Sustainable Education and Development –
Sustainable Industrialization and Innovation

Clinton Aigbavboa · Joseph N. Mojekwu ·
Wellington Didibhuku Thwala ·
Lawrence Atepor · Emmanuel Adinyira ·
Gabriel Nani · Emmanuel Bamfo-Agyei
Editors

Sustainable Education and Development – Sustainable Industrialization and Innovation

Proceedings of the Applied Research Conference in Africa (ARCA), 2022

Set 1

Editors
Clinton Aigbavboa
CIDB Centre of Excellence
University of Johannesburg
Johannesburg, South Africa

Wellington Didibhuku Thwala
Department of Civil Engineering
University of South Africa (UNISA)
Gauteng, South Africa

Emmanuel Adinyira
Department of Construction Technology
and Management
Kwame Nkrumah University of Science
and Technology
Kumasi, Ghana

Emmanuel Bamfo-Agyei
Cape Coast Technical University
Cape Coast, Ghana

Joseph N. Mojekwu
University of Lagos
Lagos, Nigeria

Lawrence Atepor
Cape Coast Technical University
Cape Coast, Ghana

Gabriel Nani
Department of Construction Technology
and Management
Kwame Nkrumah University of Science
and Technology
Kumasi, Ghana

ISBN 978-3-031-25997-5   ISBN 978-3-031-25998-2 (eBook)
https://doi.org/10.1007/978-3-031-25998-2

© The Editor(s) (if applicable) and The Author(s), under exclusive license
to Springer Nature Switzerland AG 2023

This work is subject to copyright. All rights are solely and exclusively licensed by the Publisher, whether the whole or part of the material is concerned, specifically the rights of translation, reprinting, reuse of illustrations, recitation, broadcasting, reproduction on microfilms or in any other physical way, and transmission or information storage and retrieval, electronic adaptation, computer software, or by similar or dissimilar methodology now known or hereafter developed.
The use of general descriptive names, registered names, trademarks, service marks, etc. in this publication does not imply, even in the absence of a specific statement, that such names are exempt from the relevant protective laws and regulations and therefore free for general use.
The publisher, the authors, and the editors are safe to assume that the advice and information in this book are believed to be true and accurate at the date of publication. Neither the publisher nor the authors or the editors give a warranty, expressed or implied, with respect to the material contained herein or for any errors or omissions that may have been made. The publisher remains neutral with regard to jurisdictional claims in published maps and institutional affiliations.

This Springer imprint is published by the registered company Springer Nature Switzerland AG
The registered company address is: Gewerbestrasse 11, 6330 Cham, Switzerland

Preface

Research's contribution to the continent's development faces a formidable obstacle. Africa appears to be a continent where research is unrelated to development. However, it is believed that research could contribute to the continent's socio-economic development.

This book contains ninety-three peer-reviewed papers on the United Nations Sustainable Development Goal 9, with a particular emphasis on the five targets listed below. i. Develop infrastructure that is high-quality, dependable, sustainable, and resilient, including regional and transborder infrastructure, to support economic development and human well-being, with an emphasis on affordable and equitable access for all. ii. Promote inclusive and sustainable industrialization and, by 2030, significantly increase the industry's share of employment and gross domestic product in accordance with national conditions and double its share in the least developed countries. Increase small-scale industrial and other enterprises' access to financial services, including affordable credit, and their integration into value chains and markets, particularly in developing nations. iv. By 2030, upgrade infrastructure and retrofit industries to make them sustainable, with increased resource-use efficiency and greater adoption of clean and environmentally sound technologies and industrial processes, with each nation acting in accordance with its own capabilities. v. Enhance scientific research and modernise the technological capabilities of industrial sectors in all nations, especially developing nations, by 2030, including fostering innovation and substantially boosting productivity.

Clinton Aigbavboa
Joseph N. Mojekwu
Wellington Didibhuku Thwala
Lawrence Atepor
Emmanuel Adinyira
Gabriel Nani
Emmanuel Bamfo-Agyei

Organisation

Conference Chair

Joseph N. Mojekwu — University of Lagos, Lagos, Nigeria

Local Organising Committee

Gabriel Nani	Kwame Nkrumah University of Science and Technology, Kumasi, Ghana
Emmanuel Adinyira	Kwame Nkrumah University of Science and Technology, Kumasi, Ghana
Cynthia Amaning Danquah	Kwame Nkrumah University of Science and Technology, Kumasi, Ghana
Rexford Assasie Oppong	Kwame Nkrumah University of Science and Technology, Kumasi, Ghana
Christiana Okai-Mensah	Kwame Nkrumah University of Science and Technology, Kumasi, Ghana
Abdulai Sulemana Fatoama	Kwame Nkrumah University of Science and Technology, Kumasi, Ghana
Lawrence Atepor	Cape Coast Technical University, Cape Coast, Ghana
Sophia Panarkie Pardie	Cape Coast Technical University, Cape Coast, Ghana
Emmanuel Bamfo-Agyei	Cape Coast Technical University, Cape Coast, Ghana

Editors

Clinton Aigbavboa	University of Johannesburg, Johannesburg, South Africa
Joseph N. Mojekwu	University of Lagos, Lagos, Nigeria
Wellington Thwala	University of South Africa (UNISA), South Africa
Lawrence Atepor	Cape Coast Technical University, Cape Coast, Ghana
Emmanuel Adinyira	Kwame Nkrumah University of Science and Technology, Kumasi, Ghana

Gabriel Nani   Kwame Nkrumah University of Science and Technology, Kumasi, Ghana

Emmanuel Bamfo-Agyei   Cape Coast Technical University, Cape Coast, Ghana

Review Panel

D. W. Thwala	University of South Africa (UNISA), South Africa
Clinton. O. Aigbavboa	University of Johannesburg, South Africa
Samuel Sackey	Kwame Nkrumah University of Science and Technology, Ghana
William Kodom Gyasi	University of Cape Coast, Ghana
Lazarus M. Ojigi	National Space Research and Development Agency, Nigeria
Rufus Adebayo Ajisafe	Obafemi Awolowo University, Ile–Ife, Nigeria
Gabriel Nani	Kwame Nkrumah University of Science and Technology, Ghana
J. Smallwood	Nelson Mandela Metropolitan University, South Africa
David J. Edwards	Birmingham City University, UK
R. Assasie Oppong	Kwame Nkrumah University of Science and Technology, Ghana
Emmanuel Adiniyra	Kwame Nkrumah University of Science and Technology, Ghana
Godfred Darko	Kwame Nkrumah University of Science and Technology, Ghana
Phil Hackney	Northumbria University, Newcastle upon Tyne NE1 8ST, UK
Richard Osae	Cape Coast Technical University, Ghana
Eric Danso	University of Leeds, UK
Eric Simpeh	Kwame Nkrumah University of Science and Technology, Ghana
Christopher Amoah	University of the Free State, South Africa
Frederick Simpeh	Appiah-Menka University of Skills Training and Entrepreneurial Development, Ghana
Emmanuel Bamfo-Agyei	Cape Coast Technical University, Ghana

Scientific Committee

Joseph N. Mojekwu	University of Lagos, Nigeria
Lawrence Atepor	Cape Coast Technical University Ghana

D.W. Thwala	University of South Africa (UNISA), South Africa
Clinton. O. Aigbavboa	University of Johannesburg, South Africa
Jim Ijenwa Unah	University of Lagos, Nigeria
L.O. Ogunsumi	Obafemi Awolowo University, Ibadan, Nigeria
Emanuel Amaniel Mjema	College of Business Education, Tanzania
Bashir Garba	Usmanu Danfodiyo University, Sokoto State, Nigeria
Samuel Sackey	Kwame Nkrumah University of Science and Technology, Ghana
Gabriel Nani	Kwame Nkrumah University of Science and Technology, Ghana
R. K. Nkum	Kwame Nkrumah University of Science and Technology, Ghana
A.N. Aniekwu	University of Benin, Nigeria
J. Smallwood	Nelson Mandela Metropolitan University, South Africa
Lazarus M. Ojigi	National Space Research and Development Agency, Nigeria
R. Assasie Oppong	Kwame Nkrumah University of Science and Technology, Ghana
Emmanuel Adinyira	Kwame Nkrumah University of Science and Technology, Ghana
Cynthia Amaning Danquah	Kwame Nkrumah University of Science and Technology, Ghana
Emmanuel Bamfo-Agyei	Cape Coast Technical University, Ghana
William Kodom Gyasi	University of Cape Coast, Ghana

Contents

Contributory Factors to Emerging Contractor's Non-compliance
with Project Quality Requirements 1
   C. Amoah and Y. Sibelekwana

Examining Awareness and Usage of Renewable Energy Technologies
in Non-electrified Farming Communities in the Eastern Region of Ghana 14
   T. A. Asiamah, G. Tettey, D. B. Boyetey, and R. T. Djimajor

Application of Situation Awareness Theory to the Development
of an Assessment Framework for Indoor Environmental Quality
in Classroom 28
   A. D. Ampadu-Asiamah, S. Amos-Abanyie, K. Abrokwah Gyimah,
   E. Ayebeng Botchway, and D. Y. A. Duah

Causes of Poor Workmanship in Low-Cost Housing Construction
in South Africa 40
   M. Maseti, E. Ayesu-Koranteng, C. Amoah, and A. Adeniran

Assessment of Refuse Shute Practices in Medium-Rise Buildings
Within the Greater Accra Region, Ghana 52
   M. Pim-Wusu, T. Adu Gyamfi, B. M. Arthur-Aidoo, and P. R. Nunoo

A Review of Frameworks for the Energy Performance Certification
of Buildings and Lessons for Ghana 63
   G. Osei-Poku, C. Koranteng, S. Amos-Abanyie, E. A. Botchway,
   and K. A. Gyimah

The Use of Building Information Modelling by Small to Medium-Sized
Enterprises: The Case of Central South Africa 81
   H. A. Deacon and H. Botha

Effects of Property Rights for Low-Income Housing in South Africa 94
   W. B. Mpingana, N. X. Mashwama, O. Akinradewo, and C. Aigbavboa

Recyclability of Construction and Demolition Waste in Ghana:
A Circular Economy Perspective 106
   M. Adesi, M. Ahiabu, D. Owusu-Manu, F. Boateng, and E. Kissi

Innovative Work Environment of an Informal Apparel Micro Enterprise
with PDCA Cycle: An Action-Oriented Case Study 121
   W. K. Senayah and D. Appiadu

Industrialization and Economic Development in Sub-Saharan Africa:
The Role of Infrastructural Investment 143
   Rachel Jolayemi Fagboyo and Rufus Adebayo Ajisafe

Greening the Circular Cities: Addressing the Challenges to Green
Infrastructure Development in Africa 153
   O. M. Owojori and C. Okoro

Properties of Clay Deposits in Selected Places in Sekondi-Takoradi
and Ahanta West, Ghana 166
   B. K. Mussey, A. Addae, G. Obeng-Agyemang, and S. Quayson Boahen

Lumped – Capacitance Design for Transient Heat Loss Prediction in Oil
and Gas Production Pipes in Various Media 177
   *R. N. A. Akoto, J. J. Owusu, B. K. Mussey, G. Obeng-Agyemang,
and L. Atepor*

Legal Immigration to the United States: A Time Series Analysis 190
   D. Akoto, R. N. A. Akoto, B. K. Mussey, and L. Atepor

Public Health Predictive Analysis of Chicago Community Areas:
A Data Mining Approach 202
   D. Akoto and R. N. A. Akoto

Simulation-Based Exploration of Daylighting Strategies for a Public
Basic School in a Hot-Dry Region of Ghana 215
   J. T. Akubah, S. Amos-Abanyie, and B. Simmons

The Effect of Building Collapse in Ghanaian Building Industry: The
Stakeholders' Perspectives 234
   M. Pim-Wusu, T. Adu Gyamfi, and K. S. Akorli

Estimation of the Most Sustainable Regional and Trans-border
Infrastructure Among Road, Rail and Seaborne Transport 243
   S. N. Dorhetso and I. K. Tefutor

Innovation Performance and Efficiency of Research and Development
Intensity as a Proportion of GDP: A Bibliometric Review 260
   S. N. Dorhetso, L. Y. Boakye, and D. N. O. Welbeck

Determinants of Small and Medium-Sized Enterprises Access
to Financial Services in Ghana 278
S. N. Dorhetso, L. Y. Boakye, and K. Amofa-Sarpong

The Effects of Electronic Taxes on Small and Medium-Sized
Enterprises' Access to Financial Services 293
S. N. Dorhetso, K. Amofa-Sarpong, and E. Osafoh

An Assessment of Practices on Disposal of Solar E-Waste in Lusaka,
Zambia 303
S. Chisumbe, E. Mwanaumo, K. Mwape, W. D. Thwala,
and A. Chilimunda

Lean Supply Chain Practices in the Zambian Construction Industry 313
E. Manda, E. Mwanaumo, W. D. Thwala, R. Kasongo, and S. Chisumbe

The Performance Assessment of Zambia Railways Transport Service
Quality 327
E. Mwanaumo, C. Bwalya, W. D. Thwala, and S. Chisumbe

Effective Cost Management Practices for Enabling Sustainable Success
Rate of Emerging Contractors in the Eastern Cape Province of South
Africa 339
A. Sogaxa and E. K. Simpeh

Reflections on Real Options Valuation Approach to Sustainable Capital
Budgeting Practice 358
S. Aro-Gordon, M. Al-Salmi, G. Chinnasamy, and G. Soundararajan

A Reconstructionist Approach to Communalism and the Idea
of Sustainable Development in Africa 375
J. O. Thomas

Provision of Digital Library, a Catalyst for Scholastic Creativity Among
Undergraduates in South-West, Nigeria 389
V. O. Amatari and I. U. Berezi

Evaluation of the Factors Influencing the Intention-To-Use Bim Among
Construction Professionals in Abuja, Nigeria 401
S. Isa and M. O. Anifowose

Determinants of Farmers' Satisfaction with Access to Irish Potato
Farmer Co-operatives' Services in Northern and Western Provinces,
Rwanda 413
C. Uwaramutse, E. N. Towo, and G. M. Machimu

Determining Factors Influencing Out-of-Pocket Health Care
Expenditures in Low- and Middle-Income Countries: A Systematic
Review 441
   R. Muremyi, D. Haughton, F. Niragire, and I. Kabano

Are They Really that Warm: A Thermal Assessment of Kiosks
and Metal Containers in a Tropical Climate? 451
   L. A. Nartey, M. Agbonani, and M. N. Addy

Fashion Transformational Synthesis Model for Beauty Pageants in Ghana 464
   S. W. Azuah, K. S. Abekah, and B. Atampugre

Design Creativity and Clothing Selection: The Central Focus in Clothing
Construction 474
   S. W. Azuah, A. S. Deikumah, and J. Tetteth

Assessment of Climate Change Mitigation Strategies in Building
Project Delivery Process 483
   A. Opawole and K. Kajimo-Shakantu

Prior Knowledge of Sustainability Among Freshmen Students
of University of Professional Studies 493
   N. A. A. Doamekpor and E. M. Abraham

Conceptual Framework on Proactive Conflict Management in Smart
Education 501
   P. Y. O. Amoako

A Gated Recurrent Unit (GRU) Model for Predicting the Popularity
of Local Musicians 514
   O. O. Ajayi, A. O. Olorunda, O. G. Aju, and A. A. Adegbite

Novel Cost-Effective Synthesis of Copper Oxide Nanostructures by The
Influence of pH in the Wet Chemical Synthesis 522
   R. B. Asamoah, A. Yaya, E. Annan, P. Nbelayim, F. Y. H. Kutsanedzie,
   P. K. Nyanor, and I. Asempah

The Synergestic Effect of Multiple Reducing Agents on the Synthesis
of Industrially Viable Mono-dispersed Silver Nanoparticles 530
   R. B. Asamoah, A. Yaya, E. Annan, P. Nbelayim, F. Y. H. Kutsanedzie,
   P. K. Nyanor, and I. Asempah

Perormance and Nutrient Values of *Clarias Gariepinus* Fed
with Powdered Mushroom *(Ganoderma Lucidum)* and Tetracycline
as Additives 539
   A. M. Adewole

Impact of Age Distribution and Health Insurance Towards Sustainable
Industrialization 557
   H. T. Williams, T. S. Afolabi, and J. N. Mojekwu

Determination of Overall Coefficient of Heat Transfer of Building Wall
Envelopes 568
   E. Baffour-Awuah, N. Y. S. Sarpong, I. N. Amanor, and E. Bentum

Post-harvest Losses of Coconut in Abura/asebu/kwamankese District,
Central Region, Ghana 589
   E. Baffour-Awuah, N. Y. S. Sarpong, and I. N. Amanor

Hazard Assessment and Resilience for Heavy Metals and Microbial
Contaminated Drinking Water in Akungba Metropolis, Nigeria 603
   T. H. T. Ogunribido

Usage Behaviour of Electronic Information Resources Among
Academicians' in Tertiary Institutions of Tanzania 611
   L. L. Nkebukwa

Factors for Information Seeking and Sharing During Accomplishig
Collaborative Activities in Vocational Schools 623
   L. L. Nkebukwa and I. Luambano

Expenditure on Education, Capital Formation and Economic Growth
in Tanzania 636
   K. M. Bwana

Mitigating the Challenges of Academic Information Systems
Implementation in Higher Education Institutions in Tanzania for Their
Continuous Development 650
   A. M. Kayanda

Examining the Challenges of Price Quotation as a Procurement Method
in Tertiary Institutions in Ghana 659
   G. Nani, S. F. Abdulai, and J. A. Ottou

Theoretical Review of Migration Theory of Consumer Switching
Behaviour 673
   L. Y. Boakye

Tertiary Students' Accommodation Affordability in Ghana 689
  C. Amoah, E. Bamfo-Agyei, and F. Simpeh

Disease Pandemics in Africa and Food Security: An Introduction 698
  E. Baffour-Awuah, I. N. Amanor, and N. Y. S. Sarpong

An Image-Based Cocoa Diseases Classification Based on an Improved
Vgg19 Model 711
  P. Y. O. Amoako, G. Cao, and J. K. Arthur

Impact of Promotional Strategies on Sustainable Business Growth
in a Selected Wine Processing Companies in Dodoma City, Tanzania 723
  K. Seme and P. Maziku

The Role of Land Demarcation in Addressing Conflicts Management
Between Farmers and Pastoralists for Sustainable Agriculture in Kiteto
District, Tanzania 731
  P. Maziku and M. Mganulwa

Users' Satisfaction of Autorickshaw Transport Operations Towards
Sustainable Intra-city Mobility, Cape Coast, Ghana 739
  S. B. Adi, C. Amoako, and D. Quartey

Acute and Sub-acute Toxicity Studies of Solvent Extracts of *Crinum
pedunculatum* Bulbs R.Br 752
  P. Doe, C. A. Danquah, K. A. Ohemeng, S. Nutakor, B. Z. Braimah,
  A. Amaglo, M. Abdul-Fatah, A. E. Tekpo, N. A. F. Boateng,
  S. N. Tetteh, O. K. Boateng, D. M. Sam, O. F. Batsa, J. T. Boateng,
  S. K. J. Gyasi, S. B. Dadson, and K. Oteng-Boahen

Effectiveness of E-Filing System on Improving Tax Collection
in Tanzania: A Case of Ilala Tax Region 763
  M. Jumanne and A. Mrindoko

Assessing the Impact of Internal Control on the Performance
of Commercial Banks in Tanzania 784
  L. T. Bilegeya and A. Mrindoko

Factors Affecting Tanzanian Small and Medium Enterprises
Performance in the East African Community Market: A Case of Dar es
Salaam Region 807
  S. S. Mtengela and A. E. Mrindoko

Effect of Service Quality on Customer Retention at Mount Kilimanjaro, Tanzania 827
   R. Delphin and R. G. Mashenene

Techno-Economic Feasibility of Hydropower Generation from Water Supply Networks in Ghana 840
   W. O. Sarkodie and E. A. Ofosu

The Influence of Cash Management on Financial Performance of Private Schools in Tanzania 854
   F. Johnson and D. Pastory

The Influence of Citizen Awareness and Willingness on Revenue Collection in Local Government Authorities: Evidence from Temeke Municipal Council, Tanzania 866
   B. Mwakyembe and D. Pastory

Exploring Sustainable Agriculture Through the Use of the Internet of Things 881
   F. O. Bamigboye and E. O. Ademola

Animal Ethics and Welfare as Practised by Small Ruminant Farmers in Ado-Ekiti, Ekiti State Nigeria 888
   F. O. Bamigboye, A. J. Amuda, J. O. Oluwasusi, and E. O. Ademola

Challenges Facing People with Disabilities in Acquiring Equitable Employment in Small and Medium Enterprises in Tanzania 902
   G. J. Mushi, A. P. Athuman, and E. J. Munishi

Competence of Traditional Automobile Practitioners in Maintenance of Automatic Transmission Drives and Implications for Transportation Planning in Ghana 913
   G. Boafo, R. S. Wireko-Gyebi, S. K. Nkrumah, and F. Davis

Enhancing Customer Satisfaction Through Listening in Tanzanian Higher Education 927
   A. K. Majenga and R. G. Mashenene

Technology Adoption and the Financial Market Performance in Nigeria and South Africa 935
   O. N. Oladunjoye and N. A. Tshidzumba

Burnt Clay Grinding Pot Waste Powder as a Partial Replacement of Ordinary Portland Cement for Concrete Production 953
   A. Nimo-Boakye, E. Nana-Addy, and K. Adinkrah-Appiah

The Effect of Covid-19 on the Teaching and Learning Process
of Entrepreneurship Education 967
   M. C. Ntimbwa and C. M. Ryakitimbo

Managing Pandemic Diseases and Food Security in Africa 974
   E. Baffour-Awuah, N. Y. S. Sarpong, and I. N. Amanor

Dividend and Share Price Behaviour: A Panacea for Sustainable
Industrialization 986
   N. M. Moseri, S. I. Owualah, P. I. Ogbebor, I. R. Akintoye,
   and H. T. Williams

Study of Social Capital and Business Performance of Micro Women
Entrepreneurs in Lagos State Nigeria: Implications for Sustainable
Development 993
   J. C. Ngwama and E. E. Omolewa

The Impact of Access to Finance on the Micro-enterprises' Growth
in Emerging Countries Towards Sustainable Industrialization 1010
   M. A. Mapunda and M. A. Tambwe

Human Capital Development and Economic Growth in Tanzania: Public
Spending Perspectives 1026
   K. M. Bwana

Coping with Crime Threat and Resilience Factors Among
the Motorcycle Taxi Operators and Customers in Dar es Salaam Tanzania 1039
   E. F. Nyange, I. M. Issa, K. Mubarack, and E. J. Munishi

Effects of Open Performance Review Appraisal System in Assessing
and Appraising Employees' Performance at First Housing Finance
Tanzania Limited 1053
   D. K. Nziku and C. B. Matogwa

Stakeholder's Intervention in Reducing Crime Threat Among
Motorcycle Taxi Riding Operators in Dar es Salaam, Tanzania 1064
   I. M. Issa, E. F. Nyange, K. Mubarack, and E. J. Munishi

E-learning of Mathematics and Students' Perceptions in Public
Secondary School, Oyo State, Nigeria 1077
   A. E. Kayode and E. O. Anwana

Dynamics of Silica Nanofluid Under Mixed Electric Field Effect 1088
   R. N. A. Akoto, H. Osei, E. N. Wiah, and S. Ntim

Production and Marketing Strategies by Youth Vegetable Farmers
in Urban Settlements, Tanzania 1099
   A. E. Maselle, D. L. Mwaseba, and C. Msuya-Bengesi

Conceptualising Technology Exchange as a Critical Gap for Higher
Education and Industry Collaborations in Ghana 1109
   M. Alhassan, W. D. Thwala, and C. O. Aigbavboa

Workplace Health and Safety Procedures and Compliance
in the Technical and Vocational Institutions Workshop in Ghana 1122
   T. Adu Gyamfi, S. K. Akorli, E. Y. Frempong-Jnr, and M. Pim-Wusu

The Effect of Magnetic Field on the Motion of Magnetic Nanoparticles
in Nanofluid 1135
   R. N. A. Akoto and L. Atepor

Ultrasound-Assisted Alkaline Treatment Effect on Antioxidant
and ACE-Inhibitory Potential of Walnut for Sustainable Industrialization 1143
   M. K. Golly, H. Ma, D. Liu, D. Yating, A. S. Amponsah,
   and K. A. Duodu

Achieving Sustainable Housing in Nigeria: A Rethink of the Strategies
and Constraints 1164
   I. R. Aliu

Minimization of Transportation Cost for Decision Making on Covid-19
Vaccines Distribution Across Cities 1180
   H. T. Williams, J. N. Mojekwu, and T. D. Ayodele

Promoting Sustainable Industrialization in Tanzanian Agro-Processing
Sector: Key Drivers and Challenges 1190
   M. A. Tambwe and M. A. Mapunda

Female Social Entrepreneurship in Male-Dominated Industries in Ghana
and Agenda 2030 1209
   S. Dzisi

Author Index 1217

Contributory Factors to Emerging Contractor's Non-compliance with Project Quality Requirements

C. Amoah[✉] and Y. Sibelekwana

Department of Quantity Surveying and Construction Management, University of the Free State, Bloemfontein, South Africa
amoahc@ufs.ac.za

Abstract. Purpose: Customer dissatisfaction due to poor quality of workmanship or non-desired specifications can cost the construction company thousands of rand in reworks. It could further potentially cost the business a client and future profits. The study aimed to investigate the factors contributing to emerging contractors' non-compliance with project quality requirements in South Africa and the associated consequences on their businesses.

Design/Methodology/Approach: A quantitative research approach was used to gather information from the respondents. Structured questionnaires were sent to 70 emerging contractors and construction professionals randomly selected in South Africa, of which 32 responded. Data received were analysed using Excel statistical tool to calculate the frequencies and mean values.

Findings: The results have shown that It was found that lack of quality management system implementation, cash flow problems, lack of adequate plant equipment, poor site coordination, substandard material, inexperience supervisors, lack of drawing specifications, the inexperience of artisans, poor work scheduling, lack of communication with consultants and poor contracts managements are significant contributory factors to non-compliance to quality standard among emerging contractors. These quality non-compliance issues have brought repercussions such as reduced profits, loss of clients, reduced firm reputation, bankruptcy, court action, reduced team morale on sites, lack of company expansion, increased insurance premiums to emerging contractors.

Research Limitations/Implications: There is an urgent need to educate the emerging contractors on the need to focus on project quality management systems and practices in managing their projects to reduce the effects thereof. This will help them avoid unnecessary expenditures, make projects profitable and sustain their businesses.

Practical Implication: Emerging contractors must have a quality management system that all personnel understand to eliminate quality deficiencies and reworks. This should be implemented from the design stages to the procurement and execution stage of the project. Construction contracts should make QMS mandatory, and contractors who do not comply to set qualities standards to be penalised or blacklisted.

© The Author(s), under exclusive license to Springer Nature Switzerland AG 2023
C. Aigbavboa et al. (Eds.): ARCA 2022, *Sustainable Education and Development – Sustainable Industrialization and Innovation*, pp. 1–13, 2023.
https://doi.org/10.1007/978-3-031-25998-2_1

Originality/Value: This research outcome assists the South African construction industry, particularly emerging contractors, in understanding the benefits of implementing a Quality Management System to grow and strengthen their companies and guarantee the success of their projects.

Keywords: Emerging contractors · Quality management system · Quality · Non-compliance

1 Introduction

In the construction industry, small to medium-sized construction companies seem to not invest enough time and costs on quality management systems and procedures, despite employers' increasing demand for "quality" (Muhammad and Muhammad, 2015). With various definitions of quality from literature, one can assume it becomes exceedingly difficult to put a clear quality management system in place for these contractors, with definitions ranging from "conformance with requirements" to "accomplishment of expectations of stakeholders" (CIDB 2011; PMI 2016). Further to the above, it is worth noting that Muhammad and Muhammad (2015) state that; quality management is increasingly becoming important for contractors to remain competitive in the ever-growing and developing construction industry. With comprehensive quality management systems in place, the client becomes comfortable that they will receive value for money, with their projects being completed on time and within budget (Muhammad and Muhammad 2015).

A substantial amount of research has been conducted concerning the causes of poor quality on site; contractor's lack of experience and inadequate resources (be it human, financial, or otherwise) as the biggest contributors to poor quality (Lewis et al. 2007; Rajendran et al. 2012). However, very little research seems to have been conducted on the effects of Quality Management Systems (QMS) in reducing these quality deficiencies on site and the extent of their use by emerging contractors in South Africa, with a primary focus on site-based use and implementation. Where the research is conducted about QMS in construction, the primary focus tends to be on the manufacturing systems and production of material in a warehouse, but very little input on the use of QMS on-site as well as in the inception stages of the project (Shieh and Wu 2002).

This is a concern because most defects and quality concerns rise as a result of clients changing the scope of works and project requirements when construction has commenced on site or due to the lack of thorough specification of requirements at the design development phase of the project (Arditi and Gunaydin 1997). The failure of project consultants and clients to give their input to eliminate unforeseen and time-consuming delays results in quality deficiencies and conflict (Arditi and Gunaydin 1997). Emerging contractors in South Africa are increasing and executing most construction projects as main and subcontractors, especially in the public sector; however, the quality of their work is becoming a concern to project stakeholders. Therefore, project stakeholders must understand the impact of quality management systems on the project's success and on construction companies. Hence this study sought to ascertain the causes and consequences of nonconformity to quality requirements and the possible solutions QMS will offer in reducing defects and increasing quality on site.

2 Literature Review

2.1 Emerging Contractors: South African Perspective

This study focuses on the emerging contractors' non-compliance with quality requirements on construction sites. Based on this, the study must define emerging contractors in the broader South African context. According to Martin and Root (2010), the South African construction industry is transforming. Part of this transformation results in several emerging contractors growing, partaking in procurement processes, and being awarded construction projects in South Africa. According to the CIDB (2011), an emerging contractor is defined as an enterprise owned, managed, and controlled by previously disadvantaged persons and is overcoming business impediments arising from the legacy of apartheid, most of which are in the grading category of Grade 1 to 4. It is suggested that these contractors tend to fail to develop into sustainable enterprises due to inadequate construction knowledge and lack of experience (Martin and Root 2010). Bikitsha and Amoah (2020) cite the government institutions' lack of financial support and late payment as one of the main challenges emerging contractors encounter in executing their projects, resulting in poor work execution and quality. Again Thwala and Phaladi (2009) state that most emerging contractors have similar features, including lack of mentors, lack of skilled labour, lack of training, lack of appropriate tools and equipment for works execution, lack of project experience, lack of resources to do large and complex projects, lack of management skills and poor project supervision. These features exhibited by the emerging contractors affect them in executing projects to acceptable standards.

2.2 Quality Management Assessments

Over the years, growing attention has been given to improving quality standards in building construction projects. As a result, many construction companies have endeavoured to implement quality management systems and techniques within their businesses (Aoieong et al. 2002). Rajendran, Clarke, and Andrews (2012) best define these quality management systems as the people and processes that have been put in place to ensure that the construction meets and, where possible, exceeds the quality requirements. Due to its success in the manufacturing industry, the concept of Total Quality Management (TQM) has now attained some awareness in the construction industry. The fundamental objectives are customer satisfaction, continuous improvement of systems and processes, lower costs and bottom-line savings, and better quality of the end product (Aoieong et al. 2002). It is a system based on the participation of all an organisation's members in improving the product, service, and culture in which the work is executed (Love et al. 1999). Based on the available literature, TQM is a collective responsibility within an organisation and its stakeholders to ensure absolute customer satisfaction and not just a role for the select few.

To better understand TQM, there is the need to understand the terms quality and terms associated with quality. According to Rajendran et al. (2012), quality is when a product or service conforms to the owner or client's plans, specifications, and standards. Whilst conformance has been defined as when a product and other physical objects produced do not violate a standard or other aspect of the product. Love et al. (1999)

define quality as the totality of a product's characteristics that handles its ability to meet the stated and implicit needs of the user. In summary, it is clear from the above that, for quality management to succeed, the client must be a significant role player to be considered in achieving quality. The Health and Safety Authority (2009) defines the client as a person or firm with the controlling interest in construction projects, as the project's owner. Without the client specifying the quality requirements, the project is set out for failure at the onset. As previously alluded to, quality is one of the fundamental terms within the construction industry and is a crucial contributor and determining factor for project success and client satisfaction (Amoah et al. 2020). As it has been established, implementing QMS reduces the number of defects drastically and the poor quality of the projects (Aoieong et al. 2002). As a result, Montgomery (2000) states that it is pivotal to benchmark quality management processes against other companies to ensure they are the best they can be. This aims to improve the quality management process, reduce waste, improve efficiency through continuous personnel training, and ensure that the product meets customer requirements and expectations. Based on Rosenfeld (2008), key points to assess quality management are the resources assigned to the project and management, the operating procedures, and quality metrics. These entail employee training programs, continuous management review meetings with staff, and internal audit programs to ensure compliance with the quality management system.

2.3 Challenges in Complying with Project Quality Requirements in the Construction Industry

For Total Quality Management Systems to succeed in an organisation, commitments to improve quality must be borne first from the higher structures of an organisation; from the top management right through to the lower structures at an operational level (Armeanu et al. 2017). The inclusive approach usually is missing in emerging contractors within the construction industry, which poses a challenge for attaining project quality requirements (Othman and Mdyin 2014). According to Lewis et al. (2007), the main objectives of the top management in this regard are:

- Strategy finalisation is the formulation of the quality goals and objectives and how feedback from the end-users and clients is monitored, and corrective action is taken. Furthermore, formulating resource-based strategies, whereby for firms to maintain and sustain a competitive advantage quality-wise, their focus should be on developing physical and human capital and organisational resources and ensuring that their resources are rare among competitors.
- Environmental focus entails understanding the business's environment and aligning its quality strategies to adapt, respond to, and/or shape the environment.
- Lastly, as suggested earlier, creating a quality culture realises a collective responsibility within the business domain. This culture ranges from top management involvement to the importance of quality to the firm, employees' involvement, and internal and external customers.

Othman and Mydin (2014) further state that some of the challenges and reasons resulting in poor project quality are problems arising in the design and construction stage of the project. These problems, amongst others, are flawed and inadequate communication and information, failure to check, coordinate, and verify details/specifications, and a lack of technical expertise.

2.4 Effects of Project Quality Non-conformance on Contractors

Failure to achieve the desired quality can lead to dire repercussions for the contractors. The costs of non-conformance are far much greater than the cost of conformance, whereby some costs cannot even be quantified (CIDB 2015). A study conducted in the United States of America established that the cost of reworks, change orders, etc., amounted to around 12.4% of the total project cost (Rosenfeld 2008). Some of these costs are hidden, intangible, and indirect quality-related costs such as failures to retain existing clients, the loss of potential clients due to bad publicity, and long-term and short-term reputational damages to the company, leading to a loss of profit generation (Rosenfeld 2008). According to the CIDB (2015), contractors should be concerned because the total cost cannot be necessarily assigned to this category of non-conformance.

Other quality non-conformance categories that hurt the bottom-line earnings of the contractor are costs of wasted time due to the contractor having to finance the reworks and all the time associated with preliminaries and labour/production costs. Loss of material related to the process of remedial works and dispatching costs were also a concern (Aoieong et al. 2002). Literature suggests an interrelationship between quality and safety management (Rajendran et al. 2012). The lack of quality management system implementation can directly or indirectly compromise site safety, leading to serious injury or death. Over and above safety, Rajendran et al. (2012) further suggest that the lack of QMS implantation on sites yields increased project costs and time and the potential for construction defects, whilst it decreases owner satisfaction. Concerning staff morale, they suggest that negative staff behaviour becomes more visible with a lack of quality management implemented, and team spirits are usually at their lowest. Thus there would be no culture of research and development and better ways of doing things through the promotion of continuous improvement.

3 Research Methodology

3.1 Research Approach

In this study, the quantitative research method was explored according to the research questions that need to be answered. According to Rutberg and Bouikidis (2018), quantitative research uses a measured design to examine a phenomenon based on precise measurements. Therefore, it gathers and analyses mathematical data to find patterns and averages for generalising results to the broader populace. Sukamolson (2007) further defines quantitative research as social research that engages empirical methods and empirical statements and further explains an empirical statement as being a statement that describes what "is" the case in the real world, rather than what "ought" to be the case. In essence, quantitative research deals with facts and not opinions or ideas. This approach is the most suitable for this study as there is a wealth of literature on the reasons for poor quality. Therefore, a questionnaire with pre-populated questions and options to choose from will be a better option to understand better what actual people in the construction field deem as the biggest contributors to quality deficiencies and their thoughts on the effects of not achieving the desired project quality.

3.2 Target Population and Sampling Method

A target population is an entire group of individuals being investigated (McLeod 2019) to gain intelligence on a particular study interest. The study selected participants in the construction industry, either working for construction companies from grade 1 to 4 CIDB or for construction consulting firms, to investigate and identify their understanding of non-compliance to quality by emerging contractors based on the pre-selected factors. According to Creswell (2017), sampling is a small portion of a statistical population whose qualities are studied to understand the whole. In this research study, a convenience sampling approach was adopted. This approach was adopted because participants are sequentially selected according to their convenient availability. The participant selection process ends typically when the total desired number of participants is reached (Acharya et al. 2013). Creswell and Creswell (2017) indicate that for non-probabilistic sampling, 20 participants are deemed adequate to conclude. The researcher distributed 78 questionnaires, of which 32 completed questionnaires were received and used for the analysis. According to Altunşık et al. (2004), a sample size between 30 and 500 is sufficient for quantitative analysis for many researchers; the decision on the size should depend on the quality of the sample or respondents (Thomson 2004). The respondents of this study are professionals working with construction firms; thus, the quality of their contributions cannot be trivialised. Hence the sample size of 32 is appropriate based on the views of Altunşık et al. (2004) and Thomson (2004). Cohen et al. (2000) also suggest that if a study uses a survey design, the sample size should not be less than 30. Respondents were selected from consultants and contractors all over South Africa based on their convenience.

3.3 Data Collection Instruments

The collection of the data is conducted by utilising a survey design. A questionnaire has been deemed one of the most cost-effective and efficient data collection tools, as very little money is used to conduct the research. Data can be collected far and wide with email and the internet, with no travel required (Debois 2019). Since data was collected from construction professionals all over South Africa, the questionnaire was deposited on the Survey website. The link was emailed or texted or WhatsApp to the respondents where necessary. The respondents filled in the questionnaire and submitted it via the survey monkey website. The questionnaire was divided into 3 sections; the first dealt with demographics, the second dealt with quality non-compliance to quality requirements (see Table 2), and the last was about the effects of quality non-compliance (see Table 3).

3.4 Data Analysis

According to Blaikie (2003), many analysis methods exist for analysing quantitative data. For this study, the researcher adopted the Bivariate Descriptive Analysis method to analyse the questionnaire, and the software used was Microsoft Excel statistical tool to generate frequencies and mean values. This analysis method assisted with responding to the research questions and identifying the magnitude of respondents' views on the variables given in the questionnaire. The demographic features of the respondents are shown in Table 1.

Table 1. Respondent's demographic features

Demographics		Respondents	Percentages
Gender	Male	18	56%
	Female	14	44%
	Total	32	100%
Age	Under 25	2	7%
	26–35 years	23	71%
	36–45 years	5	16%
	Over 45 years	2	6%
	Total	32	100%
Professional background	Implementing agent	3	9%
	Consultants	6	19%
	Contractor	18	56%
	Architect	2	6%
	Quantity surveying	3	9%
	Total	32	100%
Work experience	Less than 5 years	8	25%
	5–10 years	14	44%
	10–15 years	6	19%
	Over 15 years	4	13%
	Total	32	100%

Respondents' demographic features in Table 2 show that most (56%) of the respondents are males, indicating male dominance in the construction sector. Again, most (56%) of them are contractors, while 76% have over 5 years of work experience. This also shows that respondents are kingly aware of the questions asked as they have experienced quality issues in projects over the years.

4 Findings

4.1 Reasons for Quality Requirements Non-conformance

Respondents were asked to indicate the extent to which the factors indicated in Table 2 contribute to their inability to comply with quality standards. The findings show that the most critical factors contributing to this are lack of QMS implementation and cash flow problems, with a mean score of 3.70, ranked 1. This is followed by lack of adequate plant equipment (mean score = 3.88, ranked 2), poor site co-ordination (mean score = 3.54, ranked 3), and substandard material (mean score = 3.49, ranked 4).

Table 2. Contributory factors quality non-compliance

Contributors to poor quality	Very low	Low	Medium	High	Very high	Mean	Ranking
Lack of QMS implementation	3	8	18	17	72	3.70	1
Cash flow problems	2	12	15	17	72	3.70	1
Lack of adequate plant equipment	1	10	22	33	51	3.68	2
Poor site coordination	2	12	12	45	42	3.54	3
Substandard material	1	17	12	40	42	3.49	4
Inexperience foremen	2	12	22	41	32	3.41	5
Lack of drawing specifications	3	12	18	33	42	3.39	6
Inexperience of artisans	3	12	22	24	46	3.38	7
Poor work scheduling	2	12	28	33	32	3.35	8
Lack of communication with consultants	4	6	28	37	30	3.31	9
Poor contracts managements	2	12	37	26	26	3.21	10
The inexperience of consultants in advising contractors	3	14	25	29	30	3.19	11
Delays by consultants in approving change orders	3	12	28	37	21	3.16	12
Poorly described bill of quantities	4	14	25	29	26	3.07	13
Mistakes made in designs	4	12	31	33	16	3.01	14

(*continued*)

Table 2. (*continued*)

Contributors to poor quality	Very low	Low	Medium	High	Very high	Mean	Ranking
Late approval of designs	4	17	25	33	16	2.96	15
Unskilled labourers	4	14	40	13	21	2.88	16
Changes in the scope of works	4	15	31	41	0	2.83	17
Adverse weather conditions	3	25	37	12	5	2.56	18

The least contributory factors to non-quality compliance, according to the respondents, are Late approval of designs (mean score = 2.96, ranked 15), unskilled labourers (mean score = 2.88, ranked 16), changes in the scope of works (mean score = 2.83, ranked 17), and adverse weather conditions (mean score = 2.56, ranked 18).

4.2 Effects of Non-conformance of Quality Standards on Emerging Contractors

After finding out the causes of quality non-compliance among emerging contractors, the respondents were asked to indicate the effects of the quality non-compliance on their businesses using a very low to very large scale, as indicated in Table 3. The findings show that the prevalent effects on quality non-compliance are reduced profits (mean score = 3.84, ranked 1), loss of clients (mean score = 3.78, ranked 2), and reduced firm reputation (mean score = 3.66, ranked 3). Other effects identified are bankruptcy (mean score = 3.59, ranked 4), court action (mean score = 3.56, ranked 5), reduced team morale on sites (mean score = 3.53, ranked 6), lack of company expansion, and increased insurance premiums with a mean score = 3.41, ranked 7, respectively. Looking at the mean values, it can be concluded that all the effects of quality non-compliance are significant.

Table 3. Effects of quality non-compliance on emerging contractors

Effects of quality non-conformance	Very low	Low	Medium	High	Very high	Mean	Ranking
Reduced profits	1	8	12	52	50	3.84	1
Loss of clients	0	12	18	36	55	3.78	2
Reduced firm reputation	3	6	15	48	45	3.66	3

(*continued*)

Table 3. (*continued*)

Effects of quality non-conformance	Very low	Low	Medium	High	Very high	Mean	Ranking
Bankruptcy	3	6	27	24	55	3.59	4
Court action	1	10	27	36	40	3.56	5
Reduced team morale on sites	2	6	30	40	35	3.53	6
Lack of company expansion	2	8	30	44	25	3.41	7
Increased insurance premiums	2	10	27	40	30	3.41	7

5 Discussion of the Findings

5.1 Reasons for Emerging Contractor's Poor Quality

From the findings, the most prevalent causes of emerging contractors' non-compliance to project quality requirements are lack of QMS implementation, cash flow problems, lack of adequate plant equipment, poor site coordination, substandard material, inexperienced foremen, lack of drawing specifications, and inexperience of artisans. The lack of cash flow problems was also highly considered a significant factor of poor quality by the respondents and lack of QMS implementation. QMS is a system and culture where all stakeholders (top to bottom) understand quality as their responsibility and not just a responsibility of a select few. A system where everybody plays their role in eliminating poor quality and suggests new ways of doing things in the spirit of continuous improvement (Armeanu et al. 2017; Hedre 2010). Again, contractors' lack of experience and inadequate financial and human resources have been identified as the most significant contributors to poor quality (Lewis et al. 2007; Rajendran et al. 2012). According to Othman and Mydin (2014), some of the most significant contributors to poor quality on construction sites are inadequate communication and project coordination of works, design, and details. This is also echoed by Dai et al. (2009), who attribute poor project quality to a lack of appropriate project management, where delays in execution become an issue towards the end of the projects, resulting in rushed work to make up lost time. Othman and Mydin (2014) further state technical expertise and appropriate equipment as the common challenge that leads to poor quality. However, they were not elaborative enough regarding the expertise, be it expertise and skills of site personnel or expertise and skills of the design team. A factor that cannot be ignored, contributing to poor site quality is the contractors' procurement process. According to Rahman and Alzubi (2015), material that does not comply with the ISO standards is procured and utilised on construction sites without undergoing a rigorous quality check and compliance process before installation. These substandard materials result in poor end-product quality. Therefore, it is evident that poor quality standards among contractors are widespread in many jurisdictions, including the emerging contractors in South Africa who are the backbone of economic growth. The implication is that emerging contractors still find it challenging to comply with the quality standard in their project execution which may reduce their engagement by the established contractors' private and other government agencies in contract awards. Thus curtailing the government agenda of promoting the capacities of these contractors for economic growth.

5.2 Repercussions of Not Complying with Quality Management

According to the empirical findings, the significant effects of not conforming to quality requirements on emerging contractors include reduced profits, client loss, firm reputation, bankruptcy, court action, team morale on sites, lack of company expansion, and increased insurance premiums. In the literature, Rosenfeld (2008) states that quality non-conformance has a lot of repercussions, some tangible and calculable, whilst others are intangible and indirect, where it cannot be costed. Some intangible effects of poor quality are quality-related costs such as failure to retain existing clients, the loss of potential clients due to bad publicity, and long and short-term reputational damages to the company, leading to a loss of future income. Other quality non-conformance effects, according to Aoieong (2002), are costs of wasted time, loss of material associated with the process of remedial works, and dispatching costs, leading to a total loss of profits. Due to the interlinked relationship between quality and safety management (Rajendran et al. 2012), the lack of quality management system implementation can directly or indirectly compromise site safety, leading to severe injury or death. It has been suggested by Love and Li (2010) that customers become dissatisfied due to poor quality of workmanship and undesired specifications, which can ultimately cost the construction company thousands of rand in reworks and potentially cost the client future profits. The findings imply that emerging contractors are losing profits due to their inability to comply with the defined project quality standards. This, in effect, may threaten their business sustenance and further compromise the functionality of constructed facilities which subsequently affect the clients.

6 Conclusion and Recommendations

Complying with project quality standards by executing firms is necessary for increased profit and business sustenance. It can thus be concluded from the study as well as the empirical results that the reasons for the emerging contractor's non-compliance to quality requirements are lack of Quality Management Systems (QMS) implementation, cash flow problems, lack of adequate plant equipment, poor site coordination, substandard material, inexperience foremen, lack of drawing specifications, and inexperience of artisans among others. All these factors are within the control of the contractors and avoidable with extra care and the implementation of quality management systems, quality control, and assurance. The non-compliant with quality standards affects emerging contractors in various ways, including reduced profits, client loss, firm reputation, bankruptcy, court action, team morale on sites, lack of company expansion, and increased insurance premiums. Thus the repercussions and costs of non-conformance are far greater than the cost of conformance. The study has shown an in-depth understanding of contributing factors to emerging contractors' poor and non-compliance to quality requirements. Therefore, the study recommends that all emerging contractors be trained in quality management and issued renewable certification yearly. Construction projects should be assigned a mandatory quality compliance officer and auditor to check and verify that quality checks are performed on a construction project on an ongoing basis by competent personnel. Emerging contractors must have a quality management system that all personnel understand to eliminate quality deficiencies and reworks. This should be implemented from the design stages to the procurement and execution stage of the project. Construction contracts should make QMS mandatory, and contractors who do not comply to set qualities standards to be penalised or blacklisted.

References

Amoah, C., Van Schalkwyk, T., Kajimo-Shakantu, K.: Quality management of RDP housing construction: myth or reality? J. Eng. Des. Technol. (2020)
Aoieong, R.T., Tang, S.L., Ahmed, S.M.: A process approach in measuring quality costs of construction projects: model development. Constr. Manag. Econ. **20**(2), 179–192 (2002). https://doi.org/10.1080/01446190110109157
Acharya, A.S., Prakash, A., Saxena, P., Nigam, A.: Sampling: why and how of it. Indian J. Med. Specialties. **4**(2), 330–333 (2013)
Altunşık, R., Coşkun, R., Bayraktaroğlu, S., Yıldırım, E.: Sosyal bilimlerde araştırma ymleri (3. bs). Sakarya Kitabevi, İstanbul (2004)
Arditi, D., Gunaydin, H.M.: Total quality management in the construction process. Int. J. Project Manage. **15**(4), 235–243 (1997)
Armeanu, S.D., Vintila, G., Gherghina, S.C.: A cross-country empirical study towards the impact of following ISO management system standards on Euro-area economic confidence. Amfiteatru Econ. **19**(44), 144–165 (2017)
Bikitsha, L., Amoah, C.: Assessment of challenges and risk factors influencing the operation of emerging contractors in the Gauteng Province, South Africa. Int. J. Constr. Manage. 1–10 (2020). https://doi.org/10.1080/15623599.2020.1763050
Blaikie, N.: Analysing Quantitative Data: From Description to Explanation. Sage (2003)
CIDB (Construction Industry Development Board): Framework: National Contractor Development Programme. CIDB, Pretoria (2011)
Cohen, L., Manion, L., Morrison, K.: Reserch Methods in Education, 5th edn. Routledge/Falmer, London (2000)
Construction Industry Development Board. The CIDB Construction Industry Indicators: Summary Results. CIDB, South Africa (2015)
Creswell, J.W., Creswell, J.D.: Research Design: Qualitative, Quantitative, and Mixed Methods Approaches. Sage publications (2017)
Dai, J., Goodrum, P.M., Maloney, W.F.: Construction craft workers' perceptions of the factors affecting their productivity. J. Constr. Eng. Manag. **135**(3), 217–226 (2009)
Debois, S.: 10 Advantages and Disadvantages of Questionnaires [online] (2019). Available from https://surveyanyplace.com/blog/questionnaire-pros-and-cons/. Accessed 8 Oct. 2021]
Health and Safety Authority: Annual Report 2009. Metropolitan Building, Dublin (2009)
Hedre, L.V.: Quality of construction activity. Univ. Petrosani-Econ. **10**(3), 183–188 (2010)
Lewis, W.G., Pun, K.F., Lalla, T.R.M.: The effect of ISO 9001 on TQM implementation in SME in Trinidad. West Indian J. Eng. **30**(1), 1–16 (2007)

Love, P.E., Mandal, P., Li, H.: Determining the causal structure of rework influences in construction. Constr. Manag. Econ. **17**(4), 505–517 (1999)

Martin, L., Root, D.: Emerging contractors in South Africa: interactions and learning. J. Eng. Des. Technol. **8**(1), 64–79 (2010). https://doi.org/10.1108/17260531011034655

McLeod, S.A.: Sampling methods. Types and Techniques explained [online]. Available from https://www.simplypsychology.org/sampling.html. Accessed 2 Oct 2021 (2019)

Montgomery, J.D.: Quality assessment and improvement processes and techniques. Paper presented at Project Management Institute Annual Seminars & Symposium, Houston, TX. Project Management Institute, Newtown Square, PA (2000)

Muhammad, Y.S., Muhammad, M.S.:Perception based definition of construction quality in Pakistan. KICEM J. Constr. Eng. Project Manag. 2s4–34 (2015). https://doi.org/10.6106/JCEPM.2015.5.2.024

Othman, N.A., Mydin, M.A.: Poor workmanship in construction of low cost housing. Analele Universitatii.Eftimie Murgu. **21**(1), 300–305 (2014)

Project management institute. 2016. *Construction Extension to the PMBOK Guide*. Pennsylvania

Rahman, A., Alzubi, Y.: Exploring key contractor factors influencing client satisfaction level in dealing with construction project: an empirical study in Jordan. Int. J. Acad. Res. Business Soc. Sci. **5**(12), 109–126 (2015)

Rajendran, S., Clarke, B., Andrews, R.: Quality management in construction: an expanding role for SH&E professionals. Prof. Saf. **57**(11), 37–42 (2012)

Rosenfeld, Y.: Cost of quality versus cost of non-quality in construction: the crucial balance. Constr. Manag. Econ. **27**(2), 107–117 (2008)

Rutberg, S., Bouikidis, C.D.: Focusing on the fundamentals: a simplistic differentiation between qualitative and quantitative research. Nephrol. Nurs. J. **45**(2), 209–213 (2018)

Shieh, H.M., Wu, K.Y.: The relationship between total quality management and project performance in building planning phase: an empirical study of real estate industries in Taiwan. Total Qual. Manag. **13**(1), 133–151 (2002)

Sukamolson, S.: Fundamentals of quantitative research. Lang. Inst. Chulalongkorn Univ. **1**(3), 1–20 (2007)

Thomson, S.B.: Qualitative research: grounded theory – sample size and validity. Paper presented at the meeting of the 8th Faculty Research Conference. Monash University, Marysville, Victoria (2004). Available from http://www.buseco.monash.edu.au/research/studentdocs/mgt.pdf. Accessed on 25 Jul 2022

Thwala, W.D., Phaladi, M.J.: An exploratory study of problems facing small contractors in the North West Province of South Africa. Afr. J. Bus. Manage. **3**(10), 533–539 (2009)

Examining Awareness and Usage of Renewable Energy Technologies in Non-electrified Farming Communities in the Eastern Region of Ghana

T. A. Asiamah[1(✉)], G. Tettey[2], D. B. Boyetey[1], and R. T. Djimajor[3]

[1] School of Sustainable Development, University of Environment and Sustainable Development, Somanya, Ghana
taasiamah@uesd.edu.gh
[2] Bui Power Authority, Accra, Ghana
[3] Department of Social Welfare and Community Development, Eastern Regional Office, Koforidua, Ghana

Abstract. Purpose. This paper examines the level of awareness and usage of various renewable energy technologies among rural farming communities in the Eastern Region of Ghana, that are not connected to the national electricity grid. It further constructs an awareness index and examines the correlates of awareness of renewable energy technologies.

Design/Methodology/Approach. A mixed method approach is employed in this study. A total of 214 respondents were interviewed. Six in-depth interviews and 4 focus group discussions were carried out. The study employed Principal Component Analysis and Chi-square Analysis in the estimations while Content Analysis was employed in the qualitative aspects. Five districts of the Eastern Region that were connected to the national grid were purosively sampled for the study.

Findings. Preliminary findings indicate that awareness and usage of renewable energy resources is highly skewed to solar energy. The correlates of awareness include marital status, education and number of children.

Implications/Research Limitation. The study has implications for the dissemination of knowledge on various renewable energy technologies. It also directs policy on achievement of energy security and SDG 7.

Practical Implication. The study has implications for the marketing and scaling up of various renewable energy technologies to non-electrified communities.

Social Implication. The study has implications for the Government's commitment on renewable energy technology use in non-electrified communities.

Originality/Value. This study contributes to the literature on renewable energy and energy security. It also provides empirical evidence on the correlates of awareness and provides policy guidance on the supply of renewable energy technologies to meet the energy needs of non-electrified farming communities.

Keywords: Renewable energy · Renewable energy technology · Energy security · Renewable energy awareness index · Ghana

© The Author(s), under exclusive license to Springer Nature Switzerland AG 2023
C. Aigbavboa et al. (Eds.): ARCA 2022, *Sustainable Education and Development – Sustainable Industrialization and Innovation*, pp. 14–27, 2023.
https://doi.org/10.1007/978-3-031-25998-2_2

1 Introduction

Energy consumption is globally indispensable in the lives of mankind and is needed for daily activities. At the meso and macroeconomic levels, it facilitates growth and development by creating livelihood opportunities (Zakaria et al. 2019; Ningi et al. 2020; Derasid et al. 2021). From prehistoric times, conventional sources of energy such as coal, fossil fuels and gas are processed daily to provide energy and electricity for both domestic and commercial activities. These resources are not infinite and can be depleted. Fossil fuels are noted for the emission of carbon dioxide, which contributes highly to climate change and global warming (Çelikler 2013; Zakaria et al. 2019; Kuamoah 2020). There are also concerns over the scarcity and increasingly expensive supplies of fossil fuels. Consequently, there is a global call to reduce over-dependence on conventional sources and adopt alternative energy sources that are replaceable (Devine-Wright 2011). These concerns juxtapose the increasing demand in energy as a result of rising populations, have stirred up discussions among academics and stakeholders in energy policy on two themes namely renewable energy and energy security.

Renewable energy is a naturally replaceable and sustainable energy source which can be used over and over again (Çelikler 2013) and has the potential to decrease and solve the global warming problem (Ahmad et al. 2014). Renewable energy is considered safe for consumption as it is devoid of toxic gases that pollute the atmosphere and the environment when compared to non-renewable sources. Furthermore, they are limitless, clean, economical, sensible and environmentally friendly (Zeray 2010; Çelikler 2013). Globally, the potential of renewable energy in increasing access to energy is overemphasised. Studies on rural electrification extension strategies have proposed a mixture of rural electrification as well as alternative electrification sources such as renewable energy sources to increase access to energy in remote and rural communities to reduce energy insecurity (Kemausuor and Ackom 2017; Wang et al. 2018).

Energy security, on the other hand, is a multidimensional concept which concerns risks and threats to efficient and effective energy demand and supply (Winzer 2012; Ningi et al. 2020). It is one of the main targets of energy policy and reflects the continuity of energy supplies relative to demand (Winzer 2012). The fluidity and dynamic nature of the concept energy security makes it resonate with the concept of renewable energy and a departure from fossil fuel technologies (Valentin 2011; Winzer 2012). A strong synergy between renewable energy and energy security is recommended for sustainable energy development (Gyamfi et al. 2015; Wang et al. 2018).

Although much progress has been made to improve access to affordable, reliable and modern energy, some segments of the country, notably rural communities still remain non-electrified. Non-electrified communities are faced with the problem of energy security and unable to achieve development goals such as eradicating extreme poverty, increase food production and access quality healthcare and good drinking water, as these require energy. Lack of access to reliable and affordable energy sources to meet basic energy needs tend to increase the poverty status of energy poor communities (Terrapon-Pfaff et al. 2014; López-González et al. 2019). These have other trickling effects on communities such as out-migration and limited livelihood opportunities. Studies indicate a great potential to increase universal access to afforble, reliable and modern energy sources through renewable energy technologies. The awareness level of rural communities is necessary for the adoption and use of renewable energy technologies.

Li et al. (2021) confirmed that the role of awareness is important to establish the consumption of renewable energy consumption and pro-environmental choices. The consumption of pro-environmental products is in turn influenced by awareness and other subjective factors (Li et al. 2021; Rizzi et al. 2020). The awareness and usage level of renewable energy technologies by rural farming communities that are non-electrified is unknown. Previous studies on awareness of renewable energy focus on students and teachers (Derasid et al. 2021; Eshiemogie et al. 2022) and the general public (Zaaria et al. 2019; Oluoch et al. 2020). Knowledge of the awareness and usage of renewable energy among rural farming communities is imperative for energy security policies for this population. This paper constructs a renewable energy technology awareness index and examines its correlates. This paper contributes to the literature on renewable energy by focusing on non-electrified communities.

2 Literature Review

2.1 Overview of Ghana's Energy Context and Efforts in Achieving Universal Access to Affordable, Reliable and Modern Energy

For sometime, Ghana's power supply has depended on abundant inexpensive hydropower. The electricity consumption has been relatively low, as a developing country. However, in recent times due to increasing urbanizations, economic growth, associated with industrial growth and increasing economic activities, the demand and consumption of electricity has increased rapidly. Alongside the increased demand in consumption of electricity, the experience of periodic hydrological shocks has driven the use of expensive oil and gas resources in electricity generation plants (Gyamfi et al. 2015). With the new resources for energy generation, the full capacity utilization of generation plants became unachievable due to the high cost of these resources. Consequently, the demand in electricity exceeded the supply of electricity, resulting in a shortfall in supply. While the high cost of oil and gas resources has high financial consequences for the economy, the supply situation also has energy security implications for households firms in Ghana (Gyamfi et al. 2018).

Meanwhile, as at 1989, Ghana had in place an electrification plan that sought to provide universal access to electricity from 1990 to 2020 (Kemausuor and Ackom 2017). Although successes were made to increase access to electricity to many rural and newly urbanised areas, the setbacks of high cost of oil and gas affected the efficiency of energy supply. More so, many remote rural communities remained non-electrified. Experts and academics in energy policy advocate the diversification and expansion of the energy generation capacity to meet the increased demand in energy. The possible options for exploration put across include community mini grids and off-grid systems that are fuelled by renewable energy resources (Gyamfi et al. 2015; Kemausuor and Ackom 2017).

Coincidenally, the UN Agenda 2030, which contained 17 Sustainable Development goals (SDGs), with goal 7 designated to ensure universal access to affordable, reliable and modern energy, was launched in 2015. Ghana signed unto the UN Agenda to consolidate its energy mission (Kuamoah 2020). Several programs and intiatives have been made towards achieving the energy mission. The rural electrification program and the renewable energy development program are two of the major programs that worthy of noting in the context of this study.

The National Medium Term Development Policy Framework 2022–2025 contains some policies that address energy security. This includes the availability of clean, affordable and accessible energy. The energy sector of Ghana seeks to achieve universal accessibility of competitively priced energy that is produced in an environmentally sustainable manner. Interestingly, some of the outcome indicators of the energy sector includes the increased penetration of renewable energy in the national energy supply supply mix from a baseline of 1.5% to 26.4% by 2025. Yet another policy indicator is to encourage mini-grid electrification using renewable energy technologies in island and lakeside communities (Government of Ghana, 2022). In order to actualise the renewable energy use dimension of Ghana's energy policy, the Ghana Scale-Up Renewable Energy Program (SREP) is structured to spearhead the transition of the sector from a state of overdependence on fossil fuels to diversified energy sources focussing on renewable energy technologies (Government of Ghana, 2022).

2.2 Theoretical Framework

Technology Acceptance Model (TAM)

The acceptance and use of technologies is widely explained by the Technology acceptance Model (TAM) propounded by (Davis 1989). This model is a derivative of the Theory of Reasoned Action (TRA) propounded by (Fishbein and Ajzen 1975). Although versions of this model have been proposed, it is still relevant in explaining the behaviour of acceptance and use of technologies. TAM explains the motivation of consumers in using technologies by considering three factors. These are the perceived usefulness, perceived ease of use and also the attitude toward the use of the technology. Usage of technologies emanates from awareness which encompasses the perceived usefulness and ease of usefulness. Attitude toward the use of technology mediates the motivation provided by the perception on usefulness and ease of use.

Theory of Planned Behaviour (TPB)

In the usage of renewable energy technologies, the availability of the technologies as well as the financial resources to aquire these technologies is very important. Although renewable energy is derived from natural resources which are regenerative, their availability in space and time is relevant for their use. More so the perceived significance of these natural resources, which is related to awareness is also relevant for usage. In this study, the TAM is supported by the Theory of planned Behaviour which provides the dimension of perceived behavioural control (PBC) which is determined by the availability of resources, opportunities and skills, as well as the perceived significance of the resources. The availability of renewable energy resources and the capacity to aquire these resources are considerable factors in awareness and usage. These are absorbed in the study by the inclusion of the Theory of Planned Behaviour.

2.3 Empirical Studies

Existing empirical studies focus on teachers, students and the general public and households. These studies reveal disparate correlates of renewable energy awareness. There is a gap in knowledge of the awareness of renewable energy among rural farming populations that are not connected to the national electricity grid and for that matter will require alternative sources of energy such as renewable energy.

Çelikler (2013) studied awareness of renewable energy among pre-service science teachers and found no gender-based difference in awareness. However there were meaningful differences in awareness based on the year-of-study of the respondents. In this study which employed independent t-test in the analysis, other variables aside gender and year-of study were not tested. Among a selected group of teachers in secondary schools and polytechnic educators involved in the teaching of renewable energy-related courses, there is a good and positive awareness and knowledge of renewable energy and an in-depth understanding of government policies in this space. However, factors such as gender and teaching experience were not significant (Derasid et al. 2021).

Students are visualised as current and future custodians of the environment and are the targets of environmental education. Consequently, several studies on renewable energy awareness focus on students. In one of such studies, an intermediate level of awareness of renewable energy was observed among secondary school students and the correlates of awareness include grade or level, discipline of study and source of information on awareness (Altuntas and Turan 2018). Among Palestinian University students, awareness of renewable energy is not dependent on gender or educational level. However, it is dependent of choice of program pursued by students. Students pursueing vocational programs tend to have significant awareness of renewable energy resources. (Assali et al. 2019). Among Nigerian students, Eshiemogie et al. (2022), noted the need to include renewable energy knowledge in the curricula of University students. There was no gender differentials in the understanding of renewable energy.

The awareness and usage of renewable energy resources by households and the general public has been examined by previous studies. Ahmad et al. (2014), validated the Theory of Behavioural Control in identifying the factors that predict the intention to use renewable energy technologies in Malaysian households. The study finds that relative advantage and perceived behavioural controls, positively mediates attitude towards renewable energy use, while attitude positively mediates intention to use renewable energy. Li et al. (2021), identified awareness and happiness as determinants of use of renewable energy. They project that economic factors are also important for the uptake of pro-environmental consumption options.

Akinwale et al. (2014) identified a fair knowledge of renewable energy technologies such as biomass, firewood, hydropower, wind energy and solar energy in South Western Nigeria, specifically, Lagos, Oyo and Osun states. However, there is a lack of in-depth understanding of the technologies and the most known technology by the public is the solar technology. In Kenya, Oluoch et al. (2020), found a majority of their sample made up of rural and urban citizens having a high level of knowledge on renewable energy technologies. In another study, Ismail and Khembo (2015), found the determinants of energy poverty among households using a national survey data. Education, location, household size and connection to national electricity grid are some of the determinant of energy poverty in South Africa. Interestingly, the households connected to the national electricity grid were energy poor. The highly educated, those residing in urban communities and sizeable households were less energy poor. However, residents of rural communities, the less educated, very small-sized and large-sized households were energy poor.

3 Methodology

3.1 Study Area, Population and Sampling

Figure 1 shows a map of a section of the Eastern Region of Ghana, where the study was conducted. The population consists of all households in communities not connected to the national electricity grid. Five districts out of 33 districts in the Eastern Region were purposively selected as they were not fully connected to the national electricity grid. The districts selected were the Fanteakwa North and South, Suhum Kraboa Coaltar, Denkyembuor and Ayensuano Districts (Fig. 1). Two communities were randomly selected from each district and the respondents were sampled using the accidental sampling technique. These are rural communities that are not connected to the national electricity grid. A total of 214 respondents were interviewed for the survey, while 6 indepth interviews and 4 Focus group discussions were made. The sample for the focus group discussions with an average of 10 participants, were randomly selected while two females and 4 males were purposively sampled for the in-depth interviews.

Fig. 1. Map of a section of the Eastern Region showing the study area

3.2 Research Design and Data Analysis

The study adopts the cross-sectional survey design with the mixed method approach of data collection. The narrative research design was adopted for the qualitative dimension of the study. Principal components analysis was employed in the construction of the renwable energy technology awareness index. This consisted a composite of the awareness of various renewable energy products and technologies such as the solar light/lamp, solar panel, solar water pump, biogas, crop residues, biomass, briquette, improved stove, microhydro electric power and wind energy. A chi-square analysis was also used to establish the association between renewable energy awareness and some characteristics of the respondents. Stata version 15 was employed in running the quantitative analyses. Qualitative content analysis was employed in analyzing the qualitative data.

4 Results and Discussion

4.1 Summary Statistics

Two hundred and fourteen (214) individuals responded to the survey. Of this sample, one out of four persons was found to be an indigene that has lived in the community since birth. We also found that males constitute majority of the sample (54.2%). We also observed that 82% of the respondents are married but the level of education seems low, as only 6.7% had some tertiary education. Moreover, the dependency ratio was observed to be high as the average number of children and household size was found to be 4 and 6 respectively. These values are consistent with the recent findings of the rural population of the Ghana population and housing census conducted in the year 2020 (Ghana Statistical Service, 2021). This has an implication for additional cost to the family and poverty tendencies.

4.2 Awareness of Renewable Energy Technologies

We analysed the awareness of RET by testing the hypothesis that the distribution of awareness index is the same across the different RET understudy. Prior to the hypothesis testing, we constructed a composite index using Principal Component Analysis. This index, known as the RET awareness index was constructed using items for each of the 10 RETs that affirms that a respondent has ever heard of any RET, seen any RET, knows how any RET works and have ever used any of the RETs under study. A respondent who score 0 for the index has no awareness of any of the RETs understudy. A respondent whose score sums up to 4 has maximum awareness of all the RETs understudy. From Table 1, the average awareness index was 1.215 and the maximum was 3.222. Several studies have also constructed indices to measure various constructs using principal component analysis (see Kalra et al. 2015; Asiamah et al. 2017) The categorical construct of the awareness index also shows 35% of the sample demonstrating low level of awareness of the various RETs understudy. A majority (57.9%) demonstrate intermediate or fair knowledge of all the RETs put together, while a little proportion (7%) demonstrate high level of knowledge (Table 1). The study reveals low to intermediate knowledge of RETs among rural populations that are not electrified. This finding differs from that of Oluoch

Table 1. Summary statistics of the respondents

Variables	Measure	N	%
Sex	Sex of respondent Male Female	116 98	54.2 45.8
Marital Status	Married Single Divorced/Separated/Widowed	172 14 24	81.9 6.7 11.4
Household size	1–4 persons 5–8 persons 9–12 persons >12 persons	60 130 19 5	28.04 60.75 8.88 2.34
Age groups	30–39 years 40–49 years 50–59 years 60 years and above	27 44 57 51	13.04 21.26 27.54 24.64
Level of Education	None Basic Secondary Tertiary	43 90 63 14	20.5 42.8 30.0 6.7
Employment	Unemployed Farming Trading Other employment	6 153 35 6	3.0 76.5 17.5 3.0
RET awareness level	Low level Medium level High level	75 124 15	35.05 57.94 7.01

	N	Min.	Max.	Mean	Std. Dev.
Number of children	212	0	14	4.32	2.6910
Household size	214	1	15	5.96	2.5460
Age of respondent	207	18	78	45.39	12.5550
RE Awareness Index	214	0	3.222	1.215	0.572

This reports the composite awareness level for each of the 10 renewable energy technologies understudy.

et al. (2020) who identified a high level of knowledge among urban and rural population in Kenya. In contrast, a fair level of awareness of renewable energy technologies was found among citizens in South Western Nigeria. The relatively low level of awareness can be attributed to the unavailability of RETs in rural locations.

The voices of participants when asked what they know about renewable energy suggest fair knowledge of the general concept and technologies with a skewness to solar resources, emanating from the supply of the RETs. This finding has implications for education and outreach of RET options for rural communities

I know that renewable energy is energy that we can replace
- Female respondent, Denkyembuor Towoboase

Somebody talks about the solar lamp, solar panel. We have electricity but it is not part of RETs.

Male Respondent 1

I know that renewable energy is energy that we can replace
- Female respondent, Denkyembuor Towoboase

Somebody talks about the solar lamp, solar panel. We have electricity but it is not part of RETs.

Male Respondent

All we know about RET is the solar power. All the companies that came around are solar companies. It is good but we wish we can get connected to the main grid.
- A male respondent at Fanteakwa South, Abodobi

I have heard of biogas and biomass but I have not seen any of them before and the way it works. So I know of the solar panel we use in charging our phones and playing music. Some people are able to use it to watch TV.
- A male respondent at Fanteakwa South, Abodobi

The cost of the solar power is expensive and discourages many people from using.
- Female respondent at Ayensuano

4.3 Usage of Renewable Energy Technologies

Significantly, we examined the usage of 10 renewable energy technologies in the study area. These are solar light or lamps (32%), crop residues (29%) and solar panels (22%). Other forms of RETs such as solar water pump, biogas, biomass have limited usage (Fig. 2). A similar study in Nigeria by Akinwale et al. (2014), also illustrates solar resources as one of the leading technologies of usage beside hydropower and fuelwood. In this study, hydropower was excluded as the focus of study is non-electrified communities. Fuelwood was also also excluded due to its negative impact in the contributing to greenhouse gas (GHG) emission. The focus of this study is clean renewable energy.

The voices of the respondents also affirms the skewness in the use of RETs towards solar resources.

This is the first time I am hearing of renewable energy technology. I know of solar panel but I don't know it's also called renewable energy technology

-Male respondent, Denkyembuor

Here, we only use the solar panel that gives us some solar lamp. We also use the LPG, and plant residue and "bobo".

- Male respondent at Ayensuano

In this community, the only source of power is the Solar, touch and lamp. Some houses are using solar panels. This allows them to do a lot of activities even in the evening, listen to the radio and watch TVs.

Male respondent at Abodobi

Somebody talks about the solar lamp, solar panel. We have electricity but it is not part of renewable energy technologies.

-Male respondent at Denkyembuor

We have seen solar panel. Many companies supplied us with solar products and that is all we have seen when it comes to renewable energy technology. So, apart from the solar panel, I heard about other renewable energies but I have not seen any here or before.

- Male respondent, Abodobi

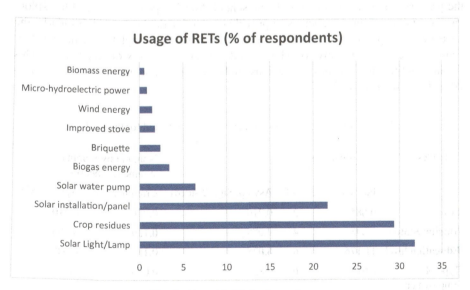

Fig. 2. Use of RETs in the study area

4.4 Determinants of Awareness of Renewable Energy Technologies

It is important to note that the distribution of the estimated indices of awareness of RET were found to be non-normal (Shapiro–Wilk statistic = 0.9659; Sig. = 0.000). Consequently we proceeded to use non parametric statistics to test the given hypothesis. Specifically, the independent samples Kruskal Wallis test was performed (X^2 = 133.184; $d.f$ = 9; $Asymp.Sig.$ = 0.000). The result led us to reject the null hypothesis and conclude at the significance level of 0.05 that the distribution of awareness index is not the same across the categories of RET.

We categorized the awareness index into three levels, namely, low awareness, medium/intermediate and high awareness. This was to enable us complete an independent Chi-square test by drawing conclusion on the hypothesis that levels of awareness of RETs are independent of nominal variables such as sex, marital status, level of education, age, number of children and household size. The results of the test are shown in Table 2. We found that different levels of awareness of RETs are not independent of marital status, level of education and the age group of respondents. Further, the estimated Crammer's V statistic for each test indicates a fairly weak association, 0.139, 0.184 and 0.224 respectively, between the awareness level and marital status, level of education and age group. Consequently, this study identifies marital status, education level and age as variables associated with awareness of RETs. Previous studies identify education as a predictor of energy security (Ismail and Khembo 2015). The explanation to this finding is that the educated tend to be exposed to various technologies and are able to gain information about the various RETs from school and media sources. Altuntas and Turan (2018), also studied the source of information on renewable energy among students and the primary source of information is the school. Married people also tend to explore various options of energy sources to provide energy for the household. Similarly, the aged are also exposed and gain knowledge of various RETs through the media and other sources. The awareness level of RETs is independent of the sex of respondents. This finding confirms the findings of previous studies among students (Altuntas and Turan 2018; Eshiemogie et al. 2022).

Table 2 Results of the non-parametric tests conducted for the study

Variables	Chi square test			Nominal by nominal symmetric measure	
	Pearson value	d.f	Asymp. Sig. (2 sided)	Crammer's V	Approx. Sig
Sex	0.685	2	0.710	0.056	0.710
Marital Status	**8.098**	4	**0.088**	**0.139**	**0.088**
Education level	**11.176**	6	**0.028**	**0.184**	**0.028**
Type of Employment	2.913	6	0.820	0.085	0.820
Age Groups	**20.691**	8	**0.008**	**0.224**	**0.008**

(*continued*)

Table 2 (*continued*)

Variables	Chi square test			Nominal by nominal symmetric measure	
	Pearson value	d.f	Asymp. Sig. (2 sided)	Crammer's V	Approx. Sig
Number of children	7.037	8	0.533	0.129	0.533
Household size groups	3.047	6	0.803	0.084	0.803
Kruskal–Wallis Test	Chi-Square	d.f	Asymp. Sig		
	133.184	9	0.000		

5 Conclusion, Policy Implications and Recommendation

This study sought to examine the awareness of RETS by residents in non-electrified rural farming communites and further identify the correlates of awareness. The various RETs covered by the study include solar light/lamp, solar panel, solar water pump, biogas, crop residues, biomass, briquette, improved stove, microhydro electric power and wind energy. Since the study area is unelectrified, hydropower was excluded from the study. Fuelwood was also excluded from the list of RETs due to its contribution to greenhouse gas emission. The average awareness index was 1.215 on a scale of 0 to 4. The maximum awareness index was 3.222 and the minimum was 0. A majority of respondents (58%) had an intermediate level of awareness of all the RETs understudy, while a lesser proportion (35%) had low level of awareness of all the RETs. Only 7% of repondents had high level of awareness of the RETs. Awareness and usage of RETs is skewed toward solar light/lamp, solar panels and crop residues. Respondents lack knowledge on technologies such as briquettes, biomass, biogas, wind energy and microhydro power. Consequently, respondents use less of these technologies. This has implications for energy poverty as there is lack of knowledge on alternative options of RETs aside solar sources and crop residues. Hence there is limited benefit to users. It also suggests limited availability of the less common technologies in the communites understudy. Efforts to increase education on the various options of RETs and scale up production to meet the needs of non-eletrified communities is strongly needed.

The study also found education level, marital status and age as the correlates of awareness. This has implications for intensification of vulnerability among groups that are already vulnerable such as widows and widowers and the less educated. An intensed education of all socio-economic groups and policies to enhance access to RETs by vulnerable groups is recommended. The lack of gender disparity in awareness also has an implication for outreach to both men and women. Men and women have similar tendencies to understand renewable energy issues. Consquently, efforts to reach men and women equally on sensitisation programs is strongly recommended.

References

Ahmad, A., Rashid, M., Omar, N.A., Alam, S.S.: Perceptions on renewable energy use in malaysia: mediating role of attitude. Pengurusan J. **41**, 123–131 (2014)

Akinwale, Y.O., Ogundari, I.O., Ilevbare, O.I., Adepoju, A.O.: A descriptive analysis of public understanding and attitudes of renewable energy resources towards energy access and development in Nigeria. Int. J. Energy Econ. Policy **4**(4), 636–646 (2014)

Altuntas, E.Ç., Turan, S.L.: Awareness of secondary school students about renewable energy sources. Renewable Energy **116**, 741–748 (2018). https://doi.org/10.1016/j.renene.2017.09.034

Asiamah, T.A., Ackah, C., Steel, W.F., Osei-Akoto, I.: Trends and Determinants of Household Use of Financial Services in Ghana. PhD Thesis Submitted to the University of Ghana (2017)

Assali, A., Khatib, T., Najjar, A.: Renewable energy awareness among future generation of Palestine. Renewable Energy **136**, 254–263 (2019). https://doi.org/10.1016/j.renene.2019.01.007

Çelikler, D.: Awareness about renewable energy of pre-service science teachers in Turkey. Renewable Energy **60**, 343–348 (2013). https://doi.org/10.1016/j.renene.2013.05.034

Davis, F.D.: Perceived usefulness, perceived ease of use and user acceptance of information technology. MIS Q. **13**(3), 319–33 (1989)

Derasid, N.C., Tahir, L.M., Musta'amal, A.H., Bakar, Z.A., Mohtaram, N., Rosmin, N., Ali, M.F.: Knowledge, awareness and understanding of the practice and support policies on renewable energy: Exploring the perspectives of in-service teachers and polytechnic lecturers. Energy Rep. **7**, 3410–3427 (2021). https://doi.org/10.1016/j.egyr.2021.05.031

Devine-Wright, H.: Envisioning the publoic engagement with renewable energy: an empirical analysis of images within the UK National Press 2006/2007. In: Devine-Wright, P. (ed.) Renewable Energy and the Public : From NIMBY to Participation, pp. 101–113. Earthscan, London (2011)

Eshiemogie, S.O., Ighalo, J.O., Banji, T.I.: Knowledge, perception and awareness of renewable energy by engineering students in Nigeria: a need for the undergraduate engineering program adjustment. Clean. Eng. Technol. **6**(100388), 1–11 (2022)

Fishbein, M., Ajzen, I.: Belief, Attitude, Intention and Behavior: An Introduction to Theory and Research. Addison-Wesley, Reading, MA (1975)

Ghana Statistical Service. Population and Housing Census General Report Volume 3A. Population of Regions and Districts (2021). https://flatprofile.com/ghana-statistical-service-released-2021-phc-general-report

Government of Ghana. Ministry of Energy Medium Term Expenditure Framework (MTEF) for 2022–2025 (2022, July 24). Retrieved from Ministry of Financec: https://mofep.gov.gh/

Gyamfi, S., Diawuo, F.A., Kumi, E.N., Sika, F., Modjinou, M.: The energy efficiency situation in Ghana. Renewable Sustainable Energy Rev. **82**, 1415–1423 (2018). https://doi.org/10.1016/j.rser.2017.05.007

Gyamfi, S., Midjinou, M., Djordjevic, S.: Improving electricity supply security in Ghana – the potential of renewable energy. Renewable Sustainable Energy Rev. **43**, 1035–1045 (2015). https://doi.org/10.1016/j.rser.2014.11.102

Ismail, Z., Khembo, P.: Determinants of energy poverty in South Africa. J. Energy South. Afr. **26**(3), 66–78 (2015)

Kalra, V., Mathur, H.P., Rajeev, P.V.: Microfinance clients' awareness index: a measure of awareness and skills of microfinance clients. IIMB Manage. Rev. **27**(4), 252–266 (2015)

Kemausuor, F., Ackom, E.: Towards universal electrification in Ghana. WIREs Energy Environ. **6**(e225), 1–14 (2017). https://doi.org/10.1002/wene.225

Kuamoah, C.: Renewable energy deployment in Ghana: the hype, hope and reality. Insight Afr. **12**(1), 45–64 (2020)

Li, X., Zhang, D., Zhang, T., Ji, Q., Lucey, B.: Awareness, energy consumption and pro-environmental choices of Chinese households. J. Clean. Prod. **279**(123734), 1–11 (2021). https://doi.org/10.1016/j.jclepro.2020.123734

López-González, A., Ferrer-Martí, L., Domenech, B.: Sustainable rural electrification planning in developing countries: a proposal for electrification of isolated communities of Venezuela. Energy Policy **129**, 327–338 (2019). https://doi.org/10.1016/j.enpol.2019.02.041

Ningi, T., Taruvinga, A., Zhou, L.: Determinants of energy security for rural households: the case of Melani and Hamburg communities, Eastern Cape, South Africa. Afr. Secur. Rev. (2020). https://doi.org/10.1080/10246029.2020.1843509

Oluoch, S., Lal, P., Susaeta, A., Vedwan, N.: Assessment of public awareness, acceptance and attitudes towards renewable energy in Kenya. Sci. Afr. **9**(e00512), 1–13 (2020). https://doi.org/10.1016/j.sciaf.2020.e00512

Rizzi, F., Annunziata, E., Contini, M., Frey, M.: On the effect of exposure to information and self-benefit appeals on consumer's intention to perform pro-environmental behaviours: A focus on energy conservation behaviours. J. Cleaner Prod. **270**, 122039 (2020)

Terrapon-Pfaff, J., Dienst, C., König, J., Ortiz, W.: A cross-sectional review: Impacts and sustainability of small-scale renewable energy products in developing countries. Renewable Sustainable Energy Rev. **20**, 1–10 (2014). https://doi.org/10.1016/j.rser.2014.07.161

Valentin, S.V.: Emerging symbiosis: renewable energy and energy security. Renewable Sustainable Energy Rev. **15**(9), 4572–4578 (2011). https://doi.org/10.1016/j.rser.2011.07.095

Wang, B., Wang, Q., Wei, Y.-M., Li, Z.-P.: Role of renewable energy in China's energy security and climate change: mitigation: an index decomposition analysis. Renewable Sustainable Energy Rev. **90**, 187–194 (2018)

Winzer, C.: Conceptualizing energy security. Energy Policy **46**, 36–48 (2012). https://doi.org/10.1016/j.enpol.2012.02.067

Zakaria, S.U., Basri, S., Kamaruddin, S.K., Majid, N.A.: Public awareness analysis on renewable energy in Malaysia. IOP Conf. Ser. Earth Environ. Sci. **268**, 1–10 (2019). https://doi.org/10.1088/1755-1315/268/1/012105

Zeray, C.: Renewable Energy Sources. Msc. Thesis, University of Çukurova, Turkey (2010)

Application of Situation Awareness Theory to the Development of an Assessment Framework for Indoor Environmental Quality in Classroom

A. D. Ampadu-Asiamah(✉), S. Amos-Abanyie, K. Abrokwah Gyimah, E. Ayebeng Botchway, and D. Y. A. Duah

Department of Architecture, Kwame Nkrumah University of Science and Technology, Kumasi, Ghana
nana.difie@gmail.com

Abstract. Purpose: Situation Awareness (SA) is the level of awareness an individual has of the environment in relation to an occurrence. This paper seeks to establish a working relationship between the SA theory and the assessment of IEQ in the development of an IEQ assessment framework.

Design/Methodology/Approach: The research design was primarily qualitative which involved the use of content analysis from articles in journals, books, reports. Search was conducted in PUBMED, ScienceDirect, Google Scholar and ResearchGate to find articles with Keywords: Situation Awareness, Indoor Environmental Quality, Classrooms, IEQ Assessment Frameworks. Articles were sieved and those that had the required information were used for this paper.

Findings: Findings from this study indicates that just as SA is used in other fields, it can be applied in the assessment of IEQ in classrooms of basic schools using the three levels of SA: Level 1 SA – Perception of the elements in the environment; Level 2 SA – Comprehension of Current situation; Level 3 SA – Projection of future status.

Research Limitations/Implications: Endsley's three levels of Situation Awareness was examined in this paper.

Practical Application: Application of Situation Awareness in the assessment of IEQ will facilitate the awareness and understanding of the internal environments of classrooms in basic schools by occupants of buildings and other stakeholders.

Social Implication: The application of Situation Awareness in IEQ assessments promote the the solicitation of occupants views in the establishment of IEQ conditions in buildings.

Originality/Value: Application of Situation awareness in the assessment of IEQ will help assessors to focus on parameters and factors in the environment that are relevant to the the health and well-being of occupants.

Keywords: Assessment · Assessment frameworks · Classrooms · Indoor environmental quality · Situation awareness

1 Introduction

Indoor environmental quality (IEQ) of a building refers to the environmental conditions within the building that promotes comfort and well-being of occupants, which consists of the air quality, lighting, acoustic and thermal conditions. (Mujeebu 2019; Toyinbo 2019; USGBC 2014). Research has established that IEQ within a space can affect the health, well-being, and performance of occupants therein (Al Horr et al. 2016; Kamaruzzaman et al. 2011). This makes the regular assessment of buildings to establish the level of safety of the internal environment so far as occupants are concerned important. Assessment of Indoor Environmental Quality in buildings is conducted through objective and subjective methods (Bluyssen 2014; Peretti and Schiavon 2011; Mahdavi et al. 2020). The predominant four IEQ components (air quality, visual comfort, acoustic comfort, and thermal comfort) are usually assessed through measurement, observation, and interviews with and questionnaires to stakeholders of the facilities being assessed. According to Mahdavi et al. (2020), the assessment of IEQ depends on the level of awareness and understanding of the factors that determine IEQ conditions by assessors, occupants and facility managers, their behaviour and interaction with each other to determine the level of performance of buildings.

In the creation of an assessment framework for IEQ, there is the need for the developer to have a working knowledge of what constitutes IEQ, how it is measured and how it affects the performance of buildings (Mahdavi et al. 2020). There is also the need for the developer to know the effects IEQ has on occupants and how buildings affect IEQ. The SA model sheds light on the need for awareness of a phenomenon in order to make correct decisions regarding that phenomenon. A person's SA is an important element in the determination of the success of decision processes in most real-world decision making (Endsley and Jones 2004) This paper seeks to establish how SA can be applied in the development of an assessment framework for indoor environmental quality.

2 Methodology

A desktop search through Google Scholar, PubMed, and ScienceDirect search engines with the keywords, 'Situational Awareness and Indoor Environmental Quality assessment frameworks' was conducted. All articles that came up were not related to the search. More articles in relation to IEQ assessment came up. As a result, separate searches were made for 'Situation awareness', and 'Indoor environmental Quality assessment', 'post-occupancy evaluation of Classrooms'. Titles of articles were perused as the results of these searches were in thousands. Articles with titles that were close to what was being searched for were selected and further filtered through the reading of abstracts. Finally, articles that dealt directly with the theory of SA and IEQ assessment frameworks of classrooms were selected. In all twenty-seven (27) articles from various sources were reviewed for this study. There were no restrictions on the years of publication for articles on SA because it was found that many of the SA theory papers dated as far as 1988 and had relevance to the study. Articles and books by Mira Endsley, the originator of the theory of SA were predominantly used. Years of publication for Articles on IEQ assessment were restricted from 2015 to the present because articles on IEQ assessment in classrooms of basic or elementary schools and IEQ assessment frameworks have received more attention within the past decade.

3 Situation Awareness

3.1 Introduction

Situation Awareness has been defined generally by authors as follows:

'The perception of the elements in the environment within a volume of time and space, the comprehension of the meaning of the projection of their status in the near future' (Endsley 1988a; Endsley 1995a; Stanton et al. 2001; Endsley and Jones 2004). In SA the elements in the environment perceived should be relevant to the goals of a task at hand (Endsley and Jones 2004). Dominguez (1994) has also been credited to define SA as 'Continuous extraction of environmental information, integration of this knowledge to form a coherent mental picture, and the use of that picture in directing further perception and anticipating future events' (National Research Council 1998). Another definition of SA by Sarter and Woods (1991) in Salmon et al. (2008) is 'The accessibility of a comprehensive and coherent situation representation which is continuously being updated in accordance with the results of situation assessments. SA is also 'The knowledge, cognition and anticipation of events, factors and variables affecting the safe, expedient and effective conduct of a mission'. (Taylor 1990 in Salmon et al. 2008). In summary, SA simply means an awareness of the essential things in one's environment that contributes to or are essential to an activity or phenomena being undertaken.

While the individual elements of SA can vary greatly from one domain to another, because of various goals and objectives, the importance of SA as a foundation for decision making and performance applies to almost every field of endeavour. (Endsley and Jones 2004).

The concept of SA was identified during World War I by Oswald Boelke who realised 'the importance of gaining an awareness of the enemy before the enemy gained a similar awareness and devised methods for accomplishing this.' (Stanton et al. 2001) SA is applied in many industries such as the aviation industry, military, and medical field. (Endsley and Garland 2000; Stanton et al. 2001) According to Endsley and Garland (2000), 'SA is now being studied in a variety of domains, however; education, driving, train dispatching, maintenance, and weather forecasting are but a few of the newer areas in which SA has been receiving attention' (pp5).

3.2 Three-Level Model of Situation Awareness (SA)

Endsley et al.'s (2000) systematic analysis of SA, discussed a list of eight descriptive models as ways of how people can achieve SA in complex domains. Amongst these models, Endsley's (Endsley 1988b; Endsley 1995b) model stood out as the prototypical descriptive model for SA. According to Rousseau et al. (2016), other models of SA focus on specific aspects of SA but remain within the constraints of Endsley's (1995a) model. Endsley's (1995a) three-level, the information-processing-based model has received the most attention and has subsequently driven research into the construct since its introduction (Salmon et al. 2008; Banbury and Tremblay 2016). Endsley's (1995a) three-level model of SA is arranged into three ranked levels of situational assessment, each level is an essential antecedent to the next higher level (Stanton et al. 2001). This model is designed for any task that needs people to monitor events or phenomena. It follows a series of information processing from perception through interpretation to prediction. From the lowest to the highest, the three levels of SA are:

i. Level 1 SA – Perception of the elements in the environment
ii. Level 2 SA – Comprehension of current situation
iii. Level 3 SA – Projection of future status

i. Level 1 SA - Perception of Existing Elements in the Environment.

Perception of existing elements in the environment involves the identification of the key elements in the environment which when combined, serve to define a phenomenon or situation (National Research Council 1998). These include the status, attributes, and dynamics of relevant elements in the environment (Endsley and Jones 2004). There is no interpretation of the data collected at this stage, what is intended at this stage is the presentation of the initial receipt of information in its raw form (Stanton et al. 2001).

Perception of information may come through visual, aural, tangible, olfactory senses, or a combination of any of them (Endsley et al. 2000). Even though in complex systems, electronic and mechanical displays and readouts are the main focus, much of Level 1 also comes from the perception of individuals on the environment (Endsley et al 2000; Endsley and Jones 2004). Additional information sources that are relied on and contribute to SA include verbal and non-verbal communication with others such as sensors, organisations, or individuals (Endsley et al. 2000; Endsley and Jones 2004). Each of these sources of information comes with various levels of reliability such as the information gathered and the confidence placed in information gathered, form an essential component of Level 1 SA in most domains (Endsley and Jones 2004). There is the danger of false information being supplied which can lead to wrong interpretation and actions. SA oriented design demand that important information is obtained and presented in a way that makes it easier to be processed by the system users who can have competing pieces of information contending for attention. (Endsley and Jones 2004).

ii. Level 2 SA—Comprehension of the Existing Situation.

Comprehension of the existing situation involves the definition of the existing status in operationally relevant terms in support of rapid decision making and action. It is the interpretation of the combination of the raw data collected at Level 1 into a comprehensive pattern or tactical situation. (National Research Council 1998). This level entails understanding what the data and cues perceived mean in relation to relevant goals and objectives. Comprehension is based on a synthesis of disjointed Level 1 SA elements and the comparison of that info to one's goals. It involves integrating many pieces of data to form information and prioritising the importance and meaning of that combined information as it relates to achieving the present goals (Endsley and Garland 2000). By understanding the importance of the pieces of data, the individual with Level 2 has associated a specific goal-oriented meaning and significance to the information.

iii. Level 3 SA - Projection of Future Status.

Projection of future status involves the forecasting of the current situation into the future to predict the evolution of a tactical situation. This level supports short-term planning and option evaluation when time permits (National Research Council 1998). This level constitutes the ability of a person to predict what the elements identified and their meaning in relation to current goals will do in future (at least in a short term) (Endsley and Garland 2000). A person can only achieve Level 3 SA by having a good understanding of the prevailing situation and the properties of the system they are working with (Endsley and Jones 2004). The use of the prevailing situation's understanding to make forecasts requires a very good understanding of the territory. The ability to constantly project ahead results is the ability to develop ready sets of strategies and responses to events. This allows for proactiveness, avoidance of undesirable situations and a very fast response to the occurrence of various events (Endsley and Garland 2000; Endsley and Jones 2004).

3.3 Factors that Influence Situation Awareness

SA's portrayal as an individual's internal model of the state of the environment suggests that individuals can determine what to do about situations that arise in the environment and carry out required mitigating actions (Endsley and Garland 2000). Even though SA is seen as a major foundation for decision making, many other factors are involved in turning good SA into successful performance (Endsley 1995a). Three such factors are discussed below.

The Individual: The ability to acquire SA differ from person to person (Endsley 1995b). This is because individuals may possess certain pre-conceptions and objectives that can act to filter and interpret the environment in forming SA (Endsley and Garland 2000). Salmon et al. (2008) also posited that because the activity of perception is not rigorous and well-defined, the results of perception are not the same for individuals or even for an individual at different times. A person's perception of the relevant elements in the environment as determined from system displays or directly by the senses form the basis for the person's SA. This may sometimes be compromised because of an individual's choice to focus on something other than the essentials at a particular point in time. A situation such as this can result in failure (Endsley and Garland 2000).

Time: Time plays a significant role in SA. It is a strong part of level 2 and Level 3. It is crucial to understand the amount of time that is available until some event occurs or some action must take place. Individuals filter situations that are of interest to them based on how soon elements will have an impact on their goals and tasks (Endsley and Jones 2004).

Dynamism in situations: The dynamic aspect of real-world situations is another important temporal aspect of SA. An understanding of the rate at which information is changing allows for the projection of future situations (Endsley 1988a, 1995a; Endsley and Garland 2000). It is important for a person's SA to change constantly or become outdated since situations always change. (Endsley and Jones 2004). Other features of the task environment that may affect SA include workload, stress, and complexity as indicated in Fig. 1.

The model Fig. 1 shows SA as a part of information processing that follows perception and leads to the making of decisions and execution of actions. According to the model, SA acquisition and maintenance is influenced by an individual through experience, training, workload, etc.; the complexity of task and systemic factors such as interface design (Endsley 1995a; Salmon et al. 2008).

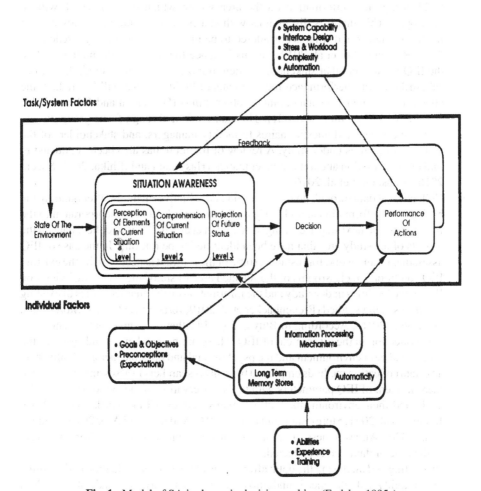

Fig. 1. Model of SA in dynamic decision making (Endsley 1995a)

4 Development Process of an IEQ Assessment Framework for Basic Schools

The development of an assessment framework for IEQ in basic schools requires steps that need to be followed. The steps involved include the need for the identification of important features of the entity being studied; collecting data through quantitative and or qualitative methods; evaluating and analysing the data, development, and validation of the framework (Ampadu-Asiamah et al. 2022) All these are done in line with the goals and objectives of the researcher (Figueirido et al. 2021).

i. Identification of important features of the entity being studied: According to Ampadu-Asiamah et al. (2022), and Figueredo et al. (2021), there is a need to build background information on the area within which the framework will be developed. This involves discussions with stakeholders within that study area or facility to understand what is considered to be important to them (Figueredo et al. 2021; Rickenbacker et al. 2019). In this instance the relevant information will be the IEQ parameters that stakeholders deem important to be measured, the technical attributes and performance requirements of buildings that will be studied, the ethical requirements involved, among other things (Hassanain and Ifthikar 2015; Soccio 2016; Ibrahim et al. 2021). This can be done through focus group discussions, interviews and questionnaires to facility managers, and stakeholders of the facility or facilities under study. A review of literature has also been conducted to find out information about the subject matter. (Hassanain and Ifthikar 2015; Soccio 2016; Amasuomo et al. 2016)

ii. Collection of data through quantitative and or qualitative methods: after building the background information and identifying the relevant factors to incorporate into the framework, the next step according to Asiamah et al. (2022), is to collect data on the aspects of the study area that have been identified to be relevant. In the case of IEQ assessment framework, for instance, some of the relevant aspects will be the existing IEQ conditions in classrooms of the selected area of study, the occupants' views on IEQ in the facilities under study and the information on the characteristics of existing structures being studied (Rickenbacker et al. 2019; Ibrahim et al. 2021; Støre-Valen and Lohne 2016). According to Bluyssen (2014), there are three main avenues for data collection in the assessment of IEQ. These are human beings and systems, the indoor and built environment and the psychosocial and possible factors of influence. The data collection methods employed by many researchers in assessing IEQ include measurement of IEQ parameters with tools, observation of the indoor spaces under study and their environments, and occupants' survey. (Rickenbacker et al. 2019; Ibrahim et al. 2021; Støre-Valen and Lohne 2016; Andargie and Azar 2019; Hameen et al. 2020). An awareness of an individual's environment plays an important role in the type of data that is collected.

iii. Evaluating and analysing the data: after data is collected, an evaluation to determine its suitability and relevance is undertaken after which analysis is made of it (Ibrahim et al. 2021; Hameen et al. 2020; Hassanain 2015). It is very essential for the evaluator to know which data is relevant or not, and also know what the data collected means for analysis to be effective. The inability to understand data collected will render the study faulty.

iv. Development and validation of the framework. Based on the information gathered from the environment and people through various means, and the evaluation and analysis of the information a framework is developed. In order to ascertain the validity of the framework, there is the need to test the framework on a few case studies. (Andargie and Azar 2019; Soccio 2016; Hameen et al. 2020)

From the above description of situation awareness and IEQ assessment development processes, the section below discusses the application of SA in the development of an IEQ assessment framework.

5 Application of Situation Awareness in the Development of an Assessment Framework IEQ of Classrooms in Basic Schools

Level 1 SA in the context of the development of an assessment framework, points out the need to identify elements in the environment which are essential to the framework's development in terms of IEQ of buildings under study and the performance of the buildings in terms of producing good IEQ. This information must be relevant to the development of the framework (Endsley 1995a, Endsley and Garland 2000).

Collection of primary data is also undertaken under Level 1. The information gathered for the development of an assessment framework for classrooms in basic schools, includes the existing environmental conditions which are determined through measurements of components of IEQ parameters; how occupants feel within the classroom through interviews or questionnaires and observation of the environment and the buildings (Ibrahim et al. 2021; Andargie and Azar 2019; Olawumi et al. 2020; Amasuomo et al. 2016; Soccio 2016). These tie in with the assertion that with Level 1, Awareness of information may come through visual, aural, tangible, olfactory senses, or a combination of any of them, as well as mechanical aids. At this stage, there is no interpretation of the data, only collection and organising. There is the need for researchers and respondents to interview and questionnaires to be aware of the environment within which they are to be able to provide the relevant data for the data collection to be successful (Endsley and Garland 2000). For instance, the operation of tools used to measure the components of IEQ parameters must be understood by the researcher, in order to know whether what is being collected is valid. In the same way, researchers must be able to identify respondents who are able to provide the relevant information to enquiries that will fit into their research objectives and not general knowledge of the environment (Endsley and Garland 2000).

Level 2 SA: The next step after the collection of primary data that is relevant to the development of the assessment framework is the evaluation and analysis of the information to understand the existing situation surrounding the development of the framework (Endsley and Jones 2004; Ibrahim et al. 2021; Hameen et al. 2020, Hassanain 2015). A synthesis of the primary data collected will be compared with the aim and objectives for the development of the framework. This step is supposed to provide a holistic view of the existing situation with an understanding of the significance of the data gathered (Endsley 1995a). Lack of clear understanding of the information gathered may affect the development of the framework. According to Hameen et al. (2020), The process of assessing IEQ in rooms goes beyond the usual performance measurements but comprehends the relationship between the components of IEQ. This realisation is a result of the understanding of critical linkages between occupant satisfaction, environmental conditions, and the technical attributes of building systems to health. There is therefore the need to pay attention to factors that influence the development of the framework (Endsley 1995a). These factors include the individual researcher and participants in the study, the time of undertaking the study and the dynamic nature of the IEQ parameters. (Endsley and Jones 2004).

Level 3 SA involves the ability to predict how the identified elements from Level 1, and their meaning from Level 2 will behave in future in relation to the aim and objectives of the development of the framework. Based on the information gathered the framework can be formulated and used to assess the indoor conditions of classrooms. Attention should be paid to how the framework will be formulated in order for it to be accepted and adapted by stakeholders in the construction and education sectors. In the case of IEQ in classrooms, the identification of relevant information, the collection and analysis thereof, will enable stakeholders to identify problems that exist in the indoor conditions of classrooms in basic schools. This will in turn enable designs of classrooms with improved IEQ. In addition to this, the creation of the awareness of IEQ and its importance through the results of the study will enable stakeholders in the education sector and built environment to pay attention to bad IEQ conditions and find solutions to them.

5.1 Theoretical Framework

Figure 2 below depicts the theoretical framework for the development of an IEQ assessment framework for classrooms in basic schools. The Level 1 SA which is the perception of environment with respect to relevant information on the environment is applied at the background information gathering and data collection stages. Level 2 SA, comprehension of existing situation applies to the evaluation and analysis of data gathered from the environments and stakeholders for easy comprehension and the last level, Level 3 SA refers to the ability to project how IEQ will be assessed in future through the developed framework. Some individual factors that can have an influence on the assessment framework development process include the dynamism of the natural environmental conditions, the type and age of buildings and the type of methodology adopted for the assessment. (Hameen et al. 2020) the ability of participants in research to focus on relevant information and the knowledge of the IEQ requirements can also affect the assessment of IEQ and thus the need to pay attention to them (Endsley and Garland 2000). Lastly, the whole process of IEQ assessment framework development must be done in relation to the aim and objectives of the study.

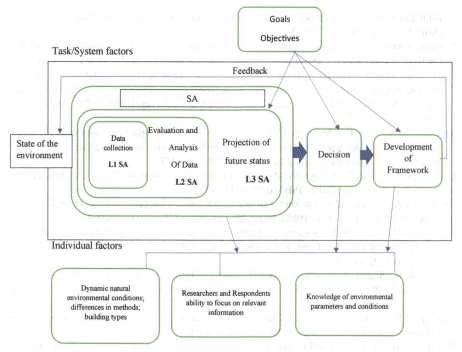

Fig. 2. Theoretical framework for development of an Assessment Framework for IEQ in classrooms in basic schools. (Endsley 1995a; Author's construct 2022)

6 Conclusion

This study set out to find out how SA can be applied in the development of IEQ assessment framework for classrooms in basic schools. It has been established that SA plays a significant role in the development of IEQ assessment framework for classrooms in basic schools. It brings to the fore the need to experience the environment and gather data in accordance with relevance to the requirements for the framework. It also sheds light on the need to understand the existing environment through synthesis and comparison of information gained with goals and objectives. The importance of the role of individuals has been highlighted and how they contribute to the SA process. Limited research exists on SA and IEQ assessment. It is important for further research to be conducted on the application of SA measurements and designs on IEQ assessments.

References

Al Horr, Y., et al.: Impact of indoor environmental quality on occupant well-being and comfort: A review of the literature. Int. J. Sustain. Built Environ. 5(1), 1–11 (2016). https://doi.org/10.1016/j.ijsbe.2016.03.006

Amasuomo, T.T., et al.: Development of a building performance assessment and design tool for residential buildings in Nigeria. Procedia Eng. 180 (2016). International High- Performance

Built Environment Conference – A Sustainable Built Environment Conference 2016 Series (SBE16), iHBE 2016, pp. 221–230

Ampadu-Asiamah, A.D., et al.: Adoption of indoor environmental quality assessment framework for naturally ventilated classrooms in basic schools in Ghana. In: Mojekwu, J.N., Thwala, W., Aigbavboa, C., Bamfo-Agyei, E., Atepor, L., Oppong, R.A. (Eds.) Sustainable Education and Development – Making Cities and Human Settlements Inclusive, Safe, Resilient, and Sustainable. ARCA 2021. Springer, Cham (2022). https://doi.org/10.1007/978-3-030-90973-4_10

Andargie, M.S., Azar, E.: 'An applied framework to evaluate the impact of indoor office environmental factors on occupants' Comfort and Working Conditions. Sustainable Cities and Society **46**, 101447 (2019) https://doi.org/10.1016/j.scs.2019.101447

Banbury, S., Tremblay S.: A Cognitive Approach to Situation Awareness: Theory and Application. Routledge, New York (2016). ISBN 0-7546-4198-8

Endsley, M.R.: Design and evaluation for situation awareness enhancement. Proc. Hum. Fact. Soc. Annu. Meet. **32**(2), 97–101 (1988). https://doi.org/10.1177/154193128803200221

Endsley, M.R., Jones, D.G.: Designing for Situation Awareness: An approach to user-centered design, 2nd edition. CRC Press, Boca Raton (2004). ISBN – 13: 978-1-4200-6358-5, https://doi.org/10.1201/b11371

Endsley, M.R.: Situation Awareness Global Assessment Technique (SAGAT). In: Proceedings of the IEEE 1988b National Aerospace and Electronics Conference (1988b). https://doi.org/10.1109/naecon.1988.195097

Endsley, M.R., Garland, D.J.: (Eds.) Situation Awareness Analysis and Measurement. Lawrence Erlbaum Associates, New Jersey (2000). ISBN 0-8058-2133-3

Endsley, M.R., et al. : Modeling and measuring Situation Awareness in an Infantry Operational Environment. (Report 1753) U.S. Army Research Institute for the Behavioral Sciences. Alexandria, VA (2000)

Endsley, M.R.: Model of situation awareness in dynamic decision making [Diagram]. In: Endsley, M.R. (Ed.) Toward a Theory of Situation Awareness in Dynamic Systems. The Journal of the Human Factors and Ergonomics Society March **37**(1), 32–64 (1995a). https://doi.org/10.1518/001872095779049543

Endsley, M.R.: Toward a theory of situation awareness in dynamic systems. The J. Hum. Fact. Ergonomi. Soc. Mar. **37**(1), 32–64 (1995b). https://doi.org/10.1518/001872095779049543

Hameen, E.C. et al.: Protocol for post occupancy evaluation in schools to improve indoor environmental quality and energy efficiency. Sustainability **12**, 3712 (2020). https://doi.org/10.3390/su12093712 https://www.mdpi.com/journal/sustainability

Hassanain, M.A., Ifthikar, A.H.: Framework model for post-occupancy evaluation of school facilities. Struct. Surv. **33**(4/5), 322–336 (2015). https://doi.org/10.1108/SS-06-2015-0029

Hofner, S., Schütze, A.: Air quality measurements and education: improving environmental awareness of high school students. Frontiers in Sensors **2** (April 2021), Article 6579202 https://doi.org/10.3389/fsens.2021.657920

Ibrahim, A.M., et al. A holistic framework for information system for the management of Indoor Environmental Qualities (IEQ) in built facilities. In: 14th IADIS International Conference Information Systems, pp. 35–44 (2021). ISBN: 978-989-8704-27-6 © 2021. https://www.researchgate.net/publication/349827977

Kamaruzzaman, S.N., et al.: The effect of indoor environmental quality on occupants' perception of performance: A case study of refurbished historic buildings in Malaysia. Energy and Buildings **43**, 407–413 (2011)

Mahdavi, A., et al.: Necessary conditions for multi-domain indoor environmental quality standards. Sustainability **12**, 8439 (2020). https://doi.org/10.3390/su12208439

Mujeebu, M.A.: Introductory chapter: indoor environmental quality. In: Indoor Environmental Quality. Intech Open (2019). https://doi.org/10.5772/intechopen.83612

National Research Council: Chapter 7: Situation Awareness. Modeling Human and Organizational Behavior: Application to Military Simulations, pp. 172–202. The National Academies Press, Washington, DC (1998). https://doi.org/10.17226/6173

Olawumi, T.O., et al.: Development of a Building Sustainability Assessment Method (BSAM) for developing countries in Sub-Saharan Africa. J. Clean. Prod. **263**, 121514 (2020). https://doi.org/10.1016/j.jclepro.2020.121514.0959-6526

Peretti, C., Schiavon, S.: Indoor environmental quality surveys. A brief literature review. Center for the Built Environment, UC Berkeley (2011). Retrieved from https://escholarship.org/uc/item/0wb1v0ss

Rousseau, R., Tremblay, S., Breton, R.: Theory: defining and modeling situation awareness: a critical review. In: Banbury, S., Tremblay, S., (Eds) A cognitive approach to situation awareness: theory and application, pp. 3-17. Routledge, New York (2016). ISBN 0-7546-4198-8

Salmon, P.M., et al.: What really is going on? review of situation awareness models for individuals and teams. Theor. Issues Ergon. Sci. **9**(4), 297–323 (2008). https://doi.org/10.1080/14639220701561775(pp297)

Soccio, P.: A new post occupancy evaluation tool for assessing the indoor environment quality of learning environments. In: Imms, W., et al. (eds.) Evaluating Learning Environments Snapshots of Emerging Issues, Methods, and Knowledge **8**(1), pp. 195–210 (2016). http://hdl.handle.net/11343/191747.SENSEPU

Stanton, N.A., et al.: Situation awareness and safety'. Safety Science **39**, 189–204 (2001). https://bura.brunel.ac.uk/bitstream/2438/1804/1/Situation_awareness_and_safety_Stanton_et_al.pdf

Støre-Valen, M., Lohne, J.: Analysis of assessment methodologies suitable for building performance. Facilities **34**(13/14), 726–747 (2016). https://doi.org/10.1108/F-12-2014-0103

USGBC: Green Building 101 (2014). What is Indoor Environmental Quality?. https://www.usgbc.org/articles/green-building-101-what-indoor-environmental-quality

Causes of Poor Workmanship in Low-Cost Housing Construction in South Africa

M. Maseti[1], E. Ayesu-Koranteng[1(✉)], C. Amoah[2], and A. Adeniran[1]

[1] Department of Building and Human Settlement Development, Nelson Mandela University, Gqeberha, South Africa
{emma.ayesu-koranteng,ayoa}@mandela.ac.za

[2] Department of Quantity Surveying and Construction Management, University of the Free State, Bloemfontein, South Africa
amoahc@ufs.ac.za

Abstract. Purpose: Poor quality workmanship is a cancer in the construction of government social housing in South Africa. The study aimed to establish the causes of poor workmanship in low-cost housing construction and determine strategies for improvement in George's local area.

Design/Methodology/Approach: A quantitative study approach was used to collect data from various stakeholders involved in constructing low-cost housing in George's local area. The survey questionnaire was distributed personally to stakeholders within the target population using simple random sampling techniques. The collected data were analysed using Excel statistical tool.

Findings: The study revealed that the causes of poor workmanship in low-cost housing construction are multifaceted. Prominent among them are cheap labour engagement, an insufficient workforce, the main contractor having limited control over sub-contractors, low subsidy from the government, lack of competency of building regulation inspectorate, scarce skills, use of unapproved materials, the poor performance of inspectors due to workload, among others.

Research Limitations/Implications: The study was limited causes of poor workmanship in low-cost housing construction.

Practical Implication: There is an urgent need to sensitise contractors on the effects of using cheap labour and unapproved building materials on construction quality. Government must also address the lack of capacity and unqualified individuals as building inspectors to ensure adequate supervision during the construction. The amount allocated to the municipalities for low-cost housing construction should be improved to encourage qualified contractors to participate in the project execution to improve quality performance.

Originality/Value: The paper has insight into the leading causes of substandard low-cost houses constructed in George's local area for policies to be implemented to curtail this phenomenon, thereby saving defect and maintenance-related costs.

Keywords: Construction · Housing quality · Labour · low-cost housing · Poor workmanship

1 Introduction

Building construction is a complex pursuit that needs trained personnel during the cycle of design and approval until the construction phase is completed. The building design commences with a site survey showing the plot's beacons and dimensions. The developer must engage the skills of industry professionals' which consist of architects, quantity surveyors and engineers among others in the process. It is widely accepted that the construction project quality should be associated with proper quality management throughout the life cycle of the project (Mallawaarachchi and Senaratne 2015). Industry experts like CESA and CIDB have argued for adopting quality principles throughout the entire project life-cycle. Mkhonto (2014) stressed the malfunctioning of the systems of Quality Management in low-cost housing developments resulted in lack of consistency and the delivery of substandard houses.

To that end, the government has legislation for example, the national constitution, which describes access to housing as a fundamental human right (Motala and Pampallis 2020) and other policies, have further been drafted, debated, and enacted in order to achieve the achievement of this fundamental right (Mabin 2020). Other scholars such as Davenport and Saunders (2000), Du Plessis (2000) and Horn (2019), among others, have further established the weight and measurement of this right. The housing legislation was enacted to guarantee that everyone's right to adequate and quality housing was gradually came to the realization (Mkuzo et al. 2019). While it is important to note that these legislations are documented, there is still something missing, as enforcing these policies and legislation leads to poorly constructed houses. Swope and Hernández (2019) argue that housing provision requires a philosophical foundation that is comprehensive and responsive enough to address these deficiencies. Taking the discussion, a step further, Oliveira and Meyfroidt (2021) proposed that the rights-based approach emphasizes a reciprocal relationship between the people as citizens and the government in the role of beneficiary and right holder respectively.

This is an old problem; it also provides them with a compelling platform to raise awareness about building sustainable, convenient houses that provide adequate employment while remaining affordable (Ntema and Marais 2013). Poor housing material indicates that the majority of the housing materials used are of poor quality, resulting in additional problems with the house's quality. Quality is a global contest for the construction industry. Because of the complexity of construction and the unique characteristics of project production, producing quality in a customer-oriented fashion presents challenges, resulting in quality defects. Poor quality is linked to human factors such as unskilled labour or inadequate supervision, as well as material, inspectorate and system failure, to name a few (Waje and Patil 2012). Quality assurance systems have been the most common quality principles applied to low-cost housing projects, including document control, audits, non-conformance tracking, CAPA and Management Review.

Furthermore, inspections have been the most dominant measure of addressing quality, which has proved ineffective in ensuring and enforcing compliance with building legislation (Rarani 2013). Rarani (2013) discovered that low-cost housing inspectors lack housing inspection training, are unaware of their roles and responsibilities, and are unfamiliar with building regulations and standards. The construction industry is a vital component of many countries' economic systems and is frequently viewed as a driver of economic growth, particularly in emerging economies. George's construction industry has recently witnessed significant growth, particularly in low-cost housing which is induced by urbanisation resulting in the increased demand for commercial and residential building spaces. Construction, because of its relatively labor-intensive nature, employs a diverse range of skilled, semi-skilled, and unskilled workers (Amra et al. 2013). However, most of these houses do not meet the quality standard leading to high maintenance costs. This study aimed to investigate why most of the low-cost houses in the George Municipality are poorly constructed.

2 Literature Review

2.1 The Quality Concept in the Construction Industry

Since its inception in the 1950s, the definition of quality has evolved, but the most common trends in defining quality are consumer satisfaction, suitability for use, and conformance to specifications or standards (KwaZulu Natal Department of Human Settlement 2010; Smallwood 2012). ISO 9000 defines quality as how inherent characteristics fulfil requirements (Grela 2015). These requirements are based on the expectations of the customer/client, which also makes consumer satisfaction a key component of how quality is defined and measured. The implication of customer satisfaction and its use for assessing the quality from their point of view is accentuated by numerous researchers in construction studies (Rashvand and Majid 2014). Duljevic and Poturak (2017) also assert that customer/client satisfaction is essential in developing the construction process and the customer relationship. Value to clients is extremely complex and frequently interpretive, but it is widely acknowledged that construction quality is a critical element of relative benefits to clients (Prior 2013). Aiyetan (2013) also emphasizes two aspects of quality in terms of material quality (which quantifies how closely the features of goods and services meet the customers' demands) and conformance quality (which pertains to a product's or service's effectiveness in accordance with design and product quality standards). Three quality "gurus", Deming, Juran and Crosby, have been profoundly influential in modern quality management practices (Neyestani 2017).

2.2 Poor Workmanship in the Construction Industry

Competent labour is considered as the most challenging to price with absolute precision. Furthermore, contractors who do not practice adequate resources for tasks will lessen labour costs in order to save fund for materials to meet their goals. Consequently, the labour provided is grossly inadequate to finish a project, and construction defects may happen. The majority of constructions use local available raw material such as timber, stone, brick, and plaster. Awareness the nature of building materials and accurate diagnosis of defects are essential in building materials management. As a result, an inadequate knowledge about proper material and structure preservation methods, as well as knowledge and understanding of standard building constituents used by consultants and contractors, can add to building deficiencies.

Conversely, contractors accountable for constructing houses utilize reduced quality items and techniques, for example Danso and Antwi (2012) indicated that the use of lower grade material against the professionals' requirements, devoid of the knowledge or approval of the client and consultants. Danso and Antwi in their investigation of time overruns in the fabrication of a telecoms tower, observed that poor workmanship resulting in modification was rated first as contractor-related issues linked with the study. According to Golchin Rad and Kim (2018), factors that determine productivity in the Iranian construction sector include an inadequate supply of equipment, supervision delays, and others. Georgiou (2010) suggested that future studies investigate the major causes of other issues affecting the construction sector, for instance a shortage of apparatus and poor workmanship, among others, primarily in emerging economies. According to Ahzahar et al. (2011), a delay is defined as work completion that is later than the planned or contract timeline. Delay occurs when a contract's progress does not meet its planned period. This can be triggered by any circumstance beyond the control of any parties to the contract. A contract rescheduling can negatively impact both the owner and the contractor, resulting in loss of revenue or additional costs. It frequently raises the controversial subject of delay responsibility, which can lead to disputes that most often result in litigation. When the real cost of a project surpasses the initial estimations, this is referred to as a cost overrun (Ahzahar et al. 2011).

Workmanship is the ability of various employees to perform a vocation or trade based on their training or profession (Ali and Wen 2011). Buys and Le Roux (2013) opined that workers are the most important factor in production because they are the only individuals who generate value and establish the overall level of efficiency. In Kenya, the contractor's expertise and qualifications have an impact on the construction projects (Nyaanga 2014). Workmanship is the most critical attribute of a project quality and poor workmanship is among the most severe challenge confronting local contractors in most emerging regions, and it has the potential to derail already-started projects (Mallawaarachchi and Senaratne 2015). A significant proportion of the contractor's payout is always retained as a warranty against hidden flaws, and other faults that may become apparent only after the completion of the project. A workforce's productivity rate is influenced by its skills domain, which means that a lower skill level personnel is a barrier to attaining full performance level. The factors that influence poor workmanship and potential solutions in Malaysia as investigated by Ali and Wen (2011) revealed issues such as lack of skill and expertise of labour, lack of communication, incompatible industrial machinery, bad weather and time and budget constraint. According to Mahamid (2016), workmanship is one of the most common causes of non-conformance to guidelines on a worksite, and as a result, parameters associated with the causes of poor construction workmanship quality have been acknowledged through literature. Ineptitude of management is widely acknowledged as a significant element in low productivity in the industry.

Ineffective site supervision adds to poor construction workmanship and it is critical because it directly influences employee's daily productivity (Forcada et al. 2016). Furthermore, Genc (2021) stated that the critical role of the subcontractor in construction work is one of the factors promoting quality deficiency. Also, Agyekum et al. (2016) observed that poor project quality leads to rework and this reduces productivity, and this arises from an insufficiency of a properly trained personnel, inadequate training as well as poor quality of training provision. Most literature acknowledges the importance of workforce skill training and development in improving construction production efficiency. Buys and Le Roux (2013) argues that contemporary literature does not go further than establishing that skill training and development are customarily valuable for the industry; it is unclear whether this perspective is evidenced by available statistical data.

Be it as it may, it has been documented that the majority of defects in construction projects are the result of human error, which can be attributed to poor workmanship in construction of housing, as well as poor control and management of building contractors, all of which have contributed to the housing crisis (Al-Tmeemy et al. 2012). Othman and Mydin (2014) also stated that poor interaction, insufficient data or failure to check data, insufficient control and tests systems, a lack of technical skills and expertise, and insufficient response all contribute to recurring errors. As a result of these, the quality of low-cost housing is becoming an issue as there are numerous protests about defects in some of the house's building components (Mkuzo et al. 2019). Poor workmanship during construction is one of the causes of this condition (Talebi et al. 2021).

3 Research Methodology

The research method involves the collection, evaluation, and explanation of the researchers aim for their investigations. The researchers collected data using a behavioural checklist (Harwell 2011). A quantitative method was employed in this study. Saunders et al. (2012) described the population as a total combination of all the elements on which the researcher focuses. Thus, it is essential to note that the term and interpreting population in research does not out-rightly refer to a group of people being considered for the study but varies depending on the nature and field of the study. For this research, the issue to be addressed was the causes of the poor quality of low-cost houses. With that in mind, the population considered were construction stakeholders (Regional Government Department overseeing housing delivery, Project Managers, Municipality, and Occupants) and contractors within Georges local area. For inclusivity, a sample of three professionals working directly with the construction of low-cost housing in the Provincial Government Department overseeing housing delivery in George Western Cape was purposefully selected. At the same time, a random sampling method was implemented for selecting other respondents (officials) from the Local Government Department overseeing housing delivery, contractors and occupants of the houses. The questionnaire, whose variables are identified from the literature, was designed to emphasise the causes of the poor quality of low-cost houses. The questionnaire was dispatched by email and manually to the relevant participants who were given 7 days to complete the questionnaire. Should the appropriate respondent not respond within 7 days, they were friendly reminded via emails and calls. Questionnaires sent out were 100, of which we retrieved 62, indicating a 62% response rate. The questionnaire was formulated using the 5 Likert scale type of questions ranging from Strongly Disagree, Disagree, Unsure, Agree, and Strongly Agree. The data collected were analysed using Microsoft Excel to determine the descriptive statistics. Table 1 shows the demographic data.

Table 1. Respondents' demographic data

	Features	Frequency	Percentages
Gender	Female	28	45%
	Male	34	55%
	Total	**62**	**100%**
Job title	Project Managers	6	10%
	Inspectors	8	13%
	Municipal Officials	4	6%
	Beneficiaries	20	32%
	Contractors	20	33%
	Officials from Government Department overseeing housing delivery	4	6%
	Total	**62**	**100%**
Age (years)	Below 25	4	6%
	26 – 35	16	26%
	36 – 45	19	31%
	45 – 55	18	29%
	above 56	5	8%
	Total	**62**	**100%**
Level of Education	Master	4	6%
	Bachelor	14	23%
	Diploma	24	39%
	Certificate	20	32%
	Total	**62**	**100%**
Work experience (years)	Below 1	2	3%
	1 – 5	6	10%
	5 – 10	30	48%
	over 10	24	39%
	Total	**62**	**100%**

From the personal data, 55% of the participants are male, 33% are contractors, and the majority (31%) are between 36 – 45 years. Again, the majority (39%) are diploma holders whilst 77% have work experience of over 5 years. This indicates that the participants are aware of the causes of poor project quality concerning low-cost housing; thus, their insights on the subject under investigation are significant.

4 Findings

The study pursued to identify the roots of the poor quality of low-cost houses. Figures below present the findings from the respondents using 5 points of the Likert Scale as follows; 1 = Strongly Disagree, 2 = Disagree, 3 = Unsure, 4 = agree, and 5 = Strongly Agree. The respondent's responses are shown in Table 2. From the findings, the lead cause of poor workmanship is cheap labour used by the contractors and insufficient manpower (Ranked 1, mean score = 4.61), respectively. The main contractor having limited control over sub-contractors was the second-ranked with a mean score of 4.32.

Table 2. Causes of poor workmanship of the low-cost housing

Variables	Strongly disagree	Disagree	Unsure	Agree	Strongly agree	Mean	S.D	Ranking
Cheap labour	0	0	2	20	40	4.61	3.71	1
The main source of concern is a lack of manpower	0	2	3	12	45	4.61	3.68	1
The main contractor has limited control over sub-contractors	1	1	10	15	35	4.32	3.52	2
Low subsidy from the government	0	6	1	25	30	4.27	3.52	3
Lack of competency of building regulation inspectorate	1	11	0	30	20	3.92	3.32	4
Scarce skills	0	12	1	36	13	3.81	3.26	5
Some materials bought on site are not SABS approved	0	12	3	35	12	3.76	3.22	6
Inspectors do not perform because of workload	3	9	12	15	23	3.74	3.15	7

(*continued*)

Table 2. (*continued*)

Variables	Strongly disagree	Disagree	Unsure	Agree	Strongly agree	Mean	S.D	Ranking
Lack of experience and level of competency	6	9	3	25	19	3.68	3.16	8
Materials bought from inferior suppliers	5	6	11	26	14	3.61	3.11	9
Site work is done by sub-contractors without proper supervision	5	10	7	24	16	3.58	3.08	10
Inspectors not enforcing orders	13	19	5	15	10	3.50	3.03	11
Contractors use defective material on low-cost houses	10	20	2	20	10	3.00	2.67	12

The next causes of poor workmanship for the low-cost housing construction identified by the respondents are low subsidy from the government (Ranked 3, mean score = 4.27), lack of competency of building regulation inspectorate (Ranked 4, mean score = 3.92) and scarce skills (Ranked 5, mean score = 3.81). Again, the usage of unapproved materials (Ranked 6, mean score = 3.76), inspectors do not perform because of workload (Ranked 7, mean score = 3.74) and lack of experience and level of competency (Ranked 8, mean score = 3.68). Other factors selected as causes of poor workmanship are materials bought from inferior suppliers (Ranked 9, mean score = 3.61), sub-contractors doing site work without proper supervision (Ranked 10, mean score = 3.58), inspectors not enforcing orders (Ranked 11, mean score = 3.50), and contractors use defective material on low-cost houses (Ranked 12, mean score = 3.00). A careful analysis of the data indicates that all the stated factors have a score of 3 and above, indicating they are significant.

5 Discussions

Contractors' using cheap labour to execute low-cost housing is a significant contributor to poor workmanship manifesting in the final deliverable. Contractors may engage unqualified workers to work on the project to maximise their profits, resulting in poor quality of low-cost housing. Likewise, contractors' usage of unapproved materials is also a significant factor stated by the respondents. The use of cheaper and inferior materials for work execution is a serious issue in the construction industry. Contractors usually look for cheaper alternative materials not approved by the client to make more profits. According to Rarani (2013), most construction industry defects are caused by the poor quality of material, e.g., bricks, cement, water, and sand. Rarani (2013) highlights the hazards which result from improper construction methods. Again, Mallawaarachchi and Senaratne (2015) mentioned that inferior material often results in repair and reconstruction costs; a certain amount of cost of construction is lost due to the alteration of substandard elements identified during construction or maintenance.

Zunguzane et al. (2012) suggest that poor management and workmanship and mechanism of building contractors have play a role to poor construction. According to Mkuzo et al. (2019), an effective construction monitoring team enhances the final deliverables. Thus, if the building inspectors are incompetent and lack capacity, there will be shortfalls in the houses' quality. Rarani (2013) states that inspections have been the most dominant measure of addressing quality, which has proved ineffective in ensuring and enforcing the compliance of building legislations. Inspectors of low-cost housing lack housing inspection training and are unmindful of their roles and responsibilities, and are unfamiliar with regulation and building standards. Therefore, it is not surprising that the lack of competency of building regulation inspectorate inspectors not enforcing orders, inspectors do not perform because of workload and site work done by sub-contractors without proper supervision were mentioned as significant factors contributing to the poor workmanship of the low-cost housing.

However, other contributory factors such as the main contractor having limited control over sub-contractors and low subsidy from the government for the building of low-cost housing may be peculiar to South Africa. In South Africa, the government policy on enhancing the participation of previously disadvantaged citizens in government work allows the government to determine the subcontractors that may work with the main contractors in the implementation of the government project without the involvement of the main contractors. This means main contractors have no control in firing and hiring subcontractors on such projects; thus, when subcontractors are not performing, the main contractors are still compelled to work with them. As observed by the respondents, this arrangement in government project execution, such as the low-cost housing, might have contributed to these subcontractors' poor quality of work. Again, the amount allocated for these low-cost housing is often insufficient compared to market-related construction costs. Contractors are thus compelled to work within this budget; thus, they look for cheaper alternative materials to executive the work to make profits, contributing to the poor quality of the houses constructed.

6 Conclusions and Recommendations

The study also concluded that poor workmanship is caused by lack of competent and experienced labour and limited resources and time. Unsuitable construction materials, poor project management and supervision of subcontractors also cause poor workmanship in low-cost houses. The research also established that many personnel in George construction are not formally educated, and they are trained on the job by starting as casual workers from where they develop an interest in construction. However, numerous projects were on very tight budgets, thus creating room for poor workmanship. Project inspectors lack capacity and often do not enforce the regulations during the inspections, allowing contractors to go away with poor quality of work. The study recommended training for construction workers and new contractors to equip them with the necessary skills. Raising the wage of skilled craftspeople will also guarantee that construction trades remain viable professional options for many people. The design-and-build construction model was also recommended by the study because it held consultants and contractors accountable for a project. Professionals should be involved in planning and project management to safeguard realistic goals, budget and time schedules. Project stakeholders must ensure project inspection and monitoring during the construction cycle. There should be a set of uniform standards that all the contractors involved in low-cost housing should adhere to. Low-cost housing projects should be awarded to competent contractors with experience and capabilities.

References

Agyekum, K., Ayarkwa, J., Amoah, P.: Built and Forgotten: Unveiling the defects associated with The Ghana Cocoa Board (Cocobod) Jubilee House in Kumasi. Journal of Building Performance 7(1) (2016)

Al-Tmeemy, S.M.H., Abdul-Rahman, H., Harun, Z.: Contractors' perception of the use of costs of quality system in Malaysian building construction projects. Int. J. Project Manage. 30(7), 827–838 (2012)

Ahzahar, N., Karim, N.A., Hassan, S.H., Eman, J.: A study of contribution factors to building failures and defects in construction industry. In: Procedia Engineering. 2nd International Building Control Conference, 11–12 July, Penang, Malaysia. Elsevier Science, Amsterdam 20, 249–255 (2011)

Ali, A.S., Wen, K.H.: Building defects: possible solution for poor construction workmanship. Journal of Building Performance 2(1), 59–69 (2011)

Aiyetan, A.O.: Causes of rework on building construction projects in Nigeria. Interim: Interdisciplinary Journal 12(3), 1–15 (2013)

Amra, R., Hlatshwayo, A., McMillan, L.:SMME employment in South Africa. In: Biennial Conference of the Economic Society of South Africa. Conference Proceedings, pp. 25–27 (2013 September)

Buys, F., Le Le Roux, M.: Causes of defects in the South African housing construction industry: perceptions of built-environment stakeholders: review articles. Acta Structilia: J. Physi. Develop. Sci. 20(2), 78–99 (2013)

Danso, H., Antwi, J.K.: Evaluation of the factors influencing time and cost overruns in telecom tower construction in Ghana. Civil and Environmental Research 2(6), 15–24 (2012)

Davenport, T., Saunders, C.: South Africa: A modern history. Springer (2000)

Du Plessis, D.J.: A critical reflection on urban spatial planning practices and outcomes in post-apartheid South Africa. In: Urban Forum, vol. 25, no. 1, pp. 69–88. Springer, Netherlands (2014 March)

Duljevic, M., Poturak, M.: Study on client-satisfaction factors in construction industry. European J. Eco. Stud. 6(2), 104–114 (2017)

Forcada, N., Macarulla, M., Gangolells, M., Casals, M.: Handover defects: comparison of construction and post-handover housing defects. Building Research & Information 44(3), 279–288 (2016)

Genc, O.: Identifying principal risk factors in Turkish construction sector according to their probability of occurrences: a relative importance index (RII) and exploratory factor analysis (EFA) approach. International Journal of Construction Management, pp. 1–9 (2021)

Georgiou, J.: Verification of a building defect classification system for housing. Struct. Surv. **28**(5), 370–383 (2010)

Grela, G.: The Framework of Quality Measurement. Management (18544223) **10**(2) (2015)

Golchin Rad, K., Kim, S.Y.: Factors affecting construction labor productivity: Iran case study. Iranian J. Sci. Technol. Trans. Civil Eng. **42**(2), 165–180 (2018)

Harwell, M.R.: Research Design in Qualitative/Quantitative/Mixed Methods, Chapter 10, the Sage Handbook for Research in Education (2011)

Horn, A.: The history of urban growth management in South Africa: tracking the origin and current status of urban edge policies in three metropolitan municipalities. Plan. Perspect. **34**(6), 959–977 (2019)

KwaZulu Natal Department of Human Settlement: Investigation into Existing Tools That Could Inform Quality assurance in Low Income Housing – November 2010, KZN Human Settlements (2010)

Mallawaarachchi, H., Senaratne, S.: Importance of quality for construction project success. In: 6th International Conference on Structural Engineering and Construction Management 2015, Kandy, Sri Lanka, 11th – 13th December 2015 (2015)

Mabin, A.: A century of South African housing acts 1920–2020. In: Urban Forum, vol. 31, no. 4, pp. 453–472. Springer Netherlands (2020 December)

Mahamid, I.: Analysis of rework in residential building projects in Palestine. Jordan Journal of Civil Engineering **10**(2) (2016)

Marais, L., Ntema, J.: The upgrading of an informal settlement in South Africa: two decades onwards. Habitat Int. **39**, 85–95 (2013)

Mkhonto, J.: An Assessment of Quality Management Practices in Low Cost Housing Projects Delivery in Mpumalanga Province, Mtech Thesis: Tshwane University of Technology (2014)

Mkuzo, T.Z., Mayekiso, T., Gwandure, C.: An observational examination of houses built under the "Breaking New Ground" housing policy of South Africa. Ergonomics SA: Journal of the Ergonomics Society of South Africa **31**(1), 1–16 (2019)

Motala, E., Pampallis, J.: Educational law and policy in post-apartheid South Africa. In: The State, Education and Equity in Post-Apartheid South Africa, pp. 14–31. Routledge (2020)

Neyestani, B.: Principles and Contributions of Total Quality Management (TQM) Gurus on Business Quality Improvement. Available at SSRN 2948946 (2017)

Nyaanga, J.K.: The effect of competence of contractors on the. Prime Journal of Social Science (PJSS) (2014)

Oliveira, E., Meyfroidt, P.: Strategic land-use planning instruments in tropical regions: state of the art and future research. J. Land Use Sci. **16**(5–6), 479–497 (2021)

Othman, N.A., Mydin, M.A.: Poor Workmanship in Construction of Low Cost Housing. Analele Universitatii'Eftimie Murgu **21**(1) (2014)

Prior, D.D.: Supplier representative activities and customer perceived value in complex industrial solutions. Ind. Mark. Manage. **42**(8), 1192–1201 (2013)

Rarani, M.: Quality Assurance in Low-Cost Housing Construction Projects in the Metropole, Mtech Thesis, Cape Peninsula University of Technology (2013)

Rashvand, P., Majid, M.Z.A.: Critical criteria on client and customer satisfaction for the issue of performance measurement. J. Manag. Eng. **30**(1), 10–18 (2014)

Saunders, M., Lewis, P., Thornhill, A.: Research Methods for Business Students, 7th edn. Pearson, London (2012)

Smallwood, J.: Quality Management in Construction, NMMU – Nelson Mandela Metropolitan University, Construction, Engineering and Public Works Inspection 2012 Conference – Cape Town, 20–21 August 2012 (2012)

Swope, C.B., Hernández, D.: Housing as a determinant of health equity: A conceptual model. Soc. Sci. Med. **243**, 112571 (2019)

Talebi, S., Koskela, L., Tzortzopoulos, P., Kagioglou, M., Rausch, C., Elghaish, F., Poshdar, M.: Causes of defects associated with tolerances in construction: A case study. Journal of Management in Engineering (2021)

Waje, V.V., Patil, V.: Cost of poor quality in construction. IOSR J. Mech. Civ. Eng. **3**(3), 16–22 (2012)

Zunguzane, N., Smallwood, J., Emuze, F.: Perceptions of the quality of low-income houses in South Africa: Defects and their causes. Acta Structilia **19**(1), 19–38 (2012)

Assessment of Refuse Shute Practices in Medium-Rise Buildings Within the Greater Accra Region, Ghana

M. Pim-Wusu[1]( ), T. Adu Gyamfi[2], B. M. Arthur-Aidoo[1], and P. R. Nunoo[1]

[1] Faculty of Built Environment, Department of Building Technology,
Accra Technical University, Accra, Ghana
mpimwusu@atu.edu.gh

[2] Faculty of Built and Natural Environment, Department of Building Technology,
Koforidua Technical University, Koforidua, Ghana

Abstract. Purpose: This study explores how waste can be effectively managed and controlled in medium-rise buildings in Ghana. This was to ease the struggle, pain, and inconveniences and safely accumulate in one discreet location that people do not inhabit. Hence the study aims to assess refuse chute practices in medium-rise buildings within the Greater Accra Region of Ghana.

Design/Methodology/Approach: A quantitative approach methodology technique was adopted for the study covering Accra Central, where people mostly live in medium-rise buildings. Primary data were collected using a questionnaire survey. The study employed a random sampling technique. A sample size of 150 was used, while data obtained from the study were analysed and presented in a frequency distribution, percentages, and inferential statistics with SPSS.

Findings: The study found that respondents have limited access to refuse chutes in their medium-rise buildings. The study reveals the benefits of using garbage chutes as convenient disposal, a hygienic environment, waste segregation for recycling, prevention of gem development, and no physical contact with refuse. The study further discovered that strategies to overcome the challenges of not using refuse chutes include checking chute design in permit approval, monitoring chute construction, and developing policies on chute construction.

Implications/Research Limitations: The revelation of this research means it is essential for an agency or authority mandated to regulate building construction activities to enforce the law regarding incorporating refuse chutes in medium-rise buildings. The present study was limited to the Greater Accra Region; however, the study could have been broadened further to cover the entire country due to the population increase where people live in medium-rise buildings.

Practical Implications: The Study's discovery is significant to building approval institutions that medium-rise building plans must incorporate refuse chutes before approval is issued.

Originality/Value: literature in Ghana indicates little or no studies on refuse chute practice in medium-rise buildings in Ghana. The outcomes of the research have proven that a refuse chute is not popular in the country, so it is imperative to incorporate one to ease the struggle, pain, and inconveniences for the occupants of the medium-rise building. Founded on previous empirical and theoretical studies, the results of this research contribute to knowledge and understanding of the refuse chute practises in medium-rise buildings in Ghana.

Keywords: Refuse chute · Disposal · Medium-rise building · Waste · Management

1 Introduction

Over the past few decades, there has been a growing recognition of the effective disposal of refuse due to the detrimental effect on humanity and the nation's economy (Owusu-Sekyere 2014). Thus, as it was, waste or refuse has ruined many countries' benefits and economic development due to its adverse impact and improper management (Macarthur 2013). Despite the country's government's numerous attempts to arrest this menace, especially in developing countries, the problem still escalates (Nwosu and Olofa 2015). The management of refuse is a key concern for the environment, and one of the major ways to address this menace is the introduction of refuse chutes in buildings (Chaban 2015). Thus, in the United States, Asia, and European countries, refuse chutes in buildings are waste management and control (Chaban 2015). Still, the situation is different among developing countries such as Nigeria, Zambia, and Zimbabwe (Chazan 2002, Mansour and Esseku 2017), as rubbish bags are a major eyesore due to the less adaption of refuse chutes in most buildings. In Ghana, the condition is worst in many towns and cities where polythene bags are scattered all over, and disposal sites overflow with filth associated with health hazards leading to uncontrolled rubbish.

Owusu-Sekyere, (2014) and Akinjare et al. (2011) postulate that the problems connected with cholera, malaria, and typhoid to residents near the dumping sites. In literal terms, a refuse chute is perpendicular or incline channel in which refuse is passed down from each floor's opening to the central refuse room on the ground floor (Hall and Greeno 2015). The chute allows the effective disposal of refuse, as it automatically directs refuse or waste to an underground storage unit (Durán and Messina 2019). It is one of the best refuse management and control methods in high and medium-rise buildings. According to Hait et al. (2021), a low-rise office or apartment building typically has fewer floors up to four, whilst a medium-rise office or apartment building raises from five to twelve floors; moreover, high-rise buildings go beyond these levels. Zahedi and Eshghi (2021) surveyed that medium-rise buildings require a lift and refuse chutes installation, which ranges from five stories in height, to ease pain and inconveniences to the occupants. Agyei-Mensah and Owusu (2009) mentioned that effective refuse disposal is like healthy life. If taken care of properly, life continues, but if not, it becomes a big problem and renders everything useless. In Ghana, refuse has been a significant problem in medium-rise buildings due to improper handling after use. It has claimed lives and properties from many individuals due to inappropriate disposal, causing floods during rainy seasons (Durán and Messina 2019). Mansour and Esseku (2017) admit that effective disposal of refuse is of key importance for reducing the destruction of properties. Institutions and organisations responsible for enforcing Ghana's building code for building design feel reluctant to institute refuse chutes for waste disposal (Macarthur 2013). The reason is that the repercussions of improper refuse disposal have no limit to the rich or poor to sickness (Gumbo et al. (2003). Therefore, this study aims to assess refuse chute practices in medium-rise buildings within the Greater Accra Region of Ghana. The study's objectives are to determine the benefits of using refuse chutes and strategies to overcome the challenges of not using refuse chutes in medium-rise buildings.

2 Literature Review

Concept of Refuse/Solid Waste: A study by Singh et al. (2016) noted that waste is not pleasant to live around. It is classified as primitive despite its critical importance within a certain environment. Boadi (2004) submitted that refuse is a material that emanates from numerous sources and is unusable or unwanted, which could be found in industries, agriculture areas, business premises, and households. Curran and Williams (2012) posit that the waste is usually classified as liquid, solid, and gaseous, depending on its premises, location,d concentration. Waste in the context of this study is not limited to solid, liquid, gaseous, or even radioactive substances that are discarded into the environment; however, any unwanted which causes a significant nuisance or adverse impact on social, economic, financial, and environmental impact on human life (Akpen and Aondoakaa (2009) (Odonkor et al. (2020)).

Regarding households, solid waste can be classified as daily disregarded products of human activities, regarded as useless in the form of refuse, garbage and sludge (Owusu et al. (2014). Gumbo et al. (2003) illustrate that many cities in developing countries have no control system for waste disposal. Gabrscek and Isljamovk (2011) posit that some parts of developing countries dispose of refuse by burning it in pits, others are dumped in random locations, whilst the majority is disposed of uncontrolled without further management. Post and Obirih-Opareh (2002) admits that solid waste is materials that are no longer used for disposal sites. Solid wastes could be materials with less liquid content that can be characterised by a reactive nature when exposed to heat. Corrode metal containers, and acid bases can be included (Owusu et al. 2014).

Refuse Management: Gravitis (2007) postulates that the industry's product no longer in use is part of the refuse waste. The technological advancement and increasing human population have contributed to waste generation resulting in environmental contamination (Lee and Min 2014). Inappropriate household refuse management in medium-rise buildings is a serious problem because it negatively affects human health and environmental discomfort. Appropriate waste handling and treatment of refuse materials will create a clean, healthy, and safe environment (Kainth 2009). Using a chute includes convenience, maintaining the apartment's cleanliness, making it a safer place to live, avoiding personal contact, and avoiding the need to submit an insurance claim (Hodcutes 2019, Valay 2022). Lawson's (2020) survey admits that the most significant advantage of refuse chutes in medium-rise buildings is that it keeps a safe distance from the refuse itself and the users making the environment and the air clean. Moreover, it prevents offensive odours in the hallways, preventing pests and insects by making it simple to dispose of refuse and better organise waste. According to Serafini (2014), having a trash chute in a high-rise is far more hygienic than having residents carry their trash out. Garbage removal is quite practical as a result. Since no one needs to touch the rubbish after it falls through the chute, less stench, bugs, rodents, and other issues arise from having garbage collected in one spot. Refuse handling as a social issue has neither spared the developed nor developing countries based on statistics showing that some developed nations are seriously grappling with this bane (Chazan 2002). According to Kwetey et al. (2014), 9 out of 10 major African cities face serious waste handling problems.

Policy Regulations to Overcome Waste Problems: According to Hall and Greeno (2015), waste disposal in medium-rise buildings can significantly impact the layout of any residential property; therefore, it is important to ensure that a refuse chute for efficient waste management is incorporated into the layout. Fulford et al. (2018) admit that it is essential that planning authorities, architects/designers as well as collection authorities collaborate to achieve the same policy regulations. This includes determining a preferred approach to residential waste management based on refuse chutes consideration that meets policies and requirements within one guidance. BS 1703 (2020) specifies that building consultants should agree with all appropriate authorities, including refuse collection agencies and local authorities, to plan for refuse chute by considering the methods of storage and collection of waste before building layout and density of waste to be adopted. Serafini (2014) posited that by integrating chutes in the design of a building, anyone residing or working there would have simple access to disposing of waste, maintaining the structure's cleanliness, and simplifying rubbish collection. According to Serafini (2014), construction planners should ensure that every high-rise they construct has at least one good trash chute because they are required for any structure greater than a story. This should incorporate the storage capacity to be provided with an allowance for the collection frequency specified by the collection authority, the volume and nature of waste material expected, and the size and type of additional containers to be used. Wiktoria (2018) conducted a survey and observed that to ensure that all relevant issues are addressed with the refuse chute, the building consultant for medium-rise buildings should consult the following bodies: local authority planning department; local authority environmental health department; waste collection authority. Chi et al. (2004) posit statutory regulations and state that it is essential to incorporate suitable and adequate provisions for the refuse chute and residual waste when designing and planning new and refurbished buildings. Chan and Lee (2006) survey and list the following considerations from a planning perspective of any high and medium-rise building. The size of the refuse chute size of additional refuse containers to be provided given more than one wheeled bin near the chute entrance to allow people to separate different materials. Siu and Xiao (2016) observe that the waste collection authority should specify waste collection frequency. However, practices vary between collection authorities, and consultants must contact the relevant local authority at an early stage.

3 Methodology

This study used a quantitative methodological approach with survey questionnaires in a closed-ended format administered to a targeted population living and working in a medium-rise building in Accra central. 150 questionnaires were distributed to residents (users/tenants) in these buildings within Accra Central; however, 125 were retrieved, representing a response rate of 83%. The sampling technique employed for the research work was a simple random technique; this ensures equal opportunities for respondents to be part of the study. The questionnaire was under three sections, with the first being sought to understand the effectiveness of refuse disposal within the Greater Accra Region. In contrast, section two adopted 5 points Likert scale comprising a range of strongly agree to disagree strongly based on respondents' knowledge of the benefits of Refuse Chute (Likert 1932). However, the third section focused on respondents' knowledge of strategies to overcome challenges related to not using refuse chutes in medium-rise buildings using 5 points Likert scale. Microsoft Excel was used to transport the data from google Forms and later transferred to the Statistical Package for Social Science (SPSS version 26.0) for data analysis to achieve the study's objectives. The appropriate statistical tools used were descriptive and inferential statistics (Kothari 2004).

4 Result and Discussion

Effectiveness of Refuse Disposal within the Greater Accra Region: Respondents was asked if they consider waste handling a struggle, pain, or inconvenience to human life in their medium-rise building. Most respondents believed that they were sure that waste handling done in medium-rise buildings was a struggle, pain or inconvenience to human life. There were one hundred and twenty-four (124) of the entire respondents. Sixty-eight (68) respondents agreed to it. However, forty-three (43) respondents were not in agreement with it. But thirteen (13) were indecisive on the matter Table 1 below illustrates the findings.

Table 1. Waste handling in medium rise building

Response	Frequency	Percentage
Yes	69	55.20
No	43	34.40
Not sure	13	10.40
Total	**125**	**100**

(Source: Field Survey).

Respondents were asked if they "have a refuse chute in their building". The majority indicated that they don't have a refuse chute in their building. Table 2 below explains the respondent's views. Eighty-five (85) stated 'NO' to having a refuse chute in their building, while twenty-seven (27) respondents stated 'YES' to having a refuse chute. However, twelve (12) respondents indicated they were not sure.

Benefits Associated with the Introduction of Refuse Chute for Waste Disposal
The research sought to find out how beneficial it is to install refuse chutes in medium-rise buildings, as displayed in Table 3. The result reveals that nine (9) items were identified to measure the benefit of using refuse chutes in medium-rise buildings. All the variables

Table 2. Refuse chute in the building

Response	Frequency	Percentage
Yes	27	21.60
No	86	68.80
Not sure	12	9.60
Total	**125**	**100**

(Source: Field Survey).

responded by the respondents had a mean score above 3 on the 5-point Likert scale, indicating that respondents agreed with the variables' importance of utilising refuse chutes in medium-rise buildings. The study further analysed the results with the Kolmogorov-Smirnov Z test to determine the significance of the variables, the outcomes indicated in Table 3 discovered that all elements were 99% significant.

Table 3. Benefits of using refuse chute in medium-rise buildings (N = 125)

Code	Chute Benefit	Mean	Standard deviation	Ranking	Kolmogorov-Smirnov Z	Sig. (2-tailed)
CB1	Effective job activities	3.7000	.99488	5th	2.111	.000
CB2	Convenient disposal	3.8000	1.16496	1st	2.306	.000
CB3	Safe environment	3.6800	.97813	7th	2.179	.000
CB4	Hygienic environment	3.7960	.89466	2nd	2.461	.000
CB5	No physical contact of refuse	3.7000	.86307	5th	2.234	.000
CB6	Avoidance of insurance claim	3.5800	.88271	9th	2.283	.000
CB7	Segregation for recycling	3.7800	.86402	3rd	2.266	.000
CB8	Quality air	3.6800	.84370	8th	2.318	.000
CB9	Gem development	3.7200	.83397	4th	2.344	.000

(Source: Field Survey 2022).

Strategies to Overcome the Challenges of Not Using Refuse Chute in Medium-Rise Buildings.

The study investigates strategies to overcome the difficulties associated with not using waste chutes in Accra's medium-rise buildings. According to Table 4, The results show that six (6) items were important in determining how to overcome the difficulties of not using garbage chutes in medium-rise buildings. All of the variables to which respondents responded had a mean score on a 5-point Likert scale greater than 3.6, indicating that respondents concurred with the significance of the variables as means of overcoming the difficulties associated with not using the refuse chute in medium-rise buildings. The Kolmogorov-Smirnov Z test was used in the studies to determine which factors were significant. According to the findings in Table 4. Each component was 99% significant.

Table 4. Strategies to overcome challenges of the refuse chute (N = 125)

Code	Strategies	Mean	Standard deviation	Ranking	Kolmogorov-Smirnov Z	Sig. (2-tailed)
SC1	Development of policy on chute construction	3.8200	.87342	3rd	2.416	.000
SC2	Implementation of chute construction	3.6800	.99877	6st	2.444	.000
SC3	Monitory of chute construction	3.9000	.78895	2nd	2.336	.000
SC4	Educating stakeholders' on chute construction	3.7200	.94847	5nd	1.952	.001
SC5	Architects design chutes in their drawings	3.7800	1.03589	4th	2.433	.000
SC6	Checking chute design in permit approval	3.9400	.89008	1st	2.028	.000

(Source: Field Survey 2022).

5 Discussions

The results from the field survey questionnaire have proven that those living in the medium-rise buildings within the Accra central have no access to refuse chutes. The consequences of not having chutes agree with Owusu-Sekyere (2014) and Akinjare et al. (2011). They postulate that not having proper refuse disposal leads to cholera, malaria, and typhoid in residents near the dumping sites. Additionally, the study discovered many benefits of using refuse chutes in the medium-rise building; the first ranked benefit of using refuse chutes is the promotion of convenient disposal, and these findings support the findings of Hodcutes (2019), Lawson (2020), and Valay (2022) who posited that having chute in the medium-rise building is a convenience for users since no one enjoys carrying refuse all the way outside the structure to be dumped. Refusing chutes has undoubtedly made life easier for those who want to remove refuse from their apartments. All contractors and builders are converting to apartment chutes because convenient waste disposal is crucial in choosing an apartment. The research found that having a refuse chute promotes a hygienic environment. This assertion agrees with Serafini (2014) and Lawson (2020) that having a trash chute in a high-rise is far more hygienic than having residents carry their trash out. The research findings reveal the benefit of using refuse chutes to offer the opportunity for refuse to be segregated for recycling. This finding is in line with Lawson (2020), who stated that refuse chutes offer better waste organising, leading to waste sorting, which contributes to waste recycling.

However, the research further finds the strategies to overcome challenges in medium-rise buildings not having to refuse chutes and reveals that there must be a policy before permit approval to ensure that medium-rise buildings are incorporated with chute design. This includes chute construction and development monitoring, which confirms Serafini's (2014) findings posited that by integrating chutes in the building design, anyone residing or working there would have simple access to disposing of waste, maintaining the structure's cleanliness, and simplifying rubbish collection. According to Serafini (2014), construction planners should ensure that every medium-rise they construct has at least one good refuse chute because they are required for any structure more significant than a few storeys. Again, Poon et al. (2004) posit statutory regulations and state that it is essential to incorporate suitable and adequate provisions for the refuse chute and residual waste when designing and planning new and refurbished buildings.

6 Conclusion

The study has explored how waste can be effectively managed and controlled by introducing refuse chutes in medium-rise buildings within the Greater Accra region of Ghana. This was to ease the struggle, pain, and inconveniences and safely accumulate waste in one discreet location that people do not inhabit. The study found that the benefits of using refuse chutes include convenient disposal, a hygienic environment, recycling segregation, gem development, effective job activities, and no physical contact with refuse. The study explores strategies to overcome the challenges of not using a refuse chute. The outcome includes checking chute design in permit approval, monitoring chute construction, developing policy on chute construction, architects designing chutes in their drawings, and educating stakeholders' on chute construction. This study contributes to the literature on refuse chute construction in a medium-rise building in Ghana.

Recommendations:

1. Regulations must ensure the full compliance of effective refuse disposal by medium-rise building users.

2. The authority who issues permits must ensure that refuse shute in all buildings is supported by fireproof, having at least two hours of fire resistance.
3. Property consultants must carefully plan to ensure adequate space for waste management trucks to turn around.
4. Regulations must be set to ensure that there is a frequency of collection to avoid contamination.
5. Regulations must ensure a front or entrance gully to drain washing up water.
6. The authority who issues permits must ensure that refuse shuts in all buildings are positioned away from habitable rooms.

References

Agyei-Mensah, S., Owusu, G.: Segregated by Neighbourhoods? A Portrait of Ethnic Diversity in the Neighbourhoods of the Accra Metropolitan Area, Ghana. Population, Space and Plac. **16**(6), 499–516 (2009)

Akinjare, O.A., Ayedun, C.A., Oluwatobi, A.O., Iroham, O.C.: Impact of sanitary landfills on urban residential property value in lagos state, Nigeria. J. Sustai. Develop. **4**(2) (April 2011)

Akpen, G.D., Aondoakaa, S.C.: Assessment of Solid Waste Management inGboko Town. Global J. Envir. Sci. **8**(2), 71–77 (2009)

Boadi, K.O.: Environment and Health in the Accra Metropolitan Area, Ghana. Jyvaskyla Studies in Biological and Environmental Science 145. Academic Dissertation (2004)

BS 1703: Refuse chutes and hoppers – Specification, 5th Edition. British Standard Institution (BSI) (2020)

Chaban, M.A.V.: The Appraisal: Garbage Collection, without the Noise or the Smell. The New York Times (2015). https://www.nytimes.com/2015/08/04/nyregion/garbage-collection-without-the-noise-or-the-smell.html. Accessed: 04 February 2021

Chan, E.H.W., Lee, G.K.L.: A review of refuse collection systems in high-rise housings in Hong Kong. Facilities **24**(9/10), 376-390 (2006). https://doi.org/10.1108/02632770610677655. Accessed: 04 February 2021

Chazan, D.: A World Drowning in Litter. BBC NEWS (2002). http://news.bbc.co.uk/2/hi/europe/1849302.stm. Accessed: 04 February 2021

Chi, S.P., Yu, A.T.W., Wong, S.W., Cheung, E.: Management of construction waste in public housing projects in Hong Kong. Constr. Manag. Econ. **22**(7), 675–689 (2004). https://doi.org/10.1080/0144619042000213292

Curran, T., Williams, I.D.: A zero-waste vision for industrial networks in Europe. J. Hazard. Mater. **207–208**, 3–7 (2012)

Durán, C.E.S., Messina, S.: Urban Management Model: Municipal Solid Waste for City Sustainability, Municipal Solid Waste Management, Hosam El-Din Mostafa Saleh, IntechOpen (2019). https://doi.org/10.5772/intechopen.82839. Available from: https://www.intechopen.com/chapters/65485

Fulford, J., Slack, A., Gillies, R.: Waste in Tall Buildings Study: Final Report for OPDC. Volume 1.0 (2018)

Gabrscek, A., Isljamovk, S.: Communal Waste Management: Case Study for Slovenia. Manage. J. Theory and Prac. Manage. **16**, 34–41 (2011)

Gravitis, J.: Zero Techniques and Systems – Zets Strength and Weakness. Journal of Cleaner Production **15**, 13–14 and 1190–1197 (2007)

Gumbo, B., Mlilo, S., Broome, J.H., Lumbroso, D.M.: Industrial water demand management and cleaner production potential: a case of three industries in bulawayo, Zimbabwe. Phys. Chem. Earth **28**, 797–804 (2003)

Hait, P., Sil, A., Choudhury, S.: Damage assessment of low to mid-rise reinforced concrete buildings considering planner irregularities. Int. J. Comput. Methods Eng. Sci. Mech. **22**(2), 150–168 (2021). https://doi.org/10.1080/15502287.2020.1856971

Hall, F., Greeno, R.: Building Services Handbook, 8th edn. Routledge (2015)

Hodcutes: Garbage Chute Benefits (2019). https://hodchutes.com/garbage-chute-benefits/#:~:text=Using%20a%20trash%20chute%20for,items%20that%20can%20harm%20them.&text=With%20construction%20trash%20chutes%20installed,another%2C%20collecting%20waste%20or%20trash. Accessed on March 2022

Kothari, C.R.: Research Methodology –Methods and Techniques. 5th Ed., New Age International (P) Ltd. New Delhi (2004)

Kwetey, S., Cobbina, S.J., Asare, W., Duwiejuah, A.B.: Household demand and willingness to pay for solid waste management service in tuobodom in the Techiman-North District. Ghana. American Journal of Environmental Protection. **2**(4), 74–78 (2014). https://doi.org/10.12691/env-2-4-3

Lawson, E.: Why Garbage Chutes Are a Must in Apartments (2020). https://constructionexec.com/article/why-garbage-chutes-are-a-must-in-apartments. Accessed on March 2022

Lee, K.-H., Min, B.: Globalisation and carbon constrained global economy: a fad or a trend? Journal of Asia-Pacific Business **15**(2), 105–121 (2014)

Likert, R.: A technique for the measurement of attitudes. Archives of Psychology **22**(140), 55 (1932)

Macarthur, E.: Towards The Circular Economy. Economic and Business Rationale for an Accelerated Transition, vol 1. Rethink The Future, Ellen MacArthur Foundation (2013)

Mansour, G., Esseku, H.: Situation analysis of the urban sanitation sector in Ghana. WSUP: Water & Sanitation for the Urban Poor (2017). Online at: https://www.wsup.com/content/uploads/2017/09/Situation-analysis-of-the-urbansanitation-sector-in-Ghana.pdf. Accessed: 19 November 2021

Nwosu, A.E., Olofa, S.A.: Effect of Waste Dumpsites on Proximate Residential Property values in Ibadan, Oyo State, Nigeria. Ethiopian J. Enviro. Stud. Manage. **8**(Suppl. 2), 976 – 982 (2015). ISSN:1998-0507

Odonkor, S.T., Frimpong, K., Kurantin, N.: An assessment of household solid waste management in a large ghanaian district. Heliyon. **6**(1), E03040 (2020)

Owusu, G., Nketiah-Amponsah, E., Codjoe, S.N.A., Afutu-Kotey, R.L.: How do Ghana's landfills affect residential property values? A case study of two sites in Accra. Urban Geography **35**(8), 1140–1155 (2014). https://doi.org/10.1080/02723638.2014.945261

Owusu-Sekyere, E.: Scavenging for wealth or death? exploring the health risk associated with waste scavenging in Kumasi, Ghana. J. Geogr. **6**, 63–80 (2014)

Post, J., Obirih-Opareh, N.: Quality assessment of public and private modes of solid waste collection in accra, Ghana. Habitat International. **26**, 95–112 (2002)

Serafini, C.: Benefits of Having a Trash Chute (2014). https://wadearch.com/trach-chute-benefits-of-having/. Accessed on March 2022

Singh, Y., Singh, A.K., Singh, R.P.: Web-GIS-based framework for solid waste complaint management for sustainable and smart city. Int. J. Adva. Remo. Sens. GIS. **5**(1), 1930–1936 (2016). https://doi.org/10.23953/cloud.ijarsg.71

Siu, K.W.M., Xiao, J.X.: Design and management of recycling facilities for household and community recycling participation. Facilities **34**(5/6), 350–374 (2016). https://doi.org/10.1108/F-08-2014-0064

Valay, P.: All You Need To Know About Garbage Chute (2022). https://www.envcure.com/all-you-need-to-know-about-garbage-chute/

Wiktoria, G.: Spaces for waste: everyday recycling and sociospatial relationships. Scottish Geographical Journal. **134**(3–4), 141–157 (2018). https://doi.org/10.1080/14702541.2018.150 0634

Zahedi, M., Eshghi, S.: A new method for seismic collapse assessment of mid-rise concrete buildings. Iranian J. Sci. Technol. Trans. Civil Eng. **45**(2), 1159–1181 (2020). https://doi.org/10.1007/s40996-020-00394-w

A Review of Frameworks for the Energy Performance Certification of Buildings and Lessons for Ghana

G. Osei-Poku[1,2(✉)], C. Koranteng[2], S. Amos-Abanyie[2], E. A. Botchway[2], and K. A. Gyimah[2]

[1] Department of Building Technology, Faculty of Built and Natural Environment, Takoradi Technical University, Takoradi, Ghana
gloseipoku@yahoo.co.uk

[2] Department of Architecture, College of Art and Built Environment, Kwame Nkrumah University of Science and Technology, Kumasi, Ghana

Abstract. Purpose: Buildings are known as a huge expender of energy especially the ones that have existed over a considerable period. However, global energy supply faces challenges, prompting the development of several frameworks to assess the performance of buildings energy-wise. This research aimed at examining energy performance certification frameworks for a possible country-specific adoption in Ghana.

Design/Methodology/Approach: To achieve this aim, scholarly publications were retrieved from the electronic search engines of Google Scholar and Research Gate. These selected papers, published in the decade between 2010 and 2020 (both years inclusive), were perused to identify the elements of the frameworks, to examine the certification process and to explore the features on Energy Performance Certificates (EPCs).

Findings: It was discovered that most energy performance assessment frameworks were developed concurrently with sustainable building rating tools. The basic features indicated on an EPC were the energy ratings and cost implications for both current and potential cases, the energy efficiency improvement recommendations and a validity period. Findings also revealed that although proposals had been made for building energy efficiency assessment, there is no framework implemented in Ghana to certify the energy performance of buildings.

Research Limitations/Implications: The study was limited to an identification of the elements in energy performance frameworks, the certification process and features on the certificate.

Practical Implication: The lessons learnt from this review would guide in the adoption of a localised EPC framework to evaluate office buildings in Ghana as a means of promoting energy efficiency from the building sector.

Social Implication: Information provided on the certificate issued would influence the decisions of several stakeholders including owners, buyers and renters of office buildings.

© The Author(s), under exclusive license to Springer Nature Switzerland AG 2023
C. Aigbavboa et al. (Eds.): ARCA 2022, *Sustainable Education and Development – Sustainable Industrialization and Innovation*, pp. 63–80, 2023.
https://doi.org/10.1007/978-3-031-25998-2_6

Originality/Value: This forms part of a broader study aimed at formulating and validating a framework for rating the energy performance of office buildings in Ghana.

Keywords: Energy efficiency · Energy performance certificate · Framework · Office building · Sustainable building

1 Introduction

Buildings provide immense benefits to occupants including shelter and comfort. Yet, they also impact negatively on the environment through the usage of resources including energy. It is well-documented that buildings are a major consumer of energy. Available statistics reveal that an estimated 40% of energy worldwide is expended by buildings from the time of construction (International Energy Agency (IEA) 2019; Baharom et al. 2015; US Department of Energy (DOE) 2011). Out of this total amount of energy consumed, about 30% to 40% is utilised for regulating the indoor environment and running appliances (Nor Azuana et al. 2019; IEA 2010). With current climate change challenges, it is envisaged that there would be an increased demand in energy consumption in the coming years. However, in spite of this need, energy provision in many countries continuously poses a challenge partly because of depletion of the sources of fossil fuels at an alarming rate (Kemausour et al. 2011; Iwaro and Mwasha 2010a; Sassi 2006). Energy is a vital ingredient towards national development albeit its sourcing and consumption should not be detrimental to the environment. Hence, worldwide efforts are being made to minimize the environmental impact of buildings on the planet Earth through the design and construction of green and sustainable buildings which are energy-efficient among other things. As such, several sustainable building rating tools have been developed by different countries to evaluate the performance of buildings from the inception stage through to the post-occupancy stage. All these rating schemes have efficient energy use as a major determinant of how sustainable a building is.

To further address the effect of buildings on the environment and on resources, a number of schemes have also been designed to assess and certify the energy consumption of buildings the world over. This is so because buildings are regarded as an important source for energy efficiency improvements since they are the largest expender of that resource globally (IEA 2019). Accordingly, energy efficiency in buildings is touted as one of the best approaches in the global quest to save the environment. In this regard, countries within the European Union (EU) have been labelled as leaders in building energy efficiency systems (Janda 2009) as they have established mandatory certification schemes since the early part of this millennium under the Energy Performance of Buildings Directive (EPBD 2002/91/EC). In making the energy performance of buildings a priority, it became compulsory for EPCs to be issued when structures were built, sold or let in Europe from 2008. The directive covers both old and new buildings (The Buildings Performance Institute Europe (BPIE) 2010). The initial directive saw some amendments over the years, namely EPBD recast (EPBD 2010/31/EC) and EPBD 2018/844/EU, all having the prime focus of achieving net zero energy buildings (NZEB) by the year 2050 (Build Up 2020).

Similarly, in the United States (US), energy efficiency assessment schemes such as Energy Star, Building Energy Quotient (BEQ) and Energy Asset Rating (AR) are operational albeit not compulsory as that of Europe (Martinez et al. 2016; Kontokosta 2015; Fuerst et al. 2012; Fuerst and McAllister 2011; Jarnagin 2009). Other countries in North and South America such as Canada and Brazil also have some building energy labelling schemes in place (Fossati et al. 2016; Qian and Chan 2010). Some Asian countries for example, China, India, Japan, Malaysia, Singapore and South Korea have also developed building energy rating schemes (Bureau of Energy Efficiency (BEE) India 2017; Jeong et al. 2017; Vyas et al. 2014; IEA 2010). Statistics from the African continent revealed that South Africa has a building energy rating scheme (Martin 2013) while others including Egypt, Ghana, Morocco, Nigeria and Tunisia have also applied some sustainable energy rating schemes to assess the performance of buildings (Anzagira et al. 2019; Amasuomo et al. 2016; Hanna 2015). There is ample evidence from the literature to support that the formulation and use of the certification schemes have helped with energy efficiency in countries which have executed same (Ministry of Power, India (MPI) 2020; BEE 2017; ECOWAS Centre for Renewable Energy and Energy Efficiency (ECREE) 2015; Ministry of Power, Ghana (MPG) 2015; Gyimah and Addo-Yobo 2014; IEA 2010; Iwaro and Mwasha 2010b; Janda 2009; Jarnagin 2009).

However, in Ghana, just like many developing countries in sub-Saharan Africa (SSA), there are no specifically developed energy efficiency frameworks to assess and certify buildings (MPG 2015; ECREE 2015; Iwaro and Mwasha 2010a). Yet, optimising efficient energy use especially from the building sector is a prime focus of the National Energy Policy of Ghana (Ministry of Energy, Ghana (MEG) 2010). As such, under the Ghana Electrical Appliance Labelling and Standards Programme (GEALSP), regulations have been put in place towards the efficient consumption of energy by household appliances such as refrigerators, air-conditioners and compact fluorescent lamps (CFL). By the Legal Instrument (LI 1932:2008), manufacturers, importers and sellers have been banned from manufacturing, importing and selling obsolete and energy inefficient products such as incandescent filament lamps, used refrigerators and used air-conditioning units. Instead, they are required to produce, import and sell only electronic appliances that meet the energy-rating standards set by the Ghana Standards Authority (MPG 2015). Additionally, all these appliances must display conspicuously the energy performance rating label (indicated by the number of stars) before they are first sold (Energy Efficiency Standards and Labelling Regulations 2005 and 2009; Energy Efficiency Regulations 2008). Reports authored by the institutions responsible for monitoring have confirmed that the compliance to these appliance regulations have helped in promoting energy conservation from the building sector (Energy Commission, Ghana 2019; MPG 2015), although the bigger potential from the buildings themselves remains to be explored.

Consequently, the lack of an energy performance certification scheme to assess and label buildings in the country presents a gap in the journey to achieve energy efficiency in Ghana. To this end, some previous studies have been undertaken by researchers which were focused on developing energy efficiency assessment and rating schemes for buildings in Ghana. Gyimah and Addo-Yobo (2014) assessed the possibility of applying EPC systems to residential buildings in the country. The authors developed and utilized the Building Energy Assessment Procedure (BEAP) model to assess the energy performance of a single case study in Tema, Ghana. Their findings revealed that buildings users would consume less energy in attempt to achieve better ratings for their structures. In the study of Addy (2016), a Building Energy Efficiency Assessment (BEEA) tool was developed for office buildings with the aim of assessing their energy performance at the design stage. Using a single existing case study building, Addy (2016) suggested that the BEEA could be applied to already constructed buildings as well. It is thus imperative to assess the components of these frameworks and test them on other buildings to ascertain their veracity. Furthermore, stakeholder consultations have been kick-started as part of efforts to implement building energy performance certification schemes in the country.

The overarching aim of this study, therefore, was to review the literature to examine energy performance certification frameworks for buildings towards a possible country-specific adoption in Ghana. The objectives were to identify the elements incorporated in an EPC framework for buildings; to examine process of certifying the energy performance of buildings; and to ascertain, by comparison, the features usually displayed on energy performance certificates globally. The significance of this review is two-fold. It primarily provides insights into the generation of an energy performance certificate by highlighting the necessary elements of frameworks that would guide the in the application of an energy rating scheme for office buildings in Ghana. Such a framework, would reveal the energy implications of the buildings and remedy any inefficiencies discovered. Secondly, this review contributes knowledge towards the implementation of EPCs in sub-Saharan Africa (SSA).

2 Literature Review

Energy performance of buildings is the measurement of the efficient use of a building to achieve its standard function while ensuring the users achieve an acceptable level of comfort (IEA 2010). A lot of countries have consequently developed tools to assess and certify the energy performance of their buildings; both at the design stage and for already constructed ones (Paterson et al. 2017; Ali and Al Nsairat 2009). This has been done concurrently with that of various sustainable and green building rating schemes since the goal of EPC is part of the sustainability agenda (Table 1). The frameworks have been developed taking into consideration the climatic and cultural factors which directly influence how diversely energy is utilised globally. As such, some of these energy performance evaluation tools are embedded within sustainable building frameworks such as the Building Research Establishment's Environmental Assessment Method (BREEAM), the Leadership in Energy and Environmental Design (LEED), the Comprehensive Assessment System for Building Environmental Efficiency (CASBEE), the Green Building Tool (GBTool), the Excellence in Design for Greater Efficiencies (EDGE), the Green Star labelling and the Sustainable Building Tool (SBTool) (Kim et al. 2020; Akbarova 2018; Nikolaou et al. 2011; Fowler and Rauch 2006). Illustrated in Table 1 are selected frameworks as evinced from literature.

Table 1. Overview of existing frameworks for evaluating building energy performance

Country	Name	Type of framework EPC	Type of framework SBR	Source
Australia	Green star		X	Sin et al. (2011)
	Energy star	X		IEA (2010)
Azerbaijan	AZERI green zoom		X	Akbarova (2018)
Canada	SB tool		X	Alyami and Rezgui (2012)
	EnerGuide	X		Qian and Chan (2010)
China	Energy label system	X		Yu et al. (2019); Zhou et al. 2013; Li and Yao (2012); IEA (2010)
	Green building label and certification		X	Li and Yao (2012)
Egypt	Green pyramid rating system		X	Hanna (2015)
France	Effinergie	X		Somuncu and Menguc (2016)
Germany	DGNB		X	Kim et al. (2020)
	Passivhaus	X		Somuncu and Menguc (2016)
Hong Kong	HK-BEAM		X	Qian and Chan (2010)
India	Star rating programme	X		BEE (2017); Vyas et al. (2014)
	Green rating for integrated habitat assessment' (GRIHA)		X	Vyas et al. (2014)
Japan	CASBEE		X	Alyami and Rezgui (2012)
Malaysia	Green building index		X	Assadi et al. (2016); Sin et al. (2011)
	EPCert scheme	X		Adam et al. (2016)
Russia	Green ZOOM		X	Akbarova (2018)
Singapore	Energy smart	X		IEA (2010)
	Green mark		X	Sin et al. (2011)
South Africa	Green star		X	
	Energy barometer	X		Martin (2013)

(*continued*)

Table 1. (*continued*)

Country	Name	Type of framework		Source
		EPC	SBR	
South Korea	Green standard for energy and environmental design		X	Kim et al. (2020)
	BECC	X		Jeong et al. (2017)
Switzerland	Minergie	X		Somuncu and Menguc (2016)
UK	BREEAM		X	Alyami and Rezgui (2012)
	EPC, DEC	X		
US	Energy star	X		Kontokosta (2015)
	LEED		X	Alyami and Rezgui (2012)

Legend: EPC – energy performance certification; SBR – sustainable building rating.

From Table 1 above, it is observed that while some countries have only one framework which holistically assesses the environmental impact of buildings with energy efficiency as a component, others concurrently have a scheme specifically for energy performance certification of buildings. These countries can be found in different continents including Africa (South Africa), Asia (India, Malaysia, and Singapore), Europe (Germany, United Kingdom), North America (Canada, USA) and Oceania (Australia).

3 Methodology

This qualitative research employed a systematic literature review strategy to gather and analyse data on EPC frameworks within the period under review. The process involves probing extant literature to identify articles with adequate information to address the set objectives of the study. The qualitative approach is appropriate as it helps one to explore salient issues from existing literature and helps to synthesise ideas (Groat and Wang 2013). Also, this methodology is recognised as a means of knowledge expansion because it reveals gaps in the findings of closely related prior studies and provides a standard to compare and contrast results to establish the importance of the present study (Creswell 2009; Aveyard 2007; Webster and Watson 2002).

Scholarly publications were selected by searching for descriptors in Google Scholar and Research Gate, with a focus on the period between 2010 and 2020 (both years inclusive). These search engines were used due to their ability to facilitate searches across different sources. The following key phrases were searched independently of each other *"building EPC frameworks"*, *"energy efficiency frameworks"*, *"features of EPCs"*, *"energy performance certification process"*. These key words are not exhaustive since it is impractical to consider all possible descriptors in a standalone study. The initial search yielded a large body of literature as energy efficiency is a widely researched area in the quest for global sustainability. To select a representative sample, a rigorous screening procedure involving the use of inclusion and exclusion criteria was adopted. All articles labelled *'editorial'*, *'in press'*, *'accepted, pending publication'* were eliminated from the selection. Furthermore, the title and abstract of each paper were skimmed to select the ones which provided specific information on EPC frameworks for buildings as well as features on EPCs. A critical systematic review and content analysis was performed on a total of 70 selected peer-reviewed publications. This number was made up journal articles, conference papers, books and technical reports.

4 Findings and Discussion

4.1 EPC Framework Development

4.1.1 Definition and Elements of a Framework

The Merriam-Webster dictionary online (2021) defines a framework as "a basic conceptual structure (as of ideas)". In essence, it is a supporting or rudimentary structure underlying a system or concept. A framework can also be described as a set of principles or ideas used when forming new decisions and judgements. When one intends to create a framework, the goal, target group and scope must be defined and properly addressed. Then the aspects required to constitute the framework are listed in a significant order after a review of literature to establish these. Afterwards, the input data required are collected and analysed in order to create the framework, which is then tested to ascertain its effectiveness on real scenarios (Ali and Al Nsairat 2009).

4.1.2 Developing an EPC Framework

It is the goal of every EPC framework to achieve efficient energy use in buildings by a determined deadline. Thus, the elements included in the framework, in order of importance, are the legislation, the administrative component, the data collection and quality assurance element, and the certification part as illustrated in Fig. 1.

The elements are described as follows:

1. *Legislation:* This refers to the legal and regulatory foundation upon which the drafting and introduction of policies and programmes on efficient energy consumption are built. In the legal framework, the schedules for setting and revising standards as well as penalties for non-compliance are defined (Zhou et al. 2013). One example of legislation is the EPBD of the EU, which mandated all member states to implement EPC schemes for buildings by 2008 (BPIE 2010). The current established legislative frameworks to reduce energy demand from buildings in the EU are the EPBD 2018/844/EU and EED 2018/2002/EU (Build Up 2020). In West Africa, all ECOWAS countries have also been required since 2013 to formulate energy efficiency standards to help in collectively achieving a regional target by 2030 through the ECOWAS Energy Efficiency Policy (EEEP) of 2013 (MPG 2015). Similar to the EPBD of Europe, the ECOWAS Directive on Energy Efficiency in Building (EDEEB)

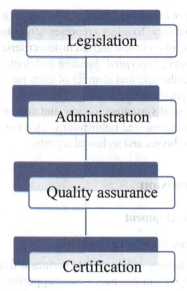

Fig. 1. Elements of EPC framework. Source: Authors' construct (2022)

has prompted the development of various policy documents by West African countries. As at 2015, the ECOWAS Centre for Renewable Energy and Energy Efficiency reported that the full operationalization of the policies in most of ECOWAS countries was yet to see the light of day (ECREE 2015).

Next under the legal aspect is for an individual country to take up the initial mandatory regional document and incorporate it within its own policies and planning documents whilst considering country-specific contexts, abilities and resources available. This involves the formulation and drafting of national legislations and regulations on building energy standards as well as national building codes with dedicated sections specifying the minimum requirements for the design and construction of energy-efficient buildings (BEE 2017; Panayiotou et al. 2010). In Ghana, for instance, a 5-year action plan, the National Energy Efficiency Action Plan (NEEAP) was drawn in 2015 in response to the ECOWAS call (MPG 2015). This action plan includes projected achievable energy efficiency targets from all sectors. Additionally, the draft new building code of Ghana (GS 1207:2018) has provisions in Part 14, Part 36 and Part 37 for mandatory energy efficiency and green building certifications respectively as part of the requirements for building permit acquisition for both new construction and renovations to existing buildings (Ghana Standards Authority (GSA) 2018). Similarly, in South Africa, the Energy Efficiency in Public Buildings and Infrastructure Programme (EEPBIP) provides a framework for all government buildings to contribute to energy efficiency and GHG mitigation targets (NAMA Facility 2021).

2. *Administration:* Under the administrative component is the implementation of legislations and regulations phase. At this stage, institutions and energy agencies responsible for the implementation of regulations and drafting of strategies to achieve same

are designated. For example, some institutions in Ghana with specific roles towards the enactment and implementation of energy efficiency policies are the Ministry of Energy, the Energy Commission, the Ghana Energy Foundation, the Ghana Standards Authority and professional bodies such as the Ghana Green Building Council (Edjekumhene 2017; MPG 2015). Afterwards, assessment methodologies are developed to conform to the resources at the disposal of the particular country. The choice of which type of rating is also established by the administrative institutions. For example, countries like Australia, China and India use energy star ratings (BEE 2017; IEA 2010) where the more stars indicate better efficiency. Such a star-rating system is currently used in the appliance labelling scheme in Ghana (Andoh 2020; Energy Commission, Ghana 2019). Similar to the star rating is the flowers icon system employed in Malaysia (Adam et al. 2016). However, the EU member states use the class band system which ranks the building's energy performance on an alphabetic scale (BPIE 2010).

3. *Quality assurance:* One of the guiding elements in developing an EPC framework is the quality assurance element. It involves controlling the quality of EPC data while ensuring that competent experts undertake the data collection and building certification (Volt et al. 2020). As averred by Li et al. (2019), the quality of the EPC is influenced by the input data, adopted methodology and the competence of the energy assessor. Consequently, crucial measures are put in place to ensure accurate and credible data of the EPC system. In this regard, the procedures involved in implementing the certification scheme such as daily administration are well-defined to ensure quality. Data collection strategies are mapped out within available resources and the methodologies for estimating the energy efficiency of buildings are defined. The characteristics of the existing building stock are noted since this plays a major role in recording credible data. Furthermore, expert stakeholders such as energy consultants and software developers are trained to handle the entire collection and processing of data as part of the control of the EPC data quality (Volt et al. 2020; BPIE 2015). To achieve this successfully, some EU countries have set minimum educational and professional requirements for certifiers (Li et al. 2019). Additionally, there are mandatory examinations of assessors' credential and recommendations for them to periodically undertake refresher courses to keep up with changing requirements in updates to directives (Cozza et al. 2020; Li et al. 2019).

4. *Certification:* This refers to issuing a certificate with the necessary details to reflect how energy efficient the assessed buildings are. After credible data has been collected by qualified experts, a certificate is given out, indicating the energy efficiency and environmental impact ratings of the buildings. All the data on individual buildings are compiled and stored in national databases for record keeping. The certification process is further described in the subsequent section.

4.2 The Energy Performance Certification Process for Existing Buildings

EPC, as a policy instrument, is an important tool aimed at stimulating market mechanisms to promote energy savings through the improvement of efficient energy consumption within the building sector (ECA 2020; IEA 2010). An EPC offers a chance to rate individual buildings by virtue of the amount of energy required to promote comfort and functionality within them (IEA 2010) and gives direction for enhancing the energy performance of buildings. The process of building energy performance certification can be grouped into three main categories of activities, which, according to the IEA (2010) and Pérez-Lombard et al. (2009) are.

- Evaluating the energy performance by a certified assessor using an approved, defined methodology
- Issuing a certificate which rates the building and displays energy classification
- Publishing the certificate to communicate the information on it to stakeholders (IEA 2010).

4.2.1 Evaluating the Building Energy Performance

A building's energy performance is defined as a measure of the total energy to be consumed by both the components and equipment within during standard use of a building (Norvaisiene et al. 2014). Evaluation of the energy performance is usually carried out by a qualified person with the relevant skill and experience to deliver reliable and accurate results (Volt et al. 2020; BPIE 2015). During the assessment process, this certifier works having an idea of what to do (scope), how to do it (method) and why it is being done (importance).

In assessing the building, the Energy Performance Index (EPI) or the Energy use intensities (EUI) is evaluated, in reference to minimum and maximum efficiency requirements as established by legislation (Pérez-Lombard et al. 2009). The terms 'EPI' as used in Europe, and 'EUI' as used in the US, refer to the numeric indicators of the energy efficiency or the annual energy use per unit area, often measured in kilowatt-hours per square meter (kWh/m^2) as indicated in Eq. 1 (BEE 2017; Kontokosta 2015).

$$EPI = \frac{\text{annual energy consumption in kWh}}{\text{total built} - \text{up area}} \quad (1)$$

The determination and selection of a methodology for assessment is a crucial portion of the certification process as the assessment method used impacts greatly on the results achieved (Panayiotou et al. 2010; Roderick et al. 2009). To this end, what parameters and data are to be collected, how accurate and comprehensive the calculation methodology employed is and the possibility of replicating the results using another method are very important factors that are considered (Harputlugil 2018; Kelly et al. 2012; BPIE 2010; IEA 2010; Ali and Al Nsairat 2009; Fowler and Rauch 2006).There are well-documented methods for calculating and reporting energy efficiency as stipulated in ISO 17743 and ISO 17742 (ISO 2016; ISO 2015). Either one of two main approaches – calculated (asset rating) or measured (operational rating) – is employed in post-design energy performance assessment and labelling (ISO 2016; ISO 2015; Li and Yao 2012). Asset rating is a theoretical simulation-based evaluation (intended for new constructions), which estimates the energy performance based on factors such as the building characteristics, energy-dependent services, age and condition of buildings while operational rating involves measurement and analysis of the real time energy consumption of existing buildings (Adam et al. 2016; Li and Yao 2012; Panayiotou et al. 2010). In some European countries, the energy ratings are calculated based on the total primary energy while others rely on annual demand for heating energy to determine the buildings' energy performance (ECA 2020).

4.3 Features of the Certificate

As part of the certification process, a tangible certificate, described by Kelly et al. (2012) as "a *necessary and important output of carrying out building performance evaluations*" is issued and published for the consumption of relevant stakeholders (IEA 2010). This section describes the nature of EPCs, presents what features are commonly displayed on the certificates and explores the benefits of the information provided on the certificate. As an output, they have one or more pages depending on what information is given on the certificate and are sometimes accompanied by evaluation reports. Usually the basic information displayed on the EPC include the energy rating, environmental impact rating, energy efficiency improvement recommendations and validity period. The certificates from different countries were compared to ascertain their features, noting similarities and differences in their appearances. Of the information, what is captured on the first pages of these certificates are presented and described below.

A. *Building information*
   Information about the type of building being assessed is a major section of the certificate. This reveals the type of structure (whether residential or commercial), the total floor area, the number of floors, the locational address and the methodology for the assessing the particular building. This section has the advantage of presenting a summary of the building's details at a glance (Fig. 2).

B. *Energy use and associated costs*
   Displayed on the certificate is the amount of energy consumed presently by the structure vis-à-vis the potential energy to be used after application of suggested remedial measures. Also, the associated cost of each section is indicated. The benefit of this information is that occupants are guided on how to use the buildings effectively in order to enhance their energy performance. Additionally, it provides the opportunity for prospective buyers or renters of the building to understand the current energy performance of the structure, what cost implications are involved to remedy any inefficiencies and how much the energy consumption could be improved in the long run. Consequently, when such information on similar buildings are compared, an informed choice can be made on which property to buy or rent.

C. *Class labels/ rating*
   This is graphically represented on the certificate using media including colour bands (in Europe), stars (in China) or other icons such as flowers (in Malaysia). In Europe, the energy efficiency rating was initially labelled on a 7-band alphabetical scale from 'A' as the most efficient to 'G' as the least efficient (BPIE 2010). However, over the period of implementation, some countries have modified the limits of the classes; some have changed the outlook of the certificates while others have rescaled the class labels to 8 or 10 bands (Heijman and Loncour 2019; Norvaisienne et al. 2014). According to Heijman and Loncour (2019), these changes were the result of updates in the directives and also because some countries wanted to increase public patronage by enhancing the appeal of the EPCs. The class labels are displayed with

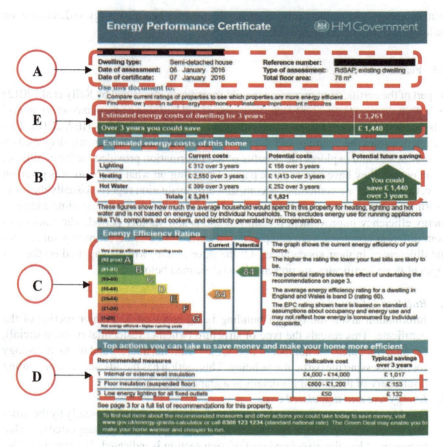

Fig. 2. Sample energy performance certificate

rainbow colour codes similar to a reverse traffic light system where the colour green represents very efficient while red indicates least efficient (illustrated in Fig. 2). In China, the most energy efficient building is rated 3 stars with the least being given 1 star (Yu et al. 2019) while in Malaysia the number of flowers indicating the highest energy efficiency is 7 and 1 star shows least efficiency (Adam et al. 2016).

Both the current and the potential energy ratings are shown; this has the advantage of easily disseminating the necessary information to occupants and other stakeholders with interest in the buildings (BPIE 2010). The downside of the ratings, however, as averred by the Energy Saving Trust (EST) is that they are based on standard assumptions about occupancy and energy use and may not reflect how energy is actually consumed by individual occupants (EST 2021). This implies that buildings may be given wrong class labels both in the present and potential sense because the real time energy use is not factored in the assessment. Thus for existing office buildings, the operational rating methodology is a better option as demonstrated by countries such as Belgium, Sweden and the UK (Volt et al. 2020).

D. *Energy efficiency improvement suggestions*

Accompanying the certificates are recommendations for enhanced energy use, listed in order of importance. This is a major part of the EPC scheme as it provides property owners and occupants with experts' opinions on what energy efficiency improvements they can undertake, how much it would cost them and the potential energy and monetary savings they could make over a given period (Li et al. 2019). However, although this is aimed at guiding building occupants on what energy and financial savings they can make, the information is not a holistic reflection on the energy usage in a building. It was discovered that the EPC presently used by most countries in Europe evaluated only the energy used by a building for heating, hot water and lighting. The energy consumed by other electrical appliances such as refrigerators, television sets and office equipment (computers, printers, scanners, etc.) are not considered in the assessment (EST 2021; Herrando et al. 2016). In such an instance, there could be misleading interpretations as the use of these appliances and equipment have a significant impact on the building's energy performance and could ultimately affect the resultant recommendations for improvement.

E. *Validity period*

A time frame is given when an issued certificate holds valid in presenting the energy performance of the assessed building. This period, Fowler and Rauch (2006) aver, is important in order to capture new developments of the rating systems and to reveal information on the building's present energy use. A re-evaluation of the energy efficiency is undertaken for an updated certificate after the period elapses or when there are dynamics in occupancy and use (Kelly et al. 2012). The maximum validity period of a certificate is 10 years in Europe (von Platten et al. 2019; BPIE 2010). However, in countries such as Sweden, a re-assessment of actual energy efficiency is required after two years of putting up and issuing an EPC for a building. This is aimed at providing up-to-date, real time information on the energy usage of a building. The validity duration in Malaysia is much less – three years – as asserted by Adam et al. (2016). In China, building energy labels for new buildings are valid for a year while operational ratings for occupied buildings expire after 5 years (Yu et al. 2019).

5 Conclusion

5.1 Summary

Present global climate challenges are driving an increase in energy demand from buildings. This has necessitated the formulation of building energy performance evaluation and certification schemes to address the environmental impact of structures. A vital output of this evaluation is the EPC. The purpose of this study was to examine EPC frameworks to identify the elements involved, to examine the certification process and explore the information provided on the certificates. A total of 70 scholarly publications comprising journal and conference papers were selected from the databases of Google

Scholar and Science Direct using a three-step selection procedure. The findings evinced that building energy performance certification (EPC) schemes and sustainable building rating (SBR) tools are linked; the former sometimes being a subset of the latter. It was discovered that all the sustainability tools assess energy performance as a core component to rate the 'greenness' of buildings, aside other elements such as indoor environmental quality (IEQ), water efficiency, site and land use, materials and resources as well as waste management (Alyami and Rezgui 2012; Ali and Al Nsairat 2009). On the other hand, there are EPC frameworks that focus solely on evaluating and labelling the energy efficiency of buildings. The elements of EPC frameworks are legislation, administration, quality assurance and certification. Furthermore, the features of the certificates were similar across various schemes globally. They indicated basically the energy consumption, cost implications and energy rating for the time of assessment and for the potential sense if recommendations were applied. Additional information included the building description, energy efficiency improvement suggestions and a validity period. Countries that had implemented the EPC schemes were backed by legislation and had adopted localised contexts for greater acceptability by their populace. Additionally, they employed expert certifiers for quality assurance. It can be concluded that EPCs make building energy efficiency transparent.

5.2 Lessons for Ghana

From the foregoing review of literature, some pertinent lessons in proposing a country-specific EPC for Ghana include the following:

- The need for a strong legal backing to ensure maximum use and compliance by stakeholders;
- It is necessary to select an appropriate national methodology which considers the culture, climatic context and resources available within the country to enhance acceptability;
- Quality assurance should be a major guiding principle in developing the frameworks to guarantee the consistency and credibility of the data presented on the certificates;
- Expert knowledge is required for an effective EPC implementation. Therefore, professional certifiers should be engaged and mandated to undergo periodic refresher training to keep up with global trends.
- The information displayed on the certificates should be as clear, consistent and credible as possible to effectively guide people to make well-informed decisions when choosing buildings.

References

Adam, S., et al.: Implementation of energy performance certification (EPCert) scheme for government buildings in melaka. In: 3rd National Conference on Knowledge Transfer. Penang (2016)

Addy, M.N.: Development of building energy efficiency assessment tool for office buildings in Ghana. PhD Thesis, Kwame Nkrumah University of Science and Technology, Department of Building Technology, Kumasi, Ghana. Retrieved June 10, 2021 (2016)

Akbarova, S.: Trends of energy performance certification of builidngs in azerbaijan. Int. J. Eng. Technol. **7**(3.2), 563–566. www.sciencepubco.com/index.php/IJET (2018). Retrieved 6 March 2021

Ali, H.H., Al Nsairat, S.F.: Developing a green building assessment tool for developing countries – case of Jordan. Build. Environ. **44**, 1053–1064 (2009). https://doi.org/10.1016/j.buildenv.2008.07.015

Alyami, S.H., Rezgui, Y.: Sustainable building assessment tool development approach. Sustain. Cities Soc. **5**, 52–62 (2012). https://doi.org/10.1016/j.scs.2012.05.004

Amasuomo, T.T., Atanda, J., Baird, G.: Development of a building performance assessment and design tool for residential buildings in Nigeria. Procedia Engineering **180**, 221–230 (2017)

Andoh, C.: Energy commission to introduce 7-star ratings on electrical appliances. Daily Graphic - Business News. https://www.graphic.com.gh/business/business-news/energy-commission-to-introduce-7-star-ratings-on-electrical-appliances.html. (14 August 2020). Retrieved 24 August 2021

Anzagira, L. F., Badu, E., Duah, D.: Towards an uptake framework for the green building concept in Ghana: a theoretical review. Int. J. Proc. Sci. Technol. 57–76 (2019)

Assadi, M.K., Zahraee, S.M., Isaabadi, M.H., Habib, K.: Evaluating different scenarios for optimizing energy consumption to achieve sustainable green building in Malaysia. ARPN J. Eng. Appl. Sci. **11**(20). Retrieved from www.arpnjournals.com (2016)

Aveyard, H.: Doing a Literature Review in Health and Social Care. McGraw Hill, London (2007)

Baharom, M., Emran, A., Rahman, A., Sulaiman, M., Bohari, Z., Jali, M.: A new construction for residential building (hostel) by focusing on green building strategies. J. Theor. Appl. Inf. Technol. **80**(1), 114–123. https://www.researchgate.net/publication/283784475_A_new_construction_for_residential_building_hostel_by_focusing_on_green_building_strategies (2015). Retrieved 21 April 2019

Build Up:.Current status of energy performance certification in Europe. The European Portal for Energy Efficiency in Buildings: https://www.buildup.eu/en/node/59101 (2020, March 12). Retrieved 28 February 2022

Bureau of Energy Efficiency: Energy Conservation Building Code. Bureau of Energy Efficiency, New Delhi. https://beeindia.gov.in/sites/default/files/advagg_js/E-Book%20ECBC%20CODE%202017/index.html (2017). Retrieved 21 March 2022

Cozza, S., Chambers, J., Brambilla, A., Patel, M.: Energy Performance certificate for buildings as a strategy for the energy transition: stakeholder insights on shortcomings. IOP Conf. Ser.: Earth Environ. Sci. **588**, 022003 (2020). https://doi.org/10.1088/1755-1315/588/2/022003

Creswell, J.: Research Design: Qualitative, Quantitative and Mixed Methods Approaches. Sage (2009)

Economic Consulting Associates: EPCs and their role in stimulating the market of building renovation and new "green" buildings. Economic Consulting Associates (EECG Meeting). www.eca-uk.com (2020). Retrieved 23 March 2022

ECOWAS Centre for Renewable Energy and Energy Efficiency (ECREEE): ECOWAS Energy Efficiency Policy. ECREEE, Praia, Cape Verde (2015)

Edjekumhene, I.: Sustainable development strategies in energy efficiency in Ghana (2017)

Energy Commission, Ghana: Report on compliance monitoring and data collection. Energy Commission of Ghana.Renewable Energy, Energy Efficiency & Climate Change Directorate, Accra (2019). Retrieved 31 January 2022

Energy Efficiency Standards and Labelling (Household Refrigerating Appliances) Regulations LI 1958. Accra, Ghana (2009)

Energy Efficiency Standards and Labelling Regulations. Accra, Ghana (2005)

Energy saving trust: energy at home: guide to energy performance certificates, Energy Saving Trust: https://energysavingtrust.org.uk/advice/guide-to-energy-performance-certificates-epcs/ (2021 July) . Retrieved 5 March 2022

Fowler, K., Rauch, E.:. Sustainable Building Rating Systems. Pacific Northwest National Laboratory, U.S. Department of Energy. Contract DE-AC05–76RL061830 (2006)

Fuerst, F., McAllister, P.: The impact of energy performance certificates on the rental and capital values of commercial property assets: some preliminary evidence from the UK. Energy Policy **39**(10), 6608–6614 (2011). https://doi.org/10.1016/j.enpol.2011.08.005

Fuerst, F., van de Wetering, J., Wyatt, P.: Is intrinsic energy efficiency reflected in the pricing of UK office leases? Building Research and Information (2012). https://doi.org/10.1080/09613218.2013.780229

Ghana Standards Authority: Draft Building Code of the Republic of Ghana (GS 1207:2018).Ghana Standards Authority, Accra, Ghana (2018)

Groat, L., Wang, D.: Architectural Research Methods. John Wiley and Sons, Hoboken, New Jersey (2013)

Gyimah, K., Addo-Yobo, F.: Energy performance certificate of buildings as a tool for sustainability of energy and environment in Ghana. Int. J. Res. **1**(6), 757–762 (2014)

Hanna, G.: Energy efficiency building codes and green pyramid rating system. Int. J. Sci. Res. **4**(5), 3055–3060. www.ijsr.net (2015). Retrieved 22 March 2022

Harputlugil, T.: Conceptual framework for developing next generation of Energy Performance Certificates (EPC) systems. In: Beyond all Limits: International Congress on Sustainability in Architecture, Planning, and Design, Ankara, Turkey (2018)

Heijmans, N., Loncour, X.: Changes in EPCs scales and layouts: experiences and best practices. Concerted Action EPBD (2019)

Herrando, M., Cambra, D., Navarro, M., de la Cruz, L., Millán, G., Zabalza, I.: Energy performance certification of faculty buildings in Spain: the gap between estimated and real energy consumption. Energy Convers. Manage. **125**, 141–153 (2016). https://doi.org/10.1016/j.enconman.2016.04.037

International Energy Agency: energy performance certification of buildings: a policy tool to improve energy efficiency. IEA Policy Pathway. International Energy Agency, Paris, France (2010)

International Energy Agency: Energy Performanc Certificate (EPC) System. Retrieved from IEA.org: https://www.iea.org/policies/1991-energy-performance-certificate-epc-system (18 November 2019)

International Organization for Standardization: ISO 17743:2016(E) Energy savings - Definition of a methodological framework applicable to calculation and reporting on energy savings. ISO, Switzerland (2016)

International Organization for Standardization: ISO 17742:2015 Energy efficiency and savings calculation for countries, regions and cities (2015)

Iwaro, J., Mwasha, A.: A review of building energy regulation and policy for energy conservation in developing countries. Energy Policy **38**, 7744–7755 (2010). https://doi.org/10.1016/j.enpol.2010.08.027

Iwaro, J., Mwasha, A.: Implications of building energy standard for sustainable energy efficient design in buildings. Int. J. Energy Environ. **1**(5), 745–756. Retrieved from www.IJEE.IEEFoundation.org (2010b)

Janda, K.B.: World status of energy standards for buildings. In: Proceedings of the Fifth Annual IEECB, (pp. 1761–1769). Frankfurt, Germany. Retrieved from http://www.eci.ox.ac.uk/publications/downloads/janda09worldwidestatus/cited 2008 (2008)

Jarnagin, R.E.: ASHRAE building EQ. ASHRAE J. **51**(12), 18–19 (2009)

Jeong, J., et al.: Development of a prediction model for the cost saving potentials in implementing the building energy efficiency rating certification. Appl. Energy **189**, 257–270 (2017). https://doi.org/10.1016/j.apenergy.2016.12.024

Kelly, S., Crawford-Brown, D., Pollitt, M.: Building performance evaluation and certification in the UK: is SAP fit for purpose? Renew. Sustain. Energy Rev. (2012). https://doi.org/10.1016/j.rser.2012.07.018

Kemausuor, F., Obeng, G.Y., Brew-Hammond, A., Duker, A.: A review of trends, policies and plans for increasing energy access in Ghana. Renew. Sustain. Energy Rev. **15**(9), 5143–5154 (2011). https://doi.org/10.1016/j.rser.2011.07.041

Kim, K.H., Chae, C.-U., Cho, D.: Development of an assessment method for energy performance of residential buildings using G-SEED in South Korea. J. Asian Archit. Build. Eng. (2020). https://doi.org/10.1080/13467581.2020.1838286

Kontokosta, C.E.: A market-specific methodology for a commercial building energy performance index. J. Real Estate Finance Econ. **51**(2), 288–316 (2014). https://doi.org/10.1007/s11146-014-9481-0

Li, B., Yao, R.: Building energy efficiency for sustainable development in China: challenges and opportunities. Build. Res. Inf. **40**(4), 417–431 (2012). https://doi.org/10.1080/09613218.2012.682419

Li, Y., Kubicki, S., Guerriero, A., Rezgui, Y.: Review of building energy performance certification schemes towards future improvements. Renew. Sustain. Energy Rev. **113**. https://doi.org/10.1016/j.rser.2019.109244

Martin, C.: Generating low-cost national energy benchmarks: a case study in commercial buildings in Cape Town, South Africa. Energy Build. **64**, 26–31 (2013). https://doi.org/10.1016/j.enbuild.2013.04.008

Martinez, M.A., Tort Ausina, I., Cho, S., Vivancos, J.: Energy efficiency and thermal comfort in historic building: a review. Renew. Sustain. Energy Rev. **61**, 70–85 (2016). https://doi.org/10.1016/j.rser.2016.03.018

Merriam-Webster Dictionary online. http://merriam-webster.com/dictionary/framework (2021). Retrieved 5 August 2021

Ministry of Power, Ghana: National Energy Efficiency Action Plan of Ghana. Ministry of Power, Republic of Ghana, Accra, Ghana (2015) . Retrieved 10 January 2022

Ministry of Power, Government of India: ECBC Commercial. Retrieved from Bureau of Energy Efficiecny: https://beeindia.gov.in/content/ecbc-commercial (2020)

Minsitry of Energy, Republic of Ghana: National Energy Policy. Accra: Ministry of Energy, Ghana (2010)

Nikolaou, T., Kolokotsa, D., Stavrakakis, G.: Review on methodologies for energy benchmarking, rating and classification of buildings. Adv. Build. Energy Res. **5**(1), 53–70 (2011). https://doi.org/10.1080/17512549.2011.582340

Nor Azuana, R., Mohd Fairuz, A., Harison, R.G.: Energy efficiency measures on two different commercial buildings in Malaysia. Journal of Building Performance **10**(1), 17–29 (2019)

Norvaisiene, R., Karbauskaite, J., Bruzgevicius, P.: Energy performance certification in Lithuanian building sector. 40th IHAS World Congress on Housing. Funchal, Portugal (2014)

Panayiotou, G., et al.: Cyprus building energy performance methodology: a comparison of the calculated and measured energy consumption results. Central Europe towards Sustainable Building (2010). Prague

Paterson, G., Mumovic, D., Das, P., Kimpian, J.: Energy use predictions with machine learning during architectural concept design. Sci. Technol. Built Environ. (2017). https://doi.org/10.1080/23744731.2017.1319176

Pérez-Lombard, L., Ortiz, J., González, R., Maestre, I.R.: A review of benchmarking, rating and labelling concepts within the framework of building energy certification schemes. Energy Build. **41**, 272–278 (2009). https://doi.org/10.1016/j.enbuild.2008.10.004

Qian, Q., Chan, E.H.: Government measures needed to promote building energy efficiency (BEE) in China. Facilities **28**(11/12), 564–589 (2010). https://doi.org/10.1108/02632771011066602

Roderick, Y., McEwan, D., Wheatley, C., Alonso, C.: Comparison of energy performance assessment between LEED, BREEAM and Green Star. In: Eleventh International IBPSA Conference, pp. 1167–1176. Glasgow, Scotland, 27–30 July 2009

Sassi, P.: Strategies for Sustainable Architecture. Taylor and Francis, Abingdon, UK (2006)

Sin, T., Sood, S.B., Peng, L.Y.: Sustainability development through energy efficiency initiatives in Malaysia (2011)

Somuncu, Y., Menguc, M.P.: Brief discussion of energy certification systems for buildings. SBE 16 (2016). Istanbul, Turkey

The Buildings Performance Institute Europe: Energy Performance Certificates across Europe: From design to implementation. The Buildings Performance Institute Europe, Brussels (2010)

The Buildings Performance Institute, Europe: Qualification and Accreditation Requirements of Building Energy Certifiers in EU28. The Building Performance Institute Europe (2015)

US Department of Energy: Buildings Energy Data Book. Office of Energy Efficiency and Renewable Energy (2011)

Volt, J., Zuhaib, S., Schmatzberger, S., Toth, Z.: Energy performance certificates: Assessing their status and potential. Buildings Performance Institute Europe. https://x-tendo.eu/ (2020). Retrieved 17 February 2022

Vyas, S., Ahmed, S., Parashar, A.: BEE (Bureau of energy efficiency) and green buildings. Int. J. Res. 23–32 (2014)

Webster, J., Watson, R.T.: Analysing the past to prepare for the future: Writing a literature review. MIS Quarterly **26**(2), xiii–xxiii (June 2002)

Yu, Y., et al.: Effect of implementing building energy efficiency labeling in China: a case study in Shanghai. Energy Policy (2019). https://doi.org/10.1016/j.enpol.2019.110898

Zhou, N., Khanna, N., Fridley, D., Romankiewicz, J.: Development and implementation of energy efficiency standards and labeling programs in China: Progress and challenges (2013)

The Use of Building Information Modelling by Small to Medium-Sized Enterprises: The Case of Central South Africa

H. A. Deacon and H. Botha

Department of Quantity Surveying and Construction Management, University of the Free State, Bloemfontein, South Africa
DeaconHA@ufs.ac.za

Abstract. Purpose: The paper identifies the benefits and challenges that impact the use of Building Information Modelling (BIM) by small to medium-sized enterprises in South Africa, as well as the organizational barriers and benefits that contribute to the utilization of BIM by these types of organizations.

Design/Methodology/Approach: The study was based on a literature review and survey, with the survey consisting of a questionnaire that was issued to 130 construction professionals. The data obtained from the questionnaires were analysed by using descriptive statistics protocols and the findings were then compared against the existing literature.

Findings: The results indicated that BIM is seldom used by small to medium-sized enterprises, with many respondents' that specifically work for such organizations having never engaged with it. The foremost barrier was a lack of support from senior management for its implementation, while the main benefit was better project planning due to visualization.

Research Limitations/Implications: The survey was geographically limited to respondents from central South Africa, namely the Free State and Gauteng provinces. The survey response rate was also only 24%, which may not be enough to generalize the findings.

Practical Implication: It is therefore evident that BIM will only be successfully adopted by small and medium-sized firms when the government supports it, and clients request its use. If this is not addressed, the AEC industry will further enforce its current underlying 'BIM imbalance' in favour of large practices.

Social Implication: Culturally, firms that focus on smaller projects do not generally use BIM as they feel it is an overkill for the nature of their work.

Originality/Value: The study enhances the current body of knowledge on BIM by enabling practitioners in the Architecture, Engineering, and Construction (AEC) industry to determine if its use is practical for their organizations.

Keywords: Barriers · Benefits · construction · BIM · SMEs

1 Introduction

Building information modelling (BIM) has been given several definitions by numerous authors (Ullah et al. 2019; Hochscheid and Halin, 2020: 1; IStructE, 2021: 4; Sompolgrunk et al. 2022), with the definitions notably being either product- or process-based, thus positing it as both a noun and verb (Hammad et al. 2012: 645).

As a noun, BIM refers to computer software that establishes a collection of building information for a construction project. The data consists of all the interconnected physical and inherent properties of the project's structure. For example, a window specification is not only a few lines of information representing the item, but contains all necessary information regarding the location, manufacturing, installation, and maintenance. The capability of the model is therefore that all the stakeholders, who are part of the project team, can work with a common database that provides them with access to up-to-date information (Sacks et al. 2018: 2). Furthermore, as a verb, BIM implies the undertaking of replicating the real-time activity of such a construction project (Smith and Tardif, 2009: 29).

Whatever the case may be, BIM refers to 3-dimensional technology that enables the visualization and communication of design, construction, and operational processes for structural facilities. It is therefore not just a computer model, but a systems approach that considers the entire life-cycle of a constructed facility (RICS, 2015: 12; Milyutina, 2018: 1).

The capabilities and benefits of BIM has been evident on large-scale construction projects, especially international projects. BIM is however seldom used on smaller projects as the general stakeholders on such projects, which are mostly SMEs, are very wary of it. They consider it an overwhelming task to adopt the technology, while also deeming it not suitable for their projects (Burger, 2019: Online). Besides, Saka et al. (2019: 1) also highlighted that most BIM research is focussed on large practices, who are its predominant users.

Considering this, it is important to identify the primary challenges that hinders the implementation of BIM by SMEs, as well as the various benefits that could be unlocked if these obstacles could be overcome.

2 Literature Review

2.1 The Evolution of BIM Technology

BIM is technically not a new development, but it has however only recently started to gain traction in the architectural, engineering, and construction (AEC) industry (RICS, 2020: 4). Its idea originated in the primitive days of computing in the 1960's. However, the initial attempts to develop software systems utilizing its concept could not be achieved until the development of a graphical interface that allowed users to interact with building models (Bergin, 2012: Online).

The first creation of practical modelling systems, better known as solid modelling, was developed in 1973 and allowed the easy generation and modifying of 3-dimensional solid shapes. Solid modelling was therefore the foundation of the first building modelling systems that were developed in the late 1970's and early 1980's (Olawumi and Chan,

2019: 53). The Building Description System, developed by Charles Eastman, was the first of these software systems. It enabled designers to fabricate building models with an integrated database, while allowing the generation of automatically uniform plans, sections, and isometrics from the model. The software further supported the generation of basic material quantities and cost estimates (Feist, 2016: 23).

Johnson (2014: 175) highlighted that many developers subsequently imitated Eastman's invention and created similar software systems, even expanding on the conceptual framework. For example, RUCAPS (Really Universal Computer Aided Production System) presented the concept of time-based phasing of construction processes. However, these early software systems mostly operated on mainframe computers that required bulky and expensive hardware. They were thus only briefly used until personal computers and geometry-based CAD (computer-aided design) software came along in the latter half of the 1980's. Accordingly, this led to the development of the next generation of building modelling systems for the personal computer (Wierzbicki et al. 2011: 3).

The first of these new software systems was ArchiCAD, which is perceived by many as the genuine beginning of BIM. It was however only suitable for small scale projects such as houses and small office blocks. Consequently, the development of Revit, which is parametric modelling software that can handle large and complex projects, followed in the late 1990's. Revit was released in 2000 and is commonly viewed as the building industry's first real attempt at efficient BIM implementation. It has been regularly updated with new features over time, while there are nowadays also other similar software systems available on the market (Smith, 2014: 483). Ultimately, although the uptake of BIM has been slow in the construction industry, the technology that underpins it has been around for more than 30 years.

2.2 The Adoption of BIM by Small and Medium-Sized Enterprises

Gledson et al. (2012) and Makowski et al. (2019: 267) studied the adoption of BIM in the United Kingdom (UK), whereby they emphasized that SMEs are aware of BIM but rarely use or engage with it. They respectively found that only 23% and 25% of the SMEs had exposure to BIM. Vidalakis et al. (2020: 140) similarly studied the adoption of BIM by SMEs in the UK, whereby they found that only 14% of the respondents had moderate to good knowledge of BIM. Tezel et al. (2020: 187) also highlighted that only 41% of small- and medium-enterprises in North America have used or engaged with BIM.

Rodgers et al. (2015: 698) further studied the awareness of BIM in the Australian construction industry, concluding that its adoption was very low among SMEs and that it was generally perceived negatively by the enterprises. Li et al. (2019: 11) investigated the use of BIM among SMEs in the Chinese AEC industry, remarking that the enterprises were mostfully unaware of its existence. Al Awad (2015: 207) similarly investigated the adoption of BIM in the construction sector of Jordan, finding that its use was basically non-existent among SMEs. Saka and Chan (2020: 263) also studied the adoption of BIM by SMEs in Nigeria and likewise concluded that these organizations are non-adopters of the technology because they largely deem themselves to have insufficient capabilities and resources to embrace it.

Regarding South Africa, Chimhundu (2015: 59) noted that the uptake of BIM was really low in the construction industry. This was attributed to the construction industry not yet being mature enough for its effective adoption, which is due to the country being a developing nation with a struggling construction sector that has mostly SMEs operating in it (CIDB, 2020).

2.3 The Benefits of BIM Implementation

Several studies have been conducted recently on the effect of the implementation of BIM in the AEC industry. Johansson et al. (2015: 81) studied the real-time visualization of large building projects, whereby they highlighted the ability of BIM to simplify sizeable projects through visualization and real-time depiction. Likewise, Kim et al. (2015: 95) pointed out the effectiveness of BIM regarding the master planning of large-scale development projects, which are long-term projects that include infrastructure and multiple facilities in a particular area.

Matthews et al. (2015) and Fan et al. (2014) independently investigated the impact of BIM during the construction phase of a project. The findings of Matthews et al. (2015: 38) revealed the timely delivery of information and accurate progress monitoring. Fan et al. (2014: 159) moreover noted a noteworthy decline in requests for information (RFI's), rework, and change orders. Similarly, El Hawary and Nassar (2016: 32) examined the effect of BIM on construction claims. They concluded that BIM greatly reduces the probability of claims during or after construction, as comprehensive designs are completed for the required work and there is coordinated stakeholder engagement throughout the project life-cycle.

Inyim et al. (2015: 9) researched the influence of BIM on sustainable construction, whereby they emphasized that BIM supported an eco-friendly approach because of its capability of environmental consideration during the design stages of a project. In addition, Karan and Irizarry (2015: 11) reasoned that BIM has the capacity to also incorporate geographic information system (GIS) data sets, which can further optimize its effectiveness in environmental sensitivity during the pre-construction stages of a project.

Morlhon et al. (2014: 1126) and Olatunji et al. (2017: 59) furthermore asserted that the implementation of BIM on construction projects brought about facilitated collaboration between key stakeholders, enhanced resource planning and management, and conformity to project programmes. Conclusively, it is evident that BIM has been beneficial to the construction industry in terms of enhanced information management and communications.

2.4 The Barriers of BIM Implementation

While BIM provides considerable benefits to users, its implementation in the AEC industry has generally been constrained by several barriers. According to Liu et al. (2015: 163) and Ahmed (2018: 109), extensive costs are required for its implementation, which compels potential adopters to consider their options thoroughly. The main cost of adopting BIM is the start-up costs, which includes the purchasing of software and upgrading of existing IT-infrastructure to accommodate additional data sharing and storage needs. In addition, organizations also need to ensure that their personnel are appropriately skilled, which is done by training existing staff or hiring new staff. Considering this, Enshassi et al. (2019: 190) emphasized that the collective costs of implementation can result in large amounts of financial expenditure. This deters companies from investing in the technology, especially small firms. As a result, BIM is mostly adopted by large companies that have the necessary resources and finances to incorporate the technology into their existing operations (Ganah and John, 2014: 143).

Moreover, Wu and Issa (2013a, 2013b) extensively studied the impact of BIM education and training on career opportunities in the construction industry, whereby they determined that tertiary education outcomes were not meeting industry expectations. This was due to the available education and training mostly focussing on the use of specific software packages and not the practical application of the methodology. Becerik-Gerber et al. (2012: 438) similarly determined that BIM implementation in the built environment suffered due to a shortage of sufficiently trained professionals. Smith and Tardif (2009: 171) also highlighted this skills gap and predicted it to last until 2030.

It is furthermore critical that all project role-players should be willing to share information in a collaborative manner to ensure that BIM is successfully implemented (Ghaffarianhoseini et al. 2017: 1047). However, Alreshidi et al. (2014: 151), Bataw et al. (2016: 14), and Mehran (2016: 1114) highlighted that the lack of a public standard regarding the use of BIM is concerning to some parties. This unease specifically relates to the sharing of data between project stakeholders and the legal ownership thereof in terms of licensing, access, and control. This corresponds to the findings of Won et al. (2013: 2), who noted that trust is a major issue among project participants due to their reluctance to openly share information with each other. The authors furthermore highlighted that project role-players were concerned with security risks in an electronic environment, and the subsequent implications thereof for professional liability.

Moreover, Halttula et al. (2015: 31) observed the lack of buy-in from executives regarding the implementation of BIM. Considering this, Aibinu and Venkatesh (2014) emphasized the importance of management support but remarked that senior managers were concerned with the business impact of implementing BIM. Managers were wary of the required learning curve to put BIM into service and transition from traditional work methods, while it was also not clear how to measure the impact of BIM on the bottom line of organizations.

3 Research Aim

The purpose of the research was to explore the use of BIM by small and medium-sized construction firms, while focussing on its benefits and barriers for the stakeholders of these organizations.

4 Research Methodology

The objectives of the study were achieved by using a self-administered survey. The purpose of the questionnaire was to establish the extent to which BIM was used by small and medium-sized construction firms, and if the use of the software is practical and adds value for these organizations.

A literature review identified the main benefits and challenges of using BIM on construction projects, followed by the creation of the questionnaire based on these identified factors. The questionnaire mostly consisted of closed-type questions, as they are simpler to answer due to respondents needing little skill and requiring less time to reply (Zikmund, 2002: 333). The questions regarding BIM were also mostly based on typical five-point Likert scales. The questionnaire was setup in Google Forms, a web-based survey administration application, for data collection (Brinkman, 2009: 32).

Potential respondents were identified by means of purposive sampling, with their contact information being sourced from the websites of various organizations and regulatory bodies that operate in the AEC industry in South Africa (Maree and Pietersen, 2019: 20). A participation invitation was sent via e-mail to 130 construction professionals, such as architects, engineers, and quantity surveyors, from the Free State and Gauteng provinces of South Africa. The e-mail stated the aim of the research, invited them to voluntarily partake in the study, and provided the link of the survey. The main advantage of an e-mail survey is that it only requires an e-mail address for distribution (Kierczak, n.d.: Online).

The survey response rate was however only 24%, with 31 construction professionals responding to the survey. The data obtained from the questionnaires were examined using descriptive statistics protocols to calculate the mean scores and standard deviations (Ellis, 2017: 85). The data was further filtered to exclude the responses from respondents working for large organizations, with the results subsequently being evaluated against the existing literature (Leedy and Ormrod, 2016: 216).

4.1 Characteristics of the Respondents

4.1.1 Profession of Respondents

This question aimed to establish the profession of the respondents within the AEC industry. The respondents indicated that 58% were quantity surveyors, 23% were architects, 10% were building contractors, 6% were engineers, and 3% specified 'other'.

4.1.2 Economic Sectors in Which Respondents Operate

This question aimed to establish the economic sectors in which the respondents operated within the AEC industry. The respondents indicated that 53% operated within both the private and public sectors, 40% operated only in the private sector, and 7% operated only in the public sector.

4.1.3 Years of Experience that Respondents Have in the Construction Industry

The purpose of this question was to establish how long the respondents have been working in the construction industry. The respondents indicated as follows in terms of years of working experience: 53% have worked in the industry for 21 years and more, 23% have worked in the industry for 6 to 10 years, 17% have worked in the industry for 11 to 20 years, while only 7% have worked in the industry for 1 to 5 years.

4.1.4 The Size of the Organizations for Which the Respondents Work

This question aimed to establish the size of the firms for which the respondents worked. The respondents indicated that 23% worked for small firms, 45% worked for medium firms, and 32% worked for large firms. The firms were categorized according to the national definition of small enterprise in South Africa (Department of Small Business Development, 2019). In summary, 68% of the respondents worked for SMEs.

4.1.5 Types of Projects that Respondents are Mainly Involved with

The purpose of this question was to establish the types of construction projects that the respondents were primarily involved with, and which BIM could thus potentially be used on. The respondents indicated as follows for the three prominent types of projects: 27% were mainly involved with residential building projects, 33% were mainly involved with commercial building projects, and 17% were mainly involved with industrial building projects. The rest of the respondents indicated 'other', specifying their main involvement with civil and specialty projects.

5 Discussion of the Results

5.1 Awareness of BIM

This question aimed to establish the respondents' awareness of BIM, irrespective of whether they have used it. Table 1 (below) indicates the responses and calculated mean score of 2.65, which indicates that the respondents' general knowledge and understanding of BIM ranged between little to moderate. A mean score of 3 and above, which equals to 60% or more, generally implies adequate comprehension of something (Sullivan and Artino Jr., 2013: 542). Considering this, the respondents' overall awareness of BIM was mediocre.

Table 1. The respondents' awareness of BIM

Awareness of BIM							
Rating	None	Little	Moderate	Good	Excellent	Mean score	Standard deviation
	1	2	3	4	5		
Percentage of responses	16%	26%	39%	16%	3%	2.65	1.05

5.2 Use of BIM

The aim of this question was to establish how many respondents have used or engaged with BIM on their projects. Only 42% of the respondents indicated that they have previously used or engaged with BIM, while 58% of the respondents have never used or engaged with BIM. The respondents meagre use or engagement with BIM makes sense when it is compared to their previously stated knowledge and understanding of it. It could also be a reason why many potential respondents did not bother to complete the survey, as they never use or engage with BIM.

This result further relates closely to the above-mentioned findings of Gledson et al. (2012), Rodgers et al. (2015), Al Awad (2015), Makowski et al. (2019), Tezel et al. (2020), and Vidalakis et al. (2020). Considering this, Carson (2018) noted that organizations undertaking smaller projects are reluctant to implement BIM due to the nature of these projects, which includes shorter timelines, limited budgets, and less risk. Li et al. (2019: 11) and Saka et al. (2019: 1) supported this assertion, whilst further noting that BIM also falls short to meet the needs of SMEs in developing countries due to the nature of the construction industries.

5.3 Benefits and Barriers of BIM

5.3.1 Benefits of BIM

The purpose of this question was to establish the respondents experienced and perceived benefits of BIM. The literature-identified causes were listed, and the scales were as follows: 1 = strongly disagree, 2 = disagree, 3 = neutral, 4 = agree, and 5 = strongly agree. The causes were ranked according to their respective average weightings, with a mean score of 3 or more again implying reasonable to strong support for a cause (i.e., factor). Table 2 (below) subsequently indicates the major benefits of BIM according to the participants of this study.

Table 2. Benefits of BIM

Benefits of BIM		
Rank	Cause	Mean score
1	Simplified project planning through visualization	3.67
2	Effective collaboration between project participants	3.63
3	Enhanced facilities management during operation	3.63
4	Timely delivery of information and reduction in claims during construction	3.50
5	Accurate progress reporting during construction	3.46
6	Enhancement of lean construction	3.24

These findings correspond with the findings of Hammad et al. (2012), Fan et al. (2014), Rokooei (2015), and Allen and Shakantu (2016), whose primary benefits were also better project planning and enhanced collaboration between project participants. It is therefore obvious that BIM ensures clear expectations from clients due to the visualized model, while improving team collaboration that delivers better project outcomes.

5.3.2 Barriers of BIM

The intention of this question was to establish the respondents experienced and perceived barriers of BIM. The literature-identified causes were also listed, and the scales were the same as with the benefits. Table 3 (below) subsequently indicates the major barriers of BIM according to the participants of this study.

Table 3. Barriers of BIM

Barriers of BIM		
Rank	Cause	Mean score
1	Lack of support from senior management for implementation	3.46
2	Incompatibility of project partners	3.36
3	Lack of the necessary expertise for implementation (i.e., skills gap)	3.12
4	Security risk in terms of data and information sharing	3.04
5	High cost of investing in IT-infrastructure and staff training	2.71

These findings correspond with the results of numerous studies that similarly found a lack of senior management support for the implementation of BIM. Bryde et al. (2013), Ahmed (2018) and Enshassi et al. (2019), found that senior management was not interested in BIM due to the prospect of organizational culture change. Sawhney et al. (2017) similarly highlighted that SMEs ignore BIM because its use is rarely requested by their clients. Ghaffarianhoseini et al. (2017) also concluded that firms were reluctant to invest in BIM as they perceived it to deliver a low return on investment. However, this study's respondents did not deem the cost of implementation to be a foremost barrier, which corresponds with the findings of Eadie et al. (2013). Conversely, Ghaffarianhoseini et al. (2017) and Liu et al. (2015) did deem implementation cost to be a critical barrier.

6 Conclusion and Recommendations

BIM is comprehended in many ways, which results in the numerous definitions that are currently used for it. It has nevertheless been used in the AEC industry for over 30 years, yet its implementation has not been made the most of and many construction practitioners remain happily unaware of its presence. However, it is evident from the literature that the paradigm of BIM brings about change in the work methods of industry role-players that do use it. This is due to it forcing a shift away from a static approach, which consists of two-dimensional documentation and drawings, towards a model-centric approach. It is furthermore viewed by many as a cohesive information system that assists in effectively integrating the organizational processes and functions of project delivery.

Regarding this study, it was found that BIM mostly assists with project planning during pre-construction and stakeholder collaboration throughout a project. Furthermore, the most noteworthy reason for not implementing BIM was a lack of support from the senior management of SMEs due to cultural or economic considerations. Culturally, firms that focus on smaller projects do not generally use BIM as they feel it is an overkill for the nature of their work. Economically, these firms are concerned with investing in a technology of which the return on investment is uncertain, which is an area for future research whereby the effectiveness of BIM could be quantified.

It has further previously been proven that governments can help to grow the utilization rate of BIM in the AEC industry. Countries such as Hong Kong, Australia, Singapore and the United Kingdom have begun to encourage or mandate the use of BIM on public sector projects. These countries are however developed nations with decent governance structures and matured construction industries, which is dissimilar to developing nations such as South Africa. The activity of the South African construction industry has declined sharply in recent times with many of the main role-players in the industry, namely SMEs, operating without advanced technologies and struggling to survive.

It is therefore evident that BIM will only be successfully adopted by small and medium-sized firms when the government supports it, and clients request its use. If this is not addressed, the AEC industry will further enforce its current underlying 'BIM imbalance' in favour of large practices.

References

Ahmed, S.: Barriers to implementation of building information modeling (BIM) to the construction industry: a review. J. Civ. Eng. Constr. **7**(2), 107–113 (2018)

Aibinu, A.A., Venkatesh, S.: The rocky road to BIM adoption: quantity surveyors perspectives. In: Management of Construction: Research to Practice (MCRP) Conference, pp. 539–554. In-house, 26–29 June, Rotterdam, Netherlands (2012)

Al Awad, O.: The uptake of advanced IT with specific emphasis on BIM by SMEs in the jordanian construction industry. Thesis (PhD). University of Salford, Salford 2015

Allen, C., Shakantu, W.: The BIM revolution: a literature review on rethinking the business of construction. In: 11th International Conference on Urban Regeneration and Sustainability, pp. 919–930. WIT Press, Alicante, Spain, 12–14 July 2016

Alreshidi, E., Mourshed, M., Rezgui, Y.: Exploring the need for a BIM governance model: UK construction practitioners' perceptions. International Conference on Computing in Civil and Building Engineering, pp. 151–158. American Society of Civil Engineers, Orlando, United States of America, 23–25 June 2014

Bataw, A., Kirkham, R., Lou, E.: The issues and considerations associated with BIM integration. In: 4th International Building Control Conference, pp. 13–20. EDP Sciences, Kuala Lumpur, Malaysia, 7–8 March 2016

Becerik-Gerber, B., Jazizadeh, F., Li, N., Calis, G.: Application areas and data requirements for BIM-enabled facilities management. J. Constr. Eng. Manag. **138**(3), 431–442 (2012)

Bergin, M.S.: History of BIM. Retrieved October 2, 2020 (2012). https://archinect.com/archlab/history-of-bim

Brinkman, W.: Design of a questionnaire instrument. In: Love, S. (ed.) Handbook of Mobile Technology Research Methods, pp. 31–57. Nova Science Publishers, New York (2009)

Bryde, D., Broquetas, M., Volm, J.M.: The project benefits of building information modelling (BIM). Int. J. Project Manage. **31**(7), 971–980 (2013)

Burger, R.: Should Small Construction Companies Use BIM? Retrieved October 2, 2020. https://www.thebalancesmb.com/should-small-construction-companies-use-bim-845327 (2019)

Carson, J.: Can BIM successfully deliver small construction projects?. Retrieved November 20, 2020. https://www.thenbs.com/knowledge/can-bim-successfully-deliver-small-construction-projects (2018)

Chimhundu, S.: A Study on the BIM Adoption Readiness and Possible Mandatory Initiatives for Successful Implementation in South Africa. Dissertation (M.Sc). University of the Witwatersrand, Johannesburg (2015)

CIDB: SME Business Conditions Survey Q2 2020. Construction Industry Development Board, Pretoria (2020)

Department of Small Business Development. Revised Schedule 1 of the National Definition of Small Enterprise. South African Government Printing Works, Pretoria (2019)

Eadie, R., Browne, M., Odeyinka, H., McKeown, C., McNiff, S.: BIM implementation throughout the UK construction project lifecycle: an analysis. Autom. Constr. **36**, 145–151 (2013)

El Hawary, A.N., Nassar, A.H.: The effect of building information modeling (BIM) on construction claims. Int. J. Sci. Technol. Res. **5**(12), 25–33 (2016)

Ellis, P.: The language of research (part 13): research methodologies: descriptive statistics - measures of central tendency. Wounds UK **13**(2), 84–85 (2017)

Enshassi, M.A., Al Hallaq, K.A., Tayeh, B.A.: Limitation factors of building information modeling (BIM) implementation. Open Const. Build. Technol. J. **13**, 189–196 (2019)

Fan, S., Skibniewski, M.J., Hung, T.W.: Effects of building information modeling during construction. J. Appl. Sci. Eng. **7**(2), 157–166 (2014)

Feist, S.T.: A-BIM: Algorithmic-based Building Information Modelling. Dissertation (M.Sc.). Technical University of Lisbon, Lisbon (2016)

Ganah, A.A., John, G.A.: Achieving level 2 BIM by 2016 in the UK. In: International Conference on Computing in Civil and Building Engineering, pp. 143–150. American Society of Civil Engineers, Orlando, United States of America, 23–25 June 2014

Ghaffarianhoseini, A., et al.: Building information modelling (BIM) uptake: clear benefits, understanding its implementation, risks and challenges. Renew. Sustain. Energy Rev. **75**, 1046–1053 (2017)

Gledson, B., Henry, D., Bleanch, P.: Does size matter? Experiences and perspectives of BIM implementation from large and SME construction contractors. In: 1st UK Academic Conference on Building Information Management (BIM), pp. 97–108. Newcastle upon Tyne, United Kingdom, 5–7 September 2012

Halttula, H., Haapasalo, H., Herva, M.: Barriers to achieving the benefits of BIM. International Journal of 3-D Information Modeling **4**(4), 16–33 (2015). https://doi.org/10.4018/IJ3DIM.2015100102

Hammad, D.B., Rishi, A.G., Yahaya, M.B.: Mitigating construction risk using building information modelling (BIM). In: 4th West Africa Built Environment Research (WABER) Conference, pp. 643–652. WABER Conference, Abuja, Nigeria 24–26 July 2012

Hochscheid, E., Halin, G.: Generic and SME-specific factors that influence the BIM adoption process: an overview that highlights gaps in the literature. Front. Eng. Manag. **7**(1), 119–130 (2019). https://doi.org/10.1007/s42524-019-0043-2

Inyim, P., Rivera, J., Zhu, Y.: Integration of building information modeling and economic and environmental impact analysis to support sustainable building design. J. Manag. Eng. **31**(1), 1–10 (2015)

IStructE.: An Introduction to Building Information Modelling (BIM). The Institution of Structural Engineers, London (2021)

Johansson, M., Roupé, M., Bosch-Sijtsema, P.: Real-time visualization of building information models (BIM). Autom. Constr. **54**, 69–82 (2015)

Johnson, B.R.: One BIM to rule them all: future reality or myth? In: Kensek, K.M., Noble, D.E. (eds.) Building Information Modeling: BIM in Current and Future Practice, pp. 175–184. Wiley, Hoboken (2014)

Karan, E.P., Irizarry, J.: Extending BIM interoperability to preconstruction operations using geospatial analyses and semantic web services. Autom. Constr. **53**, 1–12 (2015)

Kierczak, L.: Email surveys: how to send, questions & types [All-in Guide]. Retrieved November 17, 2020. https://survicate.com/email-survey/guide/ (n.d.)

Kim, J.I., Kim, J., Fischer, M., Orr, R.: BIM-based decision-support method for master planning of sustainable large-scale developments. Autom. Constr. **58**, 95–108 (2015)

Leedy, P.D., Ormrod, J.E.: Practical Research Planning and Design, 11th edn. Pearson, London (2016)

Li, Pengfei, Zheng, Shengqin, Si, Hongyun, Ke, Xu.: Critical challenges for BIM adoption in small and medium-sized enterprises: evidence from China. Adv. Civ. Eng. **2019**, 1–14 (2019). https://doi.org/10.1155/2019/9482350

Liu, S., Xie, B., Tivendal, L., Liu, C.: Critical barriers to BIM implementation in the AEC industry. Int. J. Mark. Stud. **7**(6), 162–171 (2015)

Makowski, P., Kamari, A., Kirkegaard, P.H.: BIM-adoption within small and medium enterprises (SMEs): an existing BIM-gap in the building sector. In: 36th CIB W78 Conference, pp. pp. 265–274. CIB, Newcastle, United Kingdom, 18–20 September 2019

Maree, K., Pietersen, J.: Sampling. In: Maree, K. (ed.) First steps in research, 3rd edn., pp. 214–224. Van Schaik Publishers, Pretoria (2019)

Matthews, J., Love, P.E., Heinemann, S., Chandler, R., Rumsey, C., Olatunji, O.: Real time progress management: Re-engineering processes for cloud-based BIM in construction. Autom. Constr. **58**, 38–47 (2015)

Mehran, D.: Exploring the adoption of BIM in the UAE construction industry for AEC firms. In: International Conference on Sustainable Design, Engineering and Construction, pp. 1110–1118. Elsevier Procedia, Tempe, United States of America, 18–20 May 2016

Milyutina, M. A. (2018). Introduction of Building Information Modeling (BIM) Technologies in Construction. *Journal of Physics: Conference Series, 1015*(4)

Morlhon, R., Pellerin, R., Bourgault, M.: Building information modeling implementation through maturity evaluation and critical success factors management. Procedia Technol. **16**, 1126–1134 (2014)

Olatunji, S.O., Olawumi, T.O., Awodele, O.A.: Achieving value for money (VFM) in construction projects. J. Civ. Environ. Res. **9**(2), 54–64 (2017)

Olawumi, T.O., Chan, D.W.: Building information modelling and project information management framework for construction projects. J. Civ. Eng. Manag. **25**(1), 53–75 (2019)

RICS: International BIM Implementation Guide. Royal Institution of Chartered Surveyors, London (2015)

RICS: The Future of BIM: Digital Transformation in the UK Construction and Infrastructure Sector. Royal Institution of Chartered Surveyors, London (2020)

Rodgers, C., Hosseini, M.R., Chileshe, N., Rameezdeen, R.: Building information modelling (BIM) within the South Australian construction related small and medium sized enterprises: Awareness, practices and drivers. In: 36th CIB W78 Conference, 18–20 September, Newcastle, United Kingdom. 31st Annual ARCOM Conference, pp. 691–700. Association of Researchers in Construction Management, Lincoln, United Kingdom, 7–9 September (2015)

Rokooei, S.: Building information modeling in project management: necessities, challenges and outcomes. Procedia. Soc. Behav. Sci. **210**, 87–95 (2015)

Sacks, R., Eastman, C., Lee, G., Teicholz, P.: BIM Handbook: A Guide to Building Information Modeling for Owners, Designers, Engineers, Contractors, and Facility Managers, 3rd edn. Wiley, Hoboken (2018)

Saka, A.B., Chan, D.W.: Profound barriers to building information modelling (BIM) adoption in construction small and medium-sized enterprises (SMEs): an interpretive structural modelling approach. Constr. Innov. **20**(2), 261–284 (2020)

Saka, A.B., Chan, D.W., Siu, F.M.: Adoption of building information modelling in small and medium-sized enterprises in developing countries: a system dynamics approach. World Building Congress. CIB, Hong Kong, China, 17–21 June 2019

Sawhney, A., Khanzode, A.R., Tiwari, S.: Building Information Modelling for Project Managers. Royal Institution of Chartered Surveyors, London (2017)

Smith, D.K., Tardif, M.: Building Information Modeling: A Strategic Implementation Guide for Architects, Engineers, Constructors, and Real Estate Asset Managers. Wiley, Hoboken (2009)

Smith, P.: BIM implementation - global strategies. Procedia Eng. **85**, 482–492 (2014)

Sompolgrunk, A., Banihashemi, S., Mohandes, S.R.: Building information modelling (BIM) and the return on investment: a systematic analysis. Constr. Innov. **22** (2022)

Sullivan, G.M., Artino, A.R., Jr.: Analyzing and interpreting data from Likert-type scales. J. Grad. Med. Educ. **5**(4), 541–542 (2013)

Tezel, A., Taggart, M., Koskela, L., Tzortzopoulos, P., Hanahoe, J., Kelly, M.: Lean construction and BIM in small and medium-sized enterprises (SMEs) in construction: a systematic literature review. Can. J. Civ. Eng. **47**(2), 186–201 (2020)

Ullah, K., Lill, I., Witt, E.: An overview of BIM adoption in the construction industry: benefits and barriers. In: 10th Nordic Conference on Construction Economics and Organization, pp. 297–303. Emerald Insight, Tallinn, Estonia, 7–8 May (2019)

Vidalakis, C., Abanda, F.H., Oti, A.H.: BIM adoption and implementation: focusing on SMEs. Constr. Innov. **20**(1), 128–147 (2020)

Wierzbicki, M., De Silva, C.W., Krug, D.H.: BIM - history and trends. In: 11th International Conference on Construction Applications of Virtual Reality. Bauhaus-Universität, Weimar, Germany, 3–4 November 2011

Won, J., Lee, G., Dossick, C., Messner, J.: Where to focus for successful adoption of building information modeling within organization. J. Constr. Eng. Manag. **139**(11) (2013)

Wu, W., Issa, R.: BIM education for new career options: an initial investigation. In: BIM Academic Symposium. The National Institute of Building Sciences, Washington, D.C., Unites States of America, 7–11 January 2013a

Wu, W., Issa, R.: Impacts of BIM on talent acquisition in the construction industry. In: 29th ARCOM Conference, pp. 35–45. Association of Researchers in Construction Management, Reading, United Kingdom, 2–4 September 2013b

Zikmund, W.G.: Business Research Methods, 7th edn. Thomson/South-Western, Nashville (2002)

Effects of Property Rights for Low-Income Housing in South Africa

W. B. Mpingana, N. X. Mashwama[✉], O. Akinradewo[✉], and C. Aigbavboa[✉]

Department of Construction Management and Quantity Surveying, University of Johannesburg, Johannesburg, South Africa
{nokulungam,caigbavboa}@uj.ac.za, opeakinradewo@gmail.com

Abstract. Purpose: The low-income housing sector plays an essential role in poverty alleviation, with the Department of Housing (DOH) being the driver of the sector. However, the DOH has been facing a massive backlog, which limits the issuance of property rights to South African citizens. Thus, the aim of the paper was to investigate the effects of property rights for low-income housing in South Africa.

Design/Methodology/Approach: The research was carried out using the quantitative approach. Primary data was gathered through a well-structured questionnaire which was distributed to Reconstruction Development Programme (RDP) occupants and the DOH personnel in Gauteng. A total of 104 responses were received from participants through simple random sampling from the online distributed close-ended questionnaires. The data gathered were analysed using descriptive statistics through the use of the Statistical Package for Social Science (SPSS) Version 26. The software provided the mean item score which was used to rank the variables in each research objective, and the Cronbach alpha, which tested the reliability of the questionnaire. The results of the analysed data were presented in the form of figures, graphs, and tables.

Findings: The study disclosed the major effects of property rights to be poverty reduction, poor quality of houses, and the enhancement of neighbouring property values.

Research Limitations/Implications: The study focused on the effects of property rights on low-income housing in South Africa.

Practical Implication: The research findings prove that the issuance of property rights to ordinary citizens can alleviate poverty even though there is still a need for a significant improvement in the design of the houses to ensure the quality of housing, which can be achieved through digitalization. Adoption of the various measures such as appropriate structures allowing effective whistleblowing, policy, and legislative reforms will reduce the occurrence of challenges of property rights for low-income housing, thereby allowing an increase in the provision of low-income housing in South Africa. This will result in an increase in property rights for ordinary citizens in South Africa.

Social Implication: It is therefore recommended that government must provide a clear and appropriate legal framework for the low-income housing sector through

© The Author(s), under exclusive license to Springer Nature Switzerland AG 2023
C. Aigbavboa et al. (Eds.): ARCA 2022, *Sustainable Education and Development – Sustainable Industrialization and Innovation*, pp. 94–105, 2023.
https://doi.org/10.1007/978-3-031-25998-2_8

the amendment of the Housing Act to have a section detailing how beneficiaries should enjoy the fruits of their property rights with no restrictions imposed.

Keywords: Land tenure · Poverty reduction · Property rights · Security · South Africa

1 Introduction

Everyone in South Africa has a constitutional right to sufficient housing, according to Sect. 26 of the Constitution Act 108 of 1996. This right was denied in the apartheid period, and the government had to come up with a solution in 1994. Although this was a challenging and daunting task for the South African government, the ultimate goal was to re-establish rapid economic development while also addressing the poverty and inequality that plagued the majority of South Africans (Blumenfeld 1997). The government offered the Reconstruction and Development Programme (RDP) to provide a broad structure for socio-economic change (Blumenfeld 1997). This housing strategy aimed to alleviate poverty, redressing the social imbalances of the apartheid government, and it was for the unemployed people and low-income earners. Low-income earners are the minimum earning wage group in SA of which, according to the National Minimum Wage Act 9 of 2018, earn a minimum of R20 per hour. However, the Statistics South Africa General Household Survey 2019 (2020) report shows that 20.4% of households depend on grants as a source of income nationally, and low-income earners are included in this group. Low-income homebuyers face complex challenges in obtaining a mortgage (Ebekozien et al. 2019). That is why the majority prefer this low-income housing strategy RDP provided by the government.

Low-income housing refers to housing units whose overall costs are considered affordable by low-income earners (Chepsiror 2013). Low-income earners are people working in a variety of low-wage jobs and who make a living in the informal as well as the formal economy (Chepsiror 2013). Moreover, Mashwama et al. (2019) further defined low-income housing as a provision of housing within a space that is privately owned and secure. Individuals and families with low yearly incomes can have their houses through the provision of low-income housing (Aigbavboa 2015). Mbatha (2018) argue that housing is the cornerstone of human dignity as it is a basic human necessity, and if access to housing is restricted, poverty thrives vigorously. Housing is viewed as a tool for poverty alleviation since it provides opportunities for households to build assets through investment and savings (Makgobi et al. 2019). Furthermore, low-income housing is intended to eradicate slums in informal settlements for people who are unable to access shelter and thereby improve the living conditions of the underprivileged (Mashwama et al. 2019).

Owners of the RDP houses are sometimes unable to get their title deeds due to a backlog in the administration of the local municipality tasked with implementing the programme. The RDP housing policy framework does not clearly define property rights, which makes the housing strategy susceptible to corruption because of greed by officials tasked with implementing the programme. Thus, the aim of the paper was to investigate the effects of property rights for low-income housing in the Gauteng province of South Africa. In the hope of highlighting the impact of the issuance of property rights to ordinary South African Citizens creating an awareness of the importance of property rights.

2 Property Rights

Property rights can be described as a claim to a thing which is a physical object against another person (Muller et al. 2016). Boshoff (2013) explains property rights as a right of ownership and limited real right in something belonging to another person. He et al. (2019) urge that a strong housing market can exist even without well-defined property rights. However, it is generally assumed that having properly explained property rights is a prerequisite for any housing market for efficient economic transactions and functionality purposes (Coase 2013). Incomplete property rights within the housing strategy have given birth to uncertainties (He et al. 2019). The RDP housing policy framework does not have a section detailing property rights, which makes the housing strategy susceptible to corruption because of greed by officials tasked with implementing the programme.

Property rights play a pivotal role in housing and land development processes (Choy et al. 2017). With property rights, the resource itself is not owned, but the portion of rights to use the resource is owned. Furthermore, Eggertsson (1990) describes property rights as a set of rights that include the ability to use an economic good, gain income from it, pass the good to others permanently, and enforce property rights. Property rights define the conceptual and constitutional ownership of things by natural and juristic persons (companies and states), in which the resource can be tangible or intangible, and where owners can acquire, retain, transfer, lease, or sell their property (Anderson 2020)

Property rights strengthen and acknowledge an individual's citizenship (Mbatha 2018:92). Where the individual citizens will have private ownership of their assets. Private ownership has various advantages, including the highest level of tenure security, increased economic activity through financing, an inducement to invest in the home, enhanced municipal tax collection, and thus improved land administration and efficient land and property markets (Barry and Roux 2014:28). Owning a home has enormous social and economic repercussions for everyone, and it is high on aspiring homeowners' personal and national priority lists, therefore buying or getting a home is a top savings reason (Moss et al. 2013:187). The legislative framework guides and provides security measures under which property rights can be transferred between diverse entities, as well as the government (von Benda-Beckmann et al. 2006). In knowing the legal frameworks grants better control in property relationships which results in property relations evolving (von Benda-Beckmann et al. 2006). Mbatha (2018:78) argues that property interactions in the RDP housing strategy relate to beneficiary functionalities and government-based views, which are socially directed since the beneficiaries did not purchase the houses.

Property rights in the low-income housing sector come as land and homeownership to ordinary citizens. Since the state acquires the land for development before the actual development can commence, they own the land of the state-subsidized houses. However, in some rural areas, the land may be registered to the royal families of kings who distribute it to the local residents and even sometimes without registering that land to that person. Homeownership is provided to beneficiaries through the issuance of title deeds to each respective citizen. The issuance of title deeds is not immediate after the house has been allocated. Depending on the Department of Housing's ability to submit all the necessary documents on time to the Deeds Registries Office, it takes some time. Sometimes beneficiaries occupy their property without the title deed for years. Hence, citizens with no property rights have no incentive to develop or invest, and they are not sure if they will be able to keep the benefits of their effort. Hence, the aim of this paper is to investigate the effects of property rights on low-income housing in SA.

3 Effects of Property Rights for Low-Income Housing in South Africa

3.1 Poverty Reduction

The sole purpose of the RDP housing strategy is to provide shelter while reducing poverty for low-income and unemployed citizens. Hence, Makgobi et al. (2019) indicate that establishing strong private property ownership, primarily in emerging economies, is a priority for reducing poverty. Meinzen-Dick et al. (2009) assert that property rights are particularly essential in establishing who has food entitlements, and they can help to prolong or eliminate poverty transmission between generations. Furthermore, property rights over houses on homestead land offer shelter, decency, and a means of development for both the poor and vulnerable citizens (Meinzen-Dick et al. 2009). Hence, property rights are essential for improving economic output and elevating the social position and respect of individuals who possess them (Meinzen-Dick et al. 2009).

Secure property rights enable the growth of functioning credit markets, strengthen the corporate environment and investment prospects, and maintain economic responsibility and openness (Boudreaux 2008). Therefore, land tenure security, the promotion of investment opportunities, and competent risk management should all be part of a reform strategy for successfully functioning property systems that elevate the poor (Boudreaux 2008). Moreover, if beneficiaries receive the title deed for their property in the RDP housing strategy, they should be made aware of the value and importance of this asset.

3.2 Tenure Security

Tenure security is a type of protection that allows you to retain or use a resource (Mbatha 2018). However, Porio and Crisol (2004) outline the security of tenure as giving property titles conferring legal ownership rights. In essence, tenure is a security of property rights that is vital for the endorsement and implementation of sustainable residential development (Arnot et al. 2011). Furthermore, tenure security is the establishment of exclusive ownership rights through a bundle of rules. Hence, secure property rights are fundamental for a country's economic growth and development (Ehrenberg 2006).

Tenure security is further described as the capacity to reap the advantages of use or transfer of an object or asset to others, as well as the guarantee of rights through a valid title deed (Brasselle et al. 2002; Smith 2004). Initially, beneficiaries of the RDP houses in Diepkloof, which is a southwest township located 15 km away from the Johannesburg

CBD in the Gauteng province, had all the expected services of an RDP house, but without ownership, this changed between the year 2006 to 2008 when the municipality was refurbishing the hostels in the township and handed over title deeds to beneficiaries (Rubin 2011:480). Even though beneficiaries can have a title deed for the houses, they still cannot alienate or lease out their RDP houses. As a result, they cannot use the house as collateral for a mortgage if they want to participate in the formal market. Therefore, the beneficiaries have insecure property rights or tenure (Ehrenberg 2006).

Tenure insecurity can also be caused by corruption characterized by poor governance, ambiguous regulatory framework, poor administering of land rights by the state, and lack of the required documentation (Elbow 2014). Insecure tenure to a resource diminishes chances to invest by lowering the advantages the asset can generate (Arnot et al. 2011:297). Therefore, in a poverty reduction program, tenure security is more significant than the functioning of a land market system (Barry and Roux 2014:27). Hence, the policies must be reformed to make the houses an asset that can function in the formal market without restrictions imposed.

3.3 Housing as an Asset

The primary goal of the low-income housing strategy (RDP housing) was to build a sustainable and viable asset for ordinary citizens in South Africa (Department of Human Settlement 2015). The main objectives of the Breaking New Ground (BNG) policy were to create an asset that recipients may transact and utilize to climb up the property market, an asset that can be used as security for financial aid, and an asset that can be used to strengthen recipients' economic conditions. This is impossible due to the restrictive clause in the Housing Act that prohibits housing recipients from selling or leasing their homes. Since an asset is a resource with an economic worth that a person, organization, or state possesses with the prospect of creating monetary cash flow in the future (Barone 2021), it is thus evident that the RDP houses are not an asset since you are prohibited from alienating or leasing the property. However, they can be assets to individuals' households in the sense that they reduce poverty and improve the standard of living of an ordinary citizen (Department of Human Settlement 2015).

Furthermore, Mbatha (2018) argues that housing can be an asset in the three following ways:

- Social asset where it guarantees one's place in the settlement, offers an address, legitimizes citizenship, and connects homes to the system of governance for easy access to basic services and other welfare benefits,
- Financial asset where the house can be used in transactions for monetary gain or used as security or collateral against loans, and
- Economic asset where the house can be used as an income-generating alternative for the household's financial plan.

According to the report on the evaluation of housing assets by the Department of Human Settlement (2015), low-income houses are an asset to the municipalities as they provide municipalities with possibilities to diversify their revenue streams while also developing infrastructure that will allow them to generate additional revenue and investment.

3.4 Negative Effects on Physical and Mental Health

Wolverton (2019) argues that fundamental human rights include access to medical care and a decent and safe place to live. Physical health is a human being's overall physical state at any given time; it is the body's healthiness, freedom from illness or abnormalities, and the state of optimal well-being (Kurtus 2017). Whereas our emotional, psychological, and sociological well-being are all affected by mental health, and it has an impact on how we perceive, feel, and behave (Mental Health 2020). The poor quality of houses delivered can have a negative influence on mental health where toxins such as lead, asbestos, and mold can be found in substandard public homes (Wolverton 2019). Despite the popularity of asbestos as a suitable material because of its versatility, low cost, and fire-retardant properties, asbestos can inflame and destroy the lining of critical organs, causing cancer in the inhabitants (Wolverton 2019). Walsh (2018) further highlights that asbestos was widely used in construction materials because of its tolerance to fires and chemical processes, sound insulation, and relatively cheap cost. Officials should check the quality of the low-income houses to avoid any health problems resulting from occupying the home.

3.5 Improved Investment

Makgobi et al. (2019) argue that current and documented evidence seems to support the notion that titling encourages investment. Furthermore, the research findings of Galiani and Schargrodsky (2004) support that establishing property rights boosts investment rates significantly. Clear and strongly safeguarded ownership rights are now widely recognized by researchers and lawmakers as critical to the achievement of the low-income housing provision (O'Driscoll and Hoskins 2003). Therefore, A better level of investments and economic growth should result from stronger property rights (Haas and Jones 2017). Since all beneficiaries will have title deeds of the RDP houses, the assets owned by the ordinary citizens increase, resulting in a proportional increase in the investment.

3.6 Increasing Neighbouring Property Values

The design of the state-subsidized housing can increase the property value if it fits within the community, or it can decrease the property value if it is unattractive and not aesthetically pleasing (Malgas 2018). The provision may also influence the investment in the market, moreover, the area of the development should not be overcrowded by low-income earners instead, the housing units given must include market-rate housing to cater for additional investment, which increases the property value of the housing units. The registration systems of rights have been found to increase the property value in the market (Choy et al. 2017).

4 Methodology

This study was conducted in the Gauteng Province of South Africa. Gauteng was chosen due to the researcher being familiar with the province, and there was easy accessibility to the research target area, which was Tembisa in this case. According to Statistics South Africa's (Stats SA 2019) General Household Survey, Gauteng has the most informal dwellings sitting at 37,5% in the metropolitan areas. The Reconstruction and Development Programme was first implemented in Gauteng, conducting research around the province will help unveil root problems with the RDP low-income housing strategy. Especially, the effects of property rights on low-cost housing in South Africa. Gauteng is the economic hub of the country and has the highest Gross Domestic Product per capita than any other province (Stats SA 2019).

The targeted groups were the people working at the Department of Housing in Gauteng as well as RDP occupants at Kaalfontein in Tembisa. A well-structured questionnaire was distributed online to the RDP occupants in Kaalfontein in Tembisa and people working in the Department of Housing in Gauteng, South Africa. The respondents were selected randomly in each extension. The workers from the Department of Housing were selected at random on the LinkedIn platform by the researcher. Data collection is the practice of accessing data for a research project by examining a variety of data resources (Makgobi et al. 2019). A total of 104 responses were received and were all useable. The results were analyzed and interpreted using a Microsoft spreadsheet and the Statistical Package for Social Science (SPSS). The SPSS program allowed for the recording of original data to determine the mean item scores, standard deviations, ranks, and Cronbach's alpha, both SPSS and Microsoft Excel were used to create the graphs and charts.

Internal consistency is commonly assessed by Cronbach's Coefficient Alpha (Cronbach 1951). The coefficient is effective in research studies where the Likert scale was used for responses. The range is from 0 to 1, and a Cronbach's Alpha score of 0.7 is considered to be an acceptable reliability coefficient. The Cronbach's Alpha value for the effects of property rights for low-income housing in South Africa was 0.833, which proves the reliability and validity of the questionnaire.

5 Research Findings

5.1 Demographic Characteristics of the Respondents

This section will present the results of the research findings of the demographic characteristics of the research participants.

Findings relating to the respondent's gender orientation represent 51.92% of respondents were male, and 48.08% were females.

Furthermore, findings relating to the age group of the respondents are as follows (Table 1):

Effects of Property Rights for Low-Income Housing in South Africa 101

Table 1. Respondents' age group

Respondents' age group	Percentage (%)
21–25 years old	13.46
26–30 years old	4.81
31–35 years old	6.73
36–40 years old	5.77
41–45 years old	18.27
46–50 years old	18.27
51–55 years old	22.12
Above 55 years old	10.58

Findings relating to the respondents' ethnic group of the assessed sample represent 84.62% of respondents were black, 13.46% of respondents were coloured, and 1.92% of respondents were white. Findings reveal that out of 104 respondents, 58.65% of the respondents were employed, while 41.35% were unemployed. The respondents who are employed are mostly contracted and do not earn more than R3000–8000 per month. These findings indicate that there is a balance of gender between respondents, with a difference of only four respondents. The majority of respondents are between the age of 51–55 years old, with the least respondents coming from an age group of 26–30 years old, so this implies that RDP participants surveyed have been beneficiaries for a very long time since the programme was implemented in the 2000s. Furthermore, the most dominant ethnic group amongst the participants was blacks, and the most prevalent qualification among the respondents was the Matric Certificate (Grade 12) which implies that the respondents are educated enough to understand the dynamics of the RDP housing policy and can provide credible and reliable information.

Findings relating to the current title deed status of respondents revealed that 72.12% of the respondents have title deeds while 27.88% do not have title deeds. However, the respondents claimed they could not secure a loan with any financial institution since the RDP house as an asset has less monetary value. Therefore, over 70% of respondents have title deeds, with 59% employed, which proves that the RDP housing strategy benefited the majority of the citizens, where these citizens were unemployed at the time of getting these subsidised houses, so these findings are in agreement with the findings of Mulok and Kogid (2008:28) who argue that low-income housing is provided for the betterment of the low-income and unemployed citizens which are the priority.

5.2 Descriptive Analysis of the Effects of Property Rights on Low-Income Housing in South Africa

The respondents were asked to indicate what they perceived as the effects of property rights for low-income housing in South Africa using a five-point Likert scale of 'Strongly disagree' to 'Strongly agree' on the questionnaire. Table 2 discloses the respondents' ranking of effects of property rights for low-income housing in South Africa. The top five effects of property rights based on the ranking are poverty reduction which was ranked first, poor quality of houses was ranked second, enhancing of the neighbouring property values was ranked third, housing as an asset was ranked fourth and improved investment was ranked fifth.

Table 2. Effects of property rights on low-income housing in South Africa

Effects of property rights on low-income housing in South Africa	MIS	SD	R
Poverty reduction	4,05	1,092	1
Poor quality of houses	3,94	0,974	2
Enhances neighbouring property values	3,90	1,000	3
Housing as an asset	3,86	1,018	4
Improved Investment	3,83	0,960	5
Insecure tenure security	3,66	1,030	6
Negative effects on mental health	3,37	1,208	7
Negative effects on physical health	3,35	1,164	8

MIS = Mean Item Score, SD = Standard Deviation, and R = Rank.

6 Discussion

The guaranteed effect of property rights on low-income housing is poverty reduction as it is the highest-ranked effect by the respondents and the findings of Makgobi et al. (2019), and Meinzen-Dick et al. (2009:2) support that securing property rights ensures poverty alleviation. Poverty reduction, poor quality of houses, and enhancement of neighbouring property values were the highest-ranked effects of property rights on low-income housing in South Africa. The poor quality of houses being second-ranked effect supports the findings of Wolverton (2019), who states that poor quality of houses results in houses having toxins including lead, asbestos, and mold, which can have a harmful impact on mental health. The respondents are in agreement with the findings of Malgas (2018) that the provision of state-subsidized houses increases the neighbouring property values through investment. The respondents believe that the state-subsidized houses come as social asset in the sense that it provides a connection between the government and the people through the provision of basic local amenities and this in agreement with the findings of Mbatha (2018).

7 Conclusion

The findings in relation to the literature revealed that the respondents believe that poverty reduction, poor quality of houses, and enhancement of neighbouring property values are the top three effects of property rights on low-income housing than insecure tenure security, negative effects on mental health, and negative effects on physical health which were the least ranked effects. This suggests that there should be an improvement in the design of the houses to ensure the quality of the product, which is the most crucial parameter of success for a project.

The housing Act must be reformed to cater for the full functioning of property rights of the beneficiaries without any restrictions but provide sustainability to the subsidised housing programmes. The use of legally binding contracts between state institutions and the community where both can be juristic persons is a new paradigm that can enforce accountability if one party defaults on their obligations. This paradigm can encourage transparency and accountability in the Department of Housing administration.

8 Recommendations

The government must provide a clear and appropriate legal framework for low-income housing through the amendment of the Housing Act to have a section detailing how beneficiaries should enjoy the fruits of their property rights with no restrictions imposed. The government should campaign for the creation of forums from each community in South Africa with no political affiliations where the needs of the community will be advocated without any preferences to certain individuals, and beneficiaries' participation to housing programmes will be ensured.

References

Aigbavboa, C.O.: An assessment of the critical factors impeding the delivery of low-income housing in South Africa. Socioeconomica 3(6), 219–232 (2015)

Anderson, S.: Property rights. https://www.investopedia.com/terms/p/property_rights.asp (2020). Accessed 11 Jul 2021

Arnot, C.D., Luckert, M.K., Boxall, P.C.: What is tenure security? Conceptual implications for empirical analysis. Land Econ. 87(2), 297–311 (2011)

Barone, A.: Asset. https://www.investopedia.com/terms/a/asset.asp (2021). Accessed 15 Jul 2021

Barry, M., Roux, L.: Perceptions of land registration in a state-subsidised housing project in South Africa. Hous. Financ. Int. 28(4), 27–33 (2014)

Blumenfeld, J.: From icon to scapegoat: the experience of South Africa's reconstruction and development programme. Dev. Policy Rev. 15(1), 65–91 (1997). https://africabib.org/htp.php?RID=157005380

Boshoff, D.: Understanding the Basic Principles of Property Law in South Africa. The South African Council for the Quantity Surveying Profession, Pretoria (2013)

Boudreaux, K.: Empowering the poor through property rights. Mercatus Center. https://www.mercatus.org/system/files/ch2.pdf (2008). Accessed 15 Jul 2021

Brasselle, A.S., Gaspart, F., Platteau, J.P.: Land tenure security and investment incentives: puzzling evidence from Burkina Faso. J. Dev. Econ. 67(2), 373–418 (2002)

Chepsiror, E.: The Challenges of Housing Development for the Low-Income Population in Kenya; A case of Eldoret Town. Dissertation, University of Nairobi, Nairobi (2013)

Choy, L.H.T., Lai, Y., Wang, J., Zheng, X.: Property rights and housing prices: an empirical study of small property rights housing in Shenzhen, China. Land Use Policy 68, 429–437 (2017). https://doi.org/10.1016/j.landusepol.2017.08.010

Coase, R.H.: The problem of social cost. J. Law Econ. **56**(4), 837–877 (2013). https://doi.org/10.1057/9780230523210_6

Cronbach, L.J.: Coefficient alpha and the internal structure of tests. Psychometrika **16**, 297–334 (1951)

Department of Human Settlement: Report on Evaluation of Housing Assests. http://www.dhs.gov.za/sites/default/files/u16/EVALUATION%20OF%20HOUSING%20ASSET%20WITH%20CPAGE.pdf (2015). Accessed 15 Jul 2021

Ebekozien, A., Abdul-Aziz, A.-R., Jaafar, M.: Housing finance inaccessibility for low-income earners in Malaysia: factors and solutions. Habitat Int. **87**, 27–35 (2019). https://doi.org/10.1016/j.habitatint.2019.03.009

Eggertsson, T.: Economic Behavior and Institutions: Principles of Neo institutional Economics. Cambridge University Press, Cambridge (1990)

Ehrenberg, D.: The importance and benefits of tenure security. Property Rights Land Tenure Issues **13**(3), 1–24 (2006)

Elbow, K.: What is Tenure Security? Why does it matter? https://www.land-links.org/wp-content/uploads/2017/02/USAID_Land_Tenure_2014_Haiti_Training_Module_1_Presentation_2_Elbow.pdf (2014). Accessed 15 Jul 2021

Galiani, S., Schargrodsky, E.: Effects of land titling on child health. Econ. Hum. Biol. **2**(3), 353–372 (2004)

Haas, A.R., Jones, P.: The importance of property rights for successful urbanisation in developing countries. Int. Growth Centre: Policy Brief **43609**, 1–8 (2017)

He, S., Wang, D., Webster, C., Wing Chau, K.: Property rights with price tags? Pricing uncertainties in the production, transaction, and consumption of China's small property housing. Land Use Policy **81**, 424–433 (2019)

Kurtus, R.: What is physical health. https://www.school-for-champions.com/health/what_is_health.htm#.YQHKDI4zZPY (2017). Accessed 15 Jul 2021

Makgobi, G., Mashwama, N., Aigbavboa, C.: Theoretical assessment of impacts of property rights on existing reconstruction and development program houses. In: Proceedings of International Structural Engineering and Construction Management, vol. 6, issue 1 (2019)

Malgas, S.M.: The effects of subsidized housing on the property values of neighbourhoods within its vicinity. Doctoral Thesis, University of Cape Town, Cape town (2018)

Mashwama, N., Thwala, D., Aigbavboa, C.: Challenges of reconstruction and development program (RDP) houses in South Africa. IEOM Society International (2019). https://www.researchgate.net/publication/333506568_Challenges_of_Reconstruction_and_Development_Program_RDP_Houses_in_South_Africa Accessed 26 Apr 2021

Mbatha, S.: Informal transactions of low-income houses in South Africa: a case study of eThekwini Municipality. Doctoral Thesis, University of Stuttgart, Stuttgart (2018). http://elib.uni-stuttgart.de/handle/11682/9816

Meinzen-Dick, R., Kameri-Mbote, P., Markelova, H.: Property Rights for Poverty Reduction? vol. 10017, p. 86. Bloomsbury Academic, New York, NY (2009)

Mental Health.gov: What is Mental Health. https://www.mentalhealth.gov/basics/what-is-mental-health (2020). Accessed 15 Jul 2021

Moss, V., Dincer, H., Hacioglu, U.: Financial regulations and standards in the low-income property market of South Africa. J. Bus. Manage. Econ. **4**(8), 187–194 (2013)

Muller, G., Brits, R., Pienaar, J.M., Boggenpoel, Z.: Silberberg and Schoeman's: The Law of Property, 6th edn. LexisNexis, Capetown (2016)

Mulok, D., Kogid, M.: Low-cost housing in Sabah, Malaysia: a regression analysis. Asian Soc. Sci. **4**(12), 27–33 (2008)

O'Driscoll, G.P., Hoskins, L.: Property Rights: The Key to Economic Development. Cato Institute, Washington, DC (2003)

Porio, E., Crisol, C.: Property rights, security of tenure and the urban poor in Metro Manila. Habitat Int. **28**(2), 203–219 (2004)

Rubin, M.: Perceptions of corruption in the South African housing allocation and delivery programme: what it may mean for accessing the state. J. Asian Afr. Stud. **46**(5), 479–490 (2011)

Smith, R.E.: Land tenure, fixed investment, and farm productivity: evidence from Zambia's southern province. World Dev. **32**(10), 1641–1661 (2004)

Statistics South Africa: Sustainable Development Goals: Country Report 2019. http://www.statssa.gov.za/MDG/SDGs_Country_Report_2019_South_Africa.pdf (2019). Accessed 22 Jun 2021

Statistics South Africa: General Household Survey 2019 Statistical Release. https://www.datafirst.uct.ac.za/dataportal/index.php/catalog/852 (2020). Accessed 04 Apr 2021

von Benda-Beckmann, F., von Benda-Beckmann, K., Wiber, M.: The properties of property. In: von Benda-Beckmann, F., von Benda-Beckmann, K., Wiber, M. (eds.) Changing Properties of Property, pp. 1–39. Berghahn, New York (2006)

Walsh, E.: The Effects of Low-income Housing on Health. https://sites.uab.edu/humanrights/2018/10/01/the-effects-of-low-income-housing-on-health/ (2018). Accessed 17 Jul 2021

Wolverton, S.: Low-income housing: the negative effects on both physical and mental health. https://ncrc.org/low-income-housing-the-negative-effects-on-both-physical-and-mental-health/ (2019). Accessed 17 Jul 2021

Recyclability of Construction and Demolition Waste in Ghana: A Circular Economy Perspective

M. Adesi[1(✉)], M. Ahiabu[2], D. Owusu-Manu[1], F. Boateng[3], and E. Kissi[1]

[1] Department of Construction Technology and Management,
Kwame Nkrumah University of Science and Technology, Kumasi, Ghana
{michael.adesi,ernestkissi}@knust.edu.gh, d.owusumanu@gmail.com

[2] Department of Building Technology, Ho Technical University, Ho, Ghana
mosesahiabu@yahoo.com

[3] Department of Management Studies, University of Mines and Technology, Tarkwa, Ghana
fboateng@umat.edu.gh

Abstract. Purpose: The aim of this paper is to investigate construction professionals' attitude towards the recyclability of construction and demolition waste through the circularity lens.

Design/Methodology/Approach: The paper adopts the quantitative approach using purposive sampling. Questionnaires were administered to 120 construction professionals on construction sites in which 92 useable questionnaires were retrieved for analysis giving a response rate of 77%.

Findings: The findings of this research revealed that construction professionals perceive construction and demolition wastes emanate from concrete, brick, roofing sheet, metals, gypsum board as recyclable. Also, the research findings demonstrates that the recyclability of construction and demolition waste is confronted with challenges such as cost, lack of confidence in the use of recycled construction materials, technological barriers, and lack of organised markets for recycled construction materials.

Research Limitation/Implications: The research focused mostly on participants working on building construction sites; hence, a future study to include professionals in other areas of the construction industry.

Practical Implication: The findings of this research have the potential to increase the awareness of top management in construction firms on the need to implement strategic approach that drives the recycling of construction and demolition wastes.

Social Implication: The research has the potential to improve community awareness on the need to recycle construction and demolition waste for construction operations.

Originality/Value: The study addressed pertinent issues regarding the perception of construction and demolition waste recyclability among construction professionals, which consequently drives the circular economy.

Keywords: Construction · Demolition · Industry · Recyclability · Waste

© The Author(s), under exclusive license to Springer Nature Switzerland AG 2023
C. Aigbavboa et al. (Eds.): ARCA 2022, *Sustainable Education and Development – Sustainable Industrialization and Innovation*, pp. 106–120, 2023.
https://doi.org/10.1007/978-3-031-25998-2_9

1 Introduction

The global extraction of materials for use in the construction industry has increased over the years. Globally, 40% of raw materials extracted for the construction has increased in-use stocks (Aguilar-Hernandez et al. 2021). For instance, the global extraction of raw materials has increased from 7 gigatonnes in 1900 to 89 gigatonnes in 2015 (Haas et al. 2020; Fishman et al. 2016). The extraction of materials from the environment contributes to about 20% of global CO_2 emissions (United Nations Environment Programme 2019). The construction industry is highly noted for construction of new buildings and demolition of old ones which leads to large amount of waste (Berge 2009). This study is important because Tatjana et al. (2021) found demolition works as a major source of construction waste. Similarly, waste reduction strategies such as recyclability leads to 2.5% saving in overall project budget (Begum et al. 2007), hence it is appropriate to pay attention to the issues of construction waste reduction through recyclability.

The construction industry uses a lot of non-renewable materials, which reduces the availability of materials for future generations and adverse impacts on the environment (Miatto et al. 2016). Construction wastes tend to have significant impact on the sustainability of the natural environment (Oteng-Ababio et al. 2013). For instance, construction wastes are dumped in water bodies and uncontrolled dump sites which lead to sanitation and health problems for society (Ferronato and Torretta 2022). Construction activities lead to massive destruction of ecosystems in the natural environment (Oke et al. 2019). In Ghana, lack an efficient system for managing construction and demolition wastes, leads to the reckless dumping of construction wastes in residential areas and on the shoulders of the roads (Amuna et al. 2021).

Studies on construction and demolition waste seldom focus on the attitude of construction professionals towards recyclability in Ghana. For instance, Frempong-Jnr et al. (2022) investigated the impact of stakeholder management on construction waste in Ghana found that improvement in stakeholder management has beneficial impact in addressing the high levels of construction waste. A study by Agyekum et al. (2012) on the perspective of consultants on materials waste reduction, in which they found eight waste minimisation measures. Again, recent study by Agyekum et al. (2012) focused on only architects and quantity surveyors without considering construction professionals such as building services engineers and construction managers. Kpamma and Adjei-Kumi (2011) explored the management of waste in the building design process in Ghana and found that there is little awareness about waste reduction tools such as design structure matrix, batch size reduction and set-based design. Though the findings of Kpamma and Adjei-Kumi (2011) adds to construction waste minimisation approaches, the study did not focus on recyclability as a means of reducing construction waste. Agyekum et al. (2013) highlights the need to reduce construction wastes emanating from timber, blocks, mortar, and concrete without emphasising construction professionals' perception on the recyclability of the material. While Agyekum et al. (2013) investigated only four materials without considering waste from brick, metals, structural steel, roofing sheets, electrical conduits, gypsum boards and PVC pipes. Recyclability is a vital component for achieving circular economy in developing countries; however, Asante et al. (2022) note that much light has not been shed on it in Ghana. Hence, there is a gap in terms of the recyclability of construction and demolition waste. The aim of this paper is to investigate the recyclability of construction and demolition wastes through the circularity lens. Three main objectives are set to drive the investigation. The first objective is to ascertain the perception of construction professionals towards recyclability of construction and demolition waste. The second objective focuses on identifying the challenges confronting use of recycled construction materials while the third objective seeks to identify the factors that drive the recyclability of construction and demolition waste in Ghana.

2 Literature Review

The literature review section of this paper focused on attitude towards recyclability of construction materials; factors influencing the use of recycled construction materials; challenges confronting use of recycled construction materials; technologies for recycling construction wastes; recyclability and circular economy.

2.1 Circular Economy

At the end of their life cycle, materials extracted from the environment and processed for human use are introduced back into the economy as secondary materials (Lanau et al. 2019). The reintroduction of processed materials at the end of their life cycle into the economy is consistent with the philosophy of circular economy (Kirchherr et al. 2017). The reintroduction of in-use stocks such as buildings and infrastructure require techniques such as recycling of the materials that would have gone back to pollute the natural environment. Thus, recyclability is an important driver in the promotion of circular economy. A circular economy underpinned by recyclability leads to the reduces the extraction of primary resources, waste generation and emission of greenhouse gas into the environment (Mayer et al. 2018).

2.2 Recyclability

Recyclability is an objective process of enhancing the design of buildings using recycled materials to close the construction materials cycle (Vefago and Avellaneda 2013). Recyclability supports the circular economy, which focuses on cradle-to-cradle consumption (Vefago and Avellaneda (2013) and has the potential to create environmental and economic benefits through cost mitigation and waste minimization. Though diverse types of materials are used in the construction industry, studies do not focus on their recyclability. For instance, Wang et al. (2021) focused on advances in photocatalytic fibres, as construction materials while Vefago and Avellaneda (2013) investigated the recyclability of steel, concrete, and wood. Also, Sourabh et al. (2021) studied the implications of generating waste and recycling of construction and demolition waste from concrete, mortar, and bricks. Earlier, Tam (2011) used case studies to investigate the rate of recyclability and reuse of waste in construction by focusing on materials such as plastic; paper; timber; metal; glass; and concrete. This study expands on the existing knowledge of construction material recyclability by including materials that hitherto have not been largely considered by previous studies.

2.3 Attitude of Construction Professionals towards Recyclability of Construction Waste

The problem of construction waste reduction through recyclability is due to the attitude of construction professionals (Tatjana et al. 2021). Over the past decades, attitude has become increasingly critical in the construction industry because Teo and Loosemore (2001) found that construction workforce has negative environmental attitude. According to Liu et al. (2022), the attitude of construction professionals on the construction site plays vital role in the recycling of construction materials. However, the investigation of Liu et al. (2022) focused on only construction professionals' attitude towards the sorting of construction waste for recycling with less attention on specific materials. Pradeep and Sarmah (2021) proposed sixteen factors of which moral norm, and attitude towards separation are potential factors linked to the attitude of construction professionals on recyclability. In using the theory of planned behaviour, Balador et al. (2020), investigated the attitude of architects, contractors and consumers for potential reuse and recycle of building materials. The study indicated that architects have better attitude towards reuse and recycle of building materials than contractors and consumers. However, the finding regarding architects is contrary to an earlier study by Osmani et al. (2008), which revealed that architects are unenthusiastic about construction waste with the argument that it is generated on the construction site. Construction professionals usually focus on cost, performance and quality of projects leading to the neglect of post construction waste reduction strategies (Tatjana et al. 2021).

2.4 Factors Influencing the use of Recycled Construction Materials

The need to reduce the depletion of natural resources influence the use of recycled construction materials (Kabir et al. 2016). Similarly, the application of recycled is driven by construction professionals is underpinned by the existence of efficient waste management systems (Corinaldesi and Moriconi 2004). Badraddin et al. (2022) classified the success factors for the use of recycled construction materials as external and internal factors. The external factors include stakeholders' positive attitude; government policy on recycling; and availability of uniform standards for recycling of construction wastes. The internal factors consist of availability of waste management plans in construction firms; effective communication among the workforce; and availability of marketing strategy for recycled construction materials. Likewise, the need to reduce the demand and the cost of managing landfill sites for construction waste has the potential to drive the use of recycled construction materials (Lamba et al. 2021) to ensure both economic and environmental sustainability, which are essential ingredients of the circular economy.

2.5 Technologies for Recycling Construction Waste

Technological advances in recycling are important for improving the gains made in construction waste minimisation. The application of technologies in recycling construction and demolition waste enhances the perception of construction professionals and the subsequent use of recycled materials in the construction industry. The technologies for the recycling of construction and demolition materials include physical enhancement technology, chemical strengthening technology and microbial-induced carbonate precipitation; and particle shaping technology (Feng et al. 2022; Qiuyi et al. 2016). In Zhang et al. (2022), four main types of technological systems for recycling concrete waste include wet processing system, advanced dry recovery system, thermal separation system, and smart crushing system. Studies have indicated the use of Industry 4.0 technologies such as artificial intelligence, blockchain, internet of things, digital twin, additive manufacturing for the recycling of construction and demolition wastes (Oluleye et al. 2022; Rajput and Singh 2019). Industry 4.0 technologies are fundamental to transforming linear economies into circular economies by integrating it into the recycling of construction materials (Norouzi et al. 2021).

2.6 Challenges Confronting use of Recycled Construction Materials

The challenges confronting the use of recycled construction materials include members of working group that have different attitude and lack of training on the need to use recycled materials to minimise waste in the construction industry (Badraddin et al. 2022). The lack of environmental education, which hitherto has not been taught in most construction educational training institutions poses challenges to the use of recycled construction and demolition material (Tatjana et al. 2021). Bolden et al. (2013) in their study identified twelve factors that prevent the use of recycled materials such as cost, education, environment, contamination, permits, separation, market, storage, and equipment.

3 Methodology

The target population consists of construction professionals in the Ghanaian construction firms. Specifically, the construction professional involved in the study consists of construction managers, quantity surveyors, building services engineers, and architects. The paper adopts the quantitative approach by using survey questionnaires to collect data from participants. The quantitative approach was chosen to ensure wider coverage of the data collection process as compared to the qualitative which focuses on the depth of information gathered. According to Williams (2007), the quantitative approach is appropriate for investigating factors influencing social behaviour. Since this study seeks to ascertain the attitude of construction professionals towards the recyclability of materials, the quantitative approach and survey were adopted for data collection. The survey questionnaire is divided into four sections. The first section of the questionnaire dealt with the respondents' profile, while the second section focused on the perception of construction professionals on the recyclability of materials that constitute construction and demolition waste. The third part of the questionnaire sought to identify the potential challenges of recycling construction and demolition waste while section four delved into the conditions or factors that improve recycling.

Two types of measurement scales were used to measure the responses of participants regarding specific questions posed to them. The use of different scales of measurement is to ensure that participants provide the requisite responses to the questions. For instance, in designing the questionnaire, the authors used the nominal scale for the question dealing with demography of participants while the question in section two, three and four adopted the 5-point Likert scale of measurement. A purposive sampling technique was adopted to administer the questionnaires to 120 participants comprising construction managers, quantity surveyors, building services engineers, and architects in small and medium size construction firms. The purposive sampling was used to enable the researchers to focus on specific characteristics of respondents such as professional background and level of experience. Overall, 92 questionnaires were returned for analysis with a response rate of 77%. The descriptive statistical tools used for the analysis included weighted mean, percentages, and standard deviation.

4 Analysis and Discussion of Results

This section of the paper analyses and discusses the results of the investigation pertaining to factors influencing the use of recycled construction materials, and the challenges confronting the use of recycled construction materials. The results regarding attitude of construction professionals towards recyclability of construction materials; and the profile of respondents were equally analysed and discussed.

4.1 Demography of Respondents

The paper focused on respondent's level of education, professional background, and work experience. Figure 1 shows majority of respondents have BSc degrees followed by MSc while few participants have educational backgrounds with PhD, diploma, and MPhil.

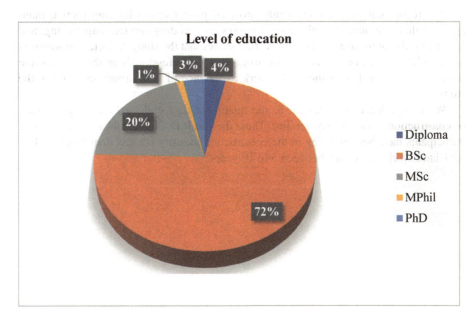

Fig. 1. Respondents' level of education

The result in Fig. 1 indicates that more people with tertiary level of education have been involved in this investigation, which is consistent with Ni et al. (2022) who found that 71% of participants involved in their study had tertiary education with bachelor's degrees and even higher. In addition to the educational level of respondents, this paper also delves into the professional background of participants shown in Fig. 2.

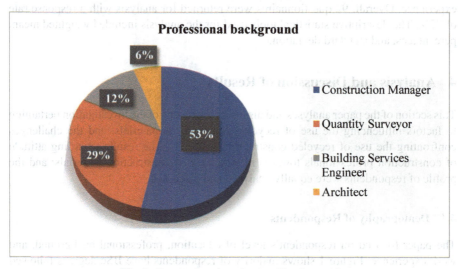

Fig. 2. Respondents' professional background

The result indicates that the study involved participants with construction management than respondents with quantity surveying, building services engineering, and architectural professional backgrounds. This shows that the study collected information on recyclability of construction waste from diverse professionals in the construction industry. Similarly, Fig. 3 shows the work experience of participants involved in the study.

Work experience is important to the quality of data collected on the perception of construction material recyclability. Thus, the result in Fig. 3 indicates that 60% of participants have been working in the construction industry for less than 5 years while 33% have work experience between 5 to 10 years.

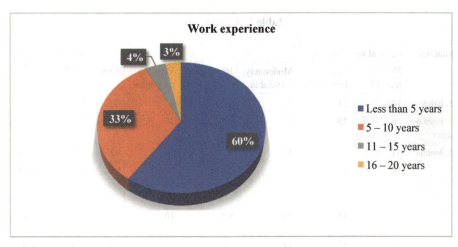

Fig. 3. Respondents' work experience

4.2 Attitude of Construction Professionals towards the Recyclability of Construction Waste

Table 1 demonstrates the results on the attitude of construction professionals towards the recyclability of nine materials. These nine materials include concrete, bricks, roofing sheet, metals, gypsum board, PVC pipes, electrical conduit, wood panelling and structural steel. The results in Table 1 show that 48 respondents involved in the study perceived construction wastes emanating from electrical conduits as recyclable. Other construction materials that participants perceived as recyclable are PVC pipes, wood panelling, structural steel, metals, and roofing sheets. However, 30 and 26 respondents perceive metals and structural steel as very recyclable.

The result also show that 31 and 25 respondents consider concrete as not recyclable and less recyclable. PVC pipes, wood panelling, structural steel, metals, and roofing sheets have their weighted mean above 3.0, indicating that the responses of the participants regarding these materials are significant. The result implies that participants involved in this study perceive construction wastes, as emanating from PVC pipes, wood panelling, structural steel, metals, and roofing sheets as recyclable. This result is consistent with Vefago et al. (2013) that identified steel profile, copper-electric cable, and hardwood without toxic preservative, as recyclable.

Table 1. Recyclability of construction materials

Materials	Level of recyclability					Total	Weighted mean	Std. Dev
	Not recyclable	Less recyclable	Moderately recyclable	Recyclable	Very recyclable			
1. Concrete	31	25	15	17	4	92	2.33	1.24

(*continued*)

Table 1. (*continued*)

Materials	Level of recyclability					Total	Weighted mean	Std. Dev
	Not recyclable	Less recyclable	Moderately recyclable	Recyclable	Very recyclable			
2. Brick	14	21	27	24	6	92	2.86	1.16
3. Roofing sheet	6	15	20	35	16	92	3.43	1.15
4. Metals	1	7	19	35	30	92	3.93	0.97
5. Gypsum board	14	17	27	27	7	92	2.96	1.18
6. PVC piping	6	11	16	43	16	92	3.57	1.11
7. Electrical conduit	5	12	18	48	9	92	3.48	1.02
8. Wood panelling	8	19	18	33	14	92	3.28	1.20
9. Structural steel	7	7	16	36	26	92	3.73	1.17

Having examined the perceptions of construction professionals towards the recyclability of construction and demolition waste, the next section of the paper focuses on the challenges facing recyclability.

4.3 Challenges Confronting use of Recycled Construction Waste Materials

The results in Table 2 show the challenges confronting recyclability, which include cost, lack of confidence, knowledge, and markets; technological barriers; and contamination of recyclable materials.

Table 2. Challenges confronting the recyclability of construction waste materials

Challenges	Weighted mean	Standard deviation
1. Cost of recycling	3.77	1.03
2. Lack of confidence	3.64	1.20
3. Technological barriers	4.04	1.12
4. Lack of knowledge	3.78	1.03
5. Lack of markets	3.72	1.05
6. Contamination of recyclable materials	3.59	1.10

The result in Table 2 indicates that all the six challenges are critical to the successful recycling of construction material because they have their weighted mean above 3.0. The lack of designated market and sectors for recycled construction materials poses threat to the recyclability of waste generated by the construction industry. Technological barriers are the most challenging issues confronting the recyclability of construction materials, as the result in Table 2 demonstrates that it has a weighted mean above 4. The inability of Ghana to use technology leads to lack of confidence in recycled materials and nonexistence of markets prevent the use of these materials in the construction industry.

Existing studies such as Feng et al. (2022); and Qiuyi et al. (2016) found the technologies for recycling of construction waste as physical enhancement technology, chemical strengthening technology and microbial-induced carbonate precipitation; and particle shaping technology. Studies by Oluleye et al. (2022); and Rajput and Singh (2019) identified industry 4.0 technologies such as artificial intelligence, blockchain, internet of things, digital twin, additive manufacturing as appropriate for the recycling of construction and demolition wastes. In developing countries and middle-income economies such as Ghana, many studies have not been undertaken on technologies for recycling construction waste. Hence, a future study on recycling technology will be appropriate in advancing the understanding of construction waste recyclability.

It is necessary to examine the factors that drive the use of recycled construction materials. The next section of the paper the factors that enhance the use of recycled construction materials in Ghana.

4.4 Factors Influencing the use of Recycled Construction Materials

Table 3 identified ten main factors that promote the reuse of recycled construction materials.

Table 3. Factor for recyclability of construction and demolition waste materials

Factors for the use recycled construction materials	Weighted mean	Standard deviation
1. The method adopted for collecting construction waste materials for recycling	3.72	0.98
2. Availability of space for storing construction waste collected for recycling	3.76	0.98
3. Availability of markets for the recycled construction materials (demand side)	3.83	0.98
4. Efficiency of construction waste collection methods (collection should not create another waste)	3.67	0.92
5. Knowledge on waste management	4.07	0.92
6. Assessment of waste stream	3.82	0.93

(*continued*)

Table 3. (*continued*)

Factors for the use recycled construction materials	Weighted mean	Standard deviation
7. Availability of statutory regulations guiding the collection, transportation and storage of construction waste	4.22	0.86
8. Community awareness	4.26	0.90
9. Innovative practices	4.29	0.79
10. Effectiveness of management	4.17	0.77

The results in Table 3 indicate that all the factors are important drivers for the use of recycled materials in the construction industry. Issues such as community awareness, regulations, innovative practices, effective management, and knowledge on waste management have their weighted mean value above 4.0. Also, the result of the study in Table 3 indicates that the availability of markets influences the use of recycled construction materials. Gunaratne et al. (2022) found that the utilisation of recycled construction materials is driven by market processes such as the use of dominant primary material, and availability of customers.

The result pertaining to community awareness has the potential to enhance the availability of customers for recycled construction materials through sensitisation programmes, which addresses the shortage of resources in the future. Community awareness improves the use of recycled construction materials, as Amuna et al. (2021) note that the efficient management of construction waste is low in Ghana. The methods for collection of construction and demolition waste enhances the approach to recyclability. In this case, the result in Table 3 demonstrate that the method adopted for collection of construction waste influences the rate of recyclability. This study also show that the availability of storage space is important to the application of recycled construction materials in Ghana. It is therefore necessary that the storage facilities for construction waste are available to strengthen the niche market for the recycled materials in the construction industry.

This paper identified three key issues, which are the identification of sources of waste materials, the awareness of the challenges to recyclability, and conditions that eliminates the challenges confronting recyclability as shown in Fig. 4.

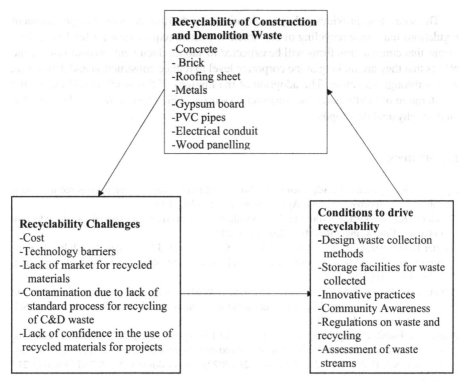

Fig. 4. Recyclability of construction and demolition waste

5 Conclusion

This paper investigates the recyclability of construction and demolition waste in Ghana. In doing so, the paper specifically focused on the attitude of construction professionals on the recyclability of construction waste through the circularity perspective. In addition, the study delved into the challenges inhibiting the recyclability of construction and demolition wastes and the conditions or factors that have the potential to drive effective recyclability of construction and demolition wastes. The paper identified key factors for driving recyclability in the construction industry, notably, the adoption of innovative practices for waste management, community awareness, and formulation of statutory regulations and policies to guide government and regulatory agencies, and construction firms in developing policies for the use of recycled construction materials in Ghana.

Though this paper identified technology as a barrier for the recyclability of construction and demolition waste, the paper is limited in terms of the key issues of technologies for recycling construction waste. Hence, a further study on technologies for separation of construction waste in terms of sorting; and technologies for improving the strength and durability of recycled construction materials will be appropriate. Again, a future study on green business models for construction and demolition waste will be appropriate in addressing the demand and marketing challenges confronting the use of recycled construction materials.

The study also provides policy implications for drafting construction procurement regulations that make recycling of construction waste a requirement for tendering. This means that construction firms will be expected to submit documents to demonstrate the efforts that they are making at the corporate level to reduce construction and demolition wastes through recycling. The adoption of the findings of this study would reduce the proliferation of landfill sites for construction wastes, which tend to reduce the available land for physical development.

References

Agyekum, K., Ayarkwa, J., Adjei-Kumi, T.: Minimizing materials wastage in construction-a lean construction approach. J. Eng. Appl. Sci. **5**(1), 125–146 (2013)

Agyekum, K., Ayarkwa, J., Adinyira, E.: Consultants' perspectives on materials waste reduction in Ghana. Eng. Manag. Res. **1**(1), 138–150 (2012)

Aguilar-Hernandez, G.A., Deetman, S., Merciai, S., Rodrigues, J.F., Tukker, A.: Global distribution of material inflows to in-use stocks in 2011 and its implications for a circularity transition. J. Ind. Ecol. **1**(1), 1–15 (2021)

Amuna, D.N., Imoro, Z.A., Cobbina, S.J., Ofori, S.A.: Waste management practices of construction companies at the airport hills and Sakumono areas in Accra, Ghana. Rwanda J. Eng. Sci. Technol. Environ. **4**(1), 1–13 (2021)

Asante, R., Faibil, D., Agyemang, M., Khan, S.A.: Life cycle stage practices and strategies for circular economy: assessment in construction and demolition industry of an emerging economy. Environ. Sci. Pollut. Res. **29**, 82110–82121 (2022). https://doi.org/10.1007/s11356-022-21470-w

Badraddin, A.K., Radzi, A.R., Almutairi, S., Rahman, R.A.: Critical success factors for concrete recycling in construction projects. Sustainability **14**(5), 1–19 (2022)

Balador, Z., Gjerde, M., Vale, B., Isaacs, N.: Towards an understanding of the potential to reuse and recycle building materials in New Zealand. In: Ghaffarianhoseini, A., et al. (eds.) Proceedings of the 54th International Conference of the Architectural Science Association, pp. 1233–1242. Architectural Science Association (ANZAScA) (2020)

Begum, R.A., Siwar, C., Pereira, J.J., Jaafar, A.H.: Factors and values of willingness to pay for improved construction waste management: a perspective of Malaysian contractors. J. Waste Manage. **27**(1), 1902–1909 (2007)

Berge, B.: Ecology of Building Materials, 2nd edn. Architectural Press, Oxford, UK (2009)

Bolden, J., Abu-Lebdeh, T., Fini, E.: Utilization of recycled and waste materials in various construction applications. Am. J. Environ. Sci. **9**(1), 14–24 (2013)

Corinaldesi, V., Moriconi, G.: Reusing and recycling C&D waste in Europe. In: Limbachiya, M.C., Roberts, J.J. (eds.) Construction Demolition Waste: Challenges and Opportunities in a Circular Economy. USA (2004)

Feng, C., Cui, B., Huang, Y., Guo, H., Zhang, W., Zhu, J.: Enhancement technologies of recycled aggregate–Enhancement mechanism, influencing factors, improvement effects, technical difficulties, life cycle assessment. Constr. Build. Mater. **317**, 126168 (2022)

Ferronato, N., Torretta, V.: Waste mismanagement in developing countries: a review of global issues. Int. J. Environ. Res. Pub. Health **16**(6), 2–28 (2019)

Fishman, T., Schandl, H., Tanikawa, H.: Stochastic analysis and forecasts of the patterns of speed, acceleration, and levels of material stock accumulation in society. Environ. Sci. Technol. **50**(7), 3729–3737 (2016)

Frempong-Jnr, E.Y., Ametepey, S.O., Cobbina, J.E.: Impact of stakeholder management on efficient construction waste management. Smart Sustain. Built Environ. (2022). https://doi.org/10.1108/SASBE-08-2021-0147

Gunaratne, T., Krook, J., Andersson, H.: Market prospects of secondary construction aggregates in Sweden. J. Clean. Prod. **360**(1), 132155 (2022)

Haas, W., Krausmann, F., Wiedenhofer, D., Lauk, C., Mayer, A.: Spaceship earth's odyssey to a circular economy – a century long perspective. Resour. Conserv. Recycl. **163**(1), 1–10 (2020)

Kabir, S., Al-Shayeb, A., Khan, I.M.: Recycled construction debris as concrete aggregate for sustainable construction materials. Procedia Eng. **145**(1), 1518–1525 (2016)

Kirchherr, J., Reike, D., Hekkert, M.: Conceptualizing the circular economy: an analysis of 114 definitions. Resour. Conserv. Recycl. **127**(1), 221–232 (2017)

Kpamma, E.Z., Adjei-Kumi, T.: Management of waste in the building design process: the Ghanaian consultants' perspective. Architectural Eng. Des. Manage. **7**(2), 102–112 (2011)

Lamba, P., Kaur, D. P., Raj, S., Sorout, J.: Recycling/reuse of plastic waste as construction material for sustainable development: a review. Environ. Sci. Pollut. Res. 1–24 (2021). https://doi.org/10.1007/s11356-021-16980-y

Lanau, M., et al.: Taking stock of built environment stock studies: progress and prospects. Environ. Sci. Technol. **53**(15), 8499–8515 (2019)

Liu, J., Chen, Y., Wang, X.: Factors driving waste sorting in construction projects in China. J. Cleaner Prod. **336**, 130397 (2022). https://doi.org/10.1016/j.jclepro.2022.130397

Mayer, A., Haas, W., Wiedenhofer, D., Nuss, P., Blengini, G.A.: Measuring progress towards a circular economy a monitoring framework for economy wide material loop closing in the EU28. J. Ind. Ecol. **23**(1), 1–15 (2018)

Miatto, A., Schandl, H., Fishman, T., Tanikawa, H.: Global patterns and trends for non-metallic minerals used for construction. J. Ind. Ecol. **21**(1), 924–937 (2016)

Ni, G., Zhang, Z., Zhou, Z., Lin, H., Fang, Y.: When and for whom organizational identification is more effective in eliciting safety voice: an empirical study from construction industry perspective. Int. J. Occup. Saf. Ergon. 1–28 (2022). https://doi.org/10.1080/10803548.2022.2081395

Norouzi, M., Chàfer, M., Cabeza, L.F., Jiménez, L., Boer, D.: Circular economy in the building and construction sector: a scientific evolution analysis. J. Build. Eng. **44**, 102704 (2021). https://doi.org/10.1016/j.jobe.2021.102704v

Oke, A., Aghimien, D., Aigbavboa C., Madonsela, Z.: Environmental sustainability: impact of construction activities. In: Proceedings of the 11th International Conference on Construction in the 21st Century, London, United Kingdom (2019)

Oluleye, B.I., Chan, D.W.M., Saka, A.B., Olawumi, T.O.: Circular economy research on building construction and demolition waste: a review of current trends and future research directions. J. Cleaner Prod. **357**, 131927 (2022). https://doi.org/10.1016/j.jclepro.2022.131927

Osmani, M., Glass, J., Price, A.D.F.: Architects' perspectives on construction waste reduction by design. Waste Manag. **28**(7), 1147–1158 (2008)

Oteng-Ababio, M., Arguello, M.J.E., Gabbay, O.: Solid waste management in African cities: sorting the facts from the fads in Accra, Ghana. Habitat Int. **39**(1), 96–104 (2013)

Pradeep, R., Sarmah, S.P.: Investigation of factors influencing source separation intention towards municipal solid waste among urban residents of India. Resour., Conserv., Recycl. **164**, 105164 (2021). https://doi.org/10.1016/j.resconrec.2020.105164

Qiuyi, L., Yunxia, L., Chongji, Z., Shuo, T.: Strengthening technique of recycled concrete aggregate. Concrete **317**(1), 74–77 (2016). https://doi.org/10.1016/j.conbuildmat.2021.126168

Rajput, S., Singh, S.P.: Connecting circular economy and industry 4.0. Int. J. Inform. Manage. **49**, 98–113 (2019). https://doi.org/10.1016/j.ijinfomgt.2019.03.002

Sourabh, J., Shaleen, S., Nikunj, K., Jain, K.: Construction and demolition waste (C&DW) in India: generation rate and implications of C&DW recycling. Int. J. Constr. Manag. **21**(3), 261–270 (2021)

Tam, V.W.Y.: Rate of reusable and recyclable waste in construction. The Open Waste Manag. J. **4**(1), 28–32 (2011)

Tatjana, T., Diana, B., Jelena, T., Irina, S.: Awareness and attitude of latvian construction companies towards sustainability and waste recycling. Humanities **14**(7), 942–955 (2021)

Teo, M.M.M., Loosemore, M.: A theory of waste behaviour in the construction industry. Constr. Manag. Econ. **19**(7), 741–751 (2001)

UNEP: Emissions Gap Report 2019. Nairobi. https://www.unep.org/resources/emissions-gap-report-2019 (2019). Accessed 8 Dec 2021

Vefago, L.H.M., Avellaneda, J.: Recycling concepts and the index of recyclability for building materials. Resour. Conserv. Recycl. **72**(1), 127–135 (2013)

Wang, W., Yang, R., Li, T., Komarneni, S., Liu, B.: Advances in recyclable and superior photocatalytic fibers: material, construction, application and future perspective. Composites Part B **205**(1), 2–21 (2021)

Williams, C.: Research methods. J. Bus. Econ. Res. **5**(3), 65–71 (2007)

Zhang, C., Hu, M., Di Maio, F., Sprecher, B., Yang, X., Tukker, A.: An overview of the waste hierarchy framework for analyzing the circularity in construction and demolition waste management in Europe. Sci. Total Environ. **803**, 149892 (2022)

Innovative Work Environment of an Informal Apparel Micro Enterprise with PDCA Cycle: An Action-Oriented Case Study

W. K. Senayah[1(✉)] and D. Appiadu[2]

[1] Department of Fashion Design and Textiles, Accra Technical University, Accra, Ghana
wksenayah@atu.edu.gh
[2] Department of Family and Consumer Science, University of Ghana, Accra, Ghana
dappiadu@ug.edu.gh

Abstract. Purpose: Working conditions are undoubtedly critical to the general well-being and performance of workers. The purpose of the study was to apply the Plan-Do-Check-Act model to improve the quality of the work environment of an informal apparel micro enterprise.

Design/Methodology/Approach: Action research with a single-case study design and qualitative approach was adopted. Direct observation and interview were used to collect the data.

Findings: Findings showed that the quality of the MSE's work environment was poor. The business operator experienced consistent body pains which affects the worker's general well-being and performance. Strategies for improving the work environment was developed based on ergonomic principles, communicated to the business operator and implemented.

Implications/Research limitations: Account of the subject implied that the improved work environment i.e. renovation of the workshop led to the worker's comfort, well-being, and increased performance.

Practical Implications: This action-oriented study can be extended to other small-scale informal apparel enterprises in poor urban areas in developing counties to improve their work environment.

Originality/Value: This would help to contribute to the attainment of SDG 3 (good health and well-being), SDG 8 (decent work and economic growth) and SDG 9 (industry, innovation, and infrastructure) set by the global community. It can also serve as case for educational purposes.

Keywords: Apparel industry · Ergonomics · Ghana · Working condition · Work environment

1 Introduction

The apparel industry in Ghana is an important part of the nation's economy. A large number of people work in the apparel industry compared with others in the manufacturing sector (The Ghana Statistical Service 2016). Majority of the firms are micro and small enterprises (MSEs) with employee size ranging from 1 to 29 (The National Board for Small Scale Industries, NBSSI 2015). The industry is dominated by businesses that produce custom-made garments. Customised garment making is a long-standing tradition in Ghana practiced by MSEs. Whereas the few medium to large enterprises are involved in mass-production. Most of these MSEs are informal businesses that mostly operate with low capital and standard equipment (Senayah 2018). MSEs in Ghana's apparel industry comprise small-scale tailors and dressmakers operating in wooden kiosks, shops made from shipping containers, brick and mortar shops, or their homes. The businesses are actively managed by the owners who are involved in the garment making and may have apprentices understudying them.

The surge in informal apparel MSEs is as a result of the relatively ease with which apparel businesses can be set up with low capital and expertise. Most of these business operators acquire the sewing skills through apprenticeship (tacit learning) whereas, few learn via the formal educational system (Sawyer 2019). Upon graduation, new trainees start business with basic sewing equipment and tools, tables, chairs and any other resource that can do the required job irrespective of whether they are ergonomically designed or contribute to a good work environment. Those who are unable to rent a space operate in their homes. Due to the low start-up capital of MSEs, the work environment are often not up to the optimum standard. The International Labour Organization (ILO) (2022) indicated that "work in the informal economy is often characterized by small or undefined work places, unsafe and unhealthy working conditions, low levels of skills and productivity, low or irregular incomes, long working hours and lack of access to information, markets, finance, training and technology".

Working conditions are undoubtedly critical to the general well-being and performance of workers. An appropriate work environment is an essential component of decent work. Three of the global sustainability agenda by the United Nations (UN) (2017a, 2017b) address work conditions, well-being, and productivity of informal MSEs. Sustainable development goal (SDG) 3: good health and well-being through which the global community has agreed to ensure healthy lives and promote well-being for all at all ages, SDG 8: decent work and economic growth, specifically target 8.3: that seeks to promote development-oriented policies that support productive activities, decent job creation, entrepreneurship, creativity and innovation, and encourage the formalization and growth of micro-, small- and medium-sized enterprises, including through access to financial services." And SDG 9: industry, innovation, and infrastructure. Specifically, target 9.4 which focuses on improving industry infrastructure and industrial processes based on respective capacities.

Occupational health and safety have been widely studied. In the case of Ghana's apparel industry, several studies on working conditions of employees in medium to large-scale apparel firms have been conducted. Although, majority of apparel businesses in Ghana are informal MSEs, there has been less attention on the working conditions of these enterprises. Poor working conditions lead to deterioration in health and well-being of workers. Ill-health can take a toll on workers' finances which can affect their livelihoods this is because the industrial processes used in the manufacture of clothes are outmoded and riddled with poor supporting infrastructure.

In spite of aforementioned issues, there is limited studies on Ghanaian apparel MSEs' working conditions. One of such studies by Vandyck and Fianu (2012) regarding Ghanaian dressmakers and tailors' work environments, showed that workers were operating in non-ergonomic conditions which affected their comfort and work productivity. They suggested that trade organizations should assist apparel producers to improve their working conditions. However, a decade has passed and a general observation in poor urban areas in Ghana shows that some of these MSEs are still working under poor environment due to lack or inadequate knowledge about ergonomics. Perhaps, this is as a result of low knowledge dissemination. Informal MSE operators hardly search for documents or attend academic presentations and thus, without interventions such as continuous occupational safety education by authorities and trade associations, the problem would still persist. In addition to that, although the government of Ghana has committed to implementing sustainable development goals, Ghana's Voluntary National Review Report on the Implementation of the 2030 Agenda for Sustainable Development, did not make any reference to the textiles and apparel industry (National Development Planning Commission 2022). It is therefore imperative for action research to be carried out. Hence, the present study applied the Plan-Do-Check-Act (PDCA) cycle to improve the work environment especially ergonomic conditions of a selected apparel MSE in a poor urban area.

The paper sheds light on how PDCA cycle was used to ascertain the selected apparel MSE's work environment specifically ergonomic issues, suggestion and implementation of strategies to improve identified shortfalls as well as evaluation to ensure continuous improvement. The rest of the paper is structured beginning with a brief review of theoretical background and understanding of working condition with emphasis on work environment and ergonomic conditions in garment industry, and the PDCA cycle. The adopted methodology is described, which is followed by presentation of results, discussion, conclusion, implications, and limitation.

2 Literature Review

2.1 The Need for Sustainable Industrialisation and Innovation: Ghana's Apparel Context

In 2015, Member States of the United Nations adopted the 2030 Agenda for Sustainable Development with the aimed of achieving peace and prosperity for people and the planet (Clark 2017). The agenda for sustainability is made up of 17 broad goals often referred to as Sustainable Development Goas (SDGs). Goal 9 of the SDG, which is the focus of this paper, focuses on Industry, innovation, and infrastructure with 8 targets (Clark 2017). Specifically, goal 9, seeks to build resilient infrastructure, promote inclusiveness and sustainable industrialisation and foster innovation. Undoubtedly, the apparel industry both in Ghana and abroad has huge sustainability issues for which sustainable industrialisation is much needed (Elrod 2017; Ertl and Schebesta 2020; Joyner Armstrong and Park 2017; Kaledzi 2022). Part of the sustainability challenge is a learning trap where MSEs do not possess the requisite knowledge on the technical know-how to increase production efficiency and make profit (Whitfield and Staritz 2021). In small scale factories, the process for enhancing sustainable manufacturing processes may require retro fitting some of the industrial processes used in the manufacturing of products. This is especially so since some of these MSEs may not have the financial resources to replace production processes with new technology and machinery. It is not always the case that sustainability requires new technology, machinery or capital. Sometimes, in the case of some MSEs, it may require the necessary knowledge that improves working environment, productivity and health. This applies in all the manufacturing sectors including the apparel sector.

In Ghana, the government considers the apparel sector as part of the strategic sectors for industrialisation. The government has identified the textiles and apparel industry as part of its industrialisation drive. Indeed, the Coordinated Programme of Economic and Social Development Policies (2017–2024) identifies the apparel industry as part of "strategic anchor industrial initiatives" aimed at economic transformation and in line with that the government has outlined steps such as lowering taxes and address structural and bottlenecks as a matter of priority. In line with the Medium-Term National Development Policy Framework (2022–2025), the government intends to promote import substitution in the sector and pursue national industrial initiatives (Government of Ghana 2016; Republic of Ghana 2017). These policy frameworks demonstrate the government's commitment to the sector in terms of national development.

Despite the policies, sustainability issues affecting the sector have not received the needed attention. Indeed, a national review of the country's implementation of the SDGs dubbed: Ghana's Voluntary National Review Report on the Implementation of the 2030 Agenda for sustainable Development does not mention any sustainable initiative for the apparel sector although the industry has sustainability challenges (National Development Planning Commission 2022). This is surprising given that Ghana is one of the highest importers of second-hand clothing (Kaledzi 2022). The country is reported to import 15 million clothing items 40% of which is discarded in landfills and water bodies (Kaledzi 2022). The apparel sector worldwide is considered as major polluter of the environment (Athreya 2022). Apart from the government, local MSEs are not left out as they mostly lack the requisite knowledge and technical know-how on sustainable industrialisation processes needed to increase efficiency and profits. A situation that has been described as a learning trap (Whitfield and Staritz 2021). Thus, whilst there ought to be government level commitment to sustainable industrialisation in the apparel sector, it does not mean that research-based practices aimed at equipping MSEs to retrofit their production processes to ensure sustainability are not important.

2.2 Work Environment and Ergonomics

Working conditions encompass physical and psycho-social aspects of work that assist employees to perform their job while providing comfort and convenience. Apart from expertise and working methods, other factors such as workplace planning, design of production devices and tools as well as physical and psycho-social environment affect work satisfaction (Kaya 2015). It is therefore important to focus on these other factors in order to achieve good work output. The apparel industry is generally viewed as a safe place to work as compared to other industries, as there are relatively few serious accidents in apparel workplace. The hazards faced by workers in the apparel industry are different. The major health risks in the industry does not result from immediate, potentially fatal hazards, but rather from more subtle hazards whose effect accumulates over time (Gunning et al. 2001).

Occupational health and safety at work has been a major source of concern. Most of the hazards in the apparel industry are ergonomic in nature resulting from poor workplace design. Most apparel firms throughout the world fail to provide quality working environments for workers. Some of the problems commonly mentioned in the literature are congested work area with non-ergonomic workstations, dust, excessive noise, improper ventilation (Parimalam et al. 2006), poor lighting, small available space, poor temperature, colour of wall surfaces, unsuitable seats, work heights, and postures (Vandyck and Fianu 2012). Workers have been reported to be experiencing numerous health problems most of which are musculoskeletal disorders. Employee in large firms usually suffered from eyes, neck, shoulders, hands, arms, and back problems (Kaya 2015). Small-scale garment producers mostly experienced frequent pains in the neck, upper back, lower back, and shoulders (Vandyck and Fianu 2012). Kaya (2015) reported that thermal comfort conditions were usually poor during summer. It is worth noting that there are few firms that are improving these poor conditions. Obeng et al. (2015) observations of the working environment in the factories including garment firms showed that occupational health and safety was quite good. Major entrances of the factory floors were opened. The factories were well-ventilated with enough windows and operational ceiling fans. Nevertheless, a good number of studies on this topic have reported poor conditions in work environment of apparel enterprises which affected workers' comfort, health, working efficiency, work performance and productivity.

The majority of these studies were conducted in large garment firms (Parimalam et al. 2006; Kaya 2015;Obeng et al. 2015). Large enterprises have been the focus of most research for a long time because of their high number of workers. It is believed that the more profound the work problems in large firms, the more people are affected. It is well-known that employees in most large firms are mostly unionised and thus, do not have a united voice which leads to some level of exploitation from management. Vandyck and Fianu's (2012) research is among the scanty studies on work environment and ergonomics that targeted apparel MSEs in Ghana. Notwithstanding, it is important that future studies target more of MSEs especially informal ones, as they make up a large chunk of the manufacturing workforce of most countries. MSEs contribute to employment growth at a higher rate than larger firms (Chakraborty 2016). More importantly, as indicated by ILO (2022), these informal MSEs do have unsafe and unhealthy working conditions and usually do not have access to information and thus, there is the need for researchers to pay attention to informal MSEs.

2.3 Workplace Improvement

Several studies have suggested ways to improve work environments in the apparel industry. However, there is lack of literature on action research which sought to assist the affected apparel firms to improve their poor work environments. Some studies sought to assess workers' perceptions of an improved environment (Lee and Park 2007; Vandyck et al. 2014) without action. It is imperative for the research to be action-oriented in other to achieve the desired improvement recommended by researchers. Different apparel enterprises may have unique work environment problems and a case by case approach would be appropriate to solve each unit's peculiar problems. Therefore, a new wave of research on work environment ought to be practiced. There should be a shift from "research to recommendation approach" to "an action-oriented case study approach" as the former usually leads to suggestions which may not been known by those affected and the latter results in real societal change.

It is in light of these arguments that the current study sought to study an informal MSE to improve its work environment. Based on the purpose of the study, Kogi's (2006) participatory methods effective for ergonomic workplace improvement of small-scale enterprises as well as the PDCA cycle, a well-established improvement model were the framework upon which the study was conducted. Highlights from Kogi's (2006) discussion on effective participatory methods for workplace improvement include the adoption of good-practice approach that can be easily modified to suit each enterprise's local situation in order to meet the specific needs of units. Methods should also focus on low-cost improvements in technical areas of physical environment, workstation design, materials handling, and work organisation.

2.4 Plan-Do-Check-Action (PDCA) Cycle

The Plan-Do-Check-Action (PDCA) cycle had its origin with Dr. W. Edwards Deming's lecture in Japan in 1950 (Moen and Norman 2009). It is called the "Model for Improvement", and has been found to support improvement efforts in a full range from the very informal to the most complex. By the 1960's the PDCA cycle in Japan had evolved into an improvement cycle and a management tool. ISO45001:2018 was founded on the concept of PDCA cycle which enables organizations to provide safe and healthy workplaces by preventing work-related injury and ill-health, and proactively improving its occupational health and safety (OHS) performance.

The four step cycle for problem solving and improvement are planning (definition of a problem and a hypothesis about possible causes and solutions), doing (implementing), checking (evaluating the results), and action (back to plan if the results are unsatisfactory or standardization if the results are satisfactory) (Moen and Norman 2009). The PDCA cycle provides a framework for the application of improvement methods and tools guided by theory of knowledge. It is used to develop, test, implement, and spread changes that result in improvement (Moen and Norman 2010). The model is widely applicable. It can be applied to the improvement of processes, products, and services in any organization, and it is applicable to all types of organizations and to all groups and levels in an organization. PDCA cycle helps to achieve measurable improvements in the performance, effectiveness, efficiency, outcomes, accountability, and other indicators of quality in services or processes (Chakraborty 2016).

Although literature has shown the effectiveness of the PDCA model for quality improvement in diverse areas (Taylor et al. 2013; Tahiduzzaman et al. 2018; Realyvásquez et al. 2018; Nguyen et al. 2020), its application in the garment industry is limited. Among the scare research is Tahiduzzaman et al.'s (2018) study that used PDCA and 5S to minimize sewing defects for knit T-Shirts produced by an apparel firm in Bangladesh. They established that PDCA cycle is good tool for improvement as the study resulted in improved quality and profit for the enterprise. Given PDCA's usefulness in improving quality in varied cases and context, it seems prudent to apply it in the present study.

2.5 Research Questions

1. What are the conditions of the work environment of the informal apparel MSE?
2. What are the ergonomic-related injuries experienced by the informal apparel MSE worker?
3. What is the level of work performance of the informal apparel MSE worker?
4. What strategies are feasible for improving the work environment of the informal apparel MSE worker?
5. What are the experiences of the apparel worker after improvement of the work environment?

3 Methodology

3.1 Study Design

An action research with a qualitative approach and case study design was employed for the study. Action research integrates theory and action with the goal of addressing important organizational, community and social issues together with those who experience them (Coghlan and Brydon-Miller 2014). It is usually concerned with improvement to practice. Thus, this method was appropriate for the study because it allowed the researcher and the apparel business operator to be involved in the change process (work environment improvement). In action research, research results are directly fed back to participants to improve the situation of interest. When those affected are involved in the project, it easily facilitates the required change. Therefore, action research is said to be practical, participatory, collaborative, and critical (Gapp and Fisher 2006).

Qualitative approach allowed for the construction of the reality and detailed understanding of the beliefs, feelings and experiences of the apparel worker. Action research has been mostly carried out using case study methods (Emerald publishing n.d.). Case study is an intensive inquiry of a single individual, group, or some other unit (Heale and Twycross 2018) in which the researcher investigates complex phenomena in the natural setting to gain in-depth understanding of them (Yin 2018). In qualitative research, up-close information is gathered by talking directly to people and observing how they behave within their natural context, i.e. the site where participants experience the problem or issue under study (Creswell and Poth 2018). Literature has shown that most PDCA cycle research used case study in order to acquire in-depth knowledge of the situation being studied. This informed appropriate change strategies for improvement. Owing to the cumbersome nature of this kind of study (where solutions to problems are implemented to effect change), the present research adopted a single-case. Knowledge generated from action research can be applied to other projects in other contexts (Marcinkoniene and Kekäle 2007). Similarly, positive outcomes from this single-case study could be replicated in other informal apparel MSEs to improve their work environment.

3.2 Instrument and Procedure

The purpose and benefits of the study were explained to the business operator. Upon acceptance to participate in the research, the researcher scheduled an appointment with the business operator. Direct observation and interview were the means by which data was collected. An observation of the work environment was done by the researcher. Field notes on the ergonomic conditions present in the workshop were taken. Interviews were conducted with the aid of a semi-structured interview guide and audio recorder. The interview sought to find out the business' profile, mode of operation as well as operator's ergonomic-related injuries and level of performance. The interview was conducted in *Twi*, the widely-spoken language in Ghana because the business operator was conversant in the *Twi* language. The data was translated, transcribed verbatim, analysed and described in detail. Since it was a single-case, analysis entailed an examination of the data in order to find feasible solutions to the identified problems. Afterwards, the results from the interview and observation, as well as suggestions for improvement were communicated to the business operator.

Next, the necessary changes were made to the work environment. Further data was collected from the business operator through interview to ascertain the experiences of the business operator after improvement of the work environment. Subsequently, the business operator was educated about the importance of a well-designed and organised work environment and its positive impact on well-being and work performance. This was to ensure the maintenance and continuous improvement of the work environment. A summary of the steps for conducting the study using PDCA model is shown in Fig. 1.

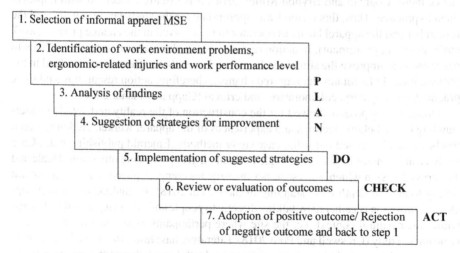

Fig. 1. Steps for conducting the study using PDCA cycle

3.3 Study Setting and Participant

In case study research, one has to be purposeful when selecting a case for a study in order to derive good and informative data (Gerring 2017). Since the study sought to improve work environment, preliminary investigation led to the selection of a case that was more likely to need improvement in its environment. The case enterprise was selected through networking, and this was necessary in order to gain the trust of the enterprise operator. The study was conducted in an informal apparel MSE in Mamprobi, a suburb of Accra, Ghana. Mamprobi is a fourth class residential area largely dominated by compound houses (single bedrooms and chamber/halls) with shared facilities (toilet, bathroom, and kitchen). The community is mostly inhabited by hawkers, carpenters, domestic workers, hairdressers, factory hands, seamstress and tailors, masons among others (Ghana Statistical Service 2014).

The apparel business has been in existence since 1992. The current shop is the second location of the business since its inception. The business was formerly operated in a wooden kiosk until 2012. Currently, it is operating in a shipping container converted into a shop, a common structure found in this type of residential area. The monthly rental charge for this structure is US$12.64 per month with 2 years of rent advance payment. The enterprise is a one-person operator business that deals in customised garment making for girls and women of all ages. It is owned and actively managed by a female. The business operator is very skilled in free-hand cutting but uses patterns where necessary. The operator acquired the skills from tacit knowledge, through apprenticeship from another informal apparel MSE. The operator's level of education is Basic Education Certificate Examination (BECE). At the time of the study, there were no apprentices although eleven apprentices have been trained since the inception of the business. Some of whom are operating their own sewing business. The business operator does not belong to any association, but has been an active member of the Ghana Tailors and Dressmakers Association (GTDA) in the past.

The business provide services to females under the following condition: Clients bring their fabrics and choose styles from available style catalogues, web photos or sample garments presented either by the clients or the operator. The measurements of the clients are taken and discussion on cost of workmanship, payment terms, and scheduled pickup are made. Afterwards, the garments are made. Clients then come for their finished products. Trying-on of finished garments take place and alterations are made where necessary to ensure fit and satisfaction of the client. This mode of operation is common among informal customised garment makers in Ghana. Garments made are usually worn to work, church, parties, marriage ceremonies, at home, and other special occasions.

4 Results

4.1 Plan- Part I (Identified Problems)

4.1.1 Work Environment

Observation of the work environment showed that the interior of the workshop was lined with plywood coated with oil paint for the walls and emulsion paint for the ceiling. The use of wood is to prevent the shop from becoming very hot when the metal used for the shipping container heats up. Broken tiles were used for the floor. Ventilation was good as the shop had four openings (two set of louvre windows, one single-door opening and one double-door opening). The two set of windows were opposite each other and the two set of doors were also slightly opposite. This allowed for free flow of air in the workshop. In terms of lighting, the openings allowed natural light to illuminate the shop very well. In the evening when the sun begins to set or on cloudy days, the intensity of the natural light reduces, and thus, artificial light from two sets of bulbs served as light source. These bulbs were fixed close to each and were positioned a little away from the middle of the ceiling. The positions of the bulbs directly illuminated the sewing and ironing areas. However, the cutting area was not well-lit since the cutting table was positioned at the other end of the workshop which was farther away from the light bulbs. The colour of the wall and ceiling surfaces was white. The white colour used for the interior coupled with the light sources contributed to good illumination of the workshop.

The workshop had one industrial single-needle sewing machine, one industrial three-thread overlocker, two small tables with domestic manual and electric sewing machines, one large table for laying, marking, and cutting, a fabric closet, a small four-drawer cabinet, an ironing board, four stools, two plastic chairs, a two-seater sofa, and a medium standing fan. Considering these facilities and their positions as well as the dimension of the shop, the available space in the workshop was small. This space was unsafe and uncomfortable as it hindered free movement in the workshop. The possibility of hitting one's leg against a stationed object especially the overlocker and small cabinet was very high. The temperature of the shop was good, but not very comfortable during hot sunny days. Even though the shop had four openings coupled with a medium standing fan which rotates sideways, it could not provide a cool temperature at the highest setting. It was also positioned close to the cutting table but far away from the main sewing area. There was no place for trying on garments. The doors are closed before finished garments are tested for fit by clients. This further increases discomfort. But since it is for a short while, the operator described the condition as bearable.

The noise level was very good. The business was operating in less busy environment. It was not located by a principal road which is characterised by noise from moving vehicles and theirs horns. However, the street was a quiet one. The noise and vibration were produced from a sewing machine in operation and the operator did not describe this as uncomfortable. The seats in the workshop were not ergonomically designed. The operator used a stool and sometimes, two stacked plastic chairs when sewing. The stool did not have a backrest and the width was unsuitable for the operator. The height of the stool was however appropriate for the industrial sewing machine and overlocker. The two plastic chairs were stacked to increase the height of the seat in order to provide comfort while sewing. But the height achieved with these two stacked chairs was still not suitable. The width and depth was however good. Although these plastic chairs have backrests, their design did not provide good back support for the operator. The operator stooped forward with raised shoulders while sewing, thereby assuming an awkward posture while sewing. Furthermore, the height of the cutting table was suitable for the height of the operator. But it was observed that there were some fabrics and finished garments packed at one side of the table, which reduced the space available for laying, marking, and cutting. The ironing board was also adjustable and the operator had fixed it at an appropriate height. Overall, the posture of the operator while laying, marking, cutting, and ironing was good.

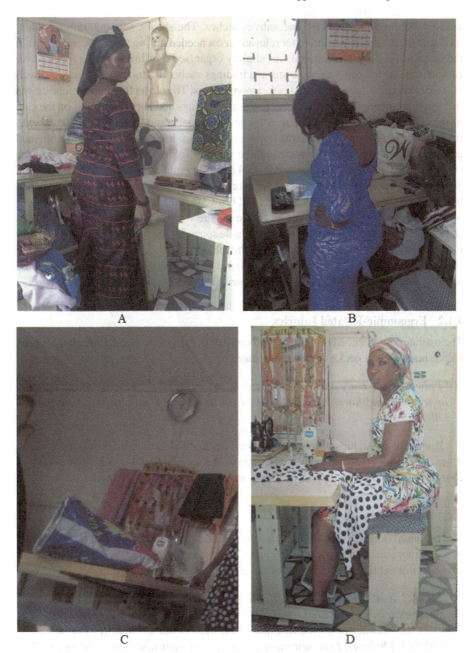

Fig. 2. Images of shop before improvement

There was limited hand movement during sewing since the frequently used machines i.e. industrial sewing machine and overlocker were operated by foot. The hand machine was used occasionally during power outage. The domestic electric machine was only used for button-holing and for decorative stitches. The shears, scissors, tailor's chalk, measuring tape, push pins and other relevant items needed during sewing were all within the reach of the operator. Storage was slightly organised. For instance, the four-drawer cabinet was use to keep notions, trims and findings such as buttons, zips, threads as well as style catalogues, measurement book, among others. The closet for keeping fabrics was full and therefore, extra fabrics were stored in sack bags and kept under, and on the side of the cutting table. Although, there was clothes hanger stand, it was placed so close to the wall due to limited space. It was impossible to place hangers on it and thus, finished garments were either placed directed on the stand or folded and kept on the side of the cutting table. Therefore, the workshop looked untidy and very busy. Nonetheless, the design of the exterior and interior of the shop was carefully planned as it was designed based on a colour scheme (white, beige, and brown theme). This gave the shop a natural woody look and feel. Figure 2 shows images of the workshop before improvement. 2a shows part of the workshop with small available space for the business operator to freely move. 2b shows small available space on the table for laying, marking, and cutting. 2c shows the position of the clothes hanger stand with finished garments. 2d shows the operator on a non-ergonomically designed seat (stool).

4.1.2 Ergonomic-Related Injuries

The operator experienced discomfort and consistent pain in the chest, lower back, upper back, shoulder, and neck. The operator stated:

"I always have body pains. I feel pains in my waist, back (pointing to the neck, upper back and lower back) as well as my shoulders. I am always taking pain killer"

Pain in the wrist, hips, thigh and feet was occasional. In addition, the operator sometimes experiences pain as a result of hitting the feet against the overlocker table or small four-drawer cabinet due to their positions and small space available in the shop.

4.1.3 Work Performance

The business' operator rated work performance as average when asked to rate on a scale of 5, ranging from extremely low (1), low (2), average (3), high (4), extremely high (5). The operator explained the reason for the average performance by stating:

"I think I do not perform to my fullest capacity. I have the sewing skills but speed is my problem. The speed at which I work is neither low nor high. I feel I can do better. The number of garments I make in a week is not the best. If I am able to work fast, I believed I can sew more garments in a week and earn more income."

The operator believed that limited space impeded free movement, decreased comfort, and thus, slowed down work activity. For instance, due to the fabric closet being full, some of the fabrics and finished garment were kept on the side of the cutting table. This reduces the space available on the table for laying, marking, and cutting. Thereby contributing to the reduced average performance.

4.2 Plan- Part II (Suggestions for Improving Work Environment)

Suggestions for improvement of the workplace based on identified problems (Table 1) were made and communicated to the business operator for onward improvement of the work environment. As proposed by Kogi (2006), education of affected enterprise is necessary to facilitate the operator's understanding of the project's importance, which is aimed at improving the well-being and performance of the worker. Additionally, new items, activities, and services that require financial commitment were selected based on the operators' budget. This is important to ensure that the operator gets on board without having to suffer financial burden after the workshop improvement project.

Table 1. Suggestions for improving work environment

Technical areas	Feedback	Changes aimed at
Ventilation	Excellent	No action
Noise level	Excellent	No action
Easy reach of items	Excellent	No action
Wall, ceiling and floor	Good	Repainting of wall and ceiling
		Installation of carpet
Lighting	Good	Spacing of light bulbs holders (one at the cutting area and the other at the sewing area)
		Change of bulbs to modern type
Temperature	Good	Replacement of standing fan with ceiling fan to provide free circulation of cooled air from the ceiling fan
Work height	Good	Introduction of large cabinet (with 4 doors and 4 drawers) at an appropriate height, width and length. Cabinet top to serve as work surface for laying, marking, cutting and ironing
Available space	Poor	*Removal, replacement, and repositioning of objects in the shop where applicable to create more space*
		Removal of cutting table, ironing board, small four-drawer cabinet, and small table carrying domestic manual sewing machine in order to create more space – Laying, marking, cutting, and ironing to be done on surface of large cabinet

(continued)

Table 1. (*continued*)

Technical areas	Feedback	Changes aimed at
		Repositioning of overlocker – To be placed at ironing board's position
		Removal of small cabinet. Items to be transferred to drawers of large cabinet
		Replacement of standing fan with ceiling fan to create more space
Seats	Poor	Introduction of ergonomically designed seat (Swivel chair with arm rest)
Organized storage		
Small items	Good	Items to be transferred from small four-drawer cabinet to 4 drawers of large cabinet
Extra fabrics	Poor	Extra clients' fabrics to be stored in large cabinet
Finished garment	Poor	Repositioning of clothes hanger stand and use of hangers to store finished garments. Extra finished garments to be stored in large cabinet
Changing room	Poor	Makeshift structure (i.e. line, pegs, and curtains should be used at the area between the fabric closet and large cabinet) when it is time for testing of finished garment by clients
Overall condition of the workshop	Poor	Implementation of above-mentioned strategies

4.3 Do (Changes Made to Improve Work Environment)

All suggested strategies for improvement were implemented within three weeks, two days. The facelift was done to maintain the white, beige and brown colour theme of the workshop. The exterior and interior of workshop was painted to improve the appearance of the shop. The white colour of the wall and ceiling was maintained. The floor was lined with a gold coloured carpet. Carpet was cheaper to acquire compared with retiling with large tiles. The initial positions of the two bulbs were changed. The bulb holders were moved apart in order to illuminate the "cutting cum ironing" area and sewing area. The bulbs were changed to modern design.

In the quest to create enough space for the business operator to freely move, the standing fan was removed from the workshop and replaced with a short blade ceiling fan. The short blade type was chosen because it was suitable for the size of the shop. Again, taking the height of the workshop into consideration, the down rod and top canopy were not used in the installation of the fan to prevent the fan from serving as a hazard.

The ceiling fan would also aid in good circulation of air to improve the temperature of the shop during hot sunny days. The large table, ironing board, and small cabinet were removed from the workshop and their roles were played by one large cabinet. The cabinet was specifically designed to have an appropriate height, width and length. The height of the business operator was taken into consideration in order to obtain a proper work height. The chosen length was to obtain a work surface with enough space for laying, marking, cutting, and ironing. The width and length was also selected in order to obtain a good storage space with a size that would not reduce the available space in the workshop. This was positioned at one corner of the shop. It had four drawers at the upper part and two big compartments each with double-door at the lower part. The drawers were used to store the items (measurement book, style catalogues, notions, trims and findings such as buttons, zips, and threads) that were previously kept in the small cabinet. The two big compartments were used to store extra fabrics and finished garments. It also became the emphasis of the shop's interior design.

In addition, the table for the domestic manual sewing machine was also removed since the machine was only used when there was power outage. The machine was kept in the one of big compartments of the large cabinet. The machine was to be replaced with the domestic electric sewing machine on the other small table when its use became necessary. The positions of the small table for the domestic electric sewing machine, fabric closet, two-seater sofa, as well as the industrial single-needle sewing machine were maintained. Lastly, the overlocker was repositioned right opposite the industrial sewing machine. The clothes hanger stand was moved to the front of the fabric closet and was furnished with hangers for storing finished garments. Although, the stand prevents easy access to the closet, it was the best option. It was decided that the fabrics that did not require immediate attention should be stored in the closet whereas, those that were to be used immediately should be kept in the cabinet. Thus, the closet was not accessed regularly. Again, one door of the closet was kept opened and covered with a curtain to make it easy to pick fabric when necessary. The operator's bodice-length mirror was hanged on the closed part of the fabric closet. These strategies (removal, replacement, repositioning of objects as well as introduction of one large multipurpose cabinet) greatly improved the space in the shop. In order to have a changing room for clients to test their garments, it was advised that a makeshift structure should be made when necessary. This should be created at the space between the fabric closet and large cabinet. A curtain should be pegged on a line, made from twine that would be fixed on nails from the wall to the top of the closet). This should be removed immediately afterwards.

A sizeable swivel chair with good backrest, arm rest with adjustable height was procured and it replaced the stool at the sewing area. Since, space was of utmost importance considering the size of the workshop, the selected swivel chair was not huge but just the right size that could accommodate the size of the business operator. This ergonomically designed chair helped to correct the poor posture of the operator while sewing. This chair also served as the seat for operating both the industrial sewing machine and the overlocker since they were opposite each other. All the operator was required to do was to turn around and move the chair closer to the machine of interest. One stool was kept to the right side of the operator at the sewing area to keep any item of necessity. A laundry basket was placed in the shop to keep large pieces of cut fabrics that can be used to make other items. Images of the renovated workshop is shown in Fig. 3. 3a shows part of the workshop with good space at the "cutting cum ironing" area and the multipurpose cabinet. 3b shows the operator on an ergonomically designed seat (swivel chair with arm rest). 3c shows the position of the clothes stand with hangers for keeping finished garments as well as the designated area for the makeshift changing room. 3d shows part of the workshop with good space at the sewing area and the two-seater sofa for clients.

Fig. 3. Images of the shop after improvement

4.4 Check (Evaluation of Business Operator's Experience after Improvement in Work Environment)

The experience of the business operator after the workshop renovation was assessed. Findings revealed that the renovated workshop improved the operator's physical and psychological well-being. It also improved performance, increased customers, enhanced customer service and client satisfaction. The worker expressed amazement after improvement of the workshop. The operator's mood was a good one. There was expression of joy, happiness and gratitude. The operator exclaimed:

"The workshop is looking very nice. I am very happy and the space in the workshop has improved."

After some months, findings from post-renovation interview showed that the worker felt very comfortable in the workshop with the introduction of ergonomically designed seats, work heights and rearrangement which resulted in good space in the workshop. This allowed the operator to move freely. The body pains had virtually stopped. The business operator recounted:

"I do not feel pains in my body like I used to before the renovation. My chair is very comfortable, I do no move much even when I want to use the overlocker. All I do is turn around and move my seat a little forward. I do not have to lift the chair because of its wheels. There is also, no object to hit my feet against."

With regards to performance the operator reported:

"I will rate my performance as high (4). These days I do not waste much time in making garment like I used to before the renovation. I have enough space for laying, marking, and cutting. The overlocker is right behind me. One turn around and I am able to use it."

Furthermore, the operator indicated that clients and onlookers gave complimentary remarks about the shop's facelift which makes the operator very happy. Clients were also impressed with the makeshift changing room. The operator added that the renovation has also earned the business new customers who confessed that they choose to engage the services of the enterprise mainly because of the shop's beautiful look.

4.5 Act (Standardisation of Outcome of Work Environment Improvement)

The experiences of the worker after improvement in work environment implies that the result of the renovation was satisfactory. Therefore, there was the need to further educate the business operator on the importance of ergonomics, a well-designed and organised workspace as well as a conducive, safe and healthy work environment. The lived experienced of the worker was emphasized in the education. It was explained that the improvement in the work environment was the reason for the improvement in the operator's well-being and performance level. The operator got to understand based on the lived experienced. It was important for the worker to understand these issues in order to maintain and strive for continuous improvement in the work environment.

5 Discussion

Although, this study is one of the few to focus on Ghanaian apparel MSE's work environment, it is the first to systematically apply PDCA cycle in an action research to explore ergonomic problems and improve the work environment of an informal apparel MSE in Ghana. The results derived confirmed numerous findings in literature. An application of the PDCA cycle resulted in an improved work environment, as well as worker well-being and performance level in the case MSE. This further confirms the effectiveness of PDCA cycle in improving the quality of systems and ensuring its continuous upgrade (Moen and Norman 2009; Moen and Norman 2010; Taylor et al. 2013; Chakraborty 2016; Tahiduzzaman et al. 2018; Realyvásquez et al. 2018; Nguyen et al. 2020). The profile characteristics of the case enterprise are consistent with past report which indicated that MSMEs are generally characterised by informality, independence, lack of external orientation, highly personalized, actively managed by the owners, limited resource, tacit knowledge, internal operational focused, and dependent on internal sources to finance growth (Ates et al. 2013; Chakraborty 2016; Senayah 2018; Sawyerr 2019; ILO 2022).

Poor conditions found in the case enterprise's work environment and its associated health problems were not different from what pertained in other context. With the exception of ventilation, noise, and easy reach of items, all other conditions (non-ergonomic workstations, bad posture resulting from inappropriate seats, small available space, poor lighting in the evening, poor comfort temperature during hot days) discovered in the present study, correspond with findings from past studies (Parimalam et al. 2006; Vandyck and Fianu 2012; Kaya 2015). The issue of poor ventilation was not found as these conditions were optimum in this case enterprise. This finding concurs with Obeng et al. (2015) report of a well-ventilated factory floor in Ghana. Vandyck and Fianu's (2012) finding of excessive noise was not observed in this case enterprise. This is because the former was conducted in an open-air market characterised by noise from traders, moving vehicles and the horns of these vehicle whereas, the latter was undertaken in a relatively quiet residential poor urban area. Similarly, issue of fabric dust (Parimalam et al. 2006) was not observed in the current study. This could be that the case enterprise operated on a small-scale compared with large enterprises that use several yards of fabrics in its operation. Another finding revealed that the case enterprise had several important but unneeded equipment considering the size of the workshop. This contributed greatly in reducing the available space in the workshop. It is worth noting that, in as much as the ironing board was required for pressing, the large cabinet played it role better and effectively since latter's surface dimension is larger than that of the former. Same can be said of the small cabinet whose function was replaced by the drawers of the large cabinet. MSEs ought to consider their workshop size when selecting equipment for the enterprise.

It is has been widely established that poor work environment with non-ergonomic designs leads to musculoskeletal disorders which affects workers' comfort and work productivity. The case worker's account of experiencing frequent pains in the neck, upper back, lower back, and shoulders as well as reduced level of work performance is consistent with Vandyck and Fianu's (2012) report of small-scale Ghanaian garment producers. Indeed, the hazards identified in the current study were not fatal just as Gunning et al. (2001) stated but rather, accumulate overtime resulting in serious health issues. Overall, the ILO's (2022) report of informal enterprises being characterized by small undefined work places with unsafe and unhealthy working conditions has been proven.

The adoption of action-oriented single-case study to improve work environment using PDCA cycle has proven to be efficient in solving one of the major problems of the apparel industry. By serving as a case tool, this study can be adopted, modified and applied to similar situations of informal apparel MSEs in developing countries. Until recommendations are directly presented to informal small-scale businesses, it would take a long time for these enterprises to improve as they do not have access to information (ILO 2022). Those who are fortunate to receive safety information may not implement them, if there are no planned participatory approach to assist them in the improvement processes (Kogi 2006). The study contributes to literature on sustainability, ergonomics, work environment, working conditions, and occupational health and safety of MSEs.

6 Conclusion, Implications and Limitations

The use of PDCA cycle guided this action research in achieving its purpose of improving the work environment of an informal apparel MSE. Identification of problems in the workplace revealed poor conditions in the physical work environment. This took a toll on the worker's well-being and the level of performance. Problems found were duly analysed and feasible strategies were developed and implemented accordingly to give the workshop a facelift. This consequently improved the physical and psychological well-being of the business operator, work performance level and customer base, service and satisfaction. It is recommended that relevant stakeholders should continually create awareness, educate and train small-scale informal apparel businesses on good working conditions and its importance in improving the safety and health of workers. This should be done through workshops and symposia tailor made to suit informal apparel MSEs' situation and needs. More action-oriented improvement projects is also recommended.

The study is limited in its use of a single case. Since it focused on one enterprise's distinctive problems, results cannot be generalised to all informal apparel MSEs in poor urban centres. Replication of the study using several garment enterprises would give a general idea of what pertains in informal apparel businesses in Ghana and other developing countries. Nonetheless, this study provided the opportunity to collect in-depth information that informed a holistic solution. It also revealed relevant issues in a less researched setting i.e. informal small-scale apparel workshop. This study can be used as a case for educational purposes. Furthermore, since it is an action-oriented study, future research, local government and non-governmental organisations can transfer this to other contexts to help in the improvement of local informal businesses. When several of these studies are undertaken, it would help to contribute to the achievement of sustainable development specifically, SDG 3: good health and well-being and SDG 8: decent work and economic growth.

References

Ates, A., Garengo, P., Cocca, P., Bititci, U.: The development of SME managerial practice for effective performance management. J. Small Bus. Enterp. Dev. **20**(1), 28–54 (2013). https://doi.org/10.1108/14626001311298402

Athreya, B.: Can fashion ever be fair? J. Fair Trade **3**(2), 16–27 (2022). https://doi.org/10.13169/jfairtrade.3.2.0016

Chakraborty, A.: Importance of PDCA cycle for SMEs. SSRG Int. J. Mech. Eng. **3**(5), 13–17 (2016). https://doi.org/10.14445/23488360/IJME-V3I5P105

Coghlan, D., Brydon-Miller, M.: The SAGE Encyclopedia of Action Research. SAGE Publications Ltd, 1 Oliver's Yard, 55 City Road, London EC1Y 1SP United Kingdom (2014). https://doi.org/10.4135/9781446294406

Clark, H.: What will it take to achieve the sustainable development goals? J. Int. Aff. 53–59 (2017). http://www.jstor.org/stable/44842600. Special 70th Anniversary Issue

Creswell, J.W., Poth, C.N.: Qualitative Inquiry & Research design: Choosing Among Five Approaches, 4th edn. Sage (2018)

Emerald Publishing: How to carry out action research. https://www.emeraldgrouppublishing.com/how-to/research-methods/carry-out-action-research - Search (bing.com) (n.d.)

Elrod, C.: The domino effect: how inadequate intellectual property rights in the fashion industry affect global sustainability. Indiana J. Glob. Legal Stud. **24**(2), 575–595 (2017). https://doi.org/10.2979/indjglolegstu.24.2.0575

Ertl, V., Schebesta, M.: Sustainability in global supply chains. Konrad-Adenauer-Stiftung e. V. **390**, 1–8 (2020)

Gapp, R., Fisher, R.: Achieving excellence through innovative approaches to student involvement in course evaluation within the tertiary education sector. Qual. Assur. Educ. **14**(2), 156–166 (2006)

Gerring, J.: Case Study Research: Principles and Practices, 2nd edn. Cambridge University Press (2019). https://doi.org/10.1017/9781316848593

Ghana Statistical Service: 2010 Population & Housing Census: Housing in Ghana. Ghana Statistical Service (2014)

Ghana Statistical Service: Integrated Business Establishment Survey: Summary Report. Ghana Statistical Service (2016)

Government of Ghana: Medium-Term National Development Policy Framework. https://ndpc.gov.gh/resource_and_publications/policy (2016)

Gunning, J., Ferrier, S., Kerr, M., King, A. Maltyby, J., Eaton, J., Frumin, E.: Ergonomic Handbook for the Clothing Industry. Union of Needletrades, Industrial and Textile Employees, Institute for Work and Health, and Occupational Health Clinics for Ontario Workers, Inc. (2001)

Heale, R., Twycross, A.: What is a case study? Evid. Based Nurs. **21**(1), 7–8 (2018)

International Labour Organization: Decent work for sustainable development (DW4SD) resource platform: Informal economy (2022)

International Organisation for Standardisation: ISO 45001:2018-Occupational health and safety management systems. https://www.iso.org/standard/63787.html (2018)

Joyner Armstrong, C.M., Park, H.: Sustainability and collaborative apparel consumption: putting the digital 'sharing' economy under the microscope. Int. J. Fashion Des., Technol. Educ. **10**(3), 276–286 (2017). https://doi.org/10.1080/17543266.2017.1346714

Kaledzi, I.: Used clothes choke both markets and environment in Ghana (2022)

Kaya, O.: Design of work place and ergonomics in garment enterprises. Procedia Manuf. **3**, 6437–6443 (2015)

Kogi, K.: Participatory methods effective for ergonomic workplace improvement. Appl. Ergon. **37**, 547–554 (2006)

Lee, Y.-K., Park, H.-S.: Workers' perception of the changes of work environment and its relation to the occurrence of work-related musculoskeletal disorders. J. Occup. Health **49**, 152–154 (2007)

Marcinkoniene, R., Kekäle, T.: Action research as culture change tool. Balt. J. Manag. **2**(1), 97–109 (2007)

Moen, R., Norman, C.: Evolution of the PDCA cycle. In: The History of the PDCA Cycle. Proceedings of the 7th ANQ Congress, Tokyo (2009)

Moen, R.D., Norman, C.L.: Circling back: clearing up the myths about the deming cycle and seeing how it keeps evolving. Qual. Prog. **42**, 23–28 (2010)

National Board for Small Scale Industries: Micro and small enterprises. www.nbssi.org (2015)

National Development Planning Commission: Ghana's Voluntary National Review Report on the Implementation of the 2030 Agenda for Sustainable Development (2022)

Nguyen, V., Nguyen, N., Schumacher, B., Tran, T.: Practical application of plan–do–check–act cycle for quality improvement of sustainable packaging: a case study. Appl. Sci. **10**, 1–15 (2020). https://doi.org/10.3390/app10186332

Obeng, F.A., Wrigley-Asante, C., Teye, J.K.: Working conditions in Ghana's export processing zone and womens empowerment. Work Organ., Labour Globalisation **9**(2), 64–78 (2015). https://doi.org/10.13169/workorgalaboglob.9.2.0064

Parimalam, P., Kamalamma, N., Ganguli, A.K.: Ergonomic interventions to improve work environment in garment manufacturing units. Indian J. Occup. Environ. Med. **10**, 74–77 (2006)

Realyvásquez, A., Arredondo-Soto, K., Carrillo-Gutiérrez, T., Ravelo, G.: Applying the Plan-Do-Check-Act (PDCA) cycle to reduce the defects in the manufacturing industry. A case study. Appl. Sci. **8**(11), 1–17 (2018). https://doi.org/10.3390/app8112181

Republic of Ghana: The Coordinated Programme of Economic and Social Development Policies (2017–2024): An Agenda for Jobs: Creating Prosperity and Equal Opportunity for All. https://s3-us-west-2.amazonaws.com/new-ndpc-static1/CACHES/PUBLICATIONS/2018/04/11/Coordinate+Programme-Final+(November+11,+2017)+cover.pdf (2017)

Sawyerr, N.O.: Consumers' evaluation of the quality of custom-made garments manufactured by micro and small scale enterprises in Sekondi/Takoradi Metropolis, Ghana. Doctoral Thesis, University of Ghana. UGSpace. http://ugspace.ug.edu.gh/handle/123456789/36363 (2019)

Senayah, W.K.: Skill-based competence and competitiveness in the garment-manufacturing firms of Ghana. Doctoral thesis, University of Ghana. UGSpace. http://ugspace.ug.edu.gh/handle/123456789/28928 (2018)

Tahiduzzaman, M., Rahman, M., Dey, S.K., Kapuria, T.K.: Minimization of sewing defects of an apparel industry in Bangladesh with 5S & PDCA. Am. J. Ind. Eng. **5**(1), 17–24 (2018). https://doi.org/10.12691/ajie-5-1-3

Taylor, M.J., McNicholas, C., Nicolay, C., Darzi, A., Bell, D., Reed, J.E.: Systematic review of the application of the plan–do–study–act method to improve quality in healthcare. BMJ Qual. Saf. **23**, 290–298 (2013)

United Nations: Sustainable development goal (SDG) 3: Good health and well-being. https://www.un.org/sustainabledevelopment/health/ (2017a)

United Nations: Sustainable development goal (SDG) 8: Decent work and economic growth. https://www.un.org/sustainabledevelopment/economic-growth/ (2017b)

Vandyck, E., Fianu, D.: The work practices and ergonomics problems experienced by garment workers in Ghana. Int. J. Consum. Stud. **36**, 486–491 (2012). https://doi.org/10.1111/j.1470-6431.2011.01066.x

Vandyck, E., Tackie-Ofosu, V., Ba-ama, M., Senayah, W.: Effects of ergonomic practices on garment production in Madina, Ghana. Int. Res. J. Arts Soc. Sci. **3**(1), 1–7 (2014). https://doi.org/10.14303/irjass.2013.070

Whitfield, L., Staritz, C.: The learning trap in late industrialisation: local firms and capability building in Ethiopia's apparel export industry. J. Dev. Stud. **57**(6), 980–1000 (2021). https://doi.org/10.1080/00220388.2020.1841169

Yin, R.K.: Case Study Research and Applications: Design and Methods, 6th edn. Sage Publication Inc. (2018)

Industrialization and Economic Development in Sub-Saharan Africa: The Role of Infrastructural Investment

Rachel Jolayemi Fagboyo[1](✉) [iD] and Rufus Adebayo Ajisafe[2]

[1] Department of Economics, Samuel Adegboyega University, Ogwa Edo State, Nigeria
rfagboyo@sau.edu.ng
[2] Department of Economics, Obafemi Awolowo University, Ile-Ife, Nigeria
rajisafe@oauife.edu.ng

Abstract. Purpose: This paper studied the role of infrastructural investment in the nexus between industrialization and Economic Development in 13 Sub-Sahara ranging from 2003 to 2020.

Design/Methodology/Approach: The study employed the panel least square technique to examine the relationship between the variables.

Findings: Findings indicates that industry value added and transport index had a significant positive effect on economic development in the Sub-Sahara Africa. Analysis suggests that low industrial activities which is largely due to poor infrastructure affect economic development negatively in the region.

Implications/Research Limitations: This study can be expanded further by adding more indicators of infrastructure investment which plays an important role in industrialization and in turn can lead to economic development.

Practical Implications: The government in these regions should invest in infrastructure as these will improve industrialization as well as expand economic growth. To achieve this, the government should sign a partnership deal with other developed countries who are vast in infrastructural development.

Originality/Value: The authenticity of this study lies on the premise that it is a cross-country study as against extant studies which focus were country specific. Also, the study examined the mediating role of infrastructural investment on the nexus between industrialization and Economic growth in an emerging economy like Sub-Saharan Africa.

Keywords: Economic Development · Industrialization · Infrastructure index · Sub-Saharan Africa · Panel least square estimation

1 Introduction

The continuous growth and development of an economy is heavily dependent on industrialization and investment in infrastructure (UN 2020). Basic infrastructure such as good roads, electricity and water supply are still a major problem in developing countries. The sustainable development goals, (SDGs) lay emphasis on industrialization, innovation and infrastructure and how it can transcend to sustainable economic development.

According to African Development Bank (2020), Economic development without investment in infrastructure to foster industrialization is unsustainable and bad for the African economy. Also, it was observed that industrialization and infrastructure is a prerequisite to ensure inclusive and sustainable economic development in Africa (ADB 2020). Industries generate about $700 of GDP per capita in Africa, which is less than a fifth of East Asia's output ($3,400) and barely a third of Latin America's output ($2,500). Africa is projected to have 2 billion people by 2050, thus basic infrastructure as well as increased industrialization will be required to cater for this increasing population.

Over the Years, Nigeria has initiated several policies with the aim of achieving economic development. The first national development plan (NDP) of 1962 to 1968 implemented the import substitution industrialization (ISI) to reduce importation and foster the growth of industries. During this period, several infrastructural projects such as the Ughelli thermal plant and Kanji dam were commissioned to improve industrialization in the country. After the civil war of 1967 to 1970, the government embarked on national rebuilding thereby introducing the second and third NDPs (1970–1974 and 1975–1980 respectively). Both plans laid emphasis on investment in public sector industry and were focused on increasing domestic production of intermediate and capital goods, while the fourth NDP (1981–1985) was centered around establishing a sector for capital goods (Bloch et al. 2015).

Generally, Sub-Saharan Africa Countries are characterized by heavy dependence on the importation of capital goods and machineries, oil price falls as well as poor economic management. African countries have over the years tried to pave the way for industrialization and economic development and as such adopted the sustainable development goals of 2015.

Numerous scholars have empirically established a relationship between industrialization and economic development (see Lionel and Enang 2014; Okezie et al. 2017; Nwogo and Orji 2019), However, less emphasis is placed on the role of infrastructural investment. It is noteworthy that the relationship between industrialization and economic development can be altered in the presence of infrastructural investment. Economic growth and investment in infrastructure are meant to complement each other when an economy pursue industrialization. According to Chukwuebuka and Jisike (2020), the infrastructure deficit in Sub-Sahara Africa pose a serious challenge to industrialization as industries strive only in economy with abundant infrastructure. One major factor retarding industrialization in Africa is poor investment on infrastructure such as electricity, water and transport services (African Development Bank 2018).

Nwogo and Oriji (2019) opined that the bedrock of economic development is industrialization, however for Industrialization to strive, basic infrastructure must be put in place to foster economic development. Ozekhome et al. (2021) opined that infrastructure is a strategic growth-driver as it constitutes a critical catalyst for economic development and trade. According to World Bank (2017), a key propeller of economic development is adequate infrastructural investment, but sub-saharan africa still performs below expectation in the Africa Infrastructure Development Index (2018). North Africa emerged as the best performing region in terms of infrastructural investment with an average index improvement of 2.23 points, followed by east Africa (1.55 points) central Africa recorded 0.64 points the lowest increase so far, while west Africa recorded no increase at all.

Extensive research has been carried out on the nexus between economic growth and infrastructure (Enimola 2010; Ogbaro and Omotoso 2017; Seidu et al. 2020), Industrialization and Economic Development (Lionel and Enang 2014; Okezie et al. 2017) and infrastructure and industrialization (Shobande and Etukomeni 2016; Udah and Ebi 2017; Prakash 2018). However, there is a paucity of literature on the role of infrastructural investment in the relationship between industrialization and Economic growth in a developing economies Sub-Saharan Africa. The recognition of this gap necessitates this study and therefore fills the gap in the literature by empirically investigating the mediating role of infrastructural investment on the nexus between industrialization and Economic development in Sub-Sahara Africa.

In addition to the introductory Section, Sect. 2 provides an overview of interrelated literature. The methodology, model specification and data are discussed in Sect. 3, while Sect. 4 presents the results and empirical analysis. Section 5 concludes the paper with some evidence-based policy recommendations.

2 Review of Related Literature

Nwogo and Oriji (2019) investigated the role of industrialization on Economic growth in Nigeria for the period 1981 to 2016, using the vector error correction mechanism. Their findings showed that industrialization indeed has a positive impact on economic growth in Nigeria. The result also showed the possibility of convergence between industrial activities and economic growth in the long run. Likewise, Okezie et al. (2017) employed ECM to examine the nexus between industrialization and economic growth from 1985 to 2015 in Nigeria. The authors found that key sectors such as Agricultural, Industrial and the manufacturing sectors are major determinant of economic growth in Nigeria.

Udah and Ebi (2017) investigated the how industrialization is affected by infrastructure and human capital from 1970 to 2014. Using the Ordinary Least Square (OLS) technique, and found out that adequate power supply, expanding public and private investment as well as openness to trade has a long-lasting effect on industrialization in Nigeria. Enimola (2010) used (VECM) to examine the effect of infrastructure investment on economic growth. The author reported that expenditure on energy and transport are significant for economic growth. In addition, the study found a positive correlation between energy consumption and economic growth.

Chukwuebuka and Jisike (2020) examined the relationship between infrastructure development and industrial sector productivity for 17 Sub-Sahara African countries from 2003 to 2018. Using panel estimation, the study argues that low productivity in the industrial sector is largely due to epileptic power supply, bad roads and poor sanitation. Seidu et al. (2020), reported the role of infrastructure on economic growth in United Kingdom. They explained that investment in infrastructure acts as a stimulant for economic growth. Ogbaro and Omotoso (2017) assessed the effect of infrastructure development on economic growth. The study employed the OLS method for the period 1980 to 2015. The result showed that transport, communication and power infrastructure have a positive impact on Economic growth.

Furthermore, Opoku and Yan (2018), found that trade stimulate industrialization impact on economic growth. They employed GMM and found that better industries will foster economic development in Africa.

2.1 Data Sources and Variables Measurement

Data were sourced from the World Development Indicator (WDI) and African Infrastructure Development Index (AIDI) of 2020 edition respectively. The data collected from these sources include Gross Domestic Product per capital growth, which is used as measure of economic development, Industrial value added which is a measure of industry value added as a percentage of GDP. Also, an index of transport infrastructure (roads, railways, and ports) data were collected which is measured by total paved roads (km per 10000 inhabitants) and total road network in km.

Data were equally collected on Electricity Infrastructure, Water and Sanitation Infrastructure, Gross capita formation, Population Growth, and Inflation.

2.2 Model Specification

Consistent with reviewed literature, economic development model can be specified as follows:

$$ED_{it} = \alpha_0 + \alpha_1 INFD_{it} + \beta \sum CV_{it} \tag{1}$$

where ED is economic development, INFD represent infrastructure investment and $\sum CV_{it}$ represent all the control variables comprising of industry value added, inflation rate, population growth and gross capital formation. Economic development is proxied by GDP per capital, infrastructure investment is proxied by transport, electricity and water supply and sanitation index. The model is thus specified as:

$$GPC_{it} = \alpha_0 + \alpha_1 IND_{it} + \alpha_2 EI_{it} + \alpha_3 WI_{it} + \alpha_4 TI_{it} + \alpha_5 POPG_{it} \\ + \alpha_6 INF_{it} + \alpha_7 GCF_{it} + \varepsilon_{it} \tag{2}$$

where GPC is GDP per capita, IND is industry value added, EI, WI and TI are electricity index, water supply and sanitation index and transport index respectively. POPG represent Population growth, INF represent inflation rate while GCF depicts gross capital formation ε_{it} = error term. An interactive term was included in Eq. (2).

$$GPC_{it} = \alpha_0 + \alpha_1 IND_{it} + \alpha_2 EI_{it} + \alpha_3 WI_{it} + \alpha_4 TI_{it} + \alpha_5 EI_{it} \\ * IND_{it} + \alpha_6 WI_{it} * IND_{it} + \alpha_7 TI_{it} * IND_{it} + \alpha_8 POPG_{it} + \alpha_9 INF_{it} + \alpha_{10} GCF_{it} + \varepsilon_{it} \tag{3}$$

The a priori expectations of the variables are as follows: $\alpha_1 - \alpha_8, \alpha_{10} > 0, \alpha_9 < 0$.

Panel data analysis is used to estimate the model, t stands for 18years from 2003 to 2020 and I stand for 13 countries. The 13 selected Sub-Saharan African countries include; Angola, Burkina-Faso, Uganda, Congo Dem Rep, Namibia, Congo Rep, Gabon, Ghana, Guinea, Mauritania, Nigeria, Tanzania and Zambia. The selection of countries was based on Sub-Saharan African countries that has industry value added to GDP at 25% and above as at 2020. The span of study period was due to the accessibility of data.

3 Empirical Findings and Discussion

3.1 Test for Multicollinearity

According to Gujarati et al. (2012), if the correlation coefficient of two variables exceeds 0.8, there is the problem of multicollinearity. The result of the correlation matrix in Table 1 shows the absence of multicollinearity, as the correlation coefficient of the explanatory variables did not exceed the bench mark. Hence, we can include all the variables in the analysis.

Table 1. Correlation matrix

	GPC	IND	EI	TI	WI	INF	POPG	GCF
GPC								
IND	−0.14							
EI	−0.25	0.29						
TI	0.11	−0.42	0.26					
WI	−0.28	0.24	0.63	0.23				
INF	0.06	0.01	−0.16	−0.13	−0.14			
POPG	0.06	0.35	−0.25	−0.64	−0.29	0.02		
GCF	−0.09	0.23	0.04	−0.16	−0.09	−0.01	0.08	

Source: Author's computation using reviews 9

3.2 Panel Unit Root Test

The study employed the Levin, Lin & Chu (LLC) and Im, Pesaran and Shin (IPS) to analyze the unit roots. The results are presented in levels and first difference forms in Table 2. Some of the variables except POPG and INF were initially non-stationary at levels. However, following Box et al. (1994) that non-stationary time series in levels can be made stationary by taking their first differences, thus the variables were first differenced and the subsequent unit root test show stationarity. Thus, the variables are a mixture of I(0) and I(1).

Table 2. Unit root stationary test for variables in levels and first difference

Variable	LLC test			IPS test		
	Level	1st difference	Remark	Level	1st difference	Remark
GPC	0.352	−7.079*	I(1)	−0.140	−7.971*	I(1)
IND	−1.357	−6.826*	I(1)	−0.125	−6.248*	I(1)
EI	−0.431	−8.343*	I(1)	0.493	−8.285*	I(1)
WI	−4.996	−10.262*	I(1)	−0.980	−6.003*	I(1)
TI	−0.314	−3.073*	I(1)	0.609	−4.103*	I(1)
POPG	−8.338*	−9.932*	I(0)	−7.371*	−9.487*	I(0)
INF	−3.584*	−8.372*	I(0)	−4.193*	−9.664*	I(0)
GCF	−2.742	−6.350*	I(1)	−2.322	−5.328*	I(1)

Source: Computed by the researchers (2022)

3.3 Descriptive Statistics

Table 3 presents the results of the descriptive analysis of the sample data of the variables used for the study. The value of the dependent variable (GDP per capita growth) ranged from a minimum of −13.02 to a maximum of 14.99 with its mean and standard deviation as 2.44 and 3.79 respectively. Industrial value added to GDP (IND). The maximum and minimum are 66.17 and 18.01 respectively. The Standard deviation suggest high volatility in the data set, indicating that industrial value added to GDP over the years has been irregular. Infrastructure investment proxied as electricity index, water and sanitation index and Transport index all have maximum values of 28.87, 84.65 and 27.18 respectively and minimum values of 0.42, 7.83 and 1.49 respectively. Their mean values were 5.42, 52.55 and 7.04 respectively, with a standard deviation of 5.77, 14.65 and 5.04 respectively. Electricity and Transport index was low (5.42, 7.04) revealing a low level of power and transport facilities in the region. The Mean value of Water and Sanitation index was high (52.5) this is true because water is in abundance in Sub-Sahara Africa.

Furthermore, the mean value of population growth is 2.81%. The maximum and minimum values are 3.78 and 1.47% respectively with a standard deviation of 0.49. Inflation (INF) has a mean value of 8.63 and a median value of 6.57. The maximum and Minimum values are 98.22 and −3.23 respectively, an indication of volatility in the data, this is further buttressed in the standard deviation value of 9.15. Gross capital Formation (GCF) has a mean value of 25.12 and a median value of 24.18. The maximum and minimum values are 117.52 and −31.11 respectively. The standard deviation value (13.43) which is lower than the mean value (25.12) indicates the absent of volatility in the data over the reference period. The Jargue-Bera statistic shows that all the variables except for Water and Sanitation index are normally distributed and its mean is asymmetric.

Table 3. Descriptive statistics of economic development, industrialization and infrastructural development

	Mean	Median	Maximum	Minimum	Std. dev.	Jarque-Bera	Probability
GPC	1.778	2.442	14.998	−13.020	3.799	49.967	0.00
IND	34.281	30.042	66.179	18.015	12.195	45.498	0.00
EI	5.424	2.537	28.874	0.427	5.774	155.74	0.00
WI	52.552	52.027	84.653	7.831	14.658	0.249	0.88
TI	7.042	5.634	27.188	1.491	5.406	190.356	0.00
POPG	2.819	2.874	3.789	1.479	0.491	8.272	0.02
INF	8.632	6.577	98.224	−3.233	9.155	15988.84	0.00
GCF	25.127	24.189	117.521	−31.111	13.434	1557.342	0.00

Source: Author's computation using reviews 9

4 Estimation of the Panel Data

The Hausman test indicates the rejection of the null hypothesis (Random effect model is appropriate) since the p-value is less than 0.05. Thus, the fixed effect model was estimated in Eq. 3 (Table 4).

Table 4. Hausman test

Test summary	Chi-Sq. statistic	Chi-Sq. d.f.	Prob.
Cross-section random	60.745	7	0.00

Source: Author's computation using reviews 9

The result of the fixed effect is presented in Table 5. The coefficient of industrial value added in each of the four equations was significant and positive. This is an indication that industrialization is very important in driving economic development in sub-Saharan Africa region. An increase in industrialization will lead to sustainable Economic Development in the region. The coefficient of electricity index is negative and it shows an insignificant effect on economic development. This implies that investment in electricity do not influence Economic Development in the region. This result is in consonance with World Bank (2017) and Chukwuebuka and Jisike (2020). This is due to the fact that in the last 20years, electricity generation in the region had witnessed very minute changes (World Bank 2017). The AfDb report of 2018 substantiate this claim that electricity generation costs thrice more in Africa than other emerging regions. Transport and water index showed a significant positive and negative effect respectively on Economic Development in the region. The negative relationship implies that despite abundant water supply in the region, the water system is underutilized and underdeveloped thereby having a negative effect on economic development. Population growth was positive and significant in each of the four equation. This means 1% increase in population will result to 2.5% increase in Economic Development. Gross capital formation was positive and insignificant. This is an indication that the level of investment in the region is relatively low and this has an insignificant influence on Economic Development. The rate of inflation was negative and significant in two out of the four equations. This depicts that higher rate of inflation discourages economic development in the region. The interaction of Electricity, transport and water and sanitation index are significant and negative to economic development. This is a clear indication that poor infrastructure and low level of industrialization are the major reason for unsustainable economic development in the region. The adjusted r-squared were 29%, 32%, 32% and 31% for each of the equations. The values are low because adjusted r-squared are usually low for cross-country regression.

Table 5. Effect of industrialization and infrastructural investment on economic development

Variables	1	2	3	4
IND	0.175***	0.269***	0.334***	0.491***
	(3.692)	(4.747)	(4.579)	(3.534)
EI	−0.112	0.804***	−0.011	−0.096
	(−0.654)	(2.250)	(−0.064)	(−0.565)
TI	0.343***	0.358***	1.319***	0.404***
	(2.401)	(2.476)	(3.566)	(0.557)
WI	−0.133***	−0.128***	−0.119***	0.045
	(−3.952)	(−3.862)	(−3.564)	(0.557)
POPG	2.365***	2.635***	2.594***	2.210*
	(2.002)	(0.239)	(2.229)	(1.891)
GCF	0.002	0.004	0.010	0.007
	(0.086)	(0.239)	(0.510)	(0.374)
INF	−0.053	−0.053*	−0.041	−0.06***
	(−1.672)	(−1.704)	(−1.315)	(−2.042)
EI*IND		−0.025***		
		(−2.906)		
TI*IND			−0.033***	
			(−2836)	
WI*IND				−0.006***
				(−2.418)
Constant	−5.294	−9.525***	−12.800**	−14.080***
	(−1.242)	(−2.151)	(2.584)	(−2.532)
Adjusted R2	0.297	0.32	0.32	0.31
Durbin-Watson Stat	1.530	1.616	1.59	1.58

Source: Reviews 9. Notes: ** and * denote 5% and 10% levels of Significance for all statistics; t-statistic in parenthesis

5 Conclusion and Recommendation

The paper examined the role of infrastructure investment and industrialization on economic development in thirteen Sub-Sahara African countries namely Angola, Burkina Faso, Congo Democratic Republic, Congo Republic, Gabon, Ghana, Guinea, Mauritania, Namibia, Nigeria, Tanzania, Uganda and Zambia from 2003 to 2020. In order to achieve the said objective, there infrastructure index, electricity, transport and water and sanitation index were used. Industrialization was proxied by industry value added to GDP, other variables such as gross capital formation, population growth and inflation were used in the model. The study made use of panel estimation and found a significant positive relationship between industry value added and Economic Development in Sub-Saharan Africa. Findings furthers showed that the interaction of infrastructure indexes with industry value add was negatively significant indicating poor infrastructure investment and low industrial activities.

Based on these results, we recommend that infrastructural investment should be improved across the countries under study. Furthermore, Sub-Saharan Africa should rigorously pursue industrialization as it has been seen to drive Sustainable Economic Development. Finally, the high level of inflation in the region should be curbed, as it negatively affects Economic Development.

References

Africa Development Bank: African Economic Outlook, African Development Bank Group, Ivory Coast, Abidjan (2018)

African Development Bank: The Africa infrastructure development index, Statistics Department (2020). http://afdb.org

Bloch, R., Makarem, N., Yunusa, M., Papachristodoulou, N., Crighton, M.: Economic Development in Urban Nigeria. Urbanisation Research Nigeria (URN) Research Report. ICF International, London. Creative Commons Attribution-Non-Commercial-ShareAlike CC BY-NC-SA (2015)

Box, G.E.P., Jenkins, G.M., Reinsel, G.C.: Time Series Analysis, Forecasting and Control, 3rd edn. Prentice Hall, Englewoods Clifs (1994)

Chukwuebuka, B.A., Jisike, J.O.: Infrastructure development and Industrial sector productivity in Sub-Saharan Africa. J. Econ. Dev. 22(1), 91–109 (2020)

Enimola, S.S.: Infrastructure and economic growth: the Nigeria experience. J. Infrastruct. Dev. 2(2), 121–133 (2010)

Gujarati, D.N., Porter, D.C., Gunasekar, S.: Basic Econometrics, 5th edn. Tata Mc Graw Hill Pvt. Ltd., New York (2012)

Lionel, E., Enang, B.U.: Industrialization and economic development in a multicultural lessons for Nigeria. Br. J. Econ. Manag. Trade 4(11), 1772–1784 (2014)

Nwogo, J.E., Orji, J.O.: Impact of industrialization on economic growth in Nigeria. IOSR J. Econ. Finan. 10(1), 42–53 (2019)

Okezie, A.C., Nwosu, C.A., Marcus, S.N.: Industrialization and economic growth in Nigeria. J. Econ. Finan. 1(1), the Nigerian Defence Academy Kaduna (2017). ISSN: 2636–5332

Ogbaro, E.O., Omotoso, D.C.: The impact of infrastructure development on economic growth in Nigeria. Niger. J. Manag. Sci. 6(1), 270–275 (2017)

Opoku, E.E., Yan, I.K.: Industrialization as driver of sustainable economic growth in Africa. J. Int. Trade Econ. Dev., 1–27 (2018). https://doi.org/10.1080/09638199.2018.1483416

Ozekhome, H.O, Fagboyo, R.J., Adesokun, A.J.: Logistics capability, public expenditure on transport infrastructure and trade performance in Nigeria: an empirical analysis in a pre-pandemic and pandemic era. In: Pandemic in the 21st Century: Multidimensional approaches, Samuel Adegboyega University, Ogwa, 21 April 2021 (2021)

Prakash, A.: Infrastructure and Industrialization: Ensuring Sustainable and Inclusive Growth in Africa. Economic Research Institute for ASEAN and East Asia, vol. 2, pp. 1–6 (2018)

The African Infrastructure Development Index: Africa Development Bank Group (2018). http://dataportal.opendataforafrica.org/AIDI

Seidu, R.D., Young, B.E., Robinson, H., Ryan, M.: The impact of infrastructure investment on economic growth in the United Kingdom. J. Infrastruct. Policy Dev. 4(2), 217–227 (2020). https://doi.org/10.24294/jipd.v4i2.1206

Shobande, A.O., Etukomeni, C.C.: Infrastructural investment and industrial growth: a private-investment led approach. Eurasian J. Bus. Econ. 6(2), 159–183 (2016)

Udah, E.B., Ebi, B.: Infrastructure, Human Capital and Industrialization in Nigeria. Nile J. Bus. Econ. 6, 58–78 (2017). https://doi.org/10.20321/nilejbe.v3i6.102

United Nations: Sustainable Development Goals (2020)

World Bank: "Africa's pulse", World Bank Group, Washington, DC (2017). http://documents.worldbank.org/curated/en/348741492463112162/Africas-pulse. Accessed 21 July 2019

Greening the Circular Cities: Addressing the Challenges to Green Infrastructure Development in Africa

O. M. Owojori and C. Okoro

Department of Finance and Investment Management, College of Business and Economics, University of Johannesburg, P.O. Box 526, Auckland Park, Johannesburg 2006, South Africa
tobiowojori@gmail.com, chiomao@uj.ac.za

Abstract. Purpose: Africa is amid a tremendous population and economic transformation with much of the growth taking place in the cities. This study explores further on the discussion on circular cities and the integration of green infrastructures (GI), as well as how they might serve as a blueprint and motivation for a revolution in African cities. The objective identified the barriers inhibiting green infrastructure development in Africa and offers a strategy for enhanced application.

Design/Methodology/Approach: A qualitative exploratory technique was adopted to collect qualitative data via an integrative literature review. The methodological approach was divided into two phases: research planning, screening and choice of publications. Utilizing content analysis, the acquired data were analysed.

Findings: The outcomes of this investigation show that scarcity of evidence to demonstrate the benefits, scarcity of data to support actions, budget constraints and lack of environmental education represent some of the barriers to green infrastructure development in Africa. This study suggests developing more innovative valuation methods, public-private partnerships, and GI marketing incentives.

Implications/Research Limitations: "The *State* of African Cities 2014" by UN-Habitat inspired this study, which focuses on African cities for a significant reimagining of current systems to promote innovation, reduce barriers, and capitalize on advantages. Thus, findings are discussed and interpreted in the light of the African environment.

Practical Implications: The results of this article will help GI decision-makers and planners in sustainable city development build appropriate strategies for how GI development could be utilised in African cities.

Originality/Value: There is a scarcity of research and lack of empirical support for the construction of green infrastructure in Sub-Saharan Africa which is a problem that has to be investigated. The global north has dominated previous studies on GI advancements, which suggests that its significance and practical use are still frequently misunderstood in the African setting.

Keywords: Africa · Barriers · Circular cities · Green infrastructure · Sustainable development

1 Introduction

Cities present a huge dilemma since they are expanding at the fastest and most unregulated rates, leading to vast variety of urban challenges. With the obvious urban growth for a better life and economic prosperity, such difficulties become chronic, particularly in the developing countries. There have been numerous efforts to solve such issues and ensure urban tenacity and vitality. Many cities, for example, have laid out strategies to become resilient eco-cities or sustainable cities (Owojori and Okoro 2022a) with sustainable housing by lowering negative effects on the environment, the economy, society and maximizing resource usage (Zhao et al. 2015).

Urbanization is pervasive throughout Africa in uncontrolled informal housing, where there is a severe paucity of infrastructures and essential services (UNDP 2017). Around 60% of African city dwellers live in these types of towns, which makes them very sensitive to climate change (Taylor and Peter 2014; Bandauko et al. 2021). This lack of sustainable infrastructure is a significant impediment to growth and progress, leading to lower intra-African and cross-regional trade. Despite making up 12% of the world's populace, Africa merely makes up 1% to world GDP and 2% of trade globally. Notwithstanding these, six of the top 10 fastest expanding economies worldwide are now found in Sub-Saharan Africa. Whilst insufficient and sustainable infrastructure may be the single biggest obstacle to Africa's long-term growth, but it also provides investors with a huge chance to support green infrastructure investments for a circular city which strengthens the case for sustainable infrastructure development (Owojori and Okoro 2022b).

Consequently, most low, and middle-income countries, particularly those in Africa, are confronted with the task of creating a more cost viable, ecologically sustainable, equitable and adaptable urban setting (Cobbinah and Darkwah 2017). This has spurred and sparked a movement to promote circular cities which is based on the circular economy theory as a significant departure from the extensive pattern of economic development. The closed loop of "resources - products - renewable resources" encourages economic society and natural ecosystem development that is both sustainable and harmonious (Ellen Macarthur foundation 2017). The shift from a linear to a circular economy paves the way for a greener, more resource-conscious future. In essence, a just transition to a circular economy must be consistent with the development of green infrastructure.

The focus of studies on green infrastructure development has inclined to be on developed countries. According to Brink et al. (2016) most investigations on the topic are undertaken in northern hemisphere cities. The majority of studies on green infrastructure in informal settlements are undertaken in Eastern Asia, North America, and Europe (Diep et al. 2019). Given that the term "green infrastructure" and its use have historically been associated with cities in the North of the world. In many circumstances, its importance and practical use are still widely misunderstood. Previous studies have looked at how GI and its ideas can be practiced in Southeast Asian states, Latin America, and Africa (e.g., Douglas 2016; Lindley et al. 2018), but there are there are still substantial gaps in the literature of how GI and its principles can be implemented in African environments. According to Herslund, Jean-Baptiste, Jalayer, Jorgensen, Kabisch, Lindley, ... Vedeld (2016), GI is not fully incorporated in African city design and management due to certain governmental structures that obstruct its development.

Regardless of the negative effects of urbanization, African cities have potent precepts and a distinctive but neglected ability to invent and explore (Nagendra et al. 2018). In particular, the massive urbanization and growth that is expected in Sub-Saharan Africa presents a significant potential to address vulnerabilities and catastrophe risk in a proactive manner. As a result, it is critical to concentrate on the possibilities and capacity for building resilience in future urban areas, thinking of metropolitan physical settings solely as "risk sites" misses out on perceiving them as environmentally relevant sources (Myers 2016).

It is realised that cities in the African Continent may have sustainability challenges because of concerns such as inequality, unemployment, and infrastructural inadequacies, but scholars should investigate how these situations can provide new avenues for urban sustainability (Fernandez 2014). This approach is now a recurrent element of futuristic ideas that strive to speed the construction of new cities, from Masdar's new towns built from the ground up to the massive development of Lusaka, and the growth of Kenyan and Ghana satellite towns, the mega-urbanization of Lusaka, Songdo, and Rajarhat (Datta and Shaban 2016). India and Indonesia both have plans to build one hundred infrastructures. National administrations in both nations have utilized large-scale initiatives as revolutionary methods to address predominant urban issues. The goal is to create structured models for upcoming urban structures or satellite cities, often known as future cities, eco or smart cities. They are designed to be brand-new constructions that utilize green technologies to facilitate effective municipal administration, improved air quality as well as industrial growth (Moser 2015; Watson 2014).

In complement to its ecological benefits, green infrastructure has economic and social benefits as a tool for planning, leading in the establishment of sustainable, adaptable, and inclusive urban settings. Despite recent advances, there has been no general agreement among scholars and practitioners on the idea of green infrastructure, with its delivery models in Africa. This review's goal is to provide knowledge into the basic barriers to African GI services considering the dearth of research that are exclusively directed at urban GI in African cities.

As a point of departure, to address the question, an integrative literature study is carried out to address the topic of what cities in Africa could do differently to achieve more circular cities via green infrastructure development according to the New Urban Agenda and Sustainable Development Goal 11, respectively.

2 Methods

There are a variety of approaches for conducting literature reviews that are useful for addressing new or existing challenges, and each one delivers unique understandings for new information, content synthesis, and reflection (Pickering et al. 2015). Green infrastructure development, as a newly emerging study field, would clearly benefit from a literature review, as there seems to be no general agreement in the literature on what to address in green infrastructure development in Africa (Gradinaru and Hersperger 2018). This study's primary objective is to conduct an integrative literature evaluation on green infrastructure. a thorough analysis of the literature on a subject that is quickly evolving like this allows for a comprehensive understanding along with a review of the existing literature. The approach used in this study is the integrative literature review, which was influenced by Tranfield et al. (2003) to ensure the review's quality and efficiency. To define the study sample, the methodology was divided into two phases: study preparation, publication screening, as well as a content analysis of the select documents.

Phase1: Planning of the Research
The first step was to select and characterize the study topic, ensuring it was unique and relevant, as well as to determine what terms to use in the query, the query string was formulated using the Boolean operator as follows; (*green AND infrastructure* AND barriers AND *green AND cities*) AND (LIMIT-TO (LANGUAGE, "English")). It is important to note that all these terms were chosen for their relevance to the research question, and the inquiry was composed to search all sections within the database, implying that the data were checked across all fields (containing title, abstract, authors, affiliations, etc.). It is pertinent to mention that the year of the publications was not filtered, therefore the sample collected from the databases contained all years of publication. The search query, which used at the time of this research, via the Scopus database produced 117 documents, which were then vetted and 60 articles were selected.

Phase Two: Document Screening and Selection
This procedure entailed the creation of criteria for scrutinizing the articles and, as a result, selecting only those that were relevant. The resulting sample's titles, abstracts, and entire text were checked for relevancy. The precondition for barriers to be in the entire text of the studies is a reasonable and crucial criterion because the major objective of this review analysis is to identify what barriers should be addressed in the development of green infrastructure. A total of eight papers were integrated with the original sample from Google Scholar, in addition to the database findings, to generate a wider range of data. In this study, only English-language documents were analyzed.

3 Results and Discussion

Green Infrastructure for Green Cities
Since the turn of the century, the term "green infrastructure" has received considerable interest from the fields of environmental protection, architecture, and construction. However, depending on the context and who uses the notion, it might have different definitions (Lennon 2015). The concept of green infrastructure is widely regarded new, despite its historical roots in the 19th and 20th centuries (Di Marino and Lapintie 2018).

Green infrastructure (GI) is gradually being acknowledged as the most cost-effective solution to mitigate and respond to social-ecological concerns through multifunctional natural ecosystems as an active physical framework. Sustaining GI connection is beneficial to nature conservation and promoting ecosystem functions, which helps to improve urban resilience and indicates that urban governance has made a concerted effort to plan for the unexpected (Lennon 2015). Grey infrastructure, on the other hand, refers to man-made, designed systems.

GI is being proposed as a a practical plan for repositioning cities in the direction of sustainability and long-term resilience, moving beyond typical green areas are critical for both health and urban ambience (Demuzere et al. 2014). Cities must develop positive connections with their structures and functions rather than develop at the expense of them to achieve a broad resilience capability even under stress (Ignatieva et al. 2011).

Green infrastructure is not just another way of describing parks and other conventionally green, open spaces. It enhances the livability and economics of cities by reducing detrimental environmental impacts and enhancing resilience (Ignatieva et al. 2011). As a result, the environment and human health are improved, and the existing built infrastructure is protected from the effects of climate change. The contribution of green infrastructure to sustainable cities in terms of social, economic, and environmental growth is depicted in Fig. 1.

Fig. 1. Contribution of green infrastructure to sustainable cities (source: Hegazy et al. 2017)

Benefits of Green Infrastructure for African Cities

The economic benefits will be discussed first, as seen in Fig. 1. When it comes to revitalizing an undeveloped city, GI can contribute significantly (Madureira and Andresen 2014). Furthermore, GI contributes to the economic development of society (Skipper et al. 2013). Value of property rises when GI enhances the social, physiological, and environmental condition of communities (Clements et al. 2013). Furthermore, GI boosts productivity and job prospects by contributing to sustainable development and design and building processes by enriching the built environment. GI reduces operating costs by improving environmental quality by enhancing water efficiency, allowing for energy savings, and offering protection from natural hazards (Skipper et al. 2013).

Following that are the sociocultural advantages of GI's functionality. GI supports recreational activities and aesthetics in communities, as well as providing benefits related to nature's pedagogical role and the preservation of historical environmental assets (Lovell and Taylor 2013). Green places that have been expanded through the use of GI and regenerating communities encourage public service accessibility, provide safety by lowering crimes and natural disasters, improve the physical environment of the community, and enhance various attributes, psychological health, and overall health (Zimmer et al. 2012). It also enables citizens to manage their own resources in order to attain sustainability, resulting in an adaptive process whereby individuals can actually learn to optimize environmental benefits (Lovell and Taylor 2013). This encourages local involvement, which develops the system, fosters a sense of belonging to a place and societal cohesion, and improves overall synergy (Wan et al. 2018; Zimmer et al. 2012). Participation also entails the society's ability to develop sustainable environments in the pursuit of climate action, as well as local residents' inclusion. Finally, the advantages will be examined from an environmental standpoint. The benefits of environmental resources include climate change adaptation and impact mitigation through creating urban locations with green spaces, which are based on considerations of precipitation emission and flood management (Clark et al. 2008). Furthermore, GI preserves biodiversity and natural environments, enhances the quality of the environment, properties, minimises carbon emissions by maintaining green areas in cities (Xiao 2018). The advantages of GI are summarized in Table 1.

Table 1. The benefits of Green infrastructures in cities

Type of benefit		Description	
Economic	Increased economic capability	– Community-based economic growth – Enhancement of marketability – Improved property values – Credits for taxes and fees – Financial incentives – Costs associated with infrastructure construction are reduced – Enhanced employee productivity – Provision of green jobs	Atkinson et al. (2010), Shakya and Ahiablame (2021), Skipper et al. (2013)
	Enhancing the built environment	– Easier access to public facilities – Noise pollution is lessened – Better-quality housing	

(*continued*)

Table 1. (*continued*)

Type of benefit		Description	
	Opportunities for education	– Improved recreational prospects – Opportunities for public education – Enhanced awareness of conservation issues	
Socio-cultural	Sustainable development	– Connections between cities and the countryside – Degraded sites are being restored – Land use efficiency	Shakya and Ahiablame (2021), Clements et al. (2013), Lovell and Taylor (2013)
	Aesthetics of the landscape	Enhanced aesthetics – Extended landscape and townscape advantages – Visual inspection of unappealing structures or infrastructure – Restoration of the landscape	Zimmer et al. (2012), Wan et al. (2018), Xiao (2018)
	Social capital development	– A higher standard of living – People's involvement – A better sense of community – Possibilities for spending time in public areas – More places for socializing – A decreases in criminal activity – Improved physical and mental well-being	
Environmental	Adaptation to climate change	– Reduced demand for grey infrastructure – Reduced ambient temperatures in cities – Resilient infrastructure – Adaptation to/mitigation of climate change	Skipper et al. (2013), Clements et al. (2013), Lovell and Taylor (2013), UNDP (2017),

(*continued*)

Table 1. (*continued*)

Type of benefit		Description	
	Improved environmental safety	– Water quality improvement and conservation – Enhanced air quality – CO2 levels in the atmosphere are lower – Conservation of biodiversity – Conserves historical sites – A lower environmental footprint	Zimmer et al. (2012), Wan et al. (2018), Xiao (2018)

4 Barriers to Green Infrastructure in Africa

The following major impediments to green infrastructure initiatives in African cities were identified based on the literature:

Dearth of Clarity and Technical Direction: Firstly, the absence of a definition for green infrastructure in systems for city administration and policy, or more generally in the context of Africa's urbanization, suggests that there isn't actually a meaningful direction for the development of green infrastructures (du Toit et al. 2018; Douglas 2016). Town planners, government officials, and those in charge of issuing permits have a distorted perception of the depth and significance of green infrastructure in the urban environment. This causes to misinterpretation or misunderstanding on what green infrastructure is and what it could offer.

Benefits Are Difficult to Quantify: Current models and procedures for categorizing and evaluating projects do not accommodate the concept of green infrastructure (du Toit et al. 2018). Because budget distribution is based on this, insufficient or incorrect substitute categorisations fail to represent the benefits of green infrastructure, implying that typical or 'grey' construction activities are more liable to be seen as easier (Shackleton et al. 2017). Because green infrastructure has not been employed as much as grey infrastructure, actual operational and maintenance expenses are unknown. Aim is to enhance or maximize the value of these assets but not being taken advantage of green infrastructure rate of return, asset life, and maintenance costs which are difficult to calculate for urban designers and finance departments.

Few Statistics Exist for Africa: Lack of African-specific data and research initiatives on green infrastructure for cities that highlight the advantages of constructing green infrastructure in African cities hinders investment in green infrastructure. Where cases are given, the data are of inadequate quality. As a result, green infrastructure is frequently regarded as a new technique with a restricted track history or as a concept that has yet to be proven in African cities (Cilliers et al. 2017). It is not that green infrastructure efforts are not being deployed across the continent; it's just that the benefits and costs of these programs, as well as their potential that would save cost and generate revenue, aren't being monitored in a methodical way.

There Is a Lack of Inter Partnerships: There are not enough cross-departmental partnerships. In accordance with the spectrum of services they provide, green infrastructure initiatives are cross-cutting and multidisciplinary in scope (Herslund et al. 2017). This necessitates interdepartmental coordination regarding the planning and design, construction, operations, and management (du Toit et al. 2018). More crucially, there is a question of shared expenses for both initial finance and public works. Different city units are focused on their own tasks and are hesitant to get active in other units' functions or allow other departments to intrude in their work. To support multifaceted and interdisciplinary planning, administration, and finance of green infrastructure, cities will have to adapt existing systems of governance amenable to inter collaboration.

The Idea of "Green Infrastructure" Is not Widely Understood: Limited awareness of the operations and principles of many types of green infrastructure among policymakers and urban dwellers leads to political and social ambiguity, which impedes green infrastructure acceptability. Green infrastructure is commonly regarded as a luxury rather than a public asset that improves municipal livability and human well-being (Niu et al. 2010, Zhao et al. 2015).

Land-Use Disputes: Pressures from increasing population and rapid urbanization raise the demand for land for development, resulting in conflicts between diverse land users and uses. Not surprisingly, when compared to more growth-oriented alternatives, green infrastructure is frequently reduced to aesthetic value and so underestimated. These issues are further classified as socioeconomic, design, policy/legislative, innovation, or financial. Table 2 lists some practical solutions to existing challenges.

Table 2. Addressing Africa's green infrastructural development barriers

Barriers	Description	Suggested solution
Technical barriers	– Dearth of clarity and technical direction – Scarcity of relevant local valuation – Green infrastructure is not well-understood – Scarce statistics for Africa	– Green infrastructure should be defined with purpose – Leverage the benefits of green infrastructure in an African setting – Develop more valuation techniques to account for the absolute and relative value of green infrastructure (Roy et al. 2018; Douglas 2016; Lindley et al. 2013

(continued)

Table 2. (*continued*)

Barriers	Description	Suggested solution
Socio-economic	– There aren't enough cross-departmental cooperation – Cooperation among stakeholder groups at diverse levels are lacking – Lack of tenure and ownership – Benefits are difficult to measure (Proof of monetary valuing being incompatible with non-monetary trading systems) – Insufficient knowledge of the benefits of green infrastructure – Limited awareness on sociocultural values	– Approaches that are inclusive and participatory in nature may be particularly beneficial – Linking the private sector via public-private partnership (Herslund et al. 2017) – A deeper and broader comprehension of the social elements of socio-ecological systems – Perception in perception and inform decision-makers, that for sustainable development, green infrastructure is not a luxury but rather a necessity (Roy et al. 2018; Lindley et al. 2013)
Innovation	– The meaning of the GI phrase is faintly explained – Green infrastructure is not well-understood – There is a scarcity of evidence to demonstrate the benefits for raising awareness – There is a scarcity of data to support actions	– Devise strategies to aid in the promotion and utilization of local ideas and experiments (du Toit et al. 2018) – Broader implementation of initiatives that garner support from a diverse range of partners – Co-creation of community learning and sustainable knowledge (Cilliers et al. 2017)
Finance	– Budget constraints – Initial infrastructure funding as well as infrastructure maintenance costs – Measurement of return on investment, asset life, and depreciation for green infrastructure is a challenge	– Budgets for green infrastructure should be prioritized (Herslund et al. 2017) – To be regarded as a need rather than a luxury – Promotional incentives should be offered for green infrastructural projects – Relaxing financial restrictions and offering enticements to promote it (lin et al. 2015; Cobbinah and Darkwah 2017)

(*continued*)

Table 2. (*continued*)

Barriers	Description	Suggested solution
Political/regulatory	– Land-use disputes – Governmental systems are frequently inadequately coordinated – Overall system delays and fragmentation – Priorities that conflicts – Environmental education on greenspaces is omitted from policies	– Coherence in policy is crucial for policy reform framework (Roy et al. 2018) – Public and private practices are being harmonized – Policy alignment and capacity building (Cilliers 2019) – Policymakers must play a role in education facilitation – Empowering builders and communities to embrace change and innovation (Demuzere et al. 2014)

5 Conclusion

Considering all of its advantages, including climate change adaption, risk prevention, community cohesion, and personal well-being enhancement, green infrastructure has the potential to lead to more circular and sustainable society transitions. Nonetheless, the highlighted constraints must be addressed since they are preventing the concept from being more widely adopted in African cities, where it has the potential to benefit both social and ecological systems. This research has significance for GI decision makers and planners, as well as sustainable development action planners, in terms of realizing the intended benefits of GI for African cities and its associated barriers and exploring how GI development could be utilized for African cities. Future research should thus analyze the cost of green infrastructure in Africa, taking into account its social, environmental, and economic advantages which may range dramatically from the western standard due to diverse contextual problems and considerations. There is a further need for Coherence in policy to institute a policy reform framework in African settings and the harmonization of public and private partnerships.

References

Atkinson, G., Brunt, A., Bryant, R., Doick, K., Lawrence, V.: Benefits of Green Infrastructure, Report to DEFRA and CLG; contract no. WC0807. Forest Research, Farnham, UK (2010)

Bandauko, E., Annan-Aggrey, E., Arku, G.: Planning and managing urbanisation in the twenty-first century: content analysis of selected African countries' national urban policies. Urban Res. Pract. **14**, 94–104 (2021)

Brink, E., et al.: Cascades of green: a review of ecosystem-based adaptation in urban areas. Glob. Environ. Chang. **36**, 111–123 (2016)

Cilliers, E.J.: Reflecting on green infrastructure and spatial planning in Africa: the complexities, perceptions, and way forward. Sustainability **11**, 455 (2019)

Cilliers, S., et al.: Health clinic gardens as nodes of social-ecological innovation to promote garden ecosystem services in Sub-Saharan Africa. Landsc. Urban Plan. **180**, 294–307 (2017)

Clark, C., Adriaens, P., Talbot, F.B.: Green roof valuation: a probabilistic economic analysis of environmental benefits. Environ. Sci. Technol. **42**, 2155–2161 (2008)

Clements, J., Juliana, A.S., Davis, P.: The Green Edge: How Commercial Property Investment in Green Infrastructure Creates Value. Natural Resources Defense Council, New York, NY, USA (2013)

Cobbinah, P.B., Darkwah, R.M.: Toward a more desirable form of sustainable urban development in Africa. Afr. Geogr. Rev. **36**, 262–285 (2017)

Datta, A., Shaban, A. (eds.): Mega-Urbanization in the Global South: Fast Cities and New Urban Utopias of the Postcolonial State. Routledge, London (2016) https://doi.org/10.4324/9781315797830

Demuzere, M., et al.: Mitigating and adapting to climate change: multi-functional and multi-scale assessment of green urban infrastructure. J. Environ. Manag. **146**, 107–115 (2014)

Diep, L., Dodman, D., Parikh, P.: Green infrastructure in informal settlements through a multiple-level perspective. Water Altern. **12**, 554–570 (2019)

Di Marino, M., Lapintie, K.: Exploring the concept of green infrastructure in urban landscape. Experiences from Italy, Canada, and Finland. Landsc. Res. **43**, 139–149 (2018)

Douglas, I.: The challenge of urban poverty for the use of green infrastructure on floodplains and wetlands to reduce flood impacts in intertropical Africa. Landsc. Urban Plan. **180**, 262–272 (2016)

Du Toit, M.J., Cilliers, S.S., Dallimer, M., Goddard, M., Guenat, S., Cornelius, S.F.: Urban green infrastructure and ecosystem services in sub-Saharan Africa. Landsc. Urban Plan. **180**, 249–261 (2018)

Ellenmacarthurfoundation.org: The Circular Economy Concept - Regenerative Economy (2017). Accessed 2 Apr 2022

Fernández, J.E.: Urban metabolism of the global south. In: Parnell, S., Oldfield, S. (eds.) The Routledge Handjournal on Cities of the Global South, pp. 597–613. Routledge, New York (2014)

Gradinaru, S.R., Hersperger, A.M.: Green infrastructure in strategic spatial plans: evidence from European urban regions. Urban for. Urban Green. **40**, 17–28 (2018)

Hegazy, I., Seddik, W., Ibrahim, H.: Towards green cities in developing countries: Egyptian new cities as a case study. Int. J. Low Carbon Technol. **12**, 358–368 (2017)

Vedeld, T.: Developing multiple-dimensional assessment of urban vulnerability 929 to climate change in Sub-Saharan Africa. Nat. Hazards **82**, 149–172 (2016)

Herslund, L., et al.: Challenges and opportunities for developing water resilient green cities in Addis Ababa and Dar es Salaam – in search of champions and paths for urban transition. Landsc. Urban Plan. **180**, 319–327 (2017)

Ignatieva, M., Stewart, G.H., Meurk, C.: Planning and design of ecological networks in urban areas. Landsc. Ecol. Eng. **7**, 17–25 (2011)

Lennon, M.: Green infrastructure and planning policy: a critical assessment. Local Environ. **20**, 957–980 (2015)

Lin, B., Meyers, J., Barnett, G.: Understanding the potential loss and inequities of green space distribution with urban densification. Urban for. Urban Green. **14**, 952–958 (2015)

Lindley, S., et al.: A GIS based assessment of the urban green infrastructure of selected case study areas and their ecosystem services (2013)

Lindley, S., Pauleit, S., Yeshitela, K., Cilliers, S., Shackleton, C.: Rethinking urban green infrastructure and ecosystem services from the perspective of sub-Saharan African cities. Landsc. Urban Plan. **180**, 328–338 (2018)

Lovell, S.T., Taylor, J.R.: Supplying urban ecosystem services through multifunctional green infrastructure in the united states. Landsc. Ecol. **28**, 1447–1463 (2013)

Madureira, H., Andresen, T.: Planning for multifunctional urban green infrastructures: promises and challenges. Urban Des. Int. **19**, 38–49 (2014)

Moser, S.: New cities: old wine in new bottles? Dialogues Hum. Geogr. **5**, 31–35 (2015)

Myers, G.: Urban Environments in Africa: A Critical Analysis of Environmental Politics. Policy Press, Bristol (2016)

Nagendra, H., Bai, X., Brondizio, E.S., Lwasa, S.: The urban south and the predicament of global sustainability. Nat. Sust. **1**, 341–349 (2018)

Owojori, O.M., Okoro, C.S.: Overcoming challenges associated with circular economy in real estate development. In: Mojekwu, J.N., Thwala, W., Aigbavboa, C., Bamfo-Agyei, E., Atepor, L., Oppong, R.A. (eds.) ARCA 2021. Springer, Cham (2022a). https://doi.org/10.1007/978-3-030-90973-4_5

Owojori, O.M., Okoro, C.S.: The private sector role as a key supporting stakeholder towards circular economy in the built environment: a scientometric and content analysis. Buildings **12**, 695 (2022b)

Niu, H., Clark, C., Zhou, J., Adriaens, P.: Scaling of economic benefits from green roof implementation in Washington, DC. Environ. Sci. Technol. **44**, 4302–4308 (2010)

Pickering, C., Grignon, J., Steven, R., Guitart, D., Byrne, J.: Publishing not perishing: How research students transition from novice to knowledgeable using systematic quantitative literature reviews. Stud. High. Educ. **40**, 1756–1769 (2015)

Roy, M., Shemdoe, R., Hulme, D., Mwageni, N., Gough, A.: Climate change and declining levels of green structures: life in informal settlements of Dar es Salaam. Tanzania. Landsc. Urban Plan. **180**, 282–293 (2018)

Shackleton, C.M., et al.: How important is green infrastructure in small and medium sized towns? Lessons from South Africa. Landsc. Urban Plan. **180**, 273–281 (2017)

Shakya, R., Ahiablame, M.: A synthesis of social and economic benefits linked to green infrastructure. Water **13**(24), 3651 (2021)

Skipper, L., Jacobson, A., Zhang, S.S., Canto, K.: Green Infrastructure Guidebook: Managing Stormwater with Green Infrastructure. University of Illinois Press, Urbana (2013)

Taylor, A., Peter, C.: Strengthening Climate Resilience in African Cities. A Framework of Working with Informality. Africa Center for Cities, Cape Town (2014)

Tranfield, D., Denyer, D., Smart, P.: Towards a methodology for developing evidence-informed management knowledge by means of systematic review. Br. J. Manag. **14**, 207–222 (2003)

UNDP: Leveraging Urbanization and Governance for Growth in Africa: A Framework for Action. UND, Addis Ababa (2017)

Wan, C., Shen, G.Q., Choi, S.: The moderating effect of subjective norm in predicting intention to use urban green spaces: a study of Hong Kong. Sustain. Cities Soc. **37**, 288–297 (2018)

Watson, V.: African urban fantasies: dreams or nightmares? Environ. Urban. **26**, 215–231 (2014)

Xiao, X.D.: The influence of the spatial characteristics of urban green space on the urban heat island in Suzhou Industrial Park. Sustain. Cities Soc. **40**, 428–439 (2018)

Zhao, D.X., He, B.J., Johnson, C., Mou, B.: Social problems of green buildings: from the humanistic needs to social acceptance. Renew. Sustain. Energy Rev. **51**, 1594–1609 (2015)

Zimmer, C., et al.: Low Impact Development Discussion Paper. ICF International, Fairfax, VA, USA (2012)

Properties of Clay Deposits in Selected Places in Sekondi-Takoradi and Ahanta West, Ghana

B. K. Mussey[1,2(✉)], A. Addae[3], G. Obeng-Agyemang[1], and S. Quayson Boahen[1]

[1] Department of Mechanical Engineering, Takoradi Technical University, Takoradi, Ghana
bernard.mussey@ttu.edu.gh
[2] Jubilee Technical Training Centre, Takoradi, Ghana
[3] Department of Ceramic Technology, Takoradi Technical University, Takoradi, Ghana

Abstract. Purpose: This paper research on properties of clay deposits in selected places in Sekondi-Takoradi. This work aims to test the properties of the identified clay and its workability to produce wares with sustainable assessment outcomes.

Design/Methodology/Approach: Physical, mechanical and chemical analyses of clay from three sites within the Sekondi Takoradi Metropolitan Assembly was performed. Using XRF and XRD (X-ray diffraction), physical, chemical and manufacturing modulus of rupture (MOR) were analysed.

Findings: Findings revealed that the Sekondi and Fijai clay deposits are rough in texture compared to the clay sample from the Apemenyin deposit site, which is fine in texture. The MOR value analysed at a temperature of 1200 °C for the Sekondi sample showed to be more robust, durable and sustainable for wares production but contains a higher percentage of oxides of aluminium, calcium, iron and silicon as compared to those of aluminium, calcium, iron and other samples.

Research Limitation/Implication: The research focused on the laboratory analysis of clay properties alone. Also, there is no scientific correlation between those properties among the clay deposit sites.

Practical/Social Implications: The study intends to provide local manufacturers with the knowledge they need to innovatively utilize nearby, sustainable clay deposits to produce durable clay products in order to establish resilient and innovative clay-related enterprises.

Originality/Value: This research will serve as an informative reference document for further research projects and industry players in STMA.

Keywords: Clay deposits · Properties · Temperature · Workability

1 Introduction

Ghana's clay industry is primarily concerned with the processing, creating and producing of various household items such as porcelain, pots and other items. Widely varying fields of usefulness, including biological, ceramics, construction, and water purification, are influenced by differences in the properties of these clays.

Rapid industrialisation, urbanisation, and the growing need for clay products have contributed to a long-standing emphasis on creating sustainable infrastructure. The primary consumers of clay reserves are the construction, ceramics, biomedical and healthcare industries. Clay can be used for various practical purposes, including synthesising materials for cutting-edge applications, food packaging, clothing, and detergents. Using clay deposits in a productive, creative and sustainable way can go a long way toward improving the economy and, obviously, the progress of the whole country.

Clay is formed mainly by volcanic deposits or marine sediments. It comes from different sources with its unique mineral composition (Chamley 1989). It contains a fine-grained earth material formed by the decomposition of igneous rock: when combined with water, clay is plastic enough to be shaped, and when dry, it is strong, and when subjected to high heat or above, it will become progressively denser and rock-like (El-Naggar et al. 2019; Yamada et al. 2019; Zhang et al. 2020). Clay is used to producing domestic appliances and can be classified into two main groups (Asamoah et al. 2018; Asamoah et al. 2020; Yaya et al. 2017). There are many different fields where clays are used, including geology, construction, environmental remediation, and the pharmaceutical and cosmetics industries field (Zhou et al. 2022).

Clays play an essential role in creating many typical dry landforms, whether as vast sheets, discrete aeolian bedforms, or as a critical component of closed basins (playas) (Annan et al. 2018; Jaskulski et al. 2020; Onwona-Agyeman et al. 2020). Clay which is the chief raw material for every ceramic and pottery work, is a natural resource made up of Alumina (Al_2O_3), Silica ($2SiO_2$), and a small amount of water (H_2O) (R. B. Asamoah et al. 2018; Dill 2020). Clay can be defined as a soil particle with a diameter less than 0.002 mm, which becomes plastic when wet and coherent when dry (El Ouahabi et al. 2019; Yamada et al. 2019). Most often, clay is formed from the disintegration of feldspathic rocks and is characterised by the presence of one or two minerals, together with a varying number of organic materials, among which quartz is predominant—field (Andrews et al. 2013; Asamoah et al. 2018).

1.1 Formation of Clay

Clay is formed mainly by volcanic deposits or marine sediments. It comes in different source that has their unique mineral composition. Clay has been used to produce cooking pots, bricks, and porcelain since the beginning of civilisation and can be classified into two main groups (Adeyemo et al. 2015; Isfahani et al. 2013; Zhou 2016).

1.2 Primary Clay

According to Wilson (1999), 'Primary clay or residual clay is a type of clay which is found around the site of the mother rock from which they are formed and have not been transported either by water, wind, or glacier'. It appears white in colour, feels rough in texture, coarse in particle size and less plastic, hence its ability to withstand high temperatures without deforming (Asante-Kyei et al. 2019; Chamley 1989; Seraj 2014). Primary clay is a type of clay formed on the site of the parent rocks. Primary clays are typically white and pure, free from organic contamination; a typical example is kaolin. Primary clay is formed on or around the site of the parent rock and has not been moved or transported either by water, ice or wind (R. B. Asamoah et al. 2018). Primary clay is pure and free from contamination. Primary clay is white in its raw and fired states and not plastic enough for pottery production methods like throwing, coiling, slabbing, and modelling.

1.3 Secondary or Sedimentary Clay

Again, Wilson (1999) further asserts that secondary clay is formed by the action of wind or running water or glacier, which transport primary clay from the site of the parent rock and deposit the grains in a valley, lake, low land sites, etc. Bediako et al., (2017) also indicated that secondary clay deposits are most commonly found in low-energy depositional environments like big lakes and marine basins. They are known as secondary clays because they are formed from primary clay and are deposited in sediments, usually in the form of layers. In the case of transportation, their particle sizes are further broken down into finer units. Thus, they tend to be more plastic than primary clays. They contain impurities such as iron, mica and other carbonaceous materials making secondary clays less refractory and having a low maturing temperature. Earthenware, stoneware, ball clays, etc., are typical examples of secondary clay and are highly plastic. Generally, secondary clays have finer particle sizes and are more malleable than primary clays, thus their preference for pottery work (Ameri et al. 2019; R. B. Asamoah et al. 2018).

1.4 Physical Characteristics of Clay

The physical characteristics of the samples under review included; plasticity, shrinkage, colour observation in the green and fired states, and porosity (Asamoah et al. 2020; Inoue 1995).

Plasticity is a property that permits moist clay to be moulded into a shape and retain the shape after drying. Clay tends to be more plastic when the particle size is very fine, a property mainly associated with secondary clays (Andrews et al. 2013).

The shrinkage property is the specific volume change of the clay relative to its water content and is mainly due to clay swelling properties (Chamley 1989; Inoue 1995; Wilson 1999). One would note that the drying shrinkage increases, as does plasticity, and more drying cracks come with that increase. This is so because plastic clays have finer particle sizes and, thus, have greater particle surface area and more inter-particle water holding the clay grains together (Dejaeghere et al. 2019; Trümer and Ludwig 2018).

Fired shrinkage results from the loss of chemically combined water during the firing of ceramic bodies or ware. As soon as the glass begins to form the body, the heat reaction between basic impurities and silica and the un-melted particles pack closer together and, overall, lose its volume (Andrews et al. 2013; Y. Zhou et al. 2022).

Porosity refers to the water absorption and the ability of clay. The porosity of clay is directly related to its hardness and verification (El-Naggar et al. 2019).

Clay contains iron to give them a slightly warm temperature when fired, of which kaolin is exempted. The presence of iron in the clay gives it a brown colour, which turns red or buff when fired and other colouring oxides such as manganese (Kirilovica et al. 2021).

2 Methodology

This entails a review of the properties of clay and analytical techniques used to perform the qualitative analysis on clay samples from selected sites within the STMA enclave.

2.1 Material

Clay from selected from places of STMA
 TOOLS
 In prospecting and testing clay, various tools and equipment were used to carry out the task; these include; a measuring cup, shovel, sack board, metal ruler, weighing scale, sieve, knife, guide sticks, rolling pin, and wheelbarrow.

2.2 Sample Preparation and Testing Procedure

Clay samples were prepared from the following sites Sekondi Clay (SC), Fijai Clay (FC), and Apemenyim Clay (AC). The clay was mined from the sites and sent to the working studio, where the physical test was conducted; the clay samples were then crushed into fine powder. The powdered clay samples were sieved through a 60 μm mesh size to separate the finer particles from unwanted materials and the larger particles. A 1 kg batch of samples was sent to the Geological Survey Department in Accra for the chemical analysis test.

2.3 Chemical Analysis

Quantitative analysis was done to identify the percentage composition of iron (Fe), Alumina (Al_2O_3), Silica (SO_2) Potash K_2O_5 in the clay samples.

Procedure
The samples from the selected sites were correctly labelled.
Samples were kept in basins and dried in the sunlight.
Soil samples were ground to obtain finer particles.
It was then sieved through 80–90 μm or microns.
The finer samples were kept in a container with a corresponding label.
Samples were then taken to the x-ray fluorescent for quantitative chemical analysis.

2.4 X-Ray Fluorescent Analysis

This is a non-destructive analytical technique used to determine the elemental composition of the materials. X-RF analysers chose the chemistry of the sample 0.900 g of wax was weighed in the electronics balance. 4.000 g of each of the samples, then topped it up.

The samples were then mixed uniformly by the use of a homogeniser.

The uniform samples were then poured into the caster to form a tablet using the Pressing machine.

The tablets were then labelled and ready for an X-RF spectrum test for the chemical elements found in the sample.

2.5 Physical Analysis and Test Procedure

1 kg of each sample was prepared dry and mixed with water to a workable state.

The wet samples were left to age for three days at the laboratory.

On the third day, the samples were kneaded to remove air pockets and other foreign materials and make them homogeneous.

The samples were rolled upon with a slab, and the formed slabs were cut into briquettes. Nine (9) briquettes with 5 cm lines measured and marked on them in the middle of each briquette were created to aid in the shrinkage test measurements.

2.6 Fire Testing

Firing Stage Firing is the process of bringing clay and glazes up to a high temperature. The final aim is to heat the object to the point that the clay and glazes are "mature"; that is, they have reached their optimal melting level (El Ouahabi et al. 2019).

The firewood kiln was for the firing test for the samples. Preheating was done for two hours; a full blast was done for about 12 h. The sample slabs were soaked in the fire for some hours at a temperature of about 600 °C. The kiln was left to cool off for three days, then the samples were removed, and the fired shrinkage was calculated using the formula below:

$$\text{Fired shrinkage} = \text{Dry Length} - \text{Fired Length}$$

2.7 Other Physical Analysis

A physical analysis is a test carried out on a clay sample to know or determine the physical status of the sample, which is the shrinkage rate, drying rate, firing rate, water adoption rate, strength of the sample, colour of the clay sample and plasticity of the clay.

2.8 Manufacturing of Modulus of Rapture (M.O.R.) or Flexural Strength

Modulus of Rapture refers to measuring the maximum load-carrying capacity or strength of the crosstie and can be defined as the stress at which the material breaks or raptures. MORs were made in bars for each sample, as shown in the figure below (Figs. 1 and 2).

Fig. 1. MOR sample

Fig. 2. Electronic MOR strength checker

MOR samples were prepared from each piece and fired at temperatures of 1000 °C, 1100 °C, and 1200 °C. An electronic MOR strength checker was used to check the strength of the sample by using force on the MOR bar until it breaks.

3 Findings

Calculation for the Dry Shrinkage
The formula for the calculation of is:

$Dry\ shrinkage\ (ds)\ =\ Wet\ Measurement\ (wm)\ -\ dry\ measurement\ (dm)$

Sekondi clay slab = wm (5 cm) − dm (4.5 cm) = 0.5 cm

Fijai\clay slab = wm (5 cm) = dm (4.6 cm) = 0.4 cm

Apemenyim clay = Wm (5 cm) − dm (4.3 cm) = 0.7 cm

Fig. 3. Sekondi clay   **Fig. 4.** Apemenyim clay   **Fig. 5.** Fijai clay

The Sekondi clay appeared brown in colour, Fijai clay was dark brown, and the Apemenyim clay looked brownish in colour in the raw state. After firing the samples, it was noted that the Sekondi clay sample fired chocolate brown, Fijai clay also came out beige colour, and Apemenyim clay fired smoke. The Sekondi and Fijai samples were rough in texture, whereas the sample from Apemenyim was fine. During the shrinkage test, it was observed that Apemenyim recorded the highest shrinkage due to its fine particle sizes. Fijai clay sample recorded minor shrinkage due to its larger particle sizes (Figs. 3, 4 and 5).

Manufacturing of Modulus of Rapture (m.o.r.) or Flexural Strength (Figs. 6, 7 and 8, Tables 1 and 2).

Table 1. Fired shrinkage = Dry Length − Fired Length

Sample	Dry Length/cm	Fired Length/cm	Fired Shrinkage/cm
Sekondi clay	4.5	4.4	0.1
Fijai clay	4.6	4.5	0.1
Apimenyim clay	4.3	4.2	0.1

$$MOR(R) = (3 * W * L)/2 * B * T2$$

Properties of Clay Deposits in Selected Places 173

At 1100°c   At 1000°c   At 1200°c

Fig. 6. Fired MOR for Sekondi samples

At 1000°C   At 1100°C   At 1200°C

Fig. 7. Fired MOR for Fijai samples

At 1000°C   At 1100°C

Fig. 8. Fired MOR for Apimenyim samples

Table 2. Shows the temperature ranges of the samples.

MOR bars			
Samples	1000 °C	1100 °C	1200 °C
Apemenyim clay	Sample 1 = 0.95 kg	Sample 1 = 1.35 kg	Sample 1 = 2.50 kg
	Sample 2 = 0.85 kg	Sample 2 = 1.35 kg	Sample 2 = 2.00 kg
	Total Weight = 1.80 kg	Total Weight = 2.70 kg	Total Weight = 4.50 kg
	Average Weight = 0.9 kg	Average Weight = 1.35 kg	Average Weight = 2,25 kg
Fijai clay	Sample 1 = 0.45 g	Sample 1 = 0.80 g	Sample 1 = 1 kg
	Sample 2 = 0.50 g	Sample 2 = 0.70 g	Sample 2 = 1 kg
	Total Weight = 0.95 kg	Total Weight = 1.50 kg	Total Weight = 2.0 kg
	Average Weight = 0.475 kg	Average Weight = 0.75 kg	Average Weight = 1.0 kg
Sekondi clay	Sample 1 = 1.35 kg	Sample 1 = 1.80 kg	Sample 1 = 2.50 kg

(*continued*)

Table 2. (*continued*)

MOR bars Samples	1000 °C	1100 °C	1200 °C
	Sample 2 = 1.20 kg	Sample 2 = 2.00 kg	Sample 2 = 2.2. kg
	Total Weight = 2.55 kg	Total Weight = 3.80 kg	Total Weight = 4.7 kg
	Average Weight = 1.275 kg	Average Weight = 1.90 kg	Average Weight = 2.35 kg

- 3 is a constant figure as well as 2
- W = weight
- L = length (7 for the MOR equipment used)
- B = breadth of the sample
- T^2 = Thickness of the bar

From analysis $2 * B * T2 = 1.458$,
L = 7, Therefore; $R = \left(\frac{3*7}{1.458}\right) * W = \underline{14.4033 * W}$ (Table 3).

Table 3. Calculation of the MOR strength

Sample	1000 °C	1100 °C	1200 °C
Apimenyim clay	R = 25.93	R = 38.89	R = 65.81
Fijai clay	R = 13.68	R = 21.60	R = 28.80
Sekondi clay	R = 36.73	R = 54.73	R = 67.70

The MOR strength for Sekondi clay samples fired at a temperature of 1200 °C was observed to be stronger and difficult to break than Apemenyim and Fijai clay samples. Moreover, all samples were observed to have an increase in MOR strength with increasing temperature (Table 4).

Table 4. Chemical analysis

Element component	Sekondi clay (%)	Fijai clay (%)	Apemenyin clay (%)
Al_2O_3	14.50	9.41	13.64
$Si O_2$	68.45	59.51	7.74
Cl	1.12	0.25	0.12
K_2O	2.42	4.32	3.21
CaO	1.42	0.94	1.01
Fe_2O_3	7.80	3.40	2.33

4 Conclusion

In conclusion, from the analysis carried out on all the samples, Apemenyim clay would be most suitable for studio-based forming techniques like throwing, coiling, pinching and slabbing due to its finer particle size and plasticity. Fijai clay, on the other hand, would be suitable for structural clay products like bricks, tiles and kiln furniture due to its refractory nature. Finally, Sekondi clay would be most appropriate for casting and modelling pinching of small handy wares due to its plasticity nature.

Conflict of Interest. The authors declare no conflict of interest.

References

Adeyemo, A.A., Adeoye, I.O., Bello, O.S.: Adsorption of dyes using different types of clay: a review. Appl. Water Sci. **7**(2), 543–568 (2015). https://doi.org/10.1007/s13201-015-0322-y

Ameri, F., Shoaei, P., Zareei, S.A., Behforouz, B.: Geopolymers vs. alkali-activated materials (AAMs): A comparative study on durability, microstructure, and resistance to elevated temperatures of lightweight mortars. Constr. Build. Mater. **222**, 49–63 (2019)

Andrews, A., Adam, J., Gawu, S.K.Y.: Development of fireclay aluminosilicate refractory from lithomargic clay deposits. Ceram. Int. **39**(1), 779–783 (2013)

Annan, E., et al.: Application of clay ceramics and nanotechnology in water treatment: a review. Cogent Eng. **5**(1), 1–35 (2018)

Asamoah, R.B., et al.: Industrial applications of clay materials from Ghana (a review). Orient. J. Chem. **34**(4), 1719–1734 (2018)

Asamoah, R.B., Yaya, A., Nbelayim, P., Annan, E., Onwona-Agyeman, B.: Development and characterization of clay-nanocomposites for water purification. Materials **13**(17), 3793 (2020)

Asante-Kyei, K., Addae, A., Abaka-Attah, M.: Production of clay containers for curbing plantain post-harvest losses in Ghana. New J. Glass Ceram. **09**(03), 50–65 (2019)

Bediako, M., Purohit, S.S., Kevern, J.T.: Investigation into Ghanaian calcined clay as supplementary cementitious material. ACI Mater. J. **114**(6), 889–896 (2017)

Chamley, H.: Clay formation through weathering. In: Chamley, H. (ed.) Clay Sedimentology, pp. 21–50. Springer, Heidelberg (1989). https://doi.org/10.1007/978-3-642-85916-8_2

Dejaeghere, I., Sonebi, M., De Schutter, G.: Influence of nano-clay on rheology, fresh properties, heat of hydration and strength of cement-based mortars. Constr. Build. Mater. **222**, 73–85 (2019)

Dill, H.G.: A geological and mineralogical review of clay mineral deposits and phyllosilicate ore guides in Central Europe – a function of geodynamics and climate change. Ore Geol. Rev. **119**, 103304 (2020)

El-Naggar, K.A.M., Amin, S.K., El-Sherbiny, S.A., Abadir, M.F.: Preparation of geopolymer insulating bricks from waste raw materials. Constr. Build. Mater. **222**, 699–705 (2019)

El Ouahabi, M., El Idrissi, H.E.B., Daoudi, L., El Halim, M., Fagel, N.:Moroccan clay deposits: Physico-chemical properties in view of provenance studies on ancient ceramics. Applied Clay Sci. **172**, 65–74 (2019)

Inoue, A.: Formation of clay minerals in hydrothermal environments. In: Velde, B. (eds.) Origin and Mineralogy of Clays. Springer, Heidelberg (1995). https://doi.org/10.1007/978-3-662-12648-6_7

Isfahani, A.P., Mehrabzadeh, M., Morshedian, J.: The effects of processing and using different types of clay on the mechanical, thermal and rheological properties of high-impact polystyrene nanocomposites. Polym. J. **45**(3), 346–353 (2013)

Jaskulski, R., Jóźwiak-Niedźwiedzka, D., Yakymechko, Y.: Calcined clay as supplementary cementitious material. Materials **13**(21), 1–36 (2020)

Kirilovica, I., Vitina, I., Grase, L.: Structural investigation of carbonation and hydration process of hydraulic dolomitic binder. Constr. Build. Mater. **275**, 122050 (2021)

Onwona-Agyeman, B., Lyczko, N., Minh, D.P., Nzihou, A., Yaya, A.: Characterization of some selected Ghanaian clay minerals for potential industrial applications. J. Ceram. Process. Res. **21**(1), 35–41 (2020)

Seraj, S.: Evaluating Natural Pozzolans for Use as Alternative Supplementary Cementitious Materials in Concrete Committee, 176 (2014)

Trümer, A., Ludwig, H.M.: Assessment of calcined clays according to the main criterions of concrete durability. RILEM Bookseries (2018)

Wilson, M.J.: The origin and formation of clay minerals in soils: past, present and future perspectives. Clay Miner. **34**(1), 7–25 (1999)

Yamada, Y., Tsuchida, T., Kyaw, N.M., Aoyama, T., Hlaing, M.M.S., Hashimoto, R.: A study on physical and mechanical properties for soft to firm clays in Yangon area – properties of clays deposit at the sedimentary basins in Myanmar. Soils Found. **59**(6), 2279–2298 (2019)

Yaya, A., Tiburu, E.K., Vickers, M.E., Efavi, J.K., Onwona-Agyeman, B., Knowles, K.M.: Characterisation and identification of local kaolin clay from Ghana: a potential material for electroporcelain insulator fabrication. Appl. Clay Sci. **150**, 125–130 (2017)

Zhang, J., Yang, J., Ying, Z.: Study on mechanical properties of metakaolin-based concretes and corrosion of carbon steel reinforcement in 3.5% NaCl. Int. J. Electrochem. Sci. **15**, 2883–2893 (2020)

Zhou, D.: Developing Supplementary Cementitious Materials from Waste London Clay, 236, October 2016

Zhou, Y., Du, H., Liu, Y., Liu, J., Liang, S.: An experimental study on mechanical, shrinkage and creep properties of early-age concrete affected by clay content on coarse aggregate. Case Stud. Constr. Mater. **16**, e01135 (2022)

Lumped – Capacitance Design for Transient Heat Loss Prediction in Oil and Gas Production Pipes in Various Media

R. N. A. Akoto[1(✉)], J. J. Owusu[2], B. K. Mussey[2], G. Obeng-Agyemang[2], and L. Atepor[3]

[1] School of Graduate Studies, University of Professional Studies, Accra, Ghana
nii.ayitey-akoto@upsamail.edu.gh
[2] Department of Mechanical Engineering, Takoradi Technical University, Takoradi, Ghana
[3] Department of Mechanical Engineering, Cape Coast Technical University, Cape Coast, Ghana

Abstract. Purpose: A Lumped Capacitance innovative approach is used here to analyze the transient temperature decay of a pipeline in water, air, and sand.

Design/Methodology/Approach: A straightforward small-scale test equipment was used to measure transient heat loss through steel pipes that were in contact with air, water and sand mediums separately. The pipe was supported in the lab using a lumped capacitance method as part of the air medium experiment. A concentric PVC pipe was used to encase the pipe in the water and sand mediums, with dry sand and water being poured into the annulus. Hot hair was generated in the pipes and transient cool-down tests were carried. Hot water replaced the hot air and another transient cool-down tests were performed.

Findings: Results obtained show experimental approach agreement. Systematic and environmental errors were attributed to a maximum deviation of 21% between design and performance.

Research Limitation/Implications: The transient temperature decay method implored in this work makes no use of the finite element method. This justified by the fact that pipes with complex pipe geometries were not used in the work.

Practical Implication: The experimental methodology presented in this work will allow the prediction of transient cool-down behavior and cool-down duration of the fluids in pipelines using an established theory during the pipelines' design. This will assist with flow assurance litigation in the oil and gas sector.

Social Implication: This work presents yet another experimental framework needed to advance the workflows within the oil and gas industry for sustainable industrialization.

Originality/Value: The novelty lies in the sustainable and innovative experimental framework for analysing transient temperature decay of oil and gas pipeline in three different media.

Keywords: Heat loss · Oil and gas · Prediction · Production · Transient

1 Introduction

Trapped fluids in oil and gas production pipelines loose heat to its surrounding during shutdown schedules. The rate of heat loss is dependent on the kind of fluid, its temperature at the time of shutdown, and the surrounding environment. This poses a threat to flow assurance since it can lead to the formation of hydrates, asphaltenes, and waxes (Sunday et al. 2021; Wang et al. 2020). The source of heat losses has been attributed largely to heat conduction and shape of the pipelines (Dayan et al. 1984; Sadegh et al. 1987).

The problem of steady state heat transfer in partially and fully buried pipelines have been extensively posited and analytical (Aja and Ramasamy 2016; Otomi et al. 2020; Ovuworie 2010; Zakarian et al. 2012) and numerical models (Enbin et al. 2013; Nyukuri et al. 2014; King et al. 2011) have been developed. These works modeled the steady state cases and that for the transient behavior was not considered. Some experimental works have been equally performed under various assumptions to explain the heat transfer problem (Král 2014; Mishra and Kumar 2019; Oh et al. 2014). An experimental approach to study the transient and steady state behaviour of looped heat pipes have been proposed using thermal systems that can be represented in networks matching to finite difference, finite element, and/or lumped parameter equations using the NASA standard thermohydraulic analyzer (Cullimore and Baumann 2000). One will however need to have a full understanding of the loop heat pipe operation and the thermophysical processes associated with those devices. Another design has made use of the macroscopic approach where a transient model for microgrooved heat pipe (Suman et al. 2005) is used. This approached cannot be used for heat pipes that are not microgrooved and polygonally shaped. Lumped parameter designs have also been introduced for capillary heat pipes and based on electrical analogy (Bernagozzi et al. 2018; Ferrandi et al. 2010, 2013). These methods ignored how gravity affected the behavior of the heat pipe and neglected to take into account how real gas would affect vapor modeling. In this study, a pipeline embedded in water, air, and sand mediums is investigated using the Lumped Capacitance technique.

2 The Lumped Capacitance Theory

A lumped capacitance method is also known as simplified analytical method. It assumes that the temperature throughout the entire hot body or system is the same, regardless of its position (Mishra and Kumar 2019). With this theoretical approach, temperature decreases over time for fluid in a pipe. In other words, the temperature gradient inside the pipe containing the fluid is minimal under temporary circumstances. In order to predict the future temperature, T of the system after a certain interval of time, we must combine Newton's cooling law with the Fourier law to arrive at

$$T(t) = T_a + (T_i - T_a)exp(-t/\psi), \qquad (1)$$

where T_i is the initial temperature, T_a is the ambient temperature and $\psi = mc_p/UA$ is known as the thermal time factor with m being the mass, c_p being the heat capacity at constant pressure, and A representing the area. U is the overall coefficient of heat transfer and its reciprocal defined without the fouling factor is

$$\frac{1}{UA} = \frac{1}{A_i h_{wi}} + \sum_j \frac{x_j}{A_j k_j} + \frac{1}{A_o h_{wo}}, \qquad (2)$$

where k_j is the material's thermal conductivity in the jth layer, $h_{wi,o}$ is the unique fluid convection heat transfer coefficient of the inside or outside walls and x_j is the thickness of the jth layer.

Certain dimensionless characteristics must be recognized and described in order to calculate the film-coefficient for convection heat transfer. The volumetric thermal expansion coefficient, β_i describes how a material volume varies as its temperature changes while maintaining a constant pressure. The ratio of buoyant force to viscosity force is captured in the Grashof number $Gr_e = \beta_i g \Delta T d^3 / \nu^2$ where ν is the kinematic viscosity and ΔT is the difference between the bulk and surface temperatures. The Rayleigh number, $Ra_e = \beta_i g \Delta T d^3 / \nu \alpha$, distinguishes laminar from turbulence flows in free convection. The thermal diffusivity $\alpha = k / \rho c_p$, measures how well a fluid transmits thermal energy relative to how well it can store thermal energy. The Nusselt number under forced or free convection indicates the amount of convectional heat transfer for a specific surface. For a transient cool-down, the Nusselt number can be exchanged for the Rayleigh number as $Nu_e = 0.55 Ra_e^{0.25}$ where $10^5 \leq Ra_e \leq 10^7$.

Equation (1) can be used to determine the cool-down time, but only if the internal heat flow resistance to external heat flow resistance ratio, otherwise known as the Biot number, is less than 0.1 (Wojtkowiak 2014). When the ratio of internal to exterior heat transfer resistance approaches this limit, temperature gradient occurs, and the lumped system method produces unreliable findings. This makes it obvious that the boundary layer conditions, spatial effect, and orientation are simplified in this method.

3 Methodology

3.1 Materials and Equipment

There are two 0.59 m long plastic pipes, with inner diameters of 0.159 m and of thicknesses 60 mm were used for the experiment. In addition, there were end covers that were drilled to fit the pipe's outer diameter, and there were covers for the medium of water that were fitted with valves to prevent leakage. Two thermocouples were used to measure the temperatures of the fluid inside each steel pipe. One thermocouple was placed inside the pipe with its protruding tip, while the other thermocouple was placed inside the pipe's outside wall to measure the temperature of the pipe's exterior surface. We used Grant Squirrel SQ400 multi-purpose data loggers, connection tubing, measuring cylinders, stop watches and thermometers to obtain this data.

3.2 Experimental Setup and Procedure

An oil and gas production scenario where the water medium cools down at the same speed as seawater flowing over and underneath a subsea pipeline is mimicked in these experiments. Pipes in air represent risers exposed to air above sea level, while pipes in sand represent production pipelines buried or covered in sand during onshore and offshore production. The experiments were carried out with the assumptions that:

i. There is very little heat transfer from steel pipe to all media by radiation.
ii. When the fluid flows in the pipe and the pipe's temperature is the same, the temperature of the pipe remains the same. Since convection is the dominant source of heat flux, diffusion is ignored along the pipe axis. Since heat flux changes over time, it only changes with time.
iii. There is a uniform flow of water above and beneath the steel pipe in the water medium that is perpendicular to it.
iv. The fluid's thermal and physical characteristics do not change throughout the pipe's axis.
v. The portion of the steel pipe and fluid outside the test length did not undergo any heat loss or gain testing.
vi. It is negligible that the covers of the PVC pipe retain or lose heat, and the fluid does not experience any effect.
vii. The fluid does not contain any air since air has different thermal properties than water, which will adversely affect the expected temperature as well as the U-value.
viii. There is uniformity in the pipe geometry.
ix. During the experiment, the fluid did not undergo any phase changes.

We performed cooling experiments in six different cases. We conducted the experiment three times for each case, including cooling hot water in water, cooling hot water in air, cooling hot water in sand, cooling hot air in water, cooling hot air in air, and cooling hot air in sand.

3.2.1 Hot Water Cooling in Water Medium

As shown in Fig. 1, a steel pipe (with thermocouples) that holds the liquid measures the rapid cooling of hot water in water medium. The test section of steel pipe is enclosed in

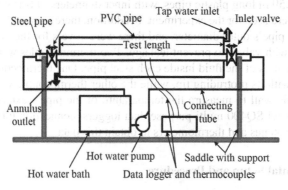

Fig. 1. Test apparatus for transient cool-down in water.

a concentric plastic pipe with valves at its ends. The temperature of the room and that of the tap water used are measured before starting the experiment.

The inlet valves on the end covers are connected to the tap, so fresh tap water can be pumped into the annulus. Water is circulated in the annulus, running over and around the steel pipe, and leaving the annulus through the outlet valve. Data loggers were connected to calibrated thermocouples on the steel pipe to ensure there was no leak from the chamber. A thermocouple measures the bulk temperature of the fluid inside the pipe, while a second thermocouple measures the temperature of the outer surface of the pipe. By selecting the appropriate temperature range, the channel temperature was determined to degrees Celsius in accordance with the instructions provided in the operating manual. Once the outer surface of the steel pipe reached a steady temperature, hot water was passed through it. By selecting each channel in the meter function of the data logger, we observed this. Following a period of time when the outer pipe surface had maintained a steady temperature, the hot water flow was stopped and the pipe ends were plugged to keep the hot water in the pipe. After this, water was circulated speedily in the annulus of the test section. The data logger recorded temperatures and times until the fluid reached the same temperature as fresh tap water - data was recorded every minute. To obtain an accurate reading of the rate of water flow leaving the annulus, three flow rates were measured using a stop watch and a measuring cylinder. The flow rate is determined by dividing the volume of water collected in the measuring cylinder by the cross-sectional area of the pipe divided by the amount of time required to collect it. In the experiment, water flowing out was measured and compared to fresh tap water in terms of temperature, T_w. Generally, the cooling down time of the water-medium experiment does not exceed thirty minutes, depending on the flow rate within the annulus and the initial temperature of the fluid. To determine the mass of the liquid in the steel pipe, a measuring cylinder is used to measure its capacity (mL) and multiply it by the density. To download the recorded data, the output of the data logger was connected to a computer at the end of the experiment.

3.2.2 Hot Water Cooling in Quiescent Air Medium

In this setup, steel pipes are connected to thermocouples, a data logger and wooden supports. In Sect. 3.2.1, an airtight pipe was mounted on wooden supports and connected to the thermocouples and data logger. A steady surface temperature was achieved through hot water running through the pipe. The pipe ends were plugged to hold the hot water in the pipe, in order to ensure no air was trapped in the pipe once the steady temperature was achieved. Temperature and length measurements were taken at this point (Fig. 2).

To avoid insulating part of the test length, one must place the wooden supports outside the test length.

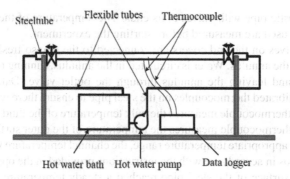

Fig. 2. Test apparatus for transient cool-down in quiescent air environment.

3.2.3 Hot Water Cooling in Soil Medium

Figure 3 illustrates the soil medium setup. In this case, steel pipes are enclosed within concentric PVC pipes and filled with sand. Dry sand was used. However, when using moist sand, the thermal conductivity might vary during the experiment. Due to heat redistribution, soil-moisture-air compositions may change continuously during the experiment (Smits et al. 2010) as heat is applied to the pipe. In addition, the packing of soil into the annulus affected how primary and secondary particles interacted. Heat flow resistance will be influenced by the compactness of the soil, and moisture and air content in the soil pack will depend on the spacing between particles. Haigh (2012), Smits et al. (2010) and Kodešová et al. (2013) reported that soil compacted closely is more thermally conductive than granular pack because it has fewer air gaps. In addition to the length of the test, and the inside and outside diameters of the PVC pipe, and the soil and room temperature, physical measurements were taken prior to the logging of the data. For some time, the steel pipe carried hot water until it reached a steady temperature outside. This was followed by the shutoff of the hot water supply and the plugging of the pipe. The temperature-time data was recorded after ten minutes by the data logging system. For this experiment to be cooled down, it usually takes about ten hours, depending on the soil thermal conductivity, ambient temperature, and fluid temperature.

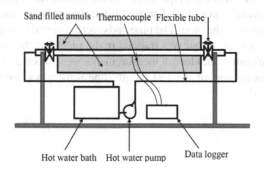

Fig. 3. Test apparatus for transient cool-down in sand.

4 Results and Discussion

Our sample calculation uses the cooling of hot water in all the media to establish cool-down temperature report of a pipeline. If we substitute the properties of hot water with those of hot air in theory, we get a hot air cool-down temperature profile. In Table 1, thermal and physical fluid properties are listed.

Table 1. Thermal properties of water and air (Beirão et al. 2012; Ramires and Nieto de Castro 1995)

Fluid	Density (kg/m^3)	Dynamic viscosity (kg/ms)	Expansion coefficient $\times 10^{-3}$ (K^{-1})	Thermal conductivity (kW/mK)	Specific heat capacity (kJ/kgK)
Water	1000	0.001	0.207	0.6000	4.187
Air	1.205		3.430	0.0257	1.005

All three media use the same inside film coefficient calculation. A kinematic viscosity was obtained by dividing the dynamic viscosity by the density, and a thermal diffusivity was obtained by substituting the appropriate values. The Rayleigh and Nusselt numbers obtained were also calculated. The inner film coefficient was obtained with the help of the characteristic length. The heat transfer of the outer film is by forced convection. We obtained the velocity of flow outside the pipe by dividing the volume flow rate by the annulus' cross-sectional area. We calculated the outside film coefficient by multiplying the pipe's thermal conductivity by its Reynolds and Prandtl numbers. The "mean pipe diameter" is the average of the pipe's inner and outer diameters. The theoretical U-value at the time the system was shut down for the various medium is shown in Table 2. It is also shown in the same table how the fluid temperature is predicted after t seconds from shutdown when U-value is computed using the outer surface area of the steel pipe.

Table 2. Measured and observed data to obtain U-values.

Media	Cool-down time after shutdown (s)	Mean diameter of pipe (m)	T_a (°C)	T_i (°C)	x/k (m^2 °C/W)	U (W/m^2 °C)	Expected temperature after shutdown (°C)
Water	20	0.0292	19	55.8	0.0006489	188.676	49.44
Air	10	0.0292	20	69.0	0.0006489	8.011	57.65
Sand	1800	0.1920	23	43.3	0.0006489	1.549	40.66

Figures 4, 5 and 6 compare experimental results with theoretically computed transient cooling down temperatures for two fluids in a steel pipe (hot water and hot air). At incremental time steps, the thermocouple readings of the bulk fluid temperature were used as experimental temperature data.

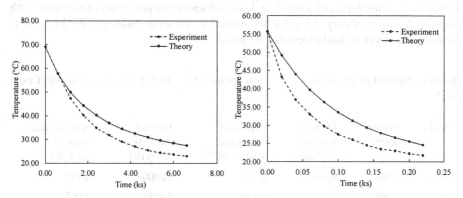

Fig. 4. Transient cool-down of hot water: (Left) in air medium (Right) in water medium.

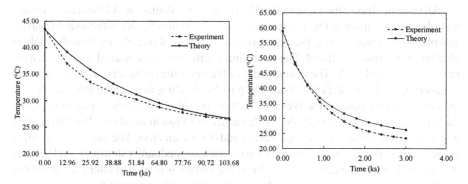

Fig. 5. Transient cool-down in sand medium: (Left) hot water (Right) hot air.

With Eq. (1), temperatures can be predicted based on the decay curves for the experiment and theory, which confirm an exponential drop in temperature under transient conditions. There may be a deviation between the tests performed and the design theory due to errors in the assumptions underpinning the theory, inaccuracies in the measurements, and temperature-dependent changes in thermophysical properties. We proceed to explain the reasons for the discrepancies in the following paragraph.

First of all, Eq. (1) assumes that the gradient of temperature in the bulk is uniformly distributed and changes over time only. Experimentally, however, the condition cannot be satisfied because at higher temperatures there is some degree of temperature variation. Three temperatures were recorded during the experiment. These temperature variations indicate that there is a temperature gradient in the inner pipe wall, where the fluid core temperature was the highest, the outer pipe surface temperature was the lowest, and the inner pipe wall temperature was in between. The readings were different at

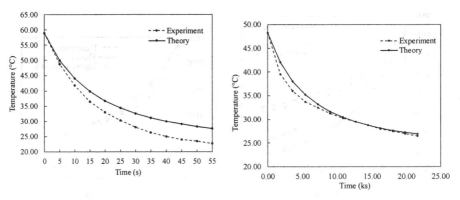

Fig. 6. Transient cool-down of hot air: (Left) in water medium (Right) sand medium.

higher temperatures, but became equal after some time and decayed at the same rate. Furthermore, Eq. (1) fails to account for heat transfer by radiation from a pipe to its surroundings. However, this assumption is reasonable, because convection heat transfer is the first step in transient heat transfer (Král 2014). A thermal resistance is introduced to pipe surfaces as a result of impurities in the fluid and/or rust formation, reducing heat transfer surface area. As well as operating temperature and chemical composition of the fluid, fouling extent depends on how long the surface is exposed to it (Awad 2011). Prior to turning off the pipe, the outside surface of the steel pipe had reached a steady temperature, so heat leakage through the ends was also ignored. Also contributing to the deviation is a non-uniform geometry of the pipe and errors made during physical measurements. Equation (1) uses fixed values for thermal and physical properties as well. These properties, however, change with temperature across a boundary layer depending on how well the heat transfer rate is accurate. The accuracy of the theory depends on how accurate these properties are. Lastly, the use of the conduction term x/k as an approximation in calculating introduces a 1% error (Lienhard and Lienhard 2020; Mcneil et al. 2013).

Based on Eq. (2), the U-value is a function of the inside and outside film coefficients and the thermal conductance of the pipe. As the flow rate of water outside the pipe increases, the Reynolds number will decay more rapidly. When the diameter of the pipe remains constant and the fluid properties remain unchanged, the Reynolds number fluctuates when the velocity is increased. The cooling to 30 °C takes approximately 2.1 min for Flow Rate 1 and 3.5 min for Flow Rate 4. In designing a subsea pipeline, it is therefore important to consider the highest fluid speed in that environment, when considering how the pipeline cools down over time under transient conditions under transient conditions (Fig. 7).

Figure 8 illustrates how inside film coefficient affects temperature decay curves theoretically. As shown in Fig. 8, the experimental curve, the theoretical curve containing the whole U-value (pipe thermal conductance, inside film coefficient, and outside film coefficient), and the theoretical curve excluding the inside film coefficient are all shown. Graphs show that the curve decays rapidly when the inside film coefficient is not present. Therefore, any factor that influences the inner film coefficient will have an effect on

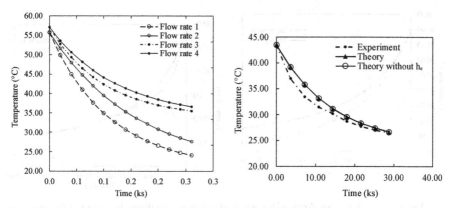

Fig. 7. Transient cool-down response to water flowrate.

Fig. 8. Effect of inside film coefficient on transient cool down.

the rate at which the system cools down. By reducing the diameter at the time, the film coefficient will increase and therefore the resistance to heat flow will decrease, assuming the heat loss was constant. Due to fouling, the surface area inside the pipe for heat transfer is reduced, so the actual inside film coefficient is higher than what would be predicted theoretically.

Under transient conditions, the higher the fluid temperature before shut down, the longer the pipeline will take to cool down. Figure 9 shows the importance of fluid temperature just before shutdown in pipeline cooling. For an industrial application, a temporary increase in temperature will lengthen the cooling process to allow the requisite work to be completed if the fluid's temperature will result in inadequate cooling time for the planned shutdown.

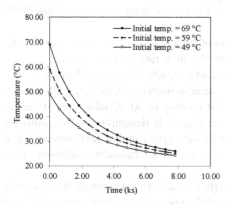
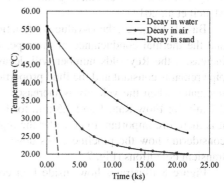

Fig. 9. Effect of initial temperature on transient cool down.

Fig. 10. Transient temperature decay in different media

As long as the fluid volume and medium remain constant, the term UA remains constant. Mass and specific heat capacity determine how fast a fluid cools. Under the same environmental conditions, water will cool down more slowly than air because of its higher density and specific heat capacity. As the seabed is usually undulating in subsea applications, understanding the topography and phase distribution of the seabed as well as the flow regime just before shutdown is necessary. It follows that the coldest part of the pipeline will influence the cool-down time. For a specific fluid, the cool-down time is inversely proportional to the characteristics of that fluid. In Fig. 10, we see that insulation extends the cooling time and that thermal time constants decrease with increasing U-values.

5 Conclusions

Pipeline transient cool-down temperature-time profiles have been analyzed based on a lumped capacitance theory. The result from the innovative setup showed that the thermal performance of a fluid in a pipe depends on the fluid's thermal mass and heat transfer coefficient. In a pipe with multiphase fluids, the temperature at which the mixture cools down will have an impact on its overall cool-down time. In order to increase the cooling duration, fluids with a shorter initial shut-down time should be warmed. The initial shutdown temperature strongly influences the cool-down period. Knowledge advanced in this work adds to the wealth of information needed to prepare a temperature decay experiment and afford sustainable industrialization.

Conflict of Interest. The authors declare that they have no conflict of interest.

References

Awad, M.M.: Fouling of heat transfer surfaces. In: Belmiloudi, A. (ed.) Heat Transfer - Theoretical Analysis, Experimental Investigations and Industrial Systems, pp. 504–542. IntechOpen, London (2011). https://doi.org/10.5772/13696

Aja, A., Ramasamy, M.: Thermal management of flow assurance challenges in offshore fields – a review. J. Eng. Appl. Sci. **11**, 6415–6422 (2016)

Beirão, S.G., Ribeiro, A.P., Lourenço, M.J., Santos, F.J., Nieto de Castro, C.A.: Thermal conductivity of humid air. Int. J. Thermophys. **33**, 1686–1703 (2012). https://doi.org/10.1007/s10765-012-1254-5

Bernagozzi, M., Charmer, S., Georgoulas, A., Malavasi, I., Michè, N., Marengo, M.: Lumped parameter network simulation of a loop heat pipe for energy management systems in full electric vehicles. Appl. Therm. Eng. **141**, 617–629 (2018). https://doi.org/10.1016/j.applthermaleng.2018.06.013

Cullimore, B., Baumann, J.: Steady state and transient loop heat pipe modeling. In: Proceedings of the International Conference on Environmental Systems, Toulouse (2000). https://doi.org/10.4271/2000-01-2316

Dayan, A., Merbaum, A., Segal, I.: Temporary distribution around buried pipe network. Int. J. Heat Mass Transf. **27**(3), 409–417 (1984). https://doi.org/10.1016/0017-9310(84)90288-6

Enbin, L., Liuting, Y., Yong, J., Ping, T., Jian, L., Yuhang, Y.: Simulation on the temperature drop rule of hot oil pipeline. Open Fuels Energy Sci. J. **6**, 55–60 (2013). https://doi.org/10.2174/1876973X01306010055

Ferrandi, C., Marengo, M., Zinna, S.: Influence of tube size on thermal behaviour of sintered heat pipe. In: Proceedings of the 2nd European Conference on Microfluidics, Toulouse, pp. 1–12 (2010)

Ferrandi, C., Iorizzo, F., Mameli, M., Zinna, S., Marengo, M.: Lumped parameter model of sintered heat pipe: transient numerical analysis and validation. Appl. Therm. Eng. **50**, 1280–1290 (2013). https://doi.org/10.1016/j.applthermaleng.2012.07.022

Haigh, S.K.: Thermal conductivity of sand. Géotechnique **62**(7), 617–625 (2012). https://doi.org/10.1680/geot.11.P.043

King, T., Phillips, R., Johansen, C.: Pipeline routing and burial depth analysis using GIS software. In: Proceedings of the Artic Technology Conference, SPE, Houston, pp. 1–11 (2011). https://doi.org/10.4043/22085-MS

Kodešová, R., et al.: Thermal properties of representative soils of the Czech Republic. Soil Water Res. **8**(4), 141–150 (2013). https://doi.org/10.17221/33/2013-SWR

Král, R.: An experimental investigation of unsteady thermal processes on a pre-cooled circular cylinder of porous material in the wind. Int. J. Heat Mass Transf. **77**, 906–914 (2014). https://doi.org/10.1016/j.ijheatmasstransfer.2014.06.045

Lienhard, J.H., Lienhard, J.H.: A Heat Transfer Textbook, 5th edn. Phlogiston Press, Cambridge (2020)

Mcneil, D.A., Raeisi, A.H., Kew, P.A., Hamed, R.S.: Flow boiling heat transfer in micro to macro transition flows. Int. J. Heat Mass Transf. **65**, 289–307 (2013). https://doi.org/10.1016/j.ijheatmasstransfer.2013.05.077

Mishra, M., Kumar, P.: Experimental lumped analysis of different solid geometries. IOP Conf. Ser. Mater. Sci. Eng. **691**, 1–9 (2019). https://doi.org/10.1088/1757-899X/691/1/012082

Nyukuri, N.W., Sigey, J.K., Okelo, J.A., Okwoyo, J.M.: Numerical study of heat transfer on fully buried pipeline under steady–periodic thermal boundary conditions. SIJ Trans. Comput. Sci. Eng. Appl. **2**(7), 229–235 (2014)

Oh, D.-W., Park, J.M., Lee, K.H., Zakarian, E., Lee, J.: Effect of buried depth on steady-state heat-transfer characteristics for pipeline-flow assurance. SPE J. **19**(6), 1162–1168 (2014). https://doi.org/10.2118/166595-PA

Otomi, O.K., Onochie, U.P., Obanor, A.I.: Steady state analysis of heat transfer in a fully buried crude oil pipeline. Int. J. Heat Mass Transf. **146**, 1–7 (2020). https://doi.org/10.1016/j.ijheatmasstransfer.2019.118893

Ovuworie, C.: Steady-state heat transfer models for fully and partially buried pipelines. In: Proceedings of the CPS/SPE International Oil and Gas Conference and Exhibition, Beijing, pp. 1–7. SPE (2010). https://doi.org/10.2523/131137-MS

Ramires, M.L., Nieto de Castro, C.A.: Standard reference data for the thermal conductivity of water. J. Phys. Chem. Ref. Data **24**, 1377 (1995). https://doi.org/10.1063/1.555963

Sadegh, A., Jiji, L., Weinbaum, S.: Boundary integral equation technique with application to freezing around a buried pipe. Int. J. Heat Mass Transf. **30**(2), 223–232 (1987). https://doi.org/10.1016/0017-9310(87)90110-4

Smits, K.M., Sakaki, T., Limsuwat, A., Illangasekare, T.H.: Thermal conductivity of sands under varying moisture and porosity in drainage–wetting cycles. Vadose Zone J. **9**(1), 172–180 (2010). https://doi.org/10.2136/vzj2009.0095

Suman, B., De, S., DasGupta, S.: Transient modeling of micro-grooved heat pipe. Int. J. Heat Mass Transf. **48**, 1633–1646 (2005). https://doi.org/10.1016/j.ijheatmasstransfer.2004.11.004

Sunday, N., Settar, A., Chetehouna, K., Gascoin, N.: An overview of flow assurance heat management systems in subsea flowlines. Energies **14**(2), 458 (2021). https://doi.org/10.3390/en14020458

Wang, H., An, C., Duan, M., Su, J.: Transient thermal analysis of multilayer pipeline with phase change material. Appl. Therm. Eng. **165**, 114512 (2020). https://doi.org/10.1016/j.applthermaleng.2019.114512

Wojtkowiak, J.: Lumped thermal capacity model. In: Hetnarski, R.B. (eds.) Encyclopedia of Thermal Stresses, pp. 2808–2817. Springer, Dordrecht (2014). https://doi.org/10.1007/978-94-007-2739-7_393

Zakarian, E., Holbeach, J., Morgan, J. E.: A holistic approach to steady-state heat transfer from partially and fully buried pipelines. In: Proceeding of the Offshore Technology Conference, Houston, pp. 1–5. SPE (2012). https://doi.org/10.4043/23033-MS

of applications denied, and the yearly total number of naturalizations. The mentioned time series data ranges from 1907 to 2019. As a result, the Legal Permanent Resident data was modified for the regression model building.

2.2 Model Specification

The plot in Fig. 1, which is a time series of Y_t = The number of immigrants entering the United States at time t (years) shows that immigration levels have gradually increased over time. From the plot, it appears that immigration levels were low in the beginning of the 19th century. According to existing literature, the late 1800s was a period when Americans rarely questioned immigration policies that were relatively free and open during the 18th and early 19th centuries. The plot also shows that immigration to the United States was summiting at the start of the 20th century. It appears that immigration levels dropped in the early years of 1900. This may be due to WWI and the Immigration Acts of 1921 and 1924 imposed numerical limits and national origins quotas. From the plot, Immigration levels tumbled more during World War II and the Great Depression.

The time series plot shows that immigration levels started to gradually increase after WW2 in 1950. A codified version of the INA was enacted in 1952, and it remains the governing law as amended to this day. A study of the graph shows that the growth in immigration since 1980 can be attributed in part to the growth in legal permanent residents as well as immediate relatives of U.S. citizens in the country. 2.1 million unauthorized aliens living in the United States as of 1982 became LPRs as a result of the Immigration Reform and Control Act (IRCA) of 1986. During the period 1981–1995, after the Refugee Act of 1980 was enacted, the number of refugees admitted increased from 1.6 million to 2.1 million, according to the Congressional Research Service (Best 2010). From the plot, it also looks like we see a big spike around 1986 due to the Refugee Act of 1980. Despite some drops after 1996, the trend of increasing continued after 2000.

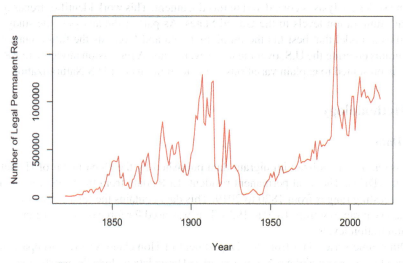

Fig. 1. Time series plot for the immigration data (1820–2019)

Overall, the immigration data seem to show there is a trend in the data. The trend seems to be a deterministic trend. In order to forecast future immigration levels, the trend will be removed from the series to achieve stationarity. Box and Cox (1964) log transformation should be used to stabilize a possible non-constant variance before taking a difference for stationary processes. An approximation of normality will also be improved by the log transformation. The plot of the transformed series where $Y_t = \ln(Y_t)$ in Fig. 3 shows an approximately increasing trend and strong momentum between observations which suggest non-stationarity. The observed increasing trend recommend taking a difference of the log-transformed data for stationarity. From the plot in Fig. 3, the log-transformed immigration data appears to be stationary after the first differencing (Fig. 2).

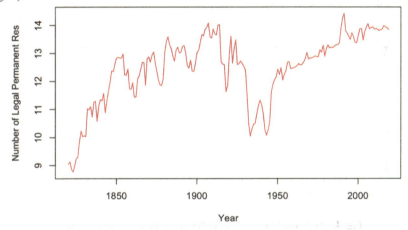

Fig. 2. Time series plot of log transformed immigration data

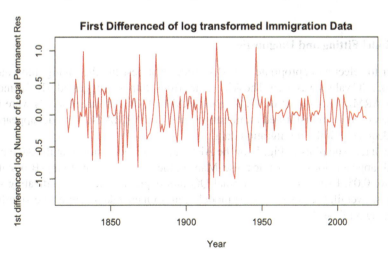

Fig. 3. Time series plot of 1st differenced log transformed immigration data

2.3 Model Specification

ACF and PACF plots were used to select the ARMA (p, q) model for the differenced data. The plot from Fig. 4 shows fluctuations of the sample autocorrelations values within the bounds suggesting a white noise process. However, there seem to be a significant spike at lag 5. This could either be by chance or a significant happening. The appropriate ARIMA model will be fit to the data in order to forecast future immigration values.

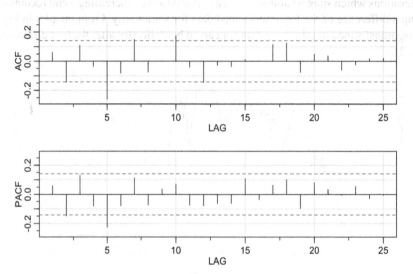

Fig. 4. ACF and PACF plot of differenced log transformed series

2.4 Model Fitting and Diagnostic

In order to select an appropriate ARIMA model that best fits the immigration data, a matrix of AIC values was used. The AIC values from Output 1 suggested ARMA models such as ARMA (3, 4), ARMA (3, 3), and ARMA (4, 3). All three models were fit to the differenced log transformed immigration data. The ARMA (4, 3) model appeared to most adequately fit the immigration data.

The diagnostic plots in Fig. 5 for the ARMA (4, 3) model show that the ACF of the residuals appear to look like white noise. Many of the p-values for the Ljung-Box statistic are above 0.05. Even though the normal QQ plot depict heavier tails indicating some skewness, overall, the ARMA (4, 3) model seem to have taken care of the significant spike at lag 5 (Table 1).

Table 1. AIC matrix for the different combinations of ARMA models (Output 1)

	[, 1]	[, 2]	[, 3]	[, 4]	[, 5]
[, 1]	152.0079	152.9501	149.2892	150.5623	151.9558
[, 2]	153.2817	150.3735	150.7229	152.4776	149.2655
[, 3]	150.8430	150.2903	151.4525	143.7939	146.0111
[, 4]	149.3918	151.0883	144.3089	143.3437	142.7381
[, 5]	150.1264	146.8614	143.5913	143.5628	146.8097

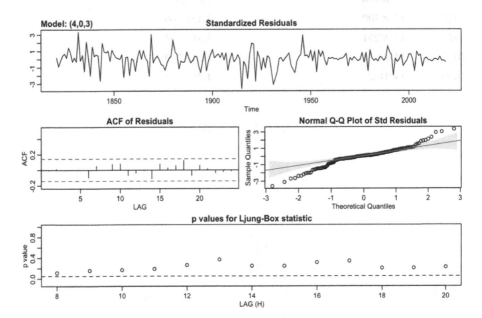

Fig. 5. Diagnostic for the ARMA (4, 3) model

3 Results and Discussion

3.1 Forecasting

The ARMA (4, 3) model was used to predict immigration levels for the years 2020 to 2026. The SARIMA for function in R (Cryer and Chan 2008; R Core Team 2020) was used to produce predicted values from the transformed series. These predicted values were then untransformed into the original scale of the original time series. The predicted values from Table 2 show that immigration will gradually keep increasing for the next seven years. However, the predicted immigration total number for 2020 is 1,029,064. This number might not be close to reality since the COVID-19 factor was not accounted for in the analysis. The predicted value also shows an immigration total of 1,263,328 by the year 2026 compared to a total of 1,031,765 in 2019. Again, an indication that immigration levels will keep increasing. The plot from Fig. 6 also shows wide confidence bounds for the predicted values.

Table 2. Predicted values for immigration levels from 2020 to 2026

Year	Log transformed predicted values	Predicted values on original scale
2020	13.84416	1,029,064
2021	13.85847	1,043,896
2022	13.90998	1,099,076
2023	13.96100	1,156,606
2024	13.98251	1,181,754
2025	14.00878	1,213,210
2026	14.04926	1,263,328

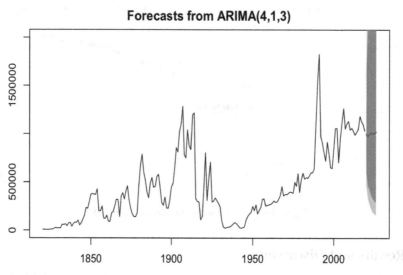

Fig. 6. Original immigration data with forecasted values from 2020–2026, ARIMA (4, 1, 3)

3.2 Regression Model

A regression model was used to establish the relationships between the differenced log transformed total yearly naturalizations, annual naturalization applications filed, annual applications denied, and the total number of legal permanent residents. The cross-correlation plots were used to check if lagged variables will be better predictors in the model. From Fig. 7, the cross correlation between transformed total naturalization and transformed total LPRs seem to show a significant spike at lag −14. However, this

Legal Immigration to the United States: A Time Series Analysis

is not what we expect since it does not take that long in reality to gaion citizenship after obtaining a "green card". The second cross-correlation plot show a significant spike at lag −1. This makes sense since on the average, it takes 6 months to a year for a naturalization application to be processed and finalized. The plot for the transformed total naturalizations and the transformed total applications denied only shows a significant spike at lag 0.

Before fitting the regression model, the model errors were modeled using ARMA. From the plot in Fig. 8, the ACF and the PACF seem to cut off at lag 1. This indicates that the errors should be modeled with an AR (1). The final regression model was then fit using AR (1) errors.

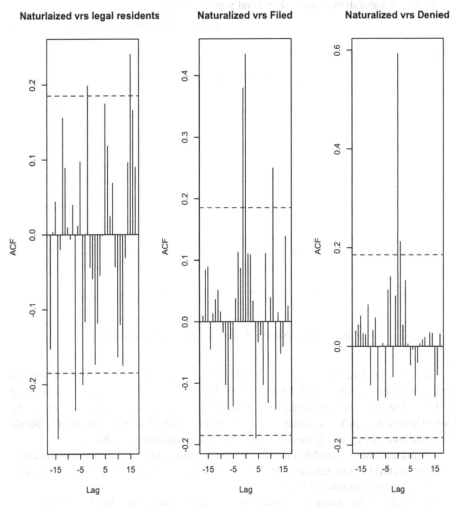

Fig. 7. Cross-correlations between the response (total naturalizations) and the predictors

Initial regression model:

$$\log(\text{Total naturalizations}) = \log(\text{LPRs}) + \log(\text{Total petitions filed})$$
$$+ \log(\text{Total petitions denied}) + \varepsilon, \quad (1)$$

where ε denote the error term. From the initial regression model, the transformed annual total number of legal permanent residents appeared to be insignificant in the model. Therefore, using p-values as the basis for variable selection, it was dropped from the model.

Final regression model:

$$\log(\text{Total naturalizations}) = \log(\text{Total petitions filed})$$
$$+ \log(\text{Total petitions denied}) + \varepsilon \quad (2)$$

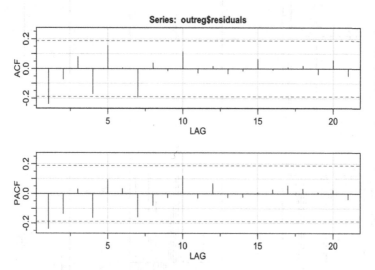

Fig. 8. ACF AND PACF plots for the model residuals

From Output 2 below, it appears that the remaining predictors in the model are all significant. The estimates both have p-values less than 0.05 presenting overwhelming evidence that they are significant in explaining the total annual naturalizations in the United States. It is quite surprising that the transformed total annual number of LPRs do not contribute in explaining the total annual U.S naturalizations (Table 3).

The present study identified that the ARMA (4, 3) model best fits the immigration data using the method of maximum likelihood. The model diagnostic plots were reasonable, and the estimated parameters were all significant.

According to the predicted values for the immigration data, the total number of immigrants will reach approximately 1,263,328 by the year 2026 which is 231, 563 greater than that of 2019. This is an indication that immigration will keep increasing for the next seven years. U.S. Census Bureau forecasts that the U.S. population will reach

Table 3. Final regression model with AR(1) errors (Output 2)

	Estimate	Std. error	z value	Pr(>|z|)
Ar1	−0.29168	0.095408	−3.0572	0.002234**
Intercept	−0.013418	0.013618	−0.9854	0.324443
TransDenied	0.305544	0.039884	7.6607	1.85E−14***
Transfiled	0.373937	0.072771	5.1386	2.77E−07***
Signif. codes:	0 '***' 0.001 '**' 0.01 '*' 0.05 '.' 0.1 ' ' 1			

392 million by 2050, with net immigration accounting for 86 percent of this growth (United States Census Bureau 2008). Also, indicator variables were created for the time periods before and after 1950 respectively in order to regress them on the total number of LPRs. However, the indicator variables all appeared insignificant, hence the output not shown in the study.

The positive coefficients of the predictors from the final regression model revealed that, controlling for the total number of applications denied, the total annual naturalizations will increase by 0.374 for each unit increase in the annual number of applications filed. Also, similar conclusion for the coefficient of applications denied with an increase of 0.306.

Some limitations for the study include not being able to find a data that involves country of origin for the immigrants for the range of years studied. This would have been a good predictor since the economic status of those countries could play a major role in immigration.

4 Conclusion

The time series analysis utilizing maximum likelihood and regression forecasted an increasing trend in America's immigration levels in the future. Also, the study established that the number of legal residents naturalized annually can be explained by the number applications filed and denied. Hence, the analysis revealed the following;

- The predicted values show that immigration in the United States will keep increasing for the next seven years. This confirms the projection from the U.S Census Bureau that immigration will make up 86% of the U.S population by year 2050.
- The regression model revealed that the number of naturalization petitions filed by legal residents and applications denied by USCIS contribute to variations in annual number of U.S naturalizations.

The findings from this study are indicators for policy decision makers to consider immigrants when making policies that affect the average resident in America. This in the long run allows for proper planning for sustainable development and industrialization. The study affirms that a significant percentage of America's population in the future will be made up of immigrants. This is an informative signal to the government to consider immigrants when drafting innovative political, socio-economic, and health policies for the country.

Conflict of Interest. Neither the authors nor any of the contributors have any conflict of interest to declare.

References

Abramitzky, R., Boustan, L.P., Eriksson, K.: A nation of immigrants: assimilation and economic outcomes in the age of mass migration. J. Polit. Econ. **122**(3), 467–506 (2014). https://doi.org/10.1086/675805

Best, R.A.: Intelligence issues for congress. CRS Report RL33539. Office of Congressional Information and Publishing, Washington, DC (2010)

Box, G.E., Cox, D.R.: An analysis of transformations. J. R. Stat. Soc. Ser. B **26**(2), 211–252 (1964). https://www.jstor.org/stable/2984418

Brunner, L., Colarelli, S.: Immigration in the twenty-first century: a personal selection approach. Indep. Rev. **14**(3), 389–413 (2010). https://www.jstor.org/stable/24562880

Chassamboulli, A., Peri, G.: The economic effect of immigration policies: analyzing and simulating the U.S. case. J. Econ. Dyn. Control **114**(1), 103898 (2020). https://doi.org/10.1016/j.jedc.2020.103898

Cryer, J., Chan, K.: Time Series Analysis with Applications in R, 2nd edn. Springer, New York (2008). https://doi.org/10.1007/978-0-387-75959-3

Gelfand, D.,Yee, B.W.K.: Trends and forces: influence of immigration, migration, and acculturation on the fabric of aging in America. Generations **15**(4), 7–10 (1991). http://www.jstor.org/stable/44877745

Grieco, E.M., Larsen, L.J., Hogan, H.: How period data influence the estimates of recently arrived immigrants in the American community survey. Int. Migr. Rev. **52**(1), 299–313 (2018). https://doi.org/10.1111/imre.12296

Fazel-Zarandi, M.M., Feinstein, J., Kaplan, E.: The number of undocumented immigrants in the United States: estimates based on demographic modeling with data from 1990 to 2016. PLoS ONE **13**(9), 1–11 (2018). https://doi.org/10.1371/journal.pone.0201193

Feigenberg, B.: Fenced out: the impact of border construction on U.S.-Mexico migration. Am. Econ. J. Appl. Econ. **12**(3), 106–139 (2020). https://doi.org/10.1257/app.20170231

Fernald, J.G., Jones, C.I.: The future of US economic growth. Am. Econ. Rev. **104**(5), 44–49 (2014). https://doi.org/10.1257/aer.104.5.44

Foged, M., Peri, G.: Immigrants' effect on native workers: new analysis on longitudinal data. Am. Econ. J. Appl. Econ. **8**(2), 1–34 (2016). https://www.jstor.org/stable/24739100

Kaushal, N., Lu, Y., Denier, N., Wang, J.-H., Trejo, S.J.: Immigrant employment and earnings growth in Canada and the USA: evidence from longitudinal data. J. Popul. Econ. **29**(4), 1249–1277 (2016). https://doi.org/10.1007/s00148-016-0600-5

Mayda, A.: International migration: a panel data analysis of the determinants of bilateral flows. J. Popul. Econ. **23**(4), 1249–1274 (2010). https://doi.org/10.1007/s00148-009-0251-x

Ortega, F., Peri, G.: The effect of income and immigration policies on international migration. Migr. Stud. **1**(1), 47–74 (2013). https://doi.org/10.1093/migration/mns004

R Core Team: R: A language and environment for statistical computing. R Foundation for Statistical Computing, Vienna (2020)

Trost, M., et al.: Immigration: analysis, trends and outlook on the global research activity. J. Glob. Health **8**(1), 1–11 (2018). https://doi.org/10.7189/jogh.08.010414

United States Census Bureau: Statistical Abstract of the United States, 127th edn. United States Census Bureau, Washington, DC (2008)

Warren, R., Passel, J.S.: A count of the uncountable: Estimates of undocumented aliens counted in the 1980 United States census. Demography **24**(3), 375–393 (1987). https://doi.org/10.2307/2061304

Warren, R., Warren, J.R.: Unauthorized immigration to the United States: annual estimates and components of change, by state, 1990 to 2010. Int. Migr. Rev. **47**(2), 296–329 (2018). https://doi.org/10.1111/imre.12022

Public Health Predictive Analysis of Chicago Community Areas: A Data Mining Approach

D. Akoto[1,2] and R. N. A. Akoto[3(✉)]

[1] Department of Mathematics and Statistics, Villanova University, Villanova, USA
[2] 421 Multi-Functional Medical Battalion, United States Army, Baumholder, Germany
[3] School of Graduate Studies, University of Professional Studies, Accra, Ghana
nii.ayitey-akoto@upsamail.edu.gh

Abstract. Purpose: This paper utilizes data mining to assist policy makers to better understand the overall health of communities in Chicago area using several public health indicators. The work utilizes regression analysis to establish relationships between social, economic and heath variables.
Design/Methodology/Approach: The main goal of the basic analysis was to identify several variables of interest for further investigation by multiple regression analysis. A correlation matrix from R was used to visualize associations between all independent variables and the dependent variables of interest for this study. The Akaike information criterion (AIC) value was then determined using backward variable selection. To classify the Chicago community areas according to similarities, k-means clustering was utilized in R. The data was transformed into a matrix and scaled.
Findings: The study found that socio-economic factors such as unemployment and crowded housing contribute to the increase in teen birth rate in Chicago community areas. This indicates that financial problems due to unemployment could lead to teenage pregnancies. The study reveals that assault, cancer, diabetes, and infant mortality all contribute to the increase in death rates in Chicago communities. In addition, unemployment and having no high school diploma is associated with communities being rated below the poverty line.
Research Limitations: The challenge in obtaining spatial statistics data for the communities was a major limitation. The spatial data affords a good way to do clustering and visualize the similarities in the communities.
Social Implications: For sustainable industrialization within the community, issues of public health are central. This will impliedly require health sector officials to focus more on the significant health issues and educate residents.
Originality: The innovation of the use of a simple data mining technique to assess the public health of communities in the Chicago area is unique to other methods employed in the literature.

Keywords: Community · Health · Mortality · Poverty · Public · Regression

1 Introduction

A person's health is determined by social and economic factors, the environment in which they live, their behavior, and the quality of health care they are able to access (Chrisman et al. 2015; Meit 2018). Several public health indicators can be used as a comprehensive measure of the overall health of a community or city (Blanchard et al. 2012; Institute of Medicine (US) 1988; Young et al. 2015). Healthy Chicago is a broad coalition led by the Chicago Department of Public Health (CDPH) that's taking important steps for a more just, equitable Chicago (Chicago Department of Public Health 2020). In order to monitor life expectancy gap, the city has been working on how to measure and monitor the city's progress. As part of achieving this goal, the CDPH acknowledged a set of principal indicators to measure significant outcomes in various communities in Chicago. In order to reach the "Healthy Chicago 2025" goal, there should exist comprehensive measures of health. The CDPH hope to measure and monitor the city's progress using public health indicators. This project utilizes data mining techniques in order to assist policy makers to better understand the overall health of communities in Chicago area using several public health indicators.

Data mining and predictive analysis is an emerging field that will assist statisticians in exploring larger amounts of data and understand the hidden patterns in the data. Data mining techniques help solve problems by obtaining relevant information from large amounts of data. It has been used in the areas of public health and medical informatics for statistical analysis, recognition of important patterns and the prediction of information as can be seen in the works of Bellinger et al. (2017), Furst et al. (2016), Lai and Stone (2020), Shirzad et al. (2021), Thongkam et al. (2015) and Wold (2015).

Application Program Interfaces (API), new digital apps, and IoT (Internet of Things) goods have emerged as relatively recent data sources for developing new applications or integrated solutions. To access, read, and handle data, nevertheless, requires highly technical abilities (Lai and Stone 2020). In the last few years, application program interfaces (API), new digital apps, and IoT (Internet of Things) products have emerged as relatively new data resources. These sources demand highly technical abilities to access, read, and handle information. There are also a wide range of survey methodologies and sources of data that are available to offer more detailed data on variables associated to public health. If you look at public health surveys at the city level, the data only includes variables related to the health of households and socioeconomic status (Lai and Stone 2020). By using data mining and analytics, public health issues can be quantified, visualized, and promoted (Ghosh and Guha 2013). For a comprehensive review, see Hawn (2009), Hou et al. (2013), Schneider et al. (2017), Seidahmed et al. (2018) and Ransome et al. (2019) for all works that involve geotagged social media data and analytics addressing citizen complaints.

To the best of our knowledge, no research work has utilized data mining technique to establish relationships between variables to analyse public health in the Chicago areas. The Chicago public health data contains several relevant social, economic, and health variables that have been of interest to the Chicago Department of Public Health. It is hypothesized that these variables are associated to each other to some degree. This paper seeks to utilize data mining techniques such as regression analysis to establish relationships between variables socio-economic and poverty level, socio-economic factors and teen birth rate, health factors and mortality.

2 Methodology

2.1 Data

The dataset contains the most recent information on 27 public health indicators by Chicago community areas. As indicators, we use rates, percentages, and other measures relating to economic status, infectious diseases, lead poisoning, mortality and natality. The rates are computed per 1000 residents. This data is obtained from the Epidemiology & Public Health Informatics Program of the Chicago Department of Public Health. The adjusted death rate dataset from the CDPH website was also merged with the initial dataset.

2.2 Multiple Linear Regression

The main goal of the basic analysis was to identify several variables of interest for further investigation by multiple regression analysis. Initially, all the missing values in the dataset were deleted and not estimated. A correlation matrix from R (R Core Team 2020) was used to visualize associations between all independent variables and the dependent variables of interest for this study. Two-dimensional scatter plots were also used to visualize correlations between variables of interest before model fitting.

For the first model, five socio-economic indicators were selected from the dataset as explanatory variables to describe teen birth rate (the dependent variable). These variables are dependency, no high school diploma, unemployment, below poverty level, and crowded housing. The associations for the independent variables were also assessed in order to prevent multicollinearity in the model. Following this, the Akaike Information Criterion (AIC) was used to select the best simple model with the lowest AIC. In order to check the model accuracy, the cross-validation method was utilized. We split the data into two categories: training and testing. Both complex and simple models were fit using the training dataset and the test dataset. The test Mean Square Error (MSE) values were computed for both models using the test dataset. The model with the lowest AIC and also the lowest test MSE was selected as the best model. The plot function in R was used to create residual plots to assess the final model accuracy.

For the second model, eight health indicators were selected from the dataset. The variables include assault resulting in homicide, birth rate, metastatic cancer, diabetes, firearm related, infant mortality rate, stroke, and tuberculosis. These variables were selected as independent variables to explain the adjusted death rate in the various Chicago communities. Again, backward variable selection and cross validation was used to select the best simple model. Residual analysis in R was also used to check model assumptions.

In addition, three socio-economic variables were selected as independent variables to explain why communities fall within the 'below poverty level' status. The independent variables selected were dependency, no high school diploma and unemployment. The AIC, R-squared, adjusted R-squared, p-values were all assessed. The plot function in R was used to create the residual plots to check model assumptions.

2.3 Clustering

An R script was used to classify Chicago community areas according to similarities based on k-means clustering. An analysis of the data was conducted by transforming it into a matrix and scaling it. The transformation reduces the impact of outliers and allows to compare a sole observation against the mean. The k-means function in R was then used with 3 clusters from the elbow point and n-start is equal to 25. The fviz_cluster() from the factoextra package in R was used to visualize k-means clusters. The cluster means and the communities that fall within each cluster were obtained.

3 Results and Discussion

3.1 First Model Results

The stepAIC function produced a simple model that contains unemployment and crowded housing as the independent variables. The two-dimensional scatterplot shows a positive correlation between unemployment and teen birth rate. There is also a positive correlation between teen birth rate and crowded housing. From Table 1, both unemployment and crowded housing appear to be significant with p-values less than 0.0001 (Fig. 1).

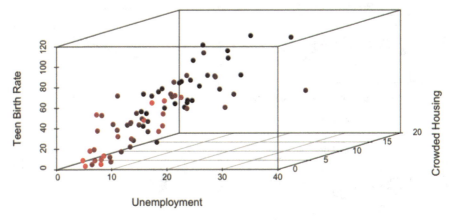

Fig. 1. Two-dimensional scatterplot of the first final model.

The residual plots appear to satisfy the normality assumption, homoscedasticity, and linearity assumptions. Also, the AIC and test MSE values for the simple model are lower than that of the complex model indicating that the simple model is a better model (Fig. 2).

Table 1. Output for first final model

Residuals:

Min	1Q	Median	3Q	Max
-50.508	-12.8540	-1.034	11.777	39.226

Coefficients:

	Estimate	Std. Error	t value	Pr(>|t|)
(Intercept)	1.6710	4.7708	0.350	0.727
unemployment	2.6562	0.2864	9.273	5.73e-14***
crowded housing	2.6200	0.5618	4.663	1.37e-05***

Signif. Codes: 0 '***' 0.001 '**' 0.01 '*' 0.05 '.' 0.1 ' ' 1

Residual standard error: 17.36 on 73 degrees of freedom
Multiple R-squared   :   0.6286
Adjusted R-squred   :   0.6185
F-statistic   :   61.78 on 2 and 73 DF
p-value   :   < 2.2e-16

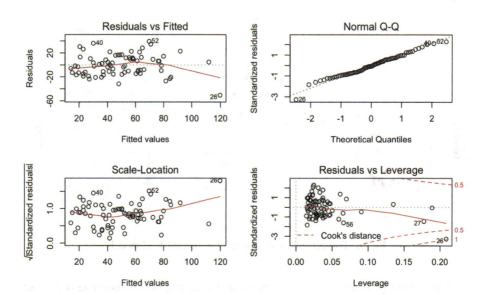

Fig. 2. Residuals for first final model

The first complex model has an AIC of 441.48 with a test error from cross validation of 509.498 and stated as

$$\text{Teen birth rate} = \alpha_0 + \alpha_1(\text{dependency}) + \alpha_2(\text{no high school diploma}) \\ + \alpha_3(\text{unemployment}) + \alpha_4(\text{below poverty level}) \\ + \alpha_5(\text{crowded housing}), \quad (1)$$

with its final model having an AIC of 436.74 with a test error from cross validation of 439.333 and stated as

$$\text{Teen birth rate} = \beta_0 + \beta_1(\text{unemployment}) + \beta_2(\text{crowded housing}), \quad (2)$$

where α_0 and β_0 are constants and α_i and β_j, are coefficients with $i = 1, 2, \ldots, 5$ and $j = 1, 2$.

3.2 Second Model Results

The stepAIC function in R produced the final model with assault, cancer, diabetes, and infant mortality as independent variables. Cancer and diabetes related have p-values less than 0.001 which indicates that they are significant in the model. Assault and infant mortality are also significant in the model. All the independent variables appear to be positively associated with adjusted death rate. The AIC and test MSE values are lower for the simple model compared to the complex model indicating that the simple model is better. The residual plots appear to satisfy the normality, homoscedasticity, and linearity assumptions.

The second complex model has an AIC of 621.77 with a test error from cross validation of 9822.347 and stated as

$$\text{Adjusted death rate} = \pi_0 +, \pi_1(\text{birth rate}) + \pi_2(\text{assult}) + \pi_3(\text{cancer}) + \pi_4(\text{diabetes}) \\ + \pi_5(\text{firearm related}) + \pi_6(\text{infant mortality}) + \pi_7(\text{stroke}) \\ + \pi_8(\text{tuberculosis}), \quad (3)$$

with its final model having an AIC of 617.84 with a test error from cross validation of 4776.304 and stated as

$$\text{Adjusted death rate} = \gamma_0 + \gamma_1(\text{assult}) + \gamma_2(\text{cancer}) + \gamma_3(\text{diabetes}) \\ + \gamma_4(\text{infant mortality}), \quad (4)$$

where π_0 and γ_0 are constants such that $\pi_0 \neq \gamma_0$ with the coefficients π_k and γ_r having indexes $k = 1, 2, \ldots, 8$ and $r = 1, 2, \ldots, 4$ (Table 2).

Table 2. Regression output for second final model.

Residuals:

Min	1Q	Median	3Q	Max
-116.214	-39.487	-2.965	38.157	139.171

Coefficients:

	Estimate	Std. Error	t value	Pr(>|t|)
(Intercept)	202.6547	36.2607	5.589	3.97e-07***
assault	2.6765	0.7229	3.703	4.19e-04***
cancer	1.6219	0.2562	6.330	1.94e-08***
diabetes	3.0667	0.469	6.539	8.13e-09***
infant mortality	5.3990	2.6779	2.016	4.7568e-02*

Signif. Codes: 0 '***' 0.001 '**' 0.01 '*' 0.05 '.' 0.1 ' ' 1

Residual standard error:	56.43 on 71 degrees of freedom
Multiple R-squared :	0.9209
Adjusted R-squred :	0.9164
F-statistic :	206.6 on 4 and 71 DF
p-value :	< 2.2e-16

The two-dimensional scatterplot in Fig. 4 depicts that there is a positive correlation between 'below poverty level' and 'no high school diploma' and also with 'unemployment'. Additionally, the scatterplot portrays a negative correlation between 'below poverty level' and 'dependency'. Table 3 shows that all the explanatory variables are significant in explaining the variations in the 'below poverty level' status by the community. The coefficients for 'unemployment' and 'no high school diploma' are positive indicating a positive association with 'below poverty level'. However, the coefficient for dependency indicates that there is a negative relationship between 'below poverty level' and 'dependency'. The residual plots also appear to satisfy the normality, homoscedasticity, and linearity assumptions. Based on correlation matrix for the explanatory variables we proceed to write a simple model for 'below poverty level' as functions of 'dependency', 'no high school diploma' and 'unemployment' as

$$\text{Below poverty level} = \phi_0 + \phi_1(\text{dependency}) + \phi_2(\text{no high school diploma}) + \phi_3(\text{unemployment}) \tag{5}$$

where ϕ_0 is a constant and the coefficients ϕ_l have index $l = 1, 2, 3$ (Fig. 3 and 5)

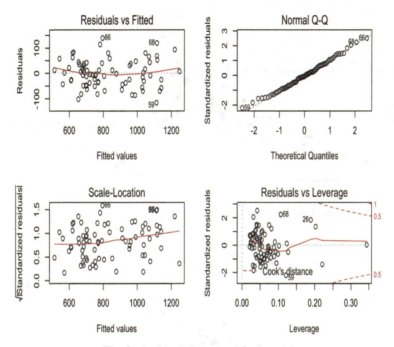

Fig. 3. Residuals for second final model

3.3 First Model Results

From the cluster analysis in Fig. 6, clusters 1, 2, and 3 contains 21, 27, and 15 observations respectively. Cluster number 2 also appear to have the largest 'within sum of squares'. From Table 4, the cluster means within the clusters for the sample variables selected appear to be close. However, the cluster means differ for the variables as we move from one cluster to the other. Some community areas that appear to be in the same cluster are Beverly, Bridgeport, and Clearing which belong to cluster 3. This indicates that these communities have some type of similarity based on the observations in the data.

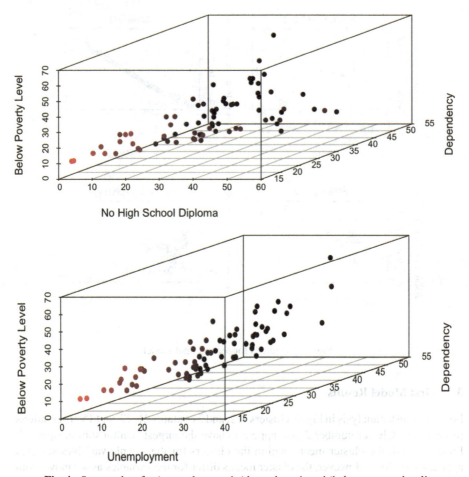

Fig. 4. Scatter plots for 'unemployment', 'dependency', and 'below poverty level'

Clusters 1, 2 and 3 have 'within sum of squares' of 243.5973, 391.5041 and 139.1965 respectively (between$_{SS}$/total$_{SS}$ = 52.0%).

The results show that socio-economic factors such as 'unemployment' and 'crowded housing' contribute to the increase in 'teen birth rate' in Chicago community areas. This indicates that financial problems due to unemployment could lead to teenage pregnancies. Also, being crowded in a household explains why teens spend more time outdoors, thereby contributing to the prevalence of teenage pregnancy. The study reveals that assault, cancer, diabetes, and infant mortality all contribute to the increase in death rates in Chicago communities. This means the health sector officials should focus more on those health issues and educate residents. In addition, unemployment and having no high school diploma is associated with communities being rated below the poverty line. Finally, communities such as Beverly, Bridgeport, and Clearing appear to have similar characteristics. Albany Park, East Side, and Hermosa have some similarities in terms of public health indicators.

Public Health Predictive Analysis of Chicago Community Areas 211

Table 3. Output from regression analysis

Residuals:

Min	1Q	Median	3Q	Max
-16.3535	-4.271	0.0415	2.872	27.6051

Coefficients:

	Estimate	Std. Error	t value	Pr(>|t|)
(Intercept)	8.6339	4.18536	2.063	4.273e-02*
dependency	-0.2940	0.14546	-2.021	4.698e-02*
no high school diploma	0.2096	0.07276	2.881	5.220e-03**
unemployment	1.3168	0.14416	9.135	1.17e-13***

Signif. Codes: 0 '***' 0.001 '**' 0.01 '*' 0.05 '.' 0.1 ' ' 1

Residual standard error: 6.992 on 72 degrees of freedom
Multiple R-squared   :   0.6441
Adjusted R-squred   :   0.6293
F-statistic   :   43.44 on 3 and 72 DF
p-value   :   3.88E-16

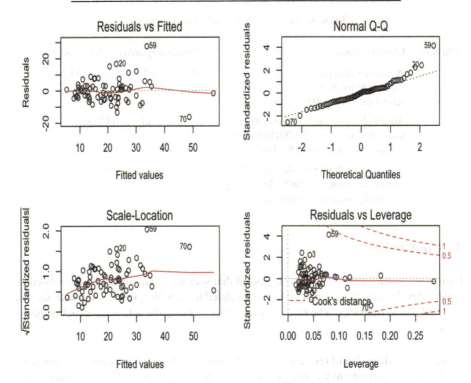

Fig. 5. Residual analysis third model.

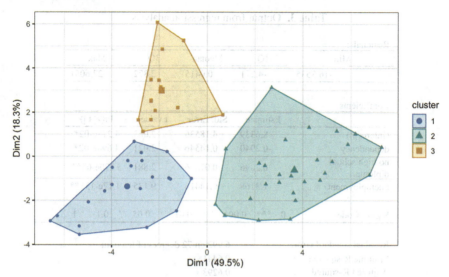

Fig. 6. Plot showing clusters

Table 4. Summary table from cluster analysis

Clusters	Cluster means	Community areas within cluster
1	Adjusted death rate = 1.1233162 unemployement = 1.0620337 no high school diploma = 0.02529301	Montclaire New city North Lawndale
2	Adjusted death rate = −0.6086666 unemployement = −0.8001857 no high school diploma = −0.73965987	Beverly Bridgeport Clearing
3	Adjusted death rate = −0.6086666 unemployement = −0.8001857 no high school diploma = 1.30862029	Albany East Side Hermosa

4 Conclusion

Using regression analysis, it was established that socioeconomic factors were related to poverty levels, socioeconomic factors were related to teen birth rates, and health factors were related to mortality. The results of the analysis bring the following conclusions to light;

- An increased number of residents without high school diploma and a high rate of unemployment contribute to an increase in poverty levels within the Chicago community areas.
- Teenage pregnancy rates and births are increasing due to high rates of unemployment and overcrowded housing.

- The study also concludes that an increase in assault cases, cancer cases, diabetes-related cases and infant mortality rate contribute to the increased reportage of deaths in the Chicago communities.

Therefore, policymakers should increase their efforts to provide employment opportunities and educate residents about the importance of completing high school. These afford sustainable industrialization and innovation within the community. Chicago's overall health depends on a combination of socio-economic and health factors, both of which were discussed in this study. In the last instance, an investigation of the spatial statistics of the communities would be interesting. This would provide a means for analysing clusters and for enhancing the visual representation of differences and similarities among the communities.

Conflict of Interest. Conflict of interest declaration: neither author declares a conflict of interest.

References

Bellinger, C., Mohomed Jabbar, M., Zaïane, O., Osornio-Vargas, A.: A systematic review of data mining and machine learning for air pollution epidemiology. BMC Public Health **17**(1), 907 (2017). https://doi.org/10.1186/s12889-017-4914-3

Blanchard, T.C., Tolbert, C., Mencken, C.: The health and wealth of US counties: how the small business environment impacts alternative measures of development. Cambridge J. Reg. Econ. Soc. **5**(1), 149–162 (2012)

Chicago Department of Public Health, 2020.Chicago Department of Public Health: Healthy Chicago 2025: closing our life expectancy gap 2020-2025. Chicago Department of Public Health, Chicago (2020)

Chrisman, M., Nothwehr, R., Yang, G., Oleson, J.: Environmental influences on physical activity in rural midwestern adults: a qualitative approach. Health Promot. Pract. **16**(1), 142–148 (2015). https://doi.org/10.1177/1524839914524958

Furst, J., Raicu, D.S., Jason, L.A.: Data mining. In: Jason, L.A., Glenwick, D.S. (eds.) Handbook of Methodological Approaches to Community-Based Research: Qualitative, Quantitative, and Mixed Methods, pp. 187–196. Oxford University Press, New York (2016). https://doi.org/10.1093/med:psych/9780190243654.003.0019

Ghosh, D., Guha, R.: What are we 'tweeting' about obesity? mapping tweets with topic modeling and geographic information system. Cartogr. Geogr. Info. Sci. **40**(2), 90–102 (2013). https://doi.org/10.1080/15230406.2013.776210

Hawn, C.: Take two aspirin and tweet me in the morning: how twitter, facebook, and other social media are reshaping health care. Health Aff. (Millwood) **28**(2), 361–368 (2009). https://doi.org/10.1377/hlthaff.28.2.36

Hou, D., Song, X., Zhang, G., Loaiciga, H.: An early warning and control system for urban, drinking water quality protection: China's experience. Environ. Sci. Pollut. Res. **20**, 4496–4508 (2013). https://doi.org/10.1007/s11356-012-1406-y

Institute of Medicine (US): The Future of Public Health. National Academies Press (US), Washington (DC) (1988)

Lai, Y., Stone, D.J.: Data integration for urban health. In: Celi, L.A., Majumder, M.S., Ordóñez, P., Osorio, J.S., Paik, K.E., Somai, M. (eds.) Leveraging data science for global health, pp. 351–363. Springer, Cham (2020). https://doi.org/10.1007/978-3-030-47994-7_21

Meit, M.: Exploring Strategies to Improved Health and Equity in Rural Communities. The Walsh Center for Rural Health Analysis, Chicago (2018)

Ransome, Y., Luan, H., Shi, X., Duncan, D.T., Subramanian, S.V.: Alcohol outlet density and area-level heavy drinking are independent risk factors for higher alcohol-related complaints. J. Urban Health **96**(6), 889–901 (2018). https://doi.org/10.1007/s11524-018-00327-z

R Core Team: R: A language and environment for statistical computing. R Foundation for Statistical Computing, Vienna (2020)

Schneider, P., Castell, N., Vogt, M., Dauge, F.R., Lahoz, W.A., Bartonova, A.: Mapping urban air quality in near real-time using observations from low-cost sensors and model information. Environ. Int. **106**, 234–247 (2017). https://doi.org/10.1016/j.envint.2017.05.005

Seidahmed, O.M., Lu, D., Chong, C.S., Ng, L.C., Eltahir, E.A.: Patterns of urban housing shape dengue distribution in Singapore at neighborhood and country scales. GeoHealth **2**(1), 54–67 (2018). https://doi.org/10.1002/2017GH000080

Shirzad, E., Ataei, G., Saadatfar, H.: Applications of data mining in healthcare area: a survey. Eng. Appl. Sci. Res. **48**(3), 314–323 (2021). https://doi.org/10.14456/easr.2021.34

Thongkam, J., Sukmak, V., Mayusiri, W.: A comparison of regression analysis for predicting the daily number of anxiety-related outpatient visits with different time series data mining. Eng. Appl. Sci. Res. **42**(3), 243–249 (2015). https://doi.org/10.14456/kkuenj.2015.26

Young, R., Willis, E., Stemmle, J., Rodgers, S.: Localized health news releases and community newspapers: a method for rural health promotion. Health Promot. Pract. **16**(4), 492–500 (2015). https://doi.org/10.1177/1524839915580538

Wold, C.: In plain sight: Is open data improving our health. California Healthcare Foundation, California (2015)

Simulation-Based Exploration of Daylighting Strategies for a Public Basic School in a Hot-Dry Region of Ghana

J. T. Akubah[1(✉)], S. Amos-Abanyie[2], and B. Simmons[3]

[1] Department of Building Technology, Tamale Technical University, Tamale, Ghana
jtdamah@gmail.com, ajtwumwaa@tatu.eu.gh

[2] Department of Architecture, Kwame Nkrumah University of Science and Technology, Kumasi, Ghana

[3] Department of Construction, Technology and Management, Kwame Nkrumah University of Science and Technology, Kumasi, Ghana

Abstract. Purpose: This study aims to evaluate a full year's indoor illuminance distribution and visual comfort conditions of a daylight public basic school located in a hot-dry region by means of an experimental simulation and a subjective response survey.

Design/Methodology/Approach: A mixed method technique was adopted. Climate-based daylighting modelling metrics for assessing daylight availability and quality were employed to assess percentage of time over the year where internationally acceptable ranges are reached (UDI $^{300lux-500lux}$; DA^{300lux}; ASE1000, 250 h of 3%). Mean values and standard deviations of questionnaire responses under different fenestrations and weather conditions were analysed at the scale of a single classroom and groups of classrooms.

Findings: UDI and DA values for the rainy season and the dry season for all fenestrations fell below the benchmark values from the simulation (UDI $^{300lux-500lux}$; DA^{300lux}). This was corroborated by the results from the subjective survey. F3 which was a casement window and had the highest WWR recorded its highest mean UDI and DA values as 43.01 lux and 38.40 lux respectively for the dry season. Similar results were recorded for F1 which recorded the lowest mean UDI (42.81 lux) and DA (42.59 lux) values.

Research Limitation: This study involved primarily three fenestration types. A further inquisition in other distinct climatic regions will provide more information on the impact of the choice of fenestration and visual comfort conditions of other public basic schools.

Practical Implication: The study highlights the poor lighting conditions in this school over a long-term assessment. A thorough assessment of annual lighting conditions to highlight conditions is required in daylight public schools. A guideline for fenestration type selection for these schools will be required based on the climatic conditions.

Social Implication: The findings of this study are essential for designers, policy makers and other stakeholder institutions to fully understand the effect of the

© The Author(s), under exclusive license to Springer Nature Switzerland AG 2023
C. Aigbavboa et al. (Eds.): ARCA 2022, *Sustainable Education and Development – Sustainable Industrialization and Innovation*, pp. 215–233, 2023.
https://doi.org/10.1007/978-3-031-25998-2_17

choice of fenestration and climatic factors on a properly daylight educational space and examine the potential of energy and cost savings.

Originality/Value: This study explored lighting quality from the perspective of simulated lighting quantities and subjective assessments from participants within the simulated classrooms on a long-term basis, a very rare type of study in Ghana. The study adds on to the body of literature in daylighting studies in the country.

Keywords: Daylight · Hot-dry · Visual comfort · Strategies · School

1 Introduction

The impact of daylighting on the educational experience and performance of students cannot be disputed. Overwhelming literature supports the fact that daylighting can provide psychological and physical healing, reduction of stresses and anxieties and many other benefits for the occupants of learning spaces (El-darwish & Gendy 2016; Fitriaty et al. 2019; Leccese et al. 2020). Other studies have also revealed a very strong impact on the human circadian system (Mien et al. 2014; Konis 2017; Spitschan 2019; Bellia et al. 2020) which is the daylight and nighttime prediction system of the human body depending mainly on the spectral distribution of the light reaching the eyes. Daylight suppresses the hormone melatonin which facilitates sleep and as the body gets filled with the presence of daylight, the effect of the melatonin is reduced, giving the body the ability to stay awake and alert, an essential attribute for students (Konis 2019).

Metrics for measuring daylight can be grouped depending on the daylighting sufficiency or insufficiency issues, their measured effects as illuminance and luminance based metrics (Costanzo et al. 2017) or in terms of the time frame of measurement as static (point-in-time) and dynamic (annual) metrics (Zomorodian and Tahsildoost 2019a, b). Static daylight metrics such as illuminance and Daylight Factor (DF) are instantaneous and are useful for assessing daylight at a single point in time but not to determine daylight distribution over a long-time span. Various authors for instance have noted that DF is static as a parameter, does not consider variation of illuminance with time, does not describe the effects of direct sunlight and because it does not consider non-overcast sky conditions, does not differentiate among different window exposures. This often leads to oversized glazing or windows which culminates in glare, thermal comfort and other issues (Moreno & Labarca 2015; Costanzo et al. 2017; Wienold et al. 2019).

Climate based daylight metrics (CBDM) for measuring are based on time series of illuminances or luminance instead of a fixed condition. They are also based on the local climate data series including daily and seasonal variations of daylight available in the area of study (Nocera et al. 2018) and considers geometry and optical properties of materials within the space under study (Mardaljevic et al. 2009). Commonly used CBDMs include Useful Daylight Illuminance (UDI), Daylight Autonomy (DA), Continuous daylight autonomy (DAcon), Spatial Daylight Autonomy (sDA) and Annual Sunlight Exposure (ASE).

UDI was originally defined as the portion of time in a year when indoor horizontal daylight illuminance at a given point fell within a given range usually at 100–2000 lux (Nabil and Mardaljevic 2005), also coined as "UDI Achieved" but this was later revised to counter for situations below 100 lux and exceeding 3000 lux (Mardaljevic et al. 2009). The ranges were divided into three bins; an upper bin (percentage of time with excessive daylight illuminance), a lower bin (percentage of time with very scarce daylight illuminance) and an intermediate bin which depicts the percentage of time when an appropriate amount of daylight illuminance was attained to ensure compliance at important set intervals (Nocera et al. 2018).

Other intervals of UDI were introduced by Mardaljevic et al. (2012) they included; UDI "fell-short" (UDI-f) when illuminance falls below 100 lux; UDI supplementary (UDI-s), when illuminance is greater than 100 lux but less than 300 lux; UDI autonomous (UDI-a) for when illuminance is greater than 300 lux but less than 3000; lux UDI combined (UDI-c), when illuminance is greater than 100 lux but less than 3000 lux; and UDI exceeded (UDI-e), when illuminance is greater than 3000 lux.

Originally initiated by the Swiss Association of Electricians in 1989 and adapted by Reinhart et al. (2006), Daylight Autonomy (DA), the first CBDM is defined as the measure of daylight availability in terms of how many hours per year a predefined minimum illuminance level is achieved (Lorenz et al. 2018). It allows for annually forecasting of daylight over a given portion of space or grid for a specific percentage of time during the day where the required visual task uses only daylight (Reinhart et al. 2006). DA provides the percentage of the occupied hours that exceeds a minimum illuminance threshold (for example, 300 lux for a classroom) so that one can determine if there is adequate daylight or supplementary light is needed for the assigned task. Results from studies (Mardaljevic et al. 2012; Sepúlveda et al. 2020a) and internationally approved standards (U.S. Green Building Council, 2017; Standard IES LM-83–1, 2012) propose a DA of 300 lux as a threshold with DA values over 300 lux counted as excessive light levels (Lee et al. 2016).

As DA does not consider times where there is too much daylight, there is the probability that a 100% autonomy could be achieved but with issues of visual discomfort. Although UDI-e, when illuminance is greater than 3000 lux can be used for the threshold for visual comfort, the IESNA further suggests an evaluation of the Annual Sun Exposure (ASE) to describe the potential for visual comfort. The ASE accounts for direct sunlight (without ambient light) which is a potential source of excessive sunlight penetration and can cause visual discomfort (Kong & Jakubiec 2021). It is the percentage of occupied area in the space of study where direct illuminance exceeds 1000 lux for a specific time (usually 250 number of hours per year) and calculated by taking into account the presence of blinds and shadings (Nezamdoost and Van Den Wymelenberg 2017b). According to IES LM-83-12 (2012), an area with an ASE1000, 250 h of 10% or greater presents "unsatisfactory visual discomfort", an ASE1000, 250 h of 7% or lower are judged to be "neutral" in terms of visual comfort, and an ASE1000, 250 h of 3% or lower are clearly acceptable.

1.1 Daylight in Public Basic Schools in Ghana

Public Basic schools in Ghana immensely depend on natural form of lighting for illumination. In terms of regulations of lighting standards for educational spaces, much of existing literature presents a long history in the United States, Europe and recently in the Asias. In Ghana, attempts have been made with regard to the regulation of visual comfort standards and energy efficiency in classrooms in some existing standards such as the Ghana National Building Regulation LI 1630 (1996) and the (Ghana-Building-Code 2018). These standards utilize point in time which are static in nature and do not present a long-term picture of daylight use in classrooms. There is also a very vast gap on research concerning dynamic daylight studies in Ghana especially with respect to how climatic conditions and fenestration types adopted affect daylighting levels in educational spaces.

The objective of this research is to evaluate a full year's indoor illuminance distribution and visual comfort conditions of a public basic school located in the Northern Region of Ghana, a hot-dry climatic region by means of an experimental simulation accompanied by a subjective response survey. There is a growing trend towards the use of simulation models to generate annual daylighting data and then use correlated occupant subjective assessments to compare with the simulation data (Kent et al. 2017; Bakmohammadi & Noorzai 2020; Leccese et al. 2020). This provides the opportunity to ascertain the results of the simulation data against the real responses of the occupants of the space. Many studies have had very effective and successful results from the use of this methodology (Mangkuto et al. 2018; Jakubiec et al. 2020; Shafavi et al. 2020; Zomorodian and Tahsildoost 2019b; Handina et al. 2017). This study will therefore adopt this methodology. The Climate based daylight metrics employed for the study include Useful Daylight Illuminance, Daylight Autonomy, continuous daylight autonomy and Annual Sunlight Exposure.

2 Methodology

The computational method for annual assessment was employed for evaluating the daylight performance of the Kanvili Roman Catholic Junior High School (KRC-JHS) block. The study also conducted field visits for observations and to conduct the subjective surveys. There were also visits for field measurements in order to collect physical parameters data such as classroom sizes, fenestration types and sizes, nature of walls, ceilings, floors and other contextual data. The study location is in the city of Tamale, Ghana which lies on latitude 9°27′17.94″N and longitude 0°50′30.00″W at an elevation of 542 feet. The area has abundant sunlight with partially cloudy skies with mean temperatures ranging from 22°C to 40°C. There are two major seasons, the wet rainy season (March-October) and the dry harmattan season (November–February).

2.1 Field Measurements

The Kanvili Roman Catholic Junior High School (KRC-JHS), Tamale is one block in a complex of school blocks housing Students from Kindergarten to Junior High School. The studied block houses only the JHS students (JHS1–JHS3). Figure 1 and 2 gives an aerial and perspective view of the school layout. A total of three classrooms out of six (JHS1A, JHS2B AND JHS3) were measured and studied. These classrooms were selected based on their location (Left End, Middle and Right end) and differences in

Fig. 1. Google Earth location of site   **Fig. 2.** Studied block

fenestration type for daylight penetration. The rest which were not selected shared similar characteristics with the selected ones and were deemed to have provided similar results.

Observation was used to collect information of some physical parameters while a tape measure was employed to determine dimensions. Characteristics that were considered include: length, breadth and height of the classrooms, wall to window ratio, dimensions of verandah and overhang, fenestration type, orientation and dimensions, presence of ceiling and presence of external features (other blocks, vegetation). According to Al-Sallal (2010) and (Atthaillah et al. 2021) these factors have significant impact on the daylight performance and quality inside classrooms. The orientation of each classroom was based on the cardinal direction of the school block which was North-South oriented with windows on the East and West sides. The block had a corridor shading each side with width 2.2 m measured from the classroom wall to an outer edge of the roof. This combination prevented direct sunlight penetration to the classrooms. There was no corridor shading or windows on the north and south sides of the block. There is a line of trees 6 m away from the fenestrations on the west side of the block. Figure 3, 4, 5, and 6 indicate the dimensions and physical characteristics of the studied blocks and classrooms.

Each classroom measured 8.14 m by 8.45 m with a double leafed door of width 1.6 m and height of 2.1 m. The different fenestration types and dimensions for each class are also indicated in Fig. 4. Students sat with their sides to the windows and facing the teaching board. For the purpose of this study, the calculation of the Wall-to-Window Ratio (WWR) included the ratio contribution of the door as it continually stood open as an extra source of daylight. JHS 3 had two sets of double leafed openable casement windows of height 1.2 m and width of 1.3 m (F3) with a door measuring 2.1 m by 1.2 m and a WWR of 3.7% for both the western and eastern sides. JHS 1A and 2B however, had one fixed metallic designed frame of height 0.9 m and width 2.1 m with circular openings within (F1). However, there were vertical and horizontal running bars around the circular openings for JHS 2B (F2). The doors to both classes which also were double leafed constantly open doors measured 2 m by 1.4 m. WWR of both JHS 2B and JHS 1A was 2.9% for both the eastern and western elevations.

The location of desk positions of students in the classroom was paramount to the study as this has great impact on the field of view and visual comfort of the student.

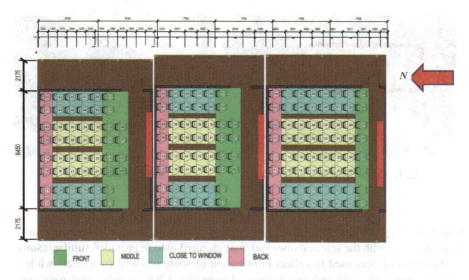

Fig. 3. Ground floor plan with seating Positions of students in studied classrooms

Classrooms were divided into four zones and labeled as; Close to window (for those on the sides), Middle, Back and Front. Figure 3 provides layouts and detailed information in this regard. Illuminance readings based on the various zones were taken at the time of responding to the survey. Though the classrooms had electrical wiring, there were no receptacles for providing light, therefore an indication that none of the studied classrooms used electrical lighting.

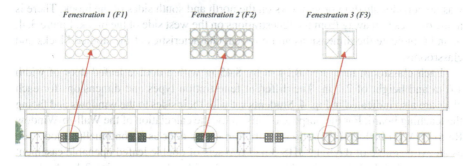

Fig. 4. Eastern elevation

2.2 Digital Modelling and Simulation

The study employed the use of Rhinoceros, a NURBS based software and Grasshopper for simulation. Accurately calibrated simulation models for the creation of annual lighting quantities within the three studied classrooms were employed. The 3D model of the whole block was built in Revit and exported to Rhino for lighting simulations.

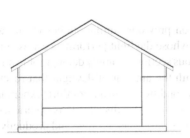
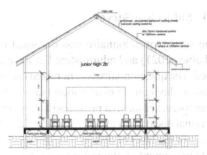

Fig. 5. View of North and South faces of school block

Fig. 6. Cross section through a classroom

The model was based on the field measurements taken including the surrounding environment. Annual climate EPW files were created for the location using parameters of latitude – 9.403; longitude - 0.842; Elevation – 151 m with the Honeybee plugin in Rhino and Grasshopper. The EPW was for the location of Tamale and the weather file was obtained from Google Earth. A horizontal grid of 1 × 1m was created in the model with the appropriate distance from the ground level of 0.5 m above the floor for the purpose of creating a work plane for suitable daylight assessment. The work plane is split into grid sensor points in which the calculation took place to create a standard model. Using the Grasshopper program, a parametric model was created, where the parameters could be easily changed in every simulation.

As reiterated by (Ayoub 2019), though the purpose of daylighting studies could differ in terms of scale and complexity, inputs for simulation to aid in evaluating and predicting outputs of the amount of daylight inside buildings are almost identical. They include weather dataset and sky models for more specific climate-based performance metrics, building geometry, daylight calculation method and daylight performance metrics. (Davoodi et al. 2017), also highlights inputs such as material properties, viewpoints/grid sensor points, space usage in terms of occupancy schedules and occupancy behaviour for a much more human centred analysis. Therefore in addition to the physical paramters, the occupancy schedule of the classrooms was taken into consideration and scheduled as 7 am to 3 pm to include times when after school classes were held. Ordinarily, classes started from 7 am and ended at 2 pm. The study was conducted from Monday to Friday as Saturdays and Sundays were weekends and was performed from January to December. The objective of this stage was to identify the daylight performance of the selected classrooms using the selected metrics UDI, DA, cDA and ASE based on acceptable levels. In addition, potential visual problems from the zoned areas and sitting positions within the classrooms were identified. The analysis was performed for the prevailing different fenestration types and climatic seasons. The simulated horizontal illuminances were compared to the measured horizontal illuminances for calibration.

2.3 Subjective Survey

The survey questionnaire was designed based on previous similar studies (Kong and Jakubiec 2021) and administered in the classes whose daylight performance was simulated. The questionnaire was itemized under various sections, namely demographic information, daylighting distribution (satisfaction with the amount of daylight that reaches your space, season that provides the best daylight conditions), visual comfort (experience of visual discomfort at certain times of the day, time of the day discomfort is experienced, rating of the level of discomfort, difficulty seeing in class when it rains) and productivity (rate agreement on the influence of the quality of daylight received during class on productivity, experience of difficulty reading in class when it rains). Students were given specific identification numbers based on their usual seating positions, which they wrote on their form to aid in the analysis of responses based on seating location. The purpose and instructions for the survey questionnaire was explained and read respectively to the students.

The survey questionnaire which took less than 10 min to fill out consisted of a mixture of multiple choice and Likert scale questions for the evaluation of their current and long-term daylighting experiences. This was to prevent disruption of classes beyond a minimal amount of time. Participants filled out the forms on hard copy printed papers as there was no access to multiple computers at the school premises which could be accessed for the purpose of this survey. There was no access to electric lighting at the time of the survey. The information was collected within the two seasons (rainy season and harmattan) under sunny and cloudy conditions during different daylit times.

3 Results and Discussion

Figures 7 and 8 present simulated outputs of UDI nad DA levels of illuminance. Table 1 highlights the summary statistics of UDI and DA in the rainy and wet seasons.

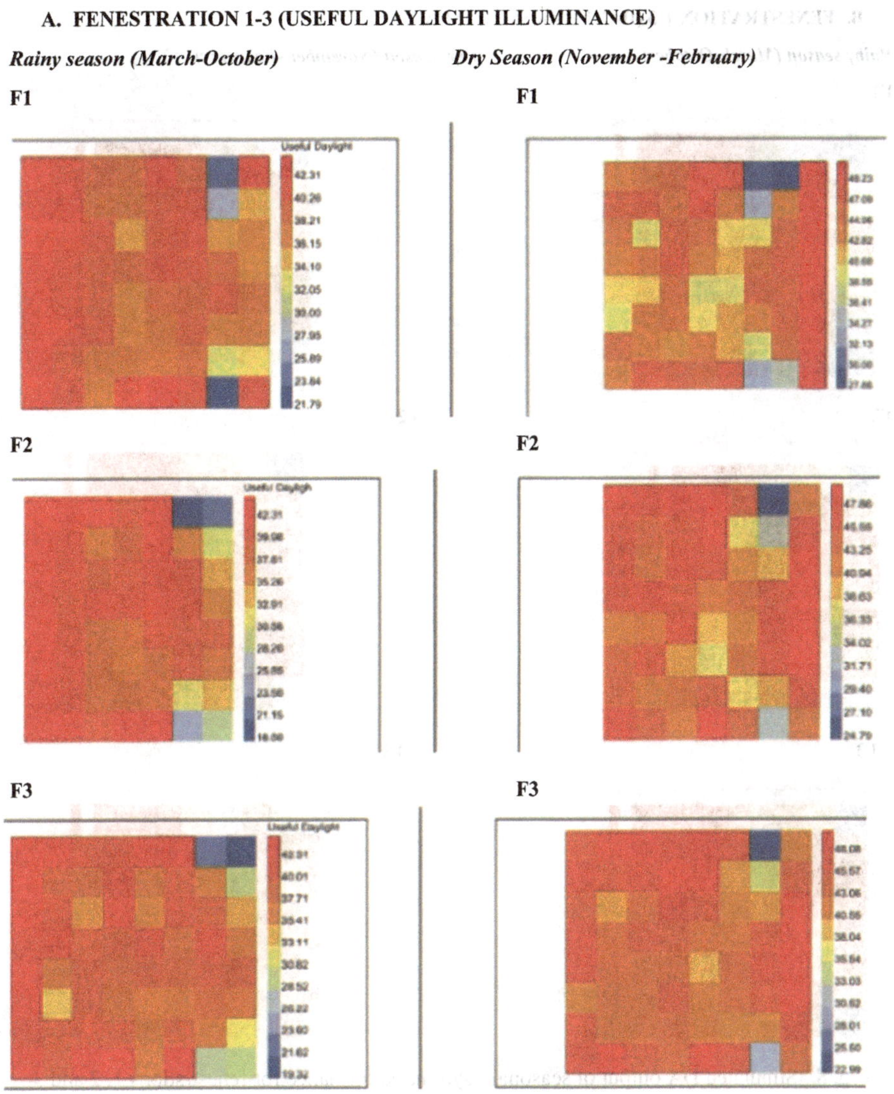

Simulation-Based Exploration of Daylighting Strategies 223

A. FENESTRATION 1-3 (USEFUL DAYLIGHT ILLUMINANCE)

Rainy season (March-October)   *Dry Season (November -February)*

F1   F1

F2   F2

F3   F3

Fig. 7. Simulated UDI output of seasonal daylight performance for fenestrstions 1, 2 and 3

The illuminance ranges for UDI for this study were set as; "too low" < 100 lux, "in range" ≥ 300 lux to < 3000 lux, and "too high" ≥ 3000 lux based on Mardaljevic et al. (2012), Nabil and Mardaljevic (2005). DA of 300 lux was counted as a threshold with DA values over 300 lux counted as excessive light levels. Table 2 presents the results of descriptive statistics of variables across the three fenestration types.

B. FENESTRATION 1-3 (DAYLIGHT AUTONOMY)

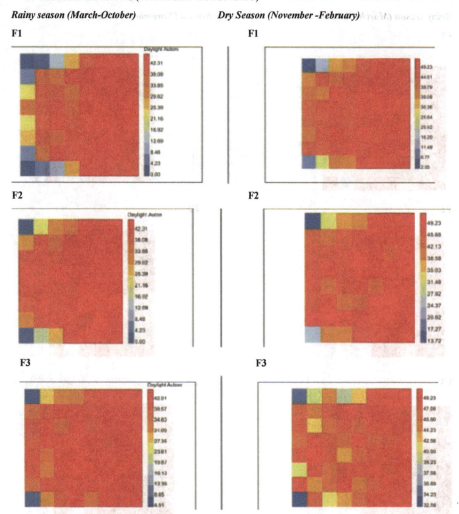

Fig. 8. Simulated DA output of seasonal daylight performance for fenestrstions 1, 2 and 3

Though, the simulated results for all the metrics fell below the threshold set by the standards adopted for the study, in terms of the subjective survey however, the variables which were analysed to represent the perception of the students on the issue of daylight distribution, revealed interesting results. Illuminance ranges for UDI and DA for all the three fenestration types, highlight, a poor distribution of daylight by the windows. This is further shown in Fig. 9 (Site recorded illuminances within the classified zones at the time of surveys). The Ghana National Building Regulation LI 1630 (1996) specifies that windows between north-east and south-west shaded by eaves or other projection of up to 1.2 m should have their entry of natural light being not less than 10% of the total floor area and the bottom of the opening not more than 1 m above floor level. F1, F2 and F3

Table 1. The result of the summary statistics of UDI and DA simulated in the rainy and dry seasons for F1, F2 and F3.

Metrics		F1	F2	F3
Useful Daylight Illuminance (UDI) in the Rainy Season	N	64	56	64
	Minimum	21.79	18.80	19.83
	Maximum	42.31	42.31	42.31
	Mean	38.1766	37.7529	37.4720
	Std. Deviation	4.07652	5.21818	4.61428
Useful Daylight Illuminance (UDI) in the Dry Season	N	64	56	64
	Minimum	27.18	24.79	22.99
	Maximum	49.23	47.86	48.08
	Mean	42.8170	42.8991	43.0148
	Std. Deviation	4.65928	4.38323	4.23052
Daylight Autonomy (DA) in the Rainy Season	N	64	56	64
	Minimum	.00	.00	1.20
	Maximum	42.31	42.31	42.31
	Mean	34.0400	37.3746	38.4028
	Std. Deviation	12.26933	9.10420	7.49313
Daylight Autonomy (DA) in the Dry Season	N	64	56	64
	Minimum	9.70	13.72	32.56
	Maximum	49.23	49.23	49.23
	Mean	42.5934	45.0495	46.3358
	Std. Deviation	9.60757	6.51861	3.50474

Table 2. Results of descriptive statistics of variables across the three fenestration types.

Fenestration type							
		F1		F2		F3	
Variables		Frequency	Valid percent	Frequency	Valid percent	Frequency	Valid percent
Access to daylight	Very dissatisfied	4	8.2	6	12.2	–	–
	Dissatisfied	2	4.1	9	18.4	5	22.7
	Neutral	2	4.1	1	2.0	–	–
	Satisfied	8	16.3	20	40.8	11	50
	Very satisfied	33	67.3	13	26.5	6	27.3
	Total	49	100.0	49	100.0	22	100
Season with the best daylight condition	Rainy season	18	36.7	25	51.0	6	27.3
	Dry season	31	63.3	24	49.0	16	72.7
	Total	49	100.0	49	100.0	22	100.0

(*continued*)

Table 2. (*continued*)

Fenestration type		F1		F2		F3	
Variables		Frequency	Valid percent	Frequency	Valid percent	Frequency	Valid percent
Visual discomfort experienced at certain time of the day	Yes	48	98.0	35	71.4	22	100.0
	No	1	2.0	14	28.6	–	–
	Total	49	100.0	49	100.0	22	100.0
Time of the day at which discomfort is experienced	Morning	12	24.5	4	8.2	4	18.2
	Mid-day	12	24.5	16	32.7	7	31.8
	Afternoon	25	51.0	29	59.2	11	50.0
	Total	49	100.0	49	100.0	22	100.0
Rating of the level of discomfort	Clearly uncomfortable	4	8.2	8	16.3	2	9.1
	Just uncomfortable	19	38.8	20	40.8	12	54.5
	Neutral	11	22.4	7	14.3	4	18.2
	Just comfortable	4	8.2	12	24.5	2	9.1
	Clearly comfortable	10	20.4	2	4.1	2	9.1
	Total	48	98.0	49	100.0	22	100.0
Difficulty seeing in class when it rains	No difficulty at all	17	34.7	17	34.7	1	4.5
	A little difficulty	16	32.7	16	32.7	4	18.2
	Some difficulty	7	14.3	2	4.1	1	4.5
	A lot of difficulty	7	14.3	9	18.4	12	54.5

(*continued*)

Table 2. (continued)

Variables		Fenestration type					
		F1		F2		F3	
		Frequency	Valid percent	Frequency	Valid percent	Frequency	Valid percent
	Completely blind	2	4.1	5	10.2	4	18.2
	Total	49	100.0	49	100.0	22	100.0
Quality of daylight influence on productivity	Strongly disagree	3	6.1	6	12.2	5	22.7
	Disagree	9	18.4	15	30.6	5	22.7
	Neutral	10	20.4	4	8.2	2	9.1
	Agree	12	24.5	20	40.8	9	40.9
	Strongly agree	15	30.6	4	8.2	1	4.5
	Total	49	100.0	49	100.0	22	100.0
Difficulty in reading when it rains	No difficulty at all	9	18.4	17	34.7	6	27.3
	A little difficulty	17	34.7	14	28.6	7	31.8
	Some difficulty	7	14.3	3	6.1	2	9.1
	A lot of difficulty	13	26.5	9	18.4	2	9.1
	Completely blind	3	6.1	6	12.2	5	22.7
	Total	49	100.0	49	100.0	22	100.0

all had WWR less than 10% of the total floor area and were more than 1 m above floor level (Table 3).

For daylight distribution, participants in F1 classroom with a WWR of 5.8% had 67% being very satisfied with the amount of daylight that reaches their space whiles for F2 with a WWR also of 5.8%, 13(26.5%) were very satisfied. For F3 with the largest WWR (7.4%), only 27.3% were very satisfied. The result in F1 which has a low WWR and which does not conform to the Ghana National Building Regulation LI 1630 (1996) was as a result of the seating positions at which the answers came from as majority of the respondents in F1, 19(38.8%) were in the front seats which had the constantly open doors allowing in light at the frontage. This was confirmed by the results of a Chi square of association between seating position and daylight satisfaction with a statistically significant association (of 0.044).

Table 3. Results of Kruskal Wallis H test of variance of mean ranks of dependent variables across the three times of the day at which discomfort were experienced

Variable	F1 Kruskal-Wallis H	df	Asymp. Sig	F2 Kruskal-Wallis H	Df	Asymp. Sig	F3 Kruskal-Wallis H	df	Asymp. Sig
Access to daylight	.397	2	.820	3.053	2	.217	.082	2	.960
Visual discomfort experienced at certain time of the day	.960	2	.619	2.193	2	.334	.000	2	1.000
Quality of daylight influence on productivity	3.917	2	141	.255	2	.881	1.384	2	.501

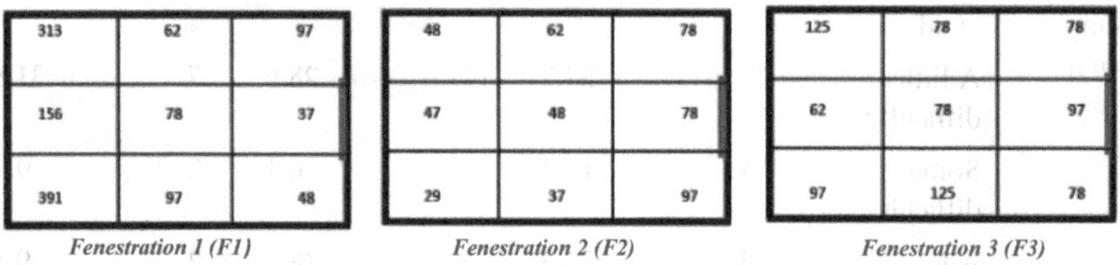

313	62	97
156	78	37
391	97	48

Fenestration 1 (F1)

48	62	78
47	48	78
29	37	97

Fenestration 2 (F2)

125	78	78
62	78	97
97	125	78

Fenestration 3 (F3)

Fig. 9. Site recorded illuminances within the classified zones at the time of surveys

Concerning the season that provides the best daylight conditions, F3 recorded the highest mean UDI (43.01 lux) and DA (38.40 lux) values for the dry season whiles with the participant survey, 72% agreed the dry season was the season with the best daylight condition. Participants in F3 also answered as having a lot of difficulty in seeing when it rains (54.5%) and 4(18.2%) were completely under that condition, a confirmation of their choice of the dry season as the most favourable season for providing the best daylight condition. F1 had 63.3% of participants also agreeing to the dry season as the season with the best daylight condition though in the simulated results it recorded the lowest mean UDI (42.81 lux) and DA (42.59 lux) values. Meanwhile 51% of respondents from F2 consider the rainy season as the season that provide best daylight conditions with a simulated result recording UDI value of 37.75 lux and a DA value of 37.37 lux, an in-between result. In contrast however, only 34.7% participants in F2 had no difficulty seeing in class when it rains but majority had some level of difficulty, either a little (32.7%) or a lot of difficulty (18.4%) in seeing in class when it rains. The results of Chi square of association (F1,0.534; F2, 0.058; F3, 0.445) implied there was no significant relationship between the seating position of the respondents and the season that provided best daylight condition. Fenestration 1 had majority of participants 17(34.7%) who had little difficulty in reading due to rain and 3(6.1%) were completely blind under that condition, 17(34.7%) of the respondents from F2 had no difficulty and 3(6.1%) had some difficulty. F3 on the other hand had majority of the respondents, 13(31.5%) having little difficulty. UDI values for the rainy season and the dry season fell below the 100 lux mark and thus classified as "too low". DA results for both seasons from all the fenestration types fell below the minimum threshold of 300 lux.

Meanwhile for issues of visual comfort, majority of respondents, 48(98%) from F1, 35(71.4%) from F2 and 22(100%) from F3 experienced visual discomfort at certain times of the day. This is matches with the results from both UDI and DA benchmark values from the simulation. Participants in all three fenestration types, experienced more discomfort in the afternoon (F1, 51%; F2, 59.2%; and F3, 50%) followed by midday and the least discomfort in the morning, most likely as a result of the East-West orientation of the windows. Out of the 49 respondents in the FI class, 19(38.8%) were just uncomfortable while 10(20.4%) were clearly comfortable in terms of the rating of visual discomfort. From F2, 20(40.8%) were just uncomfortable and only 2 participants (4.1%) were clearly comfortable. For F3, out of the 22 respondents, 12(54.5%) consider the level of discomfort just uncomfortable whiles 2(9.1) consider the level of discomfort clearly uncomfortable, just comfortable and clearly comfortable each. The highest response for the level of discomfort was "just uncomfortable" and emanated from F2 and the F3 class, with the least percentage in F1. For F2, a statistically significant association (Chi square = 0.023) was found for the position of seating and issues of visual comfort but no significant association for F1 (Chi square = 0.508) and F3 (Chi square = 0.269). A majority of participants in F1 strongly agree the quality of daylight influences their productivity, 20(40.8%) from F2 agree and 9(40.9) of participants in F3 also agree.

The results indicate (Kruskal Wallis H value = 0.297, df = 2, p = 0.820), (Kruskal Wallis H value = 3.053, df = 2, p = 0.217) and (Kruskal Wallis H value = 0.082, df = 2, p = 0.960) for F1, F2 and F3 respectively on daylight satisfaction. For visual discomfort, the results indicate (Kruskal Wallis H value = 0.960, df = 2, p = 0.619), (Kruskal Wallis H value = 2.193, df = 2, p = 0.334) and (Kruskal Wallis H value = 0.000, df = 2, p = 1.000) for F1, F2 and F3 respectively and for perceived daylight influence on productivity, the results indicate (Kruskal Wallis H value = 3.197, df = 2, p = 0.141), (Kruskal Wallis H value = 0.255, df = 2, p = 0.881) and (Kruskal Wallis H value = 1.384, df = 2, p = 0.501) for F1, F2 and F3 respectively. The results for all these variables recorded p-values more than 0.05, implying the result is statistically not significant and hence no significant difference in the mean rank of responses across the three times of the day at which discomfort was experienced in the three fenestration types. On the quality of daylight influence on productivity, results of a Chi Square of association indicates a non-significant value of 0.357 and 0.05 for F1 and F2 respectively but was significant for F3 (0.010).

3.1 Annual Sunlight Exposure

The Kanvili R.C JHS has a lot of physical characteristics that can enable adequate daylight distribution into the classrooms. The block is located in a hot dry sunny climate and has windows on the East West orientation, the point at which maximum daylight can be harnessed, though it can be excessive. However, there is adequate shading in the form of the corridor width of 2.2 m to prevent direct sunlight penetration. The major constraint as established by this study is the type of fenestrations that have been used and the WWR they occupy. Figure 10 indicates the Annual Sunlight Exposure on the block which has a range of ASE of 151 h to 1512 h for the rainy season and 173.3 h to 1733 h in the dry season, an indication of the potential for adequate daylight distribution if it the fenestrations allowed in more light.

Annual Sunlight Exposure for Rainy season (March-October)

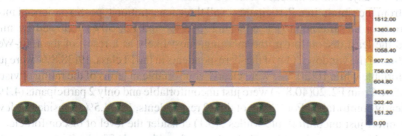

Annual Sunlight Exposure for Dry Season (November -February)

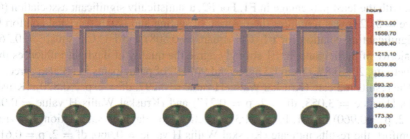

Annual Sunlight Exposure for Dry Season (whole yea

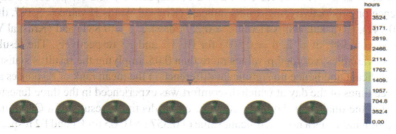

Fig. 10. Direct Sun hours of studied block with adjoining buildings

4 Conclusion

This study explores lighting quality from the perspective of simulated lighting quantities and subjective assessments from participants within the simulated classrooms on a long-term basis. In terms of predicting subjective long-term lighting assessments, UDI, DA and ASE were the most appropriate metrics for this study. The dry season presented the best daylight condition according to both the simulation and subjective survey whiles F3 which had the highest WWR recorded the highest UDI and DA values. All the three fenestration types, highlight, a poor distribution of daylight by the windows and fell below the threshold for both UDI and DA which was corroborated by the subjective study. The study recommends further research on possible modifications and retrofits of the studied fenestration types employed for all the studied classrooms.

Acknowledgements. There has been no specific funding for the conduct of this research. The authors would like to thank the reviewers for their valuable corrections, comments and time.

Author Contributions. This study is a collaborative effort by all of the authors.

Declaration of Competing Interest. The authors declare that they have no known competing financial interests or personal relationships that could have appeared to influence this study.

References

Al-Sallal, K.A.: Daylighting and visual performance : evaluation of classroom design issues in the UAE, pp. 201–209 (2010). https://doi.org/10.1093/ijlct/ctq025

Atthaillah and R. A., Koerniawan, M. D., & Soelami, F. X. N. , 2021.Atthaillah, Mangkuto, R.A., Koerniawan, M.D., Soelami, F.X.N.: Daylight annual illuminance investigation in elementary school classrooms for the tropic of lhokseumawe. Indonesia **25**(1), 129–139 (2021)

Ayoub, M.: 100 Years of daylighting: a chronological review of daylight prediction and calculation methods. Sol. Energy **194**(November), 360–390 (2019). https://doi.org/10.1016/j.solener.2019.10.072

Bakmohammadi, P., Noorzai, E.: Optimization of the design of the primary school classrooms in terms of energy and daylight performance considering occupants' thermal and visual comfort. Energy Rep. **6**, 1590–1607 (2020). https://doi.org/10.1016/j.egyr.2020.06.008

Costanzo, V., Evola, G., Marletta, L.: A Review of Daylighting Strategies in Schools: State of the Art and Expected Future Trends (2017). https://doi.org/10.3390/buildings7020041

Davoodi, A., Johansson, P., Henricson, M., Aries, M.: A conceptual framework for integration of evidence-based design with lighting simulation tools. Buildings **7** (2017). https://doi.org/10.3390/buildings7040082

El-darwish, I.I., Gendy, R.A.E.: The role of fenestration in promoting daylight performance. The mosques of Alexandria since the 19th century, pp. 3185–3193 (2016)

Ghana-Building-Code: Ministry of Water Resources, Works and Housing (MWRWH), Ghana (2018)

Handina, A., Mukarromah, N., Mangkuto, R.A., Atmodipoero, R.T.: Prediction of daylight availability in a large hall with multiple facades using computer simulation and subjective perception. Procedia Eng. **170**, 313–319 (2017). https://doi.org/10.1016/j.proeng.2017.03.037

IES LM-83-12: Approved Method : IES Spatial Daylight Autonomy (sDA) and Annual Sunlight Exposure (ASE) IES LM-83-12 IES Spatial Daylight Autonomy (sDA) (2012)

Iyendo, T.O.: Enhancing the hospital healing environment through art and day-lighting for user's Therapeutic Process, vol. 3, Issue 9, pp. 101–119 (2014)

Jakubiec, J., Quek, G., Srisamranrungruang, T.: Term visual quality evaluations correlate with climate - based daylighting metrics in tropical offices—a field study, pp. 1–27 (2020)

Kent, M.G., Altomonte, S., Wilson, R., Tregenza, P.R.: Temporal effects on glare response from daylight. Build. Environ. 113, 49–64 (2017). https://doi.org/10.1016/j.buildenv.2016.09.002

Kong, Z., Jakubiec, J.A.: Evaluations of long-term lighting quality for computer labs in Singapore Zhe Kong. Build. Environ. 194, 1–34 (2021)

Konis, K.: A circadian design assist tool to evaluate daylight access in buildings for human biological lighting needs. Sol. Energy 191(August), 449–458 (2019). https://doi.org/10.1016/j.solener.2019.09.020

Leccese, F., Salvadori, G., Rocca, M., Buratti, C., Belloni, E.: A method to assess lighting quality in educational rooms using analytic hierarchy process. Build. Environ. 168(2019), 21–24 (2020)

Lee, K.S., Han, K.J., Lee, J.W.: Feasibility study on parametric optimization of daylighting in building shading design. Sustainability 8, 1–16 (2016). https://doi.org/10.3390/su8121220

Lorenz, C., Packianather, M., Spaeth, A.B., De Souza, C.B.: Artificial Neural Network - Based Modelling for Daylight Evaluations (2018)

Mangkuto, R.A., Akbar, M., Siregar, A., Handina, A.: Determination of appropriate metrics for indicating indoor daylight availability and lighting energy demand using genetic algorithm. Solar Energy 170(2017), 1074–1086 (2018). https://doi.org/10.1016/j.solener.2018.06.025

Mardaljevic, J., Andersen, M., Roy, N., Christoffersen, J.: Daylighting metrics : is there a relation between useful daylight illuminance and daylight glare probability ? velux A/S, Adalsvej. In: Proceedings of the Building Simulation and Optimization Conference BSO12 (2012). Retrieved from https://infoscience.epfl.ch/record/179939?ln=en

Mien, I.H., et al.: Effects of exposure to intermittent versus continuous red light on human circadian rhythms, melatonin suppression , and pupillary constriction. PLoS ONE 9(5) (2014). https://doi.org/10.1371/journal.pone.0096532

Moreno, M.B.P., Labarca, C.Y.: Methodology for assessing daylighting design strategies in classroom with a climate-based method. Sustainability 7, 880–897 (2015). https://doi.org/10.3390/su7010880

Nocera, F., Faro, A. Lo, Costanzo, V.: Daylight performance of classrooms in a mediterranean school heritage building. Sustainability 10, 3705 (2018). https://doi.org/10.3390/su10103705

Fitriaty, P., Shen, Z., Achsan, A.: Daylighting Strategies in Tropical Coastal Area. International Review for Spatial Planning and Sustainable Development, vol. 7, No. 2, pp. 75–91 (2019). ISSN: 2187-3666 (Online). Http://Dx.Doi.Org/10.14246/Irspsd.7.2_75 \Copyright@SPSD Press from 2010 (SPSD Press, Kanazawa)

Sepúlveda, A., Luca, F. De, Thalfeldt, M., Kurnitski, J.: Analyzing the fulfillment of daylight and overheating requirements in residential and office buildings in Estonia. Build. Environ. 180(April), 1–12 (2020)

Shafavi, N. S., Tahsildoost, M., Zomorodian, Z.S.: Investigation of illuminance-based metrics in predicting occupants ' visual comfort (case study: Architecture design studios). Solar Energy 197(December 2019), 111–125 (2020). https://doi.org/10.1016/j.solener.2019.12.051

Spitschan, M.: ScienceDirect Melanopsin contributions to non-visual and visual function. COBEHA 30(Figure 1), 67–72 (2019). https://doi.org/10.1016/j.cobeha.2019.06.004

Wienold, J., et al.: Cross-validation and robustness of daylight glare metrics. In: Lighting Research and Technology, vol. 51 (2019). https://doi.org/10.1177/1477153519826003

Zomorodian, Z.S., Tahsildoost, M.: Assessing the effectiveness of dynamic metrics in predicting daylight availability and visual comfort in classrooms Window to Wall Ratio. Renew. Energy **134**, 669–680 (2019a)

Zomorodian, Z.S., Tahsildoost, M.: Assessing the effectiveness of dynamic metrics in predicting daylight availability and visual comfort in classrooms Window to Wall Ratio. Renew. Energy **134**, 669–680 (2019b). https://doi.org/10.1016/j.renene.2018.11.072

The Effect of Building Collapse in Ghanaian Building Industry: The Stakeholders' Perspectives

M. Pim-Wusu[1(✉)], T. Adu Gyamfi[2], and K. S. Akorli[2]

[1] Faculty of Built Environment, Department of Building Technology,
Accra Technical University, Accra, Ghana
mpimwusu@atu.edu.gh

[2] Faculty of Built and Natural Environment, Department of Building Technology,
Koforidua Technical University, Koforidua, Ghana

Abstract. Purpose: The construction industry in Ghana has experienced a rampant collapse of building over the years, and there have been various reportage in both electronic and print media. The consequences of these incidents have a detrimental effect on stakeholders' lives and properties. Hence, the study will ascertain the effect of a building collapse on the Ghanaian building industry and the stakeholders' perspectives.

Design/Methodology/Approach: The research utilised a quantitative technique by administering questionnaires to assist in attaining the aim of the study. The respondents were selected using random and purposive sampling techniques and consisted of contractors, building owners, architects, occupants, and building inspectors. A sample size of 150 was employed for the survey. Descriptive and inferential statistics were used to analyse the data gathered from the study. Again, the study achieved a reliability test with Cronbach's Alpha value of (0.870) an indication of reliable data.

Findings: The study discovered the effect of building collapse in the building industry in Ghana as leading to psychological trauma, stress and shock, loss of property, loss of valuable resources to occupants, loss of job, causes of disability, increase in the number of homeless people, injuries, and loss of trust of contractors.

Implications/Research Limitations: The present study implies that it is critical and paramount for building owners to award building contracts to qualified contractors who are tried and tested in the building industry. Also, the building industry inspection directorate should regularly inspect and monitor building projects in their jurisdictions. The study is limited to Grater Accra and the Eastern region of Ghana; therefore, a similar study can be carryout in other regions in Ghana.

Practical Implications: The outcomes of this paper would be significant to building industry stakeholders such as architects, consultants, structure engineers, project managers, quantity surveyors, clients, and MMDAs. This would strengthen the need to monitor and supervise building project delivery to avoid unexpected eventualities that may cause human lives and properties of building users.

© The Author(s), under exclusive license to Springer Nature Switzerland AG 2023
C. Aigbavboa et al. (Eds.): ARCA 2022, *Sustainable Education and Development – Sustainable Industrialization and Innovation*, pp. 234–242, 2023.
https://doi.org/10.1007/978-3-031-25998-2_18

Originality/Value: Existing literature shows no scientific inquiries into the effect of building collapse in the building industry in Ghana from stakeholders' perspectives. Grounded on previous empirical and theoretical studies, the results of this enquiry contribute to knowledge and comprehension of the effect of building collapse on building users in Ghana.

Keyword: Building · Collapse · Ghana · Industry · Stakeholders perspectives

1 Introduction

Building collapse results from conceded building components leading to low integrity of building structure, eventually resulting in the devastation of the buildings. According to Bala (2017), the failure of a building structure is due to the inability of the structural constituent to carry loads or the structure's unable to perform the expected function. In many circumstances, this failure makes the building unfit for habitation or the continuation of construction. Building collapse risk is an occurrence or course of action that may affect building occupants, investors, stakeholders, or the general public. This risk can harm success (Akande et al. 2016).

The risk of a building collapsing is significant, particularly in Ghana, where some public institutions are housed in dilapidated buildings that could fall and cause a tragedy (Asante and Sasu 2018). Ghana has experienced the collapse of several public buildings over the last ten years. Examples include the unfinished three-story Church of Prosperity construction in Akyem-Batabi in 2020, the two-story in Ashaiman in 2013, and the Achimota Melcom Shopping Mall in Accra in 2012 resulted in twenty-two fatalities (Lartey 2020).

In Ghana, the true number of building collapse cases may be more than what is publicly acknowledged and reported in the media. According to Asante and Sasu (2018), some building collapse cases go unreported. If efforts are not taken to address the root causes and triggers of the incidents, Ghana is expected to see additional building collapses (Asante and Sasu 2018). It is stated that the primary causes of building collapses in wealthy countries include terrorism, gas leak explosions, earthquakes, and global environmental changes (World Bank 2015).

In the construction industry today, particularly in Ghana, building collapse is one of the main problems facing both private and public developers. The collapse of the five-story Melcom Shopping Center building at Achimota in Accra on November 7, 2012, the collapse of a three-story building under construction in Cantonments on July 27, 2015, the collapse of a story building between the Holiday Inn and Marina Mall close to the airport in Accra on February 6, 2016, the collapse of a church building at Gbewe in Accra on May 8, 2017, and the collapse of a building at Akyem-Batabi on October 20, 2020, are just a few examples. There is a limited empirical investigation to determine how the frequent building collapses in Ghana affect the stakeholders; aside from the reports in the print media, this study fills the void. This study aims to identify characteristics that could be used to gauge how stakeholders are impacted by building collapses. Additionally, the study hypothesised that H_0, there are no divergent opinions among stakeholders regarding how a building collapse may affect Ghana's construction sector.

H₁, stakeholders' perspectives on the impact of a building collapse on Ghana's construction industry diverge significantly.

The building collapse catastrophes around the nation have claimed numerous lives and caused millions of cedis' worth of property damage. The effects of a building collapse are fatal; they include the loss of lives, the incapacity of those who were injured in the collapse, the destruction of property, financial losses, the wastage of time and valuable resources, an increase in the number of homeless persons, etc. (Akande et al. 2016; Boateng 2020a; Ayodeji 2011). The impacts of a building collapse include property loss, reputational harm, compromises to the contractors' integrity, fatalities, and legal issues between numerous parties (Obodoh et al. 2019). Ede (2013) theorised that a building collapse might cause stress, trauma, and shocks in the building owner, employees, residents, and other parties somehow connected to the structure. Additionally, an unhealthy environment increases the risk of a structure collapsing since it serves as a haven for thieves, hustlers, and potentially dangerous creatures like snakes that have made such locations their homes (Oke 2011).

Once more, building collapse causes long-term problems and has contributed to a growing loss of faith in the built environment (Boateng 2020a). Babatunde (2013) added that the decline in the market's stock of assets brought on by building collapses discourages real estate investment and leaves owners, survivor victims, and observers suffering from severe trauma, stress, and post-traumatic stress disorders. In addition, it has forced owners to declare bankruptcy (Oloke et al. 2017), forfeit contracts and commissions, revoke their practising licences, and settle insurance and compensation claims (Chendo and Obi 2015). Table 1 shows the frequency of building collapses from 2010 to 2021. This information points to the fact that all the collapsed buildings were private, and what might account for such floors from private developers call for further investigations.

Table. 1. List of selected collapsed buildings in Ghana.

Date of the collapse	Structure	Location of building	Type of building	Damage	Sources
February 11 2021	22-Story building	Airport residential area, Accra	Residential buildings	Some sustains injury	Ghana Web (2021) https://www.ghanaweb.com/GhanaHomePage/NewsArchive/Contentious-22-storey-building-at-Airport-collapses-1177381
October 20, 2020	5-story building	Akyem-Batabi	A church building	22 dead	GhanaWeb (2020) https://myinfo.com.gh/2020/11/akyem-batabi-collapsed-church-building-constructed-with-weak-concrete-expired-permit-report/
May 7 2017		Gbawe Borla in the Weija-Gbawe	A church building	20 congregants injured	Starrfm (2017) https://starrfm.com.gh/2017/05/weija-church-building-collapses-20-injured/
February 6 2016	Story building	Holiday Inn and Marina Mall around Airport in Accra	Commercial Buildings	Four people injured	Peace FM online (2016) https://www.peacefmonline.com/pages/local/social/201602/269226.php

(*continued*)

Table. 1. (*continued*)

Date of the collapse	Structure	Location of building	Type of building	Damage	Sources
July 27, 2015	3-story building under construction	Cantonments near the residence of President John Dramani Mahama"s	Residential Buildings	1 death, 20 workers were injured, Loss of property	Smith-Asante. (2015) http://www.graphic.com.gh/news/general-news/18-injured-3-dead-as-building-collapses-at-cantonments.html
March 13, 2014	Uncompleted seven-story building	Nii Boi Town, Accra	Residential Buildings	1 dead and 1 other seriously injured	https://www.graphic.com.gh/news/general-news/hotel-collapses-in-accra-killing-5-people.html
November 7 2012	5-story Melcom	Achimota, Accra	Commercial Buildings	14 people lost their lives, 78 people were injured Property damage	BBC News (2012) https://www.bbc.com/news/world-africa-20250494
January 5, 2011	Two-story building under construction	Antwirifu, Dormaa Ahenkro, Brong Shafto	Residential Buildings	2 persons died on the spot Three others were seriously injured Loss of property	Ghananewsagency2011.org/social/collapsed-building kills two-injures-others-24184
January 31, 2010	5-story hotel building under Construction	Tarkwa, Western Region	Residential Buildings	3 persons were killed Loss of property	http://www.modernghana.com/news/430078/1/melcombuilding-collapsestatementfrom-ghisep.html
June 5, 2010	4-story building	Spintex Road, Tema	Residential Buildings	Two masons were injured Damage of property	http://rosedarko.blogspot.com/2010/06/storey-buildingcollapses-1b-june-5.html

Researchers' design (2022).

2 Methodology

This study used a quantitative approach with closed-ended survey questionnaires distributed to building industry players in Accra and Koforidua in the Eastern region. 200 questionnaires were given out to stakeholders, including contractors, building owners, architects, occupants, and building inspectors, but 150 were returned, contributing a response rate of 75%. The questionnaire was divided into two sections; the first sought to solicit respondents' biodata. The second section sought data on how building collapses affected the construction sector. Both random and purposive sampling techniques were used. Random sampling offered equal chances for the respondents to be selected, whereas purposive sampling allowed the chance to involve experienced and informed people in the building sector. Descriptive and inferential statistics such as frequency, mean, and one-way ANOVA were used for data analysis using the SPSS version (26.0).

3 Results and Discussion

This section presents the respondents' demographic characteristics, as shown in Table 2. It captures the respondent's gender, age, level of education, and status/Occupation in the organisation. The study reveals that the gender of the respondent consists of 85.3% male and 14.7% female. The age of the respondents was 30.0% ranging from 18–29 years, 48.0% ranging from 30–39 years, and 22.0% ranging from 40–49 years. The study reveals the respondents' level of education as 8.7% were Technicians, 27.3% were HND holders, 44.0% were first degree holders, and 20% were master's holders. The study also investigated the Status/Occupation of the respondents; the result reveals that 24.0% of the respondents were Contractors, 23.3% were Architects, 18.7% were Occupants, 17.3% were Building inspectors, and 16.7% were building owners.

3.1 The Stakeholders' View on the Effect of a Building Collapse on the Building Industry in Ghana

The mean value of each of the five sub-groups for each item and their corresponding resultant mean rating were computed. The computed mean was compared with a theoretical mean value of 3. Any mean value less than 3 indicates disagreement with the factors, while a mean value above 3 indicates agreement. Table 3 shows that all 15 variables had a mean score higher than the theoretical mean, indicating respondents' agreement with the various elements measuring the effect of building collapse as having deadly consequences on stakeholders.

H_o There is no difference in the views of stakeholders on the effect of building collapse in the Ghanaian building industry.

H_1, there is a significant difference in the views of stakeholders on the effect of building collapse in the Ghanaian building industry.

The one-way ANOVA was used to test the hypothesis of the study. The item-by-item ANOVA results in Table 2 (last two columns) show that all the elements used were statistically insignificant regarding F-value and P-value. This indicates that stakeholders' views of the effect of building collapse are not different; therefore, the null hypothesis is accepted.

Table 2. Demographic Characteristics of Respondents (N = 150)

Variables	Frequency	Percentage (%)
Gender		
Male	128	85.3
Female	22	14.7
Age		

(continued)

Table 2. (*continued*)

Variables	Frequency	Percentage (%)
18–29 years	45	30.0
30–39 years	72	48.0
40–49 years	33	22.0
Educational level		
Technician	13	8.70
HND	41	27.3
First Degree	66	44.0
Master's Degree	30	20.0
Status/Occupation		
Contractors	36	24.0
Architects	35	23.3
Occupants	28	18.7
Building Inspectors	26	17.3
Building Owners	25	16.7

Source: Researcher's Fieldwork (2022)

Table 3. The stakeholder's view on the effect of building collapse in the building industry in Ghana (N = 150)

Variables	Contractors N = 36 $\bar{x}$	σx	Architects N = 35 $\bar{x}$	σx	Occupants N = 28 $\bar{x}$	σx	Building inspectors N = 26 $\bar{x}$	σx	Building owners N = 25 $\bar{x}$	σx	R mean	Ranking	F value	P value
1. Loss of property	4.19	.873	3.90	.737	4.43	.756	4.60	.548	4.40	.547	4.30	2	.933	.469
2. Loss of human lives	4.10	.943	3.90	.567	4.14	1.09	4.00	.707	4.00	.707	4.03	11	.168	.973
3. Loss of job	4.27	.904	4.00	.000	4.35	.633	4.20	.836	4.20	.836	4.20	3	.465	.800
4. Loss of capital	4.00	1.18	4.00	.471	4.14	.662	4.00	.707	3.80	1.64	3.99	14	.543	.742
5. Injuries	4.00	.894	4.10	.567	4.21	.893	4.40	.548	3.80	1.64	4.10	8	.329	.893
6. Loss of reputation of the contractor	4.00	.774	3.90	.875	4.14	.663	4.00	.000	4.00	1.73	4.00	13	.278	.923

(*continued*)

Table 3. (*continued*)

Variables	Contractors N = 36		Architects N = 35		Occupants N = 28		Building inspectors N = 26		Building owners N = 25		R mean	Ranking	F value	P value
	$\bar{x}$	σx	$\bar{x}$	σx	$\bar{x}$	σx	$\bar{x}$	σx	$\bar{x}$	σx				
7. Loss of trust of contractors	4.18	.603	3.80	.789	4.21	.893	3.80	.447	4.20	1.30	4.04	10	.653	.661
8. Loss of integrity of contractors	4.27	.467	3.70	.675	3.93	.829	3.80	.447	4.40	1.34	4.02	12	.653	.661
9. Loss of valuable resources to occupants	4.18	.874	4.20	.421	4.29	.825	4.20	.447	4.00	1.22	4.17	4	.164	.974
10. Loss of time	4.00	.632	4.10	.568	4.29	.611	4.00	.000	4.20	.837	4.12	7	1.07	.388
11. Leads to psychological trauma, stress and shock	4.09	1.14	4.30	.483	4.29	.726	4.60	.548	4.40	.894	4.34	1	.453	.809
12. Causes of disability	4.27	.646	4.10	.737	4.14	.663	4.20	.837	4.00	1.00	4.14	6	.347	.881
13. Loss of building materials	4.00	.774	4.40	.516	4.28	.611	3.60	.548	4.20	.837	4.09	9	3.29	.013
14. Withdrawal of licenses of construction practitioners	4.18	.750	3.50	.972	4.00	.961	3.60	1.14	4.20	.836	3.90	15	.982	.439
15. Increase the number of homeless people	4.09	1.14	4.00	.816	4.35	1.00	3.80	1.79	4.60	.548	4.17	4	.924	.475

Mean = $\bar{x}$ Standard deviation = σx Resultant = R.

3.2 Discussion

The study indicates that stakeholders rank 1st the building collapse as leading to psychological trauma, stress, and shock. This result is consistent with the body of literature; Ede (2013) asserted that building collapse contributes to stress, trauma, and shocks on the building owner and employees, residents, and other parties involved in the structure.

The second finding of the study reveals that when a building collapses, stakeholders lose their property; this finding is consistent with Boateng (2020a), Obodoh et al. (2019), Ede et al. (2014) and Akande et al. (2016); who find that the collapse of the building contributes to loss of properties of the users and stakeholders. Again, the study finds that when a building collapses, occupants lose their job and valuable resources; this outcome confirms the assertion of Ede (2010b); Ayodeji (2011); Ede et al. (2014); Akande et al. (2016) that occurrence of building collapse leads to losing a job and valuable resources of the users. In addition, the study's other outcomes reveal that building collapse leads to an increasing number of homeless occupants; this supports the findings of Akande et al. (2016); Boateng (2020a) that building collapse contributes to occupants losing their homes. The further finding of the study discovered that when a building collapses, it causes disability to some of the occupants; this finding agrees with Akande et al. (2016) and Boateng (2020a) that when a building collapses, some occupants are disabled permanently. The study's findings stipulated that there are deadly consequences associated with building collapse.

4 Conclusion

This research has clearly shown that building collapse has detrimental consequences on the occupants and stakeholders in the building industry in Ghana. The study employed quantitative research methods and a cross-sectional survey with the administration of a questionnaire. The study discovered that all fifteen (15) variables proposed to measure the effect of building collapse in the Ghanaian building industry were viewed by the stakeholders as important. The major effect of the building collapse on stakeholders includes psychological trauma, stress and shock, loss of property, loss of job, loss of valuable resources to occupants, increased number of homeless people, causes of disability, loss of time, and injuries. Also, the study further used one-way ANOVA to determine whether there were differences in the stakeholders' responses. The study's outcome revealed no differences in the stakeholders' responses, hence the acceptance of the null hypothesis. This research contributes to knowledge in the building construction industry concerning the effect of building collapse in the Ghanaian building industry. The study recommends that there should be a monitory system to check the implementation of building procedures to avoid the usage of unapproved building drawings in the building industry, which is likely to affect stakeholders. The agency to certify building materials should play its game to avoid substandard materials in the building industry. There should be a regular site investigation on each project by the inspectors from MMDAs to avoid the Absence of proper site investigation on construction sites, which puts occupants in disarray.

References

Akande, B.F., Debo-Saiye, B., Alao, T.O, Akinrogunde, O.O.: Cause, effects and remedies to incessant building collapse in Lagos State. Int. J. Basic Appl. Sci. **16**(4), 15–30 (2016)

Asante, L.A., Sasu, A.: The challenge of reducing the incidence of building collapse in ghana: analysing the perspectives of building inspectors in Kumasi. SAGE Open. April–June, pp. 1–12 (2018)

Ayodeji, O.: An examination of the causes and effects of building collapse in Nigeria. J. Design Built Environ. **9**, 37-47. (2011)

Babatunde, I.R.: Monumental effects of building collapse in Nigerian cities: the case of Lagos Island, Nigeria. Basic Res. J. Eng. Innov. **1**(2), 26–31 (2013)

Bala, K.: Building Collapse in Nigeria: challenges and remediation. In: 11th Annual Lecture/Conference. Faculty of Environmental Sciences, Nnamdi Azikiwe University, Awka, Nigeria. 5th–7th June 2017 (2017)

BBC News: Melcom shop collapse in Ghana: Negligence blamed (2012). https://www.bbc.com/news/world-africa-20250494. Accessed on 20 March 2022

Boateng, F.G.: A critique of overpopulation as a case of pathologies in African cities: evidence from building collapse in Ghana. World Dev. **37**(105161), 1–12 (2020a)

Chendo, I., Obi, N.I.: Building collapse in Nigeria; the causes, effects, consequences and remedies. Int. J. Civil Eng. Construct. Estate Manage. **3**(4), 41–49 (2015)

Darfa, F.E.: Akyem-Batabi: 5-story church building collapses and traps worshippers, 1 dead (2020)

Ede, A.N.: Structural stability in Nigeria and worsening environmental disorder: the way forward. Paper presented at the Annual Meeting of the West Africa Built Environment Research Conference. Accra, Ghana (2010b)

Ede, A.N.: Building Collapse in Nigeria: the trend of casualties in the last decade (2000–2010). Int. J. Civil Environ. Eng. **10**, 32–36 (2013)

Ede, A.N., Adebayo, S.O. Ugwu E.I., Emenike, C.P.: Life cycle assessment of environmental impacts of using concrete or timber to construct a duplex residential building. IOSR J. Mech. Civil Eng. **11**(2), Ver. I, 62–72 (2014)

Oke, A.: An Examination of the causes and effects of building collapse in Nigeria. J. Des. Built Environ. **9**, 37–47 (2011)

GhanaWeb: Akyem Batabi: Collapsed Church constructed with weak concrete, expired permit-report (2020). https://www.ghanaweb.com/GhanaHomePage/NewsArchive/Akyem-Batabi-Collapsed-church-building-constructed-with-weak-concrete-expired-permitReport-1111177#:~:text=News,Akyem%20Batabi%3A%20Collapsed%20church%20building%20constructed,weak%20concrete%2C%20expired%20permit%20%E2%80%93%20Report&text=A%20report%20by%20the%20four,quality%20workmanship%20and%20institutional%20failure. Accessed 20 March2022

Ghana Web: Contentious 22-story building at Airport residential area collapses (2021). https://www.ghanaweb.com/GhanaHomePage/NewsArchive/Contentious-22-storey-building-at-Airport-collapses-1177381. Accessed 20 March 2022

Lartey, N.L.: Akyem Batabi church collapse: original building plan was altered – Report. Citi News Report (2020). https://citinewsroom.com/2020/11/akyem-batabi-church-collapseoriginal-building-plan-was-altered-report/. Accessed 20 March 2022

Obodoh, D., Amade, B., Obodoh, C., Igwe, C.: Assessment of the effects of building collapse risks on the stakeholders in the Nigerian built environment. Niger. J. Technol. **38**(4), 822–831 (2019)

Oloke, O.C., Oni, A.S., Ogunde, A., Joshua, O., Babalola, D.O.: Incessant building collapse in Nigeria: a framework for post-development management control. J. Develop. Country Stud. **7**(3), 114–127 (2017)

Peace FM online: Four Injured in Accra Building Collapse (2016). https://www.peacefmonline.com/pages/local/social/201602/269226.php. Accessed 20 March 2022

Smith-Asante, E.: 18 Injured, 3 dead as building collapses at Cantonments (2015). https://www.graphic.com.gh/news/general-news/18-injured-3-dead-as-building-collapses-at-cantonments.html. Accessed 20 March 2022

Starrfm: 20 injured in Weija church building collapse (2017). https://starrfm.com.gh/2017/05/weija-church-building-collapses-20-injured/. Accessed on 20 March 2022

World Bank: Building Regulation for Resilience: Managing Risks for Safer Cities. The World Bank Group, Washington, DC (2015)

Estimation of the Most Sustainable Regional and Trans-border Infrastructure Among Road, Rail and Seaborne Transport

S. N. Dorhetso(✉) and I. K. Tefutor

Accra Institute of Technology (AIT), Accra, Ghana
samueldorhetso@gmail.com, dorhetso@ait.edu.gh

Abstract. Purpose: The objective of this study is to empirically determine the relative significance of the determinant factors for sustainable regional and trans-border infrastructure of road, rail and seaborne transport, based on the social, environmental and economic dimensions of sustainability.

Design/Methodology/Approach: A theoretical framework was developed from a mingle of the social, environmental and economic pillars of sustainable development, and the best-worst method was used to scrutinize and rank the identified determinant factors according to their weighted averages.

Findings: The findings of this study point to access, equity and fairness; health; cost and speed; reliability; and carbon dioxide/air emissions as the highest ranked determinant factors for sustainable regional and trans-border infrastructure of road, rail and seaborne transport.

Implications/Research Limitation: This study has significant implications for academic and organisational research and literature, and it can broaden conceptual viewpoints of implications of decisions made to adopt sustainable regional and trans-border infrastructure of road, rail and seaborne transport. However, it is limited to an extent, because it was partly founded on the opinions of transport and logistics managers in Ghana, which may be branded with biased judgment and ambiguity.

Practical Implication: In addition to its theoretical value, this study has significant implications for policy makers and practitioners, as it would expand their perspectives on choices to implement sustainable regional and trans-border infrastructure of road, rail and seaborne transport.

Originality/Value: This distinctive study forges an estimation of the most sustainable regional and trans-border infrastructure between road, rail and seaborne transport, and accentuates a critical research area that is currently understudied, using a unique framework and a different technique.

Keyword: Trans-border · Infrastructure · Road · Rail · Sea

1 Introduction

The capacity to provide safe transportation that is socially inclusive, accessible, reliable, affordable, fuel-efficient, environmentally friendly, emits low-carbon, and resilient to shocks and disruptions, including those caused by climate change and natural disasters are characteristics of a sustainable and resilient transport infrastructure (UNCTAD 2015a). Many researchers (Geurs et al. 2009; Rodrigue et al. 2013; Hanssen et al. 2012; Fan & Chan-Kang 2004) have studied the importance of a sustainable and resilient transport infrastructure. However, their studies underrated the importance of the estimation of the most sustainable regional and trans-border infrastructure between road, rail and seaborne transport. According to the report by UNCTAD (2015a), sustainable and resilient transportation can be analysed along three dimensions, namely the environment (green transport), society (inclusive transport), and the economic dimension (efficient and competitive transport). The connection among the economic, social and environmental dimensions characterizes the sustainability of the transport infrastructure. As per the report, attaining SDG Goal 9 would require that relevant sustainability and resilience criteria be integrated and mainstreamed into all modes of transport. However, it would be even more prudent to actually establish which mode of transport is linked to the most sustainable regional and trans-border infrastructure. The current momentum of sustainable regional and trans-border transport infrastructure growth has accentuated the need for a detailed study to determine the relative weights and rankings of the determinant factors for sustainable regional and trans-border infrastructure of road, rail and seaborne transport. Several studies (Fan & Chan-Kang 2004; Reis 2014; Fan et al. 2002; WHO 2015; IEA 2012a, b, 2015a, b; Hanssen et al. 2012) have discussed the determinant factors for sustainable regional and trans-border infrastructure of road, rail and seaborne transport. However, most of these studies focused on sustainability through integrated intermodal transport and logistics system (Reis 2014; Hanssen et al. 2012), and hence nothing has been found in literature about the relative significance of the determinant factors for sustainable regional and trans-border infrastructure of road, rail and seaborne transport. With impetus from this apparent research gap, the objective of this study is to empirically determine the relative significance of the determinant factors for sustainable regional and trans-border infrastructure of road, rail and seaborne transport, based on the social, environmental and economic dimensions of sustainability, using the best-worst method (Rezaei 2016; Wang et al. 2019). The Best Worst Method (BWM) would be used to estimate the relative weights and rankings of the relevant determinant factors for sustainable regional and trans-border infrastructure. The results of these analyses would bring insight into the actual level of importance of each selected determinant factor towards sustainable regional and trans-border transport infrastructure development.

2 Literature Review

The three mainstays of sustainable development need to be integrated into transport policies, planning and operation, according to the 2030 Agenda for Sustainable Development. Hence, it is imperative to identify the ways in which these three foundations of social, environmental and economic factors connect to transportation and how they are interconnected to make transport more sustainable. There have been a few studies on the determinant factors for sustainable regional and trans-border infrastructure of road, rail and seaborne transport, but none has been found from extant literature that deals with the relative level of importance of these factors. The purpose of this study is to determine the relative degree of importance of the determinant factors for sustainable regional and trans-border infrastructure of road, rail and seaborne transport. The relevant determinant factors would be identified from an extensive and meticulous literature review, and complemented by experts' opinion. The decision framework to be developed and used in this research originates from a blend of the three pillars of sustainable development. Modal preference can contribute to sustainable development and the achievement of transport-related SDGs, since each transport mode's comparative merits and demerits have impacts on these three pillars of sustainable development. In 2009, the European commission's communication on the future of transport promoted intermodal transport as a way of limiting the unsustainable expansion of the transport sector, mainly focused around road transport, without endangering the economic, social and sustainable development of the European Union (Reis 2014). However, although Reis (2014) discussed critical issues of how to best integrate the different modes of transport for sustainability, there was no mention of the relative importance of the determinant factors for sustainable regional and trans-border infrastructure of road, rail and seaborne transport. This distinctive work seeks to fill the gap by developing a theoretical framework to study the determinant factors for sustainable regional and trans-border infrastructure of road, rail and seaborne transport, with regards to the relative importance of these factors. This paper focuses on the relative importance of relevant determinant factors derived from literature, but not so much on the desk research discovery of another key determinant.

2.1 Identification of the Determinant Factors for Sustainable Regional and Trans-border Infrastructure of Road, Rail and Seaborne Transport

The decision framework developed in this research emanates from three dimensions, namely society (inclusive transport), environment (green transport) and economic (efficient and competitive transport) dimensions. The connection among the economic, social and environmental dimensions characterizes the sustainability of the transport infrastructure. This distinctive framework would be used to explore the weights and rankings of the determinant factors that significantly impact the attainment of a sustainable regional and trans-border infrastructure of road, rail and seaborne transport (see Table 1). Within each of these dimensions, there are determinant factors that were recognized by an amalgamation of literature from previous studies and opinions of transport industry experts.

2.1.1 Social Dimension

Transportation services inaccessibility is a major constraint to the reduction of poverty. The negative effects of lack of access to transport on health, education, employment and economic activities have been unveiled in several studies (WHO 2015; World Bank 2008, 2015). Transportation contributes to social development by facilitating labour mobility and enabling access to education, health care and other social benefits. Thirty-one percent of the rural population of the world, with developing nations constituting ninety-eight percent, do not have sufficient access to road transport systems (World Bank 2015). With regards to access, road networks have been found to be more open and multifaceted compared to other modes of transport. On the contrary, rail transport operates on fixed

Table 1. A theoretical framework on the dimensions and determinant factors (DFs) for sustainable regional and trans-border infrastructure of road, rail and seaborne transport.

Dimensions	Factors	References
Social (SL)	Access. Equity and fairness (SL1) Social inclusiveness and value (SL2) Community involvement (SL3) Health (SL4) Labor conditions (SL5) Safety (SL6) Congestion (SL7)	Geurs et al. (2009), World Bank (2008, 2015), WHO (2015), ATAG (2008), Rodrigue et al. (2013)
Environmental (ET)	Marine, air and soil pollution (ET1) Noise, vibrations and biodiversity (ET2) Carbon dioxide/air emissions (ET3) Climate change impacts/resilience(ET4) Energy intensity and resource depletion(ET5) Land use(ET6)	IEA (2012a, b, 2013, 2015a, b), ESCAP (2016), European Commission (2008)
Economic (EC)	Cost and speed (EC1) Reliability (EC2) Flexibility (EC3) Efficiency and productivity (EC4) Energy efficiency (EC5) Employment and revenue generation (EC6)	Hanssen et al. (2012), USCC (2006), Banomyong (2003), Fan & Chan-Kang (2004), World Travel and Tourism Council (2014), Eurostat (2009), Fan et al. (1999, 2002)

tracks; hence physical access to its service is limited. Nevertheless, it is the preferred choice of transport for the low income earners due to its low cost. According to Geurs et al. (2009), the cost of transport services is an important dimension of accessibility. Although maritime transport is less accessible than rail transport, it is one of the only means of international access for island nations. Inland waterways are also significant transport infrastructure for citizens living along water bodies or in delta regions. Air transport may be the only means of transport for very remote locations, and encourages social inclusion by linking the inhabitants of such areas with the rest of their state. Also, air transport may be deemed as the safest means of transport. With regards to safety, casualties and injuries from road accidents are possibly the most usually cited form of social impact related to transport. Road traffic injuries and fatalities cause both social and economic losses to society. Transportation by road remains the most dangerous, accounting for ninety per cent of all transport crashes on average (Rodrigue et al. 2013), although no mode of transport is completely safe. Also transportation by road creates the most intense congestion. Congestion is undesirable for sustainable regional and trans-border transport infrastructure development as it increases travel time, degrades the reliability of delivery services, decreases productivity and does not promote competitiveness.

2.1.2 Environmental Dimension

Air pollutants, such as particulate matter, nitrogen oxides, sulphur oxide, ozone and volatile organic compounds, which are by products of transportation, cause damage to human health, ecosystems, buildings and materials (European Commission 2008). The transport sector is the second leading creator of carbon dioxide emissions, making up for almost a quarter of the world's total carbon dioxide emissions in 2013 (IEA 2015a, b). Road transport is driving the growth of carbon dioxide emissions. The transport sector is one of the principal consumers of energy, and a majority of oil use is accounted for by road transport. Oil use is progressively more concentrated in just two sectors: transport and petrochemicals, according to the international energy agency. This arrangement is set to keep oil use on a rising trend to 2035 (IEA 2013). In 2013, the transport sector consumed almost sixty-six percent of the world's oil and more than twenty-seven percent of the world's total energy supply (IEA 2015a, b). Maritime transport is the most environmentally friendly means of transport, followed by rail transportation.

2.1.3 Economic Dimension

The transportation costs of cargo by rail, inland waterways, short sea shipping or ocean shipping are minor in general. Nevertheless, rail and seaborne transport modes are ranked as the slowest, whereas high-cost air transport is deemed to be the fastest (Hanssen et al. 2012). Punctuality, frequency and transport time are vital factors that affect transport users. Delivery time reliability and the average delivery time are frequently named as the most significant traits of a transportation system. Reliability of transit time is also a key variable that influences freight transport, (Banomyong 2003). Inconsistency in travel time has several adverse economic repercussions. Road transport usually has several alternative courses, unlike rail transport that is scheduled and operates on fixed tracks. As a result of this flexibility, the time reliability of road transport is lofty. Seaborne transport is also considered to have a high level of reliability (USCC 2006), although there are usually hold-ups at port. However, seaborne and rail transport modes are less competitive in terms of flexibility since they both operate on fixed routes and schedules. Road transport possess operational flexibility and a low degree of physical limitation, hence it can be a central link in cargo delivery for providing door to door services. Transport networks are critical for access to markets, public services and information at the local level. Hence, the higher the efficiency of a transport service, the higher the potential for improving production and market efficiency (Fan & Chan-Kang 2004). It is evident from extant literature that rural access to transport networks facilitates the creation of economic opportunities and employment, and reduces poverty by linking farmers to markets and suppliers to customers. Employment and incomes are essential for economic growth. The transport sector generates various job opportunities. In 2005, it was estimated that the transport sector employed 8.7 million people in the European Union (Eurostat, 2009). A study conducted by the International Food Policy Research Institute, on links between government spending and poverty in rural India, uncovered that government expenditure on rural infrastructure yielded into better chances for non-farm employment and higher wages, and this had an important effect on decreasing poverty and inequality (Fan et al. 2002). Also, transportation is indispensable to the expansion of tourism. In 2013, it was estimated that tourism contributed 2.2 trillion dollars to the world's gross domestic product and directly created 100 million jobs around the world (World Travel and Tourism Council 2014).

3 Research Methodology

The research modelling framework proposed in this study consists of the paths to prioritising the determinant factors (DFs) for sustainable regional and trans-border infrastructure of road, rail and seaborne transport, using the BWM. The BWM is employed to determine the relative significance of each dimension and determinant factor of sustainable regional and trans-border infrastructure of road, rail and seaborne transport. This is done by comparing the best dimension (most important) to the worst (least important) initially and then comparing the other dimensions to the worst afterwards using a linguistic scale for the pair wise comparison.

3.1 Best-Worst Method (BWM)

According to Wang et al. (2019), the BWM is a multi-criteria decision-making model which estimates the weights of criteria by employing two vectors of pair wise comparisons between the most important and the least important criteria. According to Rezaei (2016), the steps below are involved in determining the weights of criteria using the BWM:

Step 1: Finalisation of decision criteria.

A set of decision criteria are identified and extracted from an intensive search of literature, and experts' opinions and recorded as $\{C1, C2... Cn\}$ for n main criteria. In this study, the decision criteria are the determinant factors for sustainable regional and trans-border infrastructure of road, rail and seaborne transport.

Step 2: The best (most important) and worst (least important) criteria are selected.

At this stage, the expert selects the most important and least important criteria from the pool of identified decision criteria in Step 1 based on his/her opinion.

Step 3: A matrix is developed by determining the pair wise comparison between the most important criterion and the other decision criteria. The objective of this step is to determine the preference of the most important criterion to the other decision criteria by using a linguistic scale for the BWM having scores from 1 to 9. The outcome of the pair wise comparison of the best criterion and other decision criteria is expressed by a 'Best-to-Others' vector as follows:

$$DB = (dB1, dB2, \ldots, dBn)$$

where dBj represents the preference of the most important criterion B over a criterion j amongst the decision criteria, and $dBB = 1$.

Step 4: The 'Others –to-Worst' matrix is developed by conducting a pair-wise comparison of the other decision criteria against the least important criterion using the linguistic scale for the BWM. The outcome of comparison of the other decision criteria to the worst criterion is expressed as follows:

$$DW = (dW1, dW2, ..., dWn)^q$$

where dWj represents the preference of the criterion j amongst the decision criteria in Step 1 above the least important criterion W, and $dWW = 1$.

Step 5: Computing the optimal weights $(p1*, p2*... pn*)$.

Weights of criteria are determined such that the maximum absolute differences for all criterion j are minimised over the following set $\{|pB - dBjpj|, |pj - djW pW|\}$.

A minimax model can be formulated as:

$$\min \max j\{|pB - dBjwj|, |pj - djWwW|\}$$

Subject to :

$$\sum_j pj = 1 \qquad (1)$$

$pj \geq 0$, for all criterion j.

Model (1) can be solved by converting it into the following linear programming problem model:

$$Min\ R^L$$

Subject to :

$$|pB - dpj| \leq R^L,\ for\ all\ criterion\ j$$

$$|pj - dpW| \leq R^L,\ for\ all\ criterion\ j \qquad (2)$$

$$\sum_j pj = 1$$

$$pj \geq 0,\ for\ all\ criterion\ j$$

Solving the linear model (2), will result in optimal weights $(p1*, p2*... pn*)$ and optimal value R^L. Consistency (R^L) of comparisons also needs to be estimated. A value nearer to zero is more desired for consistency (Rezaei 2016; Wang et al. 2019).

3.2 Data Collection

For this research, questionnaires were designed and used to collect data from participants with a minimum of five years of professional management and decision-making experience in the Ghanaian transport sector. This was done to ensure accuracy of data garnered since the experts were deemed to be sufficiently knowledgeable to effectively complete the survey. The experts were purposefully selected from the fields of road transport, rail transport, maritime transport, and logistics and supply chain management in the Ghanaian transportation sector. They were assured of the confidentiality of their reports in order to allow for effective model-building and in-depth observation (Nilashi et al. 2016). Also, the participants were designated mid-level and above ranking executives, hence their responses sufficiently represent the transport sector (Fu et al. 2006).

To conduct the survey, several steps were undertaken to maximise the rate of response and minimise response bias amongst the experts from the selected transport organisations. Initially, a pilot study was done by sending the Google form questionnaires designed for this study to three researchers through emails and interviewing three participants face-to-face to review and provide feedbacks. The three researchers that participated in the pilot study were a female and two males who hold PhD degrees and have at least seven years of research experience in transportation, logistics and supply chain management. The experts who participated in the pilot study have managerial experience of at least five years in the Ghanaian transport and logistics sector. As a result of the feedbacks from the pilot study, the questionnaires were modified and emailed to thirty experts. Six experts each were selected to represent the five different roles of the participants as illustrated in Table 2. A follow- up on the respondents was done via phone conversations and personal visits (Yang et al. 2018). Eventually, twenty completed questionnaires were received out of the thirty that were emailed to the experts, a response rate of 67%. This response rate is considered suitable for efficient analysis and to yield reliable findings, according to the BWM used in this study that does not require a large sample size to provide precise and consistent results (Wang et al. 2019) (Fig. 1).

4 Results and Discussion

As the first step of the BWM, the dimensions and determinant factors for sustainable regional and trans-border infrastructure of road, rail and seaborne transport that have been derived from extant literature were evaluated by the decision-makers using questionnaires. The questionnaires were planned to require a 'YES' or 'NO' answer which indicated that a dimension or determinant factor is 'relevant' or 'not relevant' for sustainable regional and trans-border infrastructure. A simple mean method was used to select the variables that are above the arithmetic mean and the analysis of results at this stage indicated that all the identified dimensions or determinant factors were accepted with no auxiliary inclusions. Therefore, inclusiveness of relevant data was ensured and content validity was confirmed.

4.1 Calculation of the Weights of Determinant Factors (DFs) Using BWM

After the finalisation of the determinant factors for sustainable regional and trans-border infrastructure of road, rail and seaborne transport, the weights of the DFs were calculated using the BWM. For this study, twenty experts performed the identification of the best and worst criteria for the main dimensions as well as subcategory criteria. Subsequent to obtaining the best and worst criteria, all the respondents were requested to give preference ratings of the best criteria to other criteria and other criteria to worst criteria for the main dimensions' criteria DFs as well as subcategory criteria DFs. The preference ratings of the first expert, Expert 1, for the main category criteria DFs, as well as subcategory criteria DFs are illustrated in Table 3.

An identical process of the BWM survey, as described in the paragraph above was performed by all the experts in this research to estimate the performance ratings of the

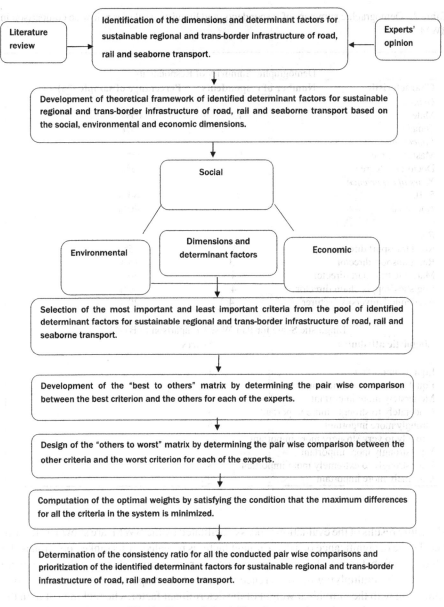

Fig. 1. Research modelling framework.

main category and subcategory DFs for sustainable regional and trans-border infrastructure of road, rail and seaborne transport. The entire weights of the DFs for both the main category and subcategory measured in this study were obtained using Eq. (1). All the aggregated weights were computed by applying the data sourced from the twenty experts of this research to Eq. (2) and estimating the mean using the simple average technique.

Table 2. Demographic summary of respondents, and linguistic scale for pair wise comparison in BWM.

Demographic Summary of Respondents		
Characteristic	**Number of respondents**	**Percentage of sample (%)**
Gender		
Male	13	65%
Female	7	35%
Education		
Master degree	11	55%
Doctorate degree	9	45%
Years of experience		
5–10	12	60%
Above 10	8	40%
Roles		
Road transport director	5	25%
Rail transport director	3	15%
Maritime transport director	4	20%
Logistics/supply chain director	4	20%
Academic/University lecturer	4	20%
Linguistic Scale for Pair Wise Comparison in BWM.		
Linguistic attributes	**Scores**	
Equally important	1	
Equal to moderately more important	2	
Moderately more important	3	
Moderately to strongly more important	4	
Strongly more important	5	
Strongly to very strongly more important	6	
Very strongly more important	7	
Very strongly to extremely more important	8	
Extremely more important	9	

The entire results of the evaluation process, facilitated by the BWM, are shown in Table 3. The degree of significance of a determinant factor is revealed by its ranked position in the table. The global ranks of the recognised determinant factors, shown in the table, were calculated by multiplying the preference weights of the respective determinant factor's dimension with the individual weight of the determinant factor. The rankings of both the main category dimensions and sub category determinant factors are discussed in detail in the next segment of this report.

4.2 Ranking of the Dimensions of the DFs for Sustainable Regional and Trans-border Infrastructure of Road, Rail and Seaborne Transport

As exuded in Table 3, the results point to the social dimension as the most substantial dimension for sustainable regional and trans-border infrastructure of road, rail and

Table 3. Pairwise comparison of main category and subcategory DFs by Expert 1, and aggregate weights of main and subcategory DFs for all the experts.

Main Category DFs			
Best to Others	Social (SL)	Environmental (ET)	Economic (EC)
Best criteria: Social (SL)	1	3	5
Others to Worst		Worst criteria: Environmental (ET)	
Social (SL)		5	
Environmental (ET)		1	
Economic (EC)		2	

Social (SL) Subcategory DFs							
Best to Others	SL 1	SL 2	SL 3	SL 4	SL 5	SL6	SL7
Best criteria: SL1	1	3	2	3	5	5	4
Others to Worst				Worst criteria: SL5			
SL1				3			
SL2				5			
SL3				3			
SL4				3			
SL5				1			
SL6				4			
SL7				5			

Environmental (ET) Subcategory DFs						
Best to Others	ET 1	ET 2	ET 3	ET 4	ET 5	ET 6
Best criteria: ET3	3	2	1	3	6	5
Others to Worst			Worst criteria: ET6			
ET1			3			
ET2			5			
ET3			2			
ET4			3			
ET5			2			
ET6			1			

Economic (EC) Subcategory DFs						
Best to Others	EC1	EC2	EC3	EC4	EC5	EC6
Best criteria: EC1	1	3	2	3	4	3
Others to Worst			Worst criteria: EC6			
EC1			3			
EC2			3			
EC3			2			
EC4			5			
EC5			4			

(*continued*)

Table 3. (*continued*)

EC6			1	

Aggregate Weights of Nain and Subcategory DFs for all the Experts.

Main category DFs	Weights of main category DFs	Subcategory DFs	Weights of subcategory DFs	Global weights	Ranking
Social (SL)	0.466	SL1	0.397	0.185	1
		SL2	0.062	0.028	13
		SL3	0.047	0.022	17
		SL4	0.255	0.119	2
		SL5	0.032	0.015	19
		SL6	0.106	0.049	7
		SL7	0.101	0.047	9
Environmental (ET)	0.259	ET1	0.137	0.035	12
		ET2	0.140	0.036	11
		ET3	0.255	0.066	5
		ET4	0.218	0.056	6
		ET5	0.152	0.039	10
		ET6	0.098	0.025	15
Economic (EC)	0.275	EC1	0.329	0.090	3
		EC2	0.241	0.069	4
		EC3	0.172	0.048	8
		EC4	0.099	0.027	14
		EC5	0.078	0.021	18
		EC6	0.081	0.023	16

seaborne transport. It is also gotten from the table that the economic and environmental dimensions follow respectively in order of importance. From the findings, it can be construed that the DFs that are related to the social context are exceedingly important and should be adequately guaranteed for sustainable regional and trans-border infrastructure development. The economically-related DFs are the next in rank of importance with regards to the development of sustainable regional and trans-border infrastructure of road, rail and maritime transport. The environmental context is ranked least amongst the main category dimensions. Transport and logistics managers must be encouraged to be in conformity with these DFs to ensure the sustainability of regional and trans-border infrastructure of road, rail and seaborne transport.

4.3 Global Ranks of the DFs

Table 3 exudes the global ranking of the DFs for sustainable regional and trans-border infrastructure of road, rail and seaborne transport. The top five DFs under the global ranks, which represent the top 25% of DFs, belong to all the three dimensions considered in this study. These top DFs are access, equity and fairness; health; cost and speed; reliability; and carbon dioxide/air emissions. The results of this study propose that access, equity and fairness is the highest ranked DFs for the attainment of the objective of developing sustainable regional and trans-border infrastructure of road, rail and seaborne transport. Access, equity and fairness are hinged on the transport network at the local level that facilitates access to markets, social services and information, particularly in remote areas. Market efficiency and productivity is improved by efficient transport services, and this has a significant impact on reducing poverty and inequality (Fan & Chan-Kang 2004; Fan et al. 2002). The second most significant DF in the ranking hierarchy is health. This denotes how a dependable, properly working and easy to use transport network facilitates access to health care. Transportation networks are predominantly essential in delivering critical health supplies and services to populations living in rural and far-flung communities. Thirty-one percent of the world's rural population, with ninety-eight percent of them in developing nations, do not have ample access to road transport networks (World Bank 2015), and this impedes the timely delivery of health services. Cost and speed are also a highly ranked DF in this study. Cost and speed designate the time cost of the freight that is being transported. Low-cost rail and seaborne transport are usually ranked as the slowest of transport modes, whilst high-cost air transport is ranked as the fastest (Hanssen et al. 2012). The next most important DF, reliability, signifies the dependability of a transport system as a result of reduced variability in travel time. According to surveys by shippers', reliability of transit time is one of the most essential factors influencing cargo transport (Banomyong, 2003). The time-reliability of road transport is lofty, seaborne transport is also deemed to have a high level of reliability (USCC 2006). Carbon dioxide/air emissions is the fifth ranked most important DFs for sustainable regional and trans-border infrastructure of road, rail and seaborne transport, according to this research. Road transport is fuelling the intensification of transport emissions. Seaborne transport is the most environmentally sustainable mode of transport, trailed by rail transport.

4.4 Ranking of the DFs Within Each Dimension

4.4.1 Social DFs

The findings of this research exude that access, equity and fairness has the highest rank in this dimension. As discussed in the preceding paragraphs, access, equity and fairness are pivoted around the transport network at the local level that facilitates access to markets, social services and information. This implies that managers and other actors in the transport sector should endeavour to improve the access of rural communities and remote areas to transport services in order to enhance equity and fairness. The next ranked in this dimension is health, which has also been widely discussed in the preceding section. The next five ranked DFs in this dimension, in order of importance, are safety; congestion; social inclusiveness and value; community involvement; and labour conditions respectively.

4.4.2 Environmental DFs

This dimension has carbon dioxide/air emissions as the highest ranked DF. The next most important DF in this context is climate change impacts/resilience. The next four DFs, ranked in order of their significance to the objective of the study, are: energy intensity and resource depletion; noise, vibrations and biodiversity; marine, air and soil pollution; and land use.

4.4.3 Economic DFs

Cost and speed is the highest ranked DF within the economic dimension. The next five DFs, ranked in order of their significance to the development of sustainable regional and trans-border infrastructure of road, rail and seaborne transport, are: reliability; flexibility; efficiency and productivity; employment and revenue generation; and energy efficiency respectively.

4.5 Theoretical and Practical Implications

This study significantly contributes to the theory and practice of developing sustainable regional and trans-border infrastructure of road, rail and seaborne transport. From a theoretical perspective, the theoretical framework based on the social, environmental and economic (SEE) dimensions of sustainability aids in understanding the limited areas that, when concentrated on, can attain sustainability of regional and trans-border infrastructure of road, rail and seaborne transport. From the SEE concepts of sustainability, the development of sustainable regional and trans-border infrastructure is influenced by determinant factors categorized into SEE dimensions. Theoretically, through the lenses of the SEE framework, this study enacts a proliferation in the level of variance explained on the drivers of sustainable regional and trans-border infrastructure of road, rail and maritime transport in a distinctive fashion by using the BWM.

The findings of this research suggest that access, equity and fairness; health; cost and speed; reliability; and carbon dioxide/air emissions have significant influence on the sustainability of regional and trans-border infrastructure of road, rail and seaborne transport. These findings are in line with the studies by Fan et al. (2002) and Fan & Chan-Kang (2004), with regards to how sustainable transport infrastructure and services at the local level can facilitate access to markets, social services, health care and information, particularly in remote areas. Furthermore, Market efficiency and productivity is improved by efficient transport services, and this has a significant impact on reducing poverty and inequality.

Succinctly, this research corroborates studies on the determinant factors for sustainable regional and trans-border infrastructure of road, rail and seaborne transport. However, this approach and context differ from previously published papers on the subject. For this distinct work, a theoretical framework based on the SEE concepts of sustainability was used to study the determinant factors which influence sustainable regional and trans-border infrastructure of road, rail and seaborne transport. The developed theoretical model for this study can be applied by any transportation and logistics firm to classify its organisational success factors according to their importance rankings.

4.6 Managerial Implications

The proceeds of this research may offer a comprehensive and in-depth understanding to transport and logistics managers on effective measures for sustainability of regional and trans-border infrastructure of road, rail and seaborne transport. The research is especially helpful for managers of major transport and logistics organisations in developing countries like Ghana, Nigeria, Ivory Coast, and other nations within the regional bloc.

Managers may adopt the modelling framework of this study, and focus more on improving the success factors for sustainable regional and trans-border infrastructure of road, rail and seaborne transport across the regional bloc. Significantly, this study would assist managers to highlight the highly ranked determinant factors for sustainability of regional and trans-border infrastructure, and focus on them with committed resources.

5 Conclusion

The application of the SEE theoretical framework for this study finely tuned the level of variance explained on the determinant factors of developing sustainable regional and trans-border infrastructure of road, rail and seaborne transport. In the shade of the SEE concept, this study proposed a comprehensive research framework that is relevant to the context of sustainability of trans-border infrastructure of road, rail and seaborne transport in the West African regional league. It is envisaged that this would afford a better understanding of a prolific mix of intermodal transport to achieve efficiency and sustainability. It is suggested that rail transport infrastructure is developed across sections of the sub regional bloc linked by land to enhance the momentum towards sustainability in 2030 as per the SDG goals. This work further recommends that efforts are also enhanced to maximise the use of inland waterways, with accessible and reliable infrastructure, for freight transport.

This study has revealed the ranked values of the determinant factors for sustainable regional and trans-border infrastructure of road, rail and seaborne transport. However, the research is partly based on the opinions of transport and logistics managers in Ghana, which may be characterised with biased judgment and ambiguity. In future research, fuzzy logic may be used to reduce uncertainty in experts' opinions (Orji and Wei 2016). Also, in future, determinant factors for sustainable regional and trans-border infrastructure of a new mingle of transportation modes that includes air transport may be investigated by using the theoretical framework developed in this work. Also, the research model may be modified to fit other multi-criteria decision methods such as the Technique for Order Preference by Similarity to Ideal Solution, and the Analytical Hierarchy Process. A wider perspective of this current work may be carried out by garnering data from a bigger pool of experts in the transport and logistics industries of other countries. Moreover, a comparative analysis may be done by either comparing diverse modelling frameworks on the research subject or comparing findings of transport and logistics institutions of different countries or comparing findings from different transport and logistics organisations locally.

References

Banomyong, R.: International freight transport choices for Lao PDR: the dilemma of a less developed and land-locked country. In: Proceedings of the 9th World Conference on Transport Research, CD-ROM (2003)

European Commission: Handbook on estimation of external costs in the transport sector: internalisation measures and policies for all external cost of transport (IMPACT) – version 1.1 (2008)

Eurostat: Panorama of Transport, 2009 ed (2009)
Fan, S., Chan-Kang, C.: Returns to investment in less-favoured areas in developing countries: a synthesis of evidence and implications for Africa. Food Policy **29**, 431–444 (2004)
Fan, S., Hazell, P., Thorat, S.: Linkages between Government Spending, Growth, and Poverty in Rural India, International Food Policy Research Institute Research Report, Washington, D.C., International Food Policy Research Institute, vol. 110 (1999)
Fan, S., Zhang, L., Zhang, X.: Growth, Inequality, and Poverty in Rural China: The Role of Public Investments, International Food Policy Research Institute Research Report, Washington, D.C., International Food Policy Research Institute, vol. 125 (2002)
Fu, J.R., Farn, C.K., Chao, W.P.: Acceptance of electronic tax filing: a study of taxpayer intentions. Inform. Manage. **43**, 109–126 (2006)
Geurs, K.T., Boon, W., & Van Wee, Bert. (2009). "Social impacts of transport: literature review and the state of the practice of transport appraisal in the Netherlands and the United Kingdom", *Transport Reviews,* 29(1)
Hanssen, T.S., Mathisen, T.A., Jørgensen, F.: Generalized transport costs in intermodal freight transport", Procedia – Social and Behavioral Sciences, In: Proceedings of the 15th meeting of the EURO Working Group on Transportation, vol. 54 (2012)
International Energy Agency: World Energy Outlook 2012 (2012a)
International Energy Agency: CO_2 Emissions from Fuel Combustion: Highlights–2015 Edition (2012b)
International Energy Agency: Executive summary: World Energy Outlook 2013 (2013)
International Energy Agency: Key world energy statistics 2015 (2015a)
International Energy Agency: CO_2 Emissions from Fuel Combustion: Highlights – 2015 Edition (2015b)
Nilashi, M., Ahmadi, H., Ahani, A., Ravangard, R., Ibrahim, O.B.: Determining the importance of hospital information system adoption factors using fuzzy analytic network process (ANP). Technol. Forecast. Soc. Chang. **111**, 244–264 (2016)
Orji, I.J., Wei, S.: A detailed calculation model for costing of green manufacturing. Ind. Manage. Data Syst. **116**(1), 65–86 (2016)
Reis, V.: Analysis of mode choice variables in short-distance intermodal freight transport using an agent-based model. Transport. Res. Part A: Policy Pract. **61**, 100–120 (2014)
Rezaei, J.: Best- worst multi- criteria decision making method: some properties and a linear model. Omega **64**, 126–130 (2016)
Rodrigue, J.P., Comtois, C., Slack, B.: The Geography of Transport Systems, 3rd edn. Oxford and New York, Routledge (2013)
The World Bank: Safe, clean, and affordable: transport for development – the World Bank Group's transport business strategy for 2008–2012 (2008)
UNCTAD: Fostering Africa's Services Trade for Sustainable Development (United Nations publication. Geneva) (2015a)
United Nations Economic and Social Council: Economic and Social Commission for Asia and the Pacific, Ministerial Conference on Transport. Sustainable Development Goals and transport E/ESCAP/MCT (3)/2 (2016)
United States Chamber of Commerce: Land transport options between Europe and Asia: commercial feasibility study (2006)
Wang, Z.G., Xu, R., Lin, H., Wang, R.J.: Energy performance contracting, risk factors, and policy implications: identification and analysis of risks based on the best- worst network method. Energy **170**, 1–13 (2019)
World Bank: Massive Drop in Number of Unbanked, says New Report (2015). Retrieved from http://www.worldbank.org/en/news/pressrelease/2015/04/15/massive-drop-in-number-of-unbanked-says-new-report

World Health Organizations: Global Status Report on Road Safety 2015 (2015)
World Travel and Tourism Council: Travel & tourism: economic impact – world (2014)
www.adb.org/sectors/transport/key-priorities/urban-transport
www.eea.europa.eu/data-and-maps/figures/specific-co2-emissions-per-tonne-2
www.env.go.jp/en/statistics/contents/index_e.html#onshitukoukagasu
www.worldbank.org/transport/transportresults/headline/rural-access/rai-updated-modelbasedscores5-20070305.pdf
Yang, Y., Lau, A.K.W., Lee, P.K.C., Yeung, A.C.L., Cheng, T.C.E.: Efficacy of China's strategic environmental management in its institutional environment. Int. J. Oper. Prod. Manage. **39**(1), 138–163 (2018)

Innovation Performance and Efficiency of Research and Development Intensity as a Proportion of GDP: A Bibliometric Review

S. N. Dorhetso[✉], L. Y. Boakye, and D. N. O. Welbeck

Accra Institute of Technology (AIT), Accra, Ghana
samueldorhetso@gmail.com, dorhetso@ait.edu.gh,
danielwelbeck@upsamail.edu.gh

Abstract. Purpose: The resolution of this study was to conduct a bibliometric analysis of innovation performance and efficiency of research and development (R&D) expenditure as a proportion of gross domestic product (GDP), to estimate its relative effects on economic growth and productivity.

Design/Methodology/Approach: Scopus database was the source of material collected to conduct this study. Selected keywords from literature were combined with Boolean operators and searched. Subsequently a rigorous screening procedure with an inclusion and exclusion criteria was adopted to select the relevant articles for the study. A four prong bibliometric analysis: co-authorship analysis; keyword co-occurrence analysis; bibliographic coupling; and co-citation analysis was then conducted.

Findings: From the co-authorship analysis based on countries, the United states emerged with the highest number of documents, followed by China and the United Kingdom, with Italy in the third position. However, China had the greatest number of citations. South Africa, Nigeria and Ghana had 3,2 and 1 documents with 14, 20 and 5 citations respectively. From the results of the keyword co-occurrence analysis, it was observed that innovation, and research and development were the most dominant keywords that reoccurred, indicating that they were the topmost research hotspots. From the bibliographic coupling analysis, it was discovered that articles themed on innovation and R&D subsidies have clouded the research space of innovation performance and efficiency of R&D intensity as a proportion of GDP since 2007 till date.

Implications/Research Limitation: It is expected that this study would significantly contribute to theoretical literature by providing a bibliometric record of innovation performance and efficiency of R&D intensity as a ratio of GDP. However, the study uses data from Scopus and limitations of the database have implications for the results. Also, although numerous manual screening criteria were set to filter documents, there may have been subjective bias.

Practical Implication: Besides the theoretical contribution of this study, the results would also help governments to plan and instigate research that is geared towards national development.

Originality/Value: This distinctive research conducted a bibliometric review of the performance of innovation and efficiency of R&D spending as a fraction of GDP, and corroborated a grave research field that is currently understudied.

Keywords: Bibliometric · Efficiency · Innovation · Performance · Subsidies

1 Introduction

Research on innovation has been deeply explored among scholars in management (Anderson et al. 2014), and served as a crucial source of competitive advantage in societies. Innovation research in organizations originated from the late 1960s, when scholars conducted research on innovation from an organizational perspective, especially in the healthcare industry, converging on innovation diffusion (Walker 1969) and centralization in organizations (Zaltman et al. 1973). The significance of innovation to economic development is of no doubt. Innovation contributes to the objectives of shared prosperity and poverty alleviation by creating gains that escalate employment, raise wages, and expand access of the underprivileged to products and services. Investing in innovation upsurges organisational abilities and enables the adoption of new technologies to increase labour productivity. Innovation is a first world activity, which is usually seen as the product of highly educated and skilled labour in R&D departments, laboratories, or research organisations. Governments can improve a country's innovation performance by rectifying the externalities and ambiguity inherent to the course of innovation. Thus, there has been a renewed focus on innovation policies in developing countries in recent decades. Nevertheless, few governments can answer with sureness basic queries of how much was spent, by whom, for what purpose, and with what fallouts. Developing economies have been paying additional attention to the influence of innovation policies to their development policies. Subsequently, investments in innovation and R&D by developing nations have soared significantly in the past decade, although governments habitually lack the utensils to properly apportion resources, ensure satisfactory returns on the expenditure, or even justify its use. The purpose of this study is to conduct a bibliometric analysis of innovation performance and efficiency of R&D spending as a proportion of GDP, to assess the influence of a well-resourced research and innovation on the economic development and productivity of countries. Taking que from a large number of articles relating intensely to innovation performance and efficiency of research and development intensity as a proportion of GDP, this research studied and produced results through the use of bibliometric analysis. This study adopted VOSviewer, a broadly used bibliometric mapping tool to analyse the dissemination of research publications, the scientific community, intellectual structure, and research hotspots.

2 Literature Review

Innovation has long been renowned as an important source of economic growth and a significant activity for addressing major development encounters, such as food scarcity;

access to services, and climate change (IEG 2013). Developing countries have been paying more attention to the contribution of innovation and R&D policies to their development strategies. Accordingly, investments in innovation and R&D by developing countries have increased substantially over the past decade. Nevertheless, governments of developing countries usually do not have the tools for proper allocation of resources, and safeguarding suitable returns on its expenditure. R&D encompass creative and systematic work carried out to amplify the stock of knowledge, including the knowledge of humankind, culture and society, and to invent new applications of available knowledge (OECD 2015). R&D intensity as a proportion of GDP was chosen by IAEG-SDG as one of two measures for its target 9.5. The other measure is the number of researchers, in full-time equivalent, per million inhabitants. According to the OECD Main Science and Technology Indicators (MSTI) database (2021), OECD R&D intensity increased from 2.4% in 2018 to nearly 2.5% in 2019. Several studies (Fagerberg 1994; Jones 1995) have postulated that R&D increases productivity by expanding reserve bases and making resource utilization more efficient. Lall and Pietrobelli (2005) reviewed the national innovation systems in Sub-Saharan and discovered that R&D institutions lacked proper infrastructure to access and diffuse their technological needs. Many of the preceding bibliometric studies of innovation research mostly focused on reviewing some sub-topics of innovation, such as inclusive innovation (Mortazavi et al. 2021), open innovation (Randhawa et al. 2016), frugal innovation (Dangelo and Magnusson 2021), and new product development (Marzi et al. 2021). Others also focused on reviewing a specific journal associated to innovation research, for example Journal of Product Innovation Management (Durisin et al. 2010; Antons et al. 2016; Sarin et al. 2018). However, gaps still remain as there has not been sufficient bibliometric studies that review papers on the performance of innovation and efficiency of R&D spending as a fraction of GDP found in extant literature. The intention of this study is to conduct a bibliometric analysis of innovation performance and efficiency of R&D expenditure as a proportion of GDP, to estimate the effects of a well-funded research and innovation on the economic growth and productivity of countries. Following a large number of documents relating keenly to innovation performance and efficiency of research and development intensity as a proportion of GDP, this work studied and produced findings via the use of bibliometric analysis. Bibliometric analysis, a more objective and efficient tool than traditional qualitative analysis, is the application of mathematics and statistical methods to study scientific publications (Leydesdorff 1995). Besides the theoretical contribution of this study to existing literature on innovation and R&D expenditure, the results would also help governments to plan on committing more funds to instigate research that is geared towards national development.

3 Research Methodology

Scopus database was used as the source of material collection to conduct this study. Scopus is a research platform containing several records. It has more than 34,385 annals of published and proceeding journals, books, conference papers, et cetera. Scopus is one of the largest multidisciplinary databases of academic journals (Norris and Oppenheim 2007). The methods used to accomplish this research are categorised into the search procedure and the screening procedure by the authors. The dichotomous methodological approach to this study is unveiled in the following sub sections.

3.1 Search Procedure

Search keywords related to innovation performance and efficiency of research and development intensity as a proportion of GDP were identified from a preliminary literature review. The retrieval search string was entered by combing the keywords with Boolean operators as exuded below:

((" Innovation performance" OR" Efficiency of R&D "OR" R&D Intensity" OR "Research and development") AND ("Proportion of GDP" OR "Portion of GDP" OR "Government expenditure" OR "Public spending "OR "Government spending"))

In March 2022, the keywords were typed in and searched and the initial results yielded 226 documents, as shown in Fig. 1. The abstracts of the documents were read to be certain that they were related to the subject of study. The result of the primary search is shown below:

TITLE-ABS-KEY ((("Innovation performance" OR " Efficiency of R&D " OR" R&D Intensity" OR" Research and development") AND ("Proportion of GDP" OR" Portion of GDP" OR" Government expenditure" OR" Public spending "OR" Government spending")))

3.2 Screening Procedure

This section explains the inclusion and exclusion criteria used to source material for the conduct of this study. The procedure for inclusion and exclusion is explained below in detail according to the order of occurrence.

The search was initially limited to articles: the results were 165 documents. Hence conference papers, book chapters, conference reviews, books, reviews, editorials, notes and undefined documents were excluded from the search. The result of these exclusions from the search is shown below:

TITLE-ABS-KEY ((("Innovation performance" OR" Efficiency of R&D "OR" R&D Intensity" OR "Research and development") AND ("Proportion of GDP" OR" Portion of GDP" OR" Government expenditure" OR" Public spending "OR" Government spending"))) AND (LIMIT-TO (DOCTYPE," ar"))

The source type of the document was limited to journals: the results were 154 documents. Hence trade journals and book series were excluded from the search. The result of these exclusions from the search is shown below:

TITLE-ABS-KEY ((("Innovation performance" OR" Efficiency of R&D "OR" R&D Intensity" OR" Research and development") AND ("Proportion of GDP" OR "Portion of GDP" OR "Government expenditure" OR "Public spending " OR "Government spending"))) AND (LIMIT-TO (DOCTYPE , "ar")) AND (LIMIT-TO (SRCTYPE , "j"))

The search was then limited to only articles published in the English language: the results were 147 documents. Hence articles in Czech, Chinese, Croatian, Malay, Portuguese and Spanish were excluded. The result of these exclusions from the search is shown below:

TITLE-ABS-KEY ((("Innovation performance" OR" Efficiency of R&D "OR " R&D Intensity" OR "Research and development") AND ("Proportion of GDP" OR "Portion of GDP" OR "Government expenditure" OR "Public spending " OR "Government spending"))) AND (LIMIT-TO (DOCTYPE , "ar")) AND (LIMIT-TO (SRCTYPE , "j")) AND (LIMIT-TO (LANGUAGE , "English"))

The search was not limited any further, since from 1962 to 2022, only 226 documents were available from the Scopus search on the subject, out of which only 147 journal articles in English were selected for this review. This means that scrupulous research on the topic is critically required. The general subject of the research is not new but little work has been done in the area of innovation performance and efficiency of research and development intensity as a proportion of GDP.

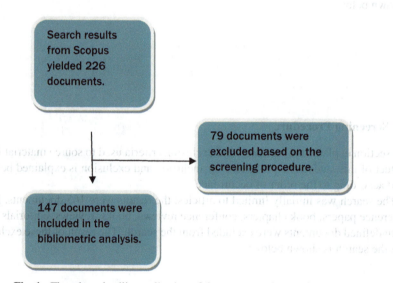

Fig. 1. Flowchart detailing collection of documents and screening process.

4 Bibliometric Analysis and Discussion

Bibliometrics, a class of scientometrics, is a tool that was developed in 1969 for library and information science. It has since been adopted by other fields of study that require a quantitative assessment of academic articles to determine trends and predict future research scenarios by compiling output and type of publication, title, keyword, author, institution, and countries data (Ho 2008; Li et al. 2017). Bibliometric analysis is a use of quantitative tools to study science. It has been applied in various fields of literature. Bibliometric analysis is simply the application of mathematics and statistical methods to study scientific publications (Leydesdorff 1995). Over time, an increasing number of indicators and techniques have been developed to quantify the research performance and contribution of authors, journals, institutions, and countries. Another feature of bibliometric analysis is that it examines scientific developments and visualizes scientific knowledge from conceptual, intellectual and social structures. Bibliometric analysis, may be combined with other useful tools to quantitatively estimate research activities from the past to the present, thus bridging a historical gap and predicting the future of research.

Bibliometric analysis, to statistically investigate and provide a comprehensive overview of a collection of related papers, is facilitated by various bibliometric matrixes such as bibliographic coupling (Kessler 1963), co-citation (Small 1973), and co-occurrence of author keywords (Nicholas and Ritchie 1978). In this study, the bibliometric analysis would be done using VOSviewer, a software tool for constructing and visualising bibliometric networks. This study contributes to literature by providing a bibliometric record of innovation performance and efficiency of research and development intensity as a proportion of GDP. For this study, a four prong bibliometric analysis would be conducted as follows: co-authorship analysis, keyword co-occurrence analysis, bibliographic coupling, and co-citation analysis.

4.1 Co-authorship Analysis

The co-authorship analysis was initially carried out based on authors. A second co-authorship analysis was performed based on countries. The subsections below would review the results of these two co-authorship analysis with different bases.

4.1.1 Co-authorship Analysis Based on Authors

The results of the co-authorship analysis based on authors were encapsulated into 138 clusters, ranking papers from old to new. The older documents were captured in the purple clusters and the newer ones were featured in the yellow clusters. The green clustered documents hovered in between the old and the new. This can be seen from the results of the overlay visualization of the analysis using VOSviewer as shown in Fig. 2. It can also be observed from the results that the maximum number of co-authored documents were two, by authors such as Alston JM, Blume-Kohout ME and Blackburn JM.

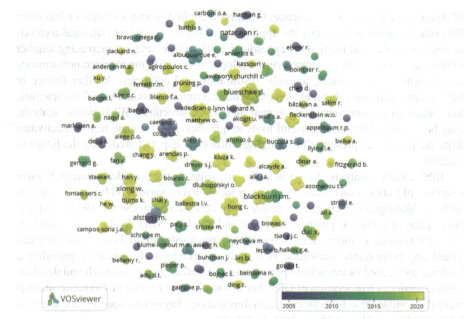

Fig. 2. VOSviewer results of overlay visualization of the co-authorship analysis based on authors

4.1.2 Co-authorship Analysis Based on Countries

The results of the co-authorship analysis based on countries were grouped into twenty-four clusters. From the results, it can be observed that the United States have the greatest number of co-authored documents amongst the countries represented by publications. The united states have 38 documents with 632 citations. This is followed by the United Kingdom and China, both having an equal number of 16 documents. However, whilst the Chinese publications have 747 citations, documents from the United Kingdom have only 386 citations. The next country in this category is Italy, which has 14 documents with 238 citations. The results of the analysis, for a few select countries in addition to the first four, is exuded in Table 1. From the overlay visualization of the results of the analysis, it is notable amongst these four countries that the papers from China are the most recent, followed by the papers from the United Kingdom and Italy, with the United States having the oldest documents. This finding can be explained to mean that authors from the United States, as well as other purple coded countries such as Norway and Japan, commenced publishing on research into innovation performance and efficiency of R&D intensity as a proportion of GDP much earlier than the others. The findings from the co-authorship analysis based on countries is consistent with studies by Peng et al. (2021) and Li et al. (2021), with regards to the United States, the United Kingdom and China been ranked as the top three countries where the majority of publications emanate from. The result of the overlay visualization of the analysis using VOSviewer is shown in Fig. 3 below:

Table 1. Results of co-authorship analysis based on countries

Country	Number of papers	Number of citations
United States	38	632
China	16	747
United Kingdom	16	386
Italy	14	238
Germany	8	29
South Korea	4	19
South Africa	3	14
Nigeria	2	20
Ghana	1	5
Check Republic	1	7

Source: authors' own construct

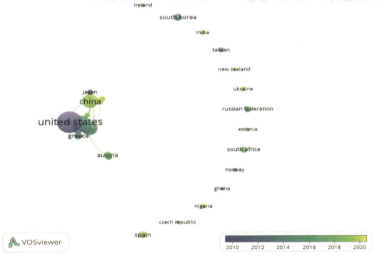

Fig. 3. VOSviewer results of overlay visualization of the co-authorship analysis based on countries

4.2 Keyword Co-occurrence Analysis Based on Author Keywords

Keywords are the convergence and generalization of the principal content of literature, and their analysis helps to recognise the research hotspots of a particular research field or discipline. Author keywords are remarkable emblems of the editorial content or its link with its research question (Strozzi et al. 2017). According to Ding et al. (2001), the co-occurrence of author keywords amongst papers might be a pointer that the documents convey a collective theme. For this study, the keyword co-occurrence analysis was hinged on author keywords, and a minimum number of three occurrences. From the results, it can be observed that innovation, and research and development were the most dominant keywords that reoccurred, with a frequency of eight occurrences. However, innovation had more links and a greater link strength than research and development. The second set of keywords with seven occurrences were R&D and economic growth. The third set with five occurrences included government spending and R&D funding. This was followed by renewable energy, fiscal policy and productivity, with four occurrences. The last set with three occurrences entailed Africa, developing countries, policy, sustainability and public expenditure. It can also be discovered from the findings that documents in the field of innovation, R&D, R&D funding, developing countries and policy are the oldest. It is further observed that papers on fiscal policy, sustainability, government spending and renewable energy are relatively the most current. From these observations made, it can be advocated that innovation, and research and development were the topmost research hotspots, followed by R&D and economic growth. Based on the findings from the keyword co-occurrence analysis. It is suggested that future research should focus more on connecting the earlier set of keywords with the most recent set, to motivate the development of sustainable fiscal policy and government spending to expand R&D and innovation in Africa and other developing countries. The result of the overlay visualization of this analysis using VOSviewer is shown in Fig. 4 below:

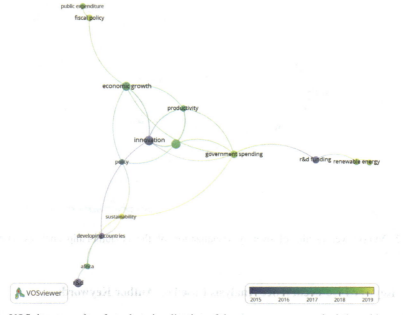

Fig. 4. VOSviewer results of overlay visualization of the co-occurrence analysis based on author keywords.

4.3 Bibliographic Coupling Analysis Based on Documents

It has been suggested by Kessler (1963) and Weinberg (1974) that two papers sharing one or more common references constitute a bibliographic couple and may lean towards similar intellectual content. Hence, bibliographic coupling is deployed in this review to identify the thematic structure of publications on innovation performance and efficiency of R&D expenditure as a proportion of GDP. It is noteworthy to state that only documents that have received at least ten citations were used for the bibliographic coupling analysis. Hence only fifty-one documents were selected. From the VOSviewer results, it was observed that the documents were categorized into seven clusters. However, only five clusters were selected for this analysis. The contents of these five clusters are briefly discussed below:

Cluster 1 (Red Cluster)

The cluster consists of eight documents. The papers in this cluster blend innovation and R&D expenditure with economic growth and government expenditure in developing countries. The link between the economy and public research was strongly exuded in most of these papers. Cruz-Castro and Sanz-Menéndez (2016) tops the citation list in this cluster with thirty-one citations. Cruz-Castro and Sanz-Menéndez (2016) published on the effects of the economic crises on public research in the Spanish budgetary policy and research organisations. They asserted that the dialogue of governments and international organizations often connect actions to cope with the crisis with reforms and alterations in many areas, such as R&D, although in practice overarching fiscal consolidation strategies could be damaging opportunities to launch government policies to change and proliferate the efficiency of the sector. Awang (2004) and Dao (2012) follow as the next papers to be cited most with eighteen citations each. Awang (2004) wrote on human capital and technology development in Malaysia. The article investigated the development of data and Communication Technology (ICT) and its relation to the development of human capital in Malaysia as a country experiencing a revolution into an ICT-driven and knowledge-based society. Education and training, was examined in terms of the government spending on education and training, years of schooling, number of enrolment and degree of education of the labour force. ICT development was estimated in terms of the personnel involvement in R&D in related areas of technology or the expansion allocation and disbursement for R&D, and the degree of ICT usage in the various sectors of the economy and amongst the general populace. Dao (2012) published on government expenditure and growth in developing countries, based on data from the World Bank and using samples of 28 developing economies. The findings of Dao (2012) revealed that per capita GDP growth is dependent on the growth of per capita public health expenditure in the GDP, growth of per capita public spending on education in the GDP, population growth, growth of the share of total health expenditure in the GDP and the share of gross capital formation in the GDP.

Cluster 2 (Green Cluster)

This cluster consists of seven items. The papers in the cluster primordially focused on innovation and R&D subsidies. Gorg and Strobl (2007) tops the citation list in this cluster with 189 citations. Gorg and Strobl (2007) published on the effects of R&D subsidies on private R&D. They investigated the relationship between government support for R&D and R&D expenditure financed privately by firms. They used a comprehensive plant level data set for the manufacturing sector in the Republic of Ireland, and found that for domestic plants small grants serve to increase private R&D expenditure, whereas too hefty a grant may crowd out private financing of R&D. Carboni (2011) follows as the next paper to be cited most with 62 citations. Carboni (2011) wrote on R&D subsidies and private R&D expenditures, citing evidence from Italian manufacturing data. The article used a comprehensive firm level data set for the manufacturing sector in Italy to study the influence of government assistance on privately financed R&D spending. Estimates from a non-parametric matching procedure suggested that public support has a positive influence on private R&D investment, in the wisdom that the beneficiary firms attain more private R&D than they would have without government assistance. This showed that the likelihood of perfect crowding out between private and public coffers can be rejected. Tax incentives appeared to be more effective than direct grants in the sample of Italian firms used for the study. The study found that grants encouraged the use of inside sources, and some evidence of positive effects on credit financing for R&D.

Cluster 3 (Blue Cluster)

This cluster encapsulates six articles. Innovation in the energy sector and R&D expenditure were the main themes of most of the papers in this cluster. Garrone and Grilli (2010) tops the citation list in this cluster with 93 citations. Garrone and Grilli (2010) published on an empirical investigation into the relationship between public expenditure in energy R&D and carbon emissions per GDP. Their paper focused on public energy R&D, an old-fashioned and controversial option amid the various climate technology guidelines, and empirically analyses its association with carbon emissions per GDP and its two components: energy intensity and the carbon factor. Indication of the causality relations that have prevailed in 13 advanced economies over the 1980–2004 period were acquired via dynamic panel models. Their findings confirmed that government R&D spending is not sufficient by itself to boost energy innovation. Bointner (2014) follows as the next paper to be cited most with 77 citations. Bointner (2014) published on innovation in the energy sector with regards to lessons learnt from R&D expenditures and patents in selected IEA countries. He conducted an extensive literature review on innovation drivers and barriers, and an examination of the knowledge brought by public R&D expenditures and patents in the energy sector. The cumulative knowledge stock induced by public R&D expenditures in 14 investigated IEA-countries was found to be 102.3bnEUR in 2013. Nuclear energy had the largest portion of 43.9bnEUR, trailed by energy efficiency contributing 14.9bnEUR, fossil fuels had 13.5bnEUR, and renewable energy had 12.1bnEUR. A linear relation between the GDP and the cumulative knowledge, with each billion EUR of GDP leading to an additional knowledge of 3.1milEUR was established from a regression analysis. Nevertheless, linearity was not found for single energy technologies, but the fallouts show that suitable public R&D funding for research and development connected with a subsequent upgrade of the market diffusion of a niche technology may lead to a development of the particular technology.

Cluster 4 (Yellow Cluster)

This cluster consists of five documents The cluster consists of papers that have similar themes, which discuses research policies and issues of the link between academia and industry. Alston et al. (1998) tops the citation list in this cluster with 67 citations. Alston et al. (1998) published on international investment patterns and policy perspectives with regards to financing agricultural research. Their study concluded that scoring is a way of developing shortcut pointers of the significances of research and possibly, for evaluating the estimated offerings of research to various specified objectives in order to obtain a concise measure of the effects of research. They suggested that sporadically a simplified scoring model could be used to assist research administrators with prioritization in situations where research funds are being apportioned across huge numbers of commodity research programs or research zones, and where resources are insufficient to afford a more complete analysis. Fernandes et al. (2010) follows as the next paper to be cited most with 47 citations. Fernandes et al. (2010) wrote on academy-industry links in Brazil, and gave evidence of channels and benefits for firms and researchers. Their study explained that knowledge flows between universities, public research organisations may adopt several conduits according to agents' motivations and expected benefits. Models were used to examine which channels of interaction led to particular benefits for firms, universities and research organisations in Brazil. The findings indicated that bi-directional channels were predominantly relevant, yielding both innovative and productive benefits for the organisations, and intellectual and economic benefits for the universities. Bi-directional channels were the most significant in terms of intellectual benefits for the researchers and innovative benefits for the organisations.

Cluster 5 (Purple Cluster)

This cluster consists of four papers. The themes of papers in this cluster revolve around R&D and productivity, as well as public spending and economic growth. Zhang et al. (2021) tops the citation list in this cluster with 79 citations. Zhang et al. (2021) wrote on the mediating role of green finance on public spending and green economic growth in Belt and Road Initiative (BRI) region. Their study analysed panel data of BRI member countries from 2008 to 2018 using the generalized method of moments (GMM) method and data envelopment analysis (DEA) to assess the association between public expenditure on R&D and green economic growth and energy efficiency. Their findings revealed a fluctuating green economic growth indicator during the research period attributed to the non-serious nature of government strategies. They found that the GMM method confirmed both composition and technique effects in the entire sample. However, the outcome of the sub-sample presented a heterogeneous effect on high GDP per capita countries. Also, their study indicated that public expenditure on human resources and R&D of green energy technologies instigates a sustainable green economy through labour and technology-oriented production events and it renders different effects in different countries. Bravo-Ortega and García Marin (2011) follows as the next paper to be cited most with 53 citations. Bravo-Ortega and García Marin (2011) published on R&D and productivity, and classified it as a two-way avenue. Their work used a 65-country panel for the period between 1965 and 2005 to study the association between R&D and productivity using several R&D indicators. Their findings revealed that per capita R&D

expenditure was strongly exogenous to productivity, and this allowed them to cultivate an advanced argument that demonstrated the high social returns to R&D expenditure.

From the findings of the bibliographic coupling analysis, it is evident that the major themes of research that have emerged in the development of research on innovation performance and efficiency of R&D intensity as a proportion of GDP are: innovation and R&D expenditure, economic growth and government expenditure in developing countries; innovation and R&D subsidies; Innovation in the energy sector and R&D expenditure; research policies and the link between academia and industry; and R&D and productivity, public spending and economic growth. It can also be observed from the results, as summarized in Table 2, that the second cluster (green cluster) has the greatest number of citations, topping the list with 392 citations and 56 cites per publication. This means that articles on innovation and R&D subsidies, especially those by Gorg and Strobl (2007), have clouded the research space of innovation performance and efficiency of R&D intensity as a proportion of GDP since 2007 till date. However, there are some theoretical gaps with the determination of innovation performance and efficiency of R&D intensity as a proportion of GDP. Henceforth, more work is required in this area to tease out the most substantial driving factors of innovation performance for the efficiency of R&D intensity as a ratio of GDP. Future studies should focus on the determination of the most significant driving factors for innovation performance and efficiency of R&D intensity as a fraction of public expenditure. Figure 5 shows the VOSviewer results of the network visualization of the bibliographic coupling analysis based on documents.

Table 2. Cluster summary

Description	Cluster 1	Cluster 2	Cluster 3	Cluster 4	Cluster 5
Period	2004:2021	2007:2021	2010:2018	1998:2011	2011:2021
Total publications (TP)	8	7	6	5	4
Total citations (TC)	129	392	266	161	187
TC/TP	16.3	56.00	22.33	26.83	46.75

Source: authors' own construct

In Table 2 above, TP denotes total publication, TC denotes total citations, and TC/TP denotes cites per publication.

4.4 Co-citation Analysis Based on Cited References

Prior to performing the co-citation analysis, the minimum number of citations of a cited reference was set to three. Hence only seven documents were selected. From the results it can be observed that the documents were categorized into two clusters. The first cluster had Romer (1986) as the most co-cited reference with 4 citations. The article showed a totally specified model of long-run growth in which knowledge was presumed to be an input in production that has increasing marginal productivity. This was fundamentally a competitive equilibrium model with endogenous technological revolution. The study

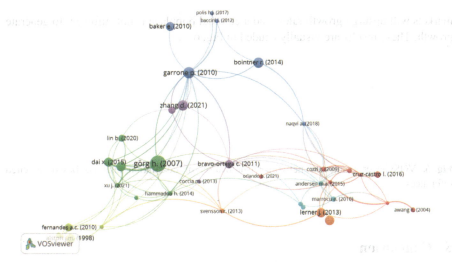

Fig. 5. VOSviewer results of network visualization of the bibliographic coupling analysis based on documents.

found that, in disparity with models grounded on diminishing returns, growth rates could be rising over time. The influences of small disturbances can be augmented by the actions of private agents, and big countries may always promulgate faster than small countries. This was followed by Lucas (1988) with 3 citations. Lucas (1988) studied the prospects for constructing a neoclassical theory of growth and international trade that is consistent with some of the main features of economic development. Three models were studied and likened to evidence: a model emphasizing physical capital growth and technological modification, a model highlighting human capital build-up through education, and a model accentuating specialized human capital accrual through learning-by-doing.

The second cluster had Aghion and Howitt (1992) as the most co-cited reference with 8 citations. Aghion and Howitt (1992) developed a model of endogenous growth in which growth is driven by vertical innovations that encompass creative destruction. Equilibrium was determined by a forward-looking difference equation, according to which the volume of research in any period depends negatively on the volume expected in the next. The study analysed positive and normative properties of stationary equilibria, and indicated conditions for the presence of cyclical equilibria and no-growth traps. The study found that the growth rate may be more or less than optimal because a business-stealing effect counteracts the usual overflow and appropriatability effects. Aghion and Howitt (1992) was followed by Romer (1990) with 3 citations. In the model by Romer (1990), growth was driven by technological change that emanates from deliberate investment decisions made by profit-maximizing agents. The unique feature of technology as an input is that it is it is a nonrival, partially excludable good, and not a conservative or public good. As a result of the nonconvexity introduced by a nonrival good, price-taking competition could not be supported, and instead the equilibrium was with monopolistic competition. Romer (1990) concluded that: the stock of human capital governs the rate of growth; too little human capital is dedicated to research in equilibrium; integration into world markets will upsurge growth rates; and a colossal populace is not sufficient to generate growth. These results are visually exuded in Fig. 6.

Fig. 6. VOSviewer results of network visualization of co-citation analysis based on cited references.

5 Conclusion

5.1 Summary

The purpose of this study was to conduct a bibliometric analysis of innovation performance and efficiency of R&D expenditure as a proportion of GDP, to estimate its relative effects on economic growth and productivity. Scopus database was the source of material collected to conduct this study. Selected keywords from literature were combined with Boolean operators and searched. Afterwards a rigorous screening procedure with an inclusion and exclusion criteria was adopted to select the relevant documents for the study. Consequently, only 147 journal articles that were published in English were selected from a total of 226 papers for this analysis. A four prong bibliometric analysis: co-authorship analysis; keyword co-occurrence analysis; bibliographic coupling; and co-citation analysis was then conducted. From the findings of the co-authorship analysis based on countries, the United states emerged with the highest number of documents, followed by China and the United Kingdom, with Italy in the third position. However, China had the highest number of citations. South Africa, Nigeria and Ghana had 3,2 and 1 documents with 14, 20 and 5 citations respectively. The findings from the co-authorship analysis based on countries is in conformity with studies by Peng et al. (2021) and Li et al. (2021), with regards to the United States, the United Kingdom and China been graded as the topmost three countries where the bulk of publications originate from. Also, from the findings of the keyword co-occurrence analysis, it was detected that innovation, and research and development were the most dominant keywords that reoccurred. It was also detected that papers on fiscal policy, sustainability, government spending and renewable energy were relatively new, and documents in the field of innovation, R&D, R&D funding, developing countries and policy were the oldest. Furthermore, from the bibliographic coupling analysis, it was discovered that articles themed on innovation and R&D subsidies had clouded the research space of innovation performance and efficiency of R&D intensity as a proportion of GDP since 2007 till date. It is anticipated that this study would significantly contribute to literature by providing a bibliometric record of innovation performance and efficiency of R&D intensity as a proportion of GDP. Besides the theoretical contribution of this study, the findings can also help governments to plan and instigate research that is geared towards national development. However, the study is limited by the inability of the researchers to capture unpublished R&D records which may have been of great implication to the study. Furthermore, the study uses data from Scopus and limitations of the database have implications for the outcomes. Also, although copious manual screening criteria were set to filter documents, there may have been subjective prejudice which limited the study. This unique research conducted a bibliometric review of the performance of innovation and efficiency of R&D expenditure as a portion of GDP, and corroborates an acute research field that is currently understudied.

5.2 Directions for Future Research

Observing from the keyword co-occurrence analysis, it was realized that documents in the field of innovation, R&D, R&D funding, developing countries and policy were the oldest. It was also observed that papers on fiscal policy, sustainability, government spending and renewable energy were relatively the most current. From these observations made, it can be recommended that future research should focus more on connecting the earlier set of keywords with the most recent set, to inspire the development of sustainable fiscal policy and government expenditure to expand R&D and innovation in Africa and other developing countries. Furthermore, from the bibliographic coupling analysis, it was evident that the major themes of research that have emerged in the development of research on innovation performance and efficiency of R&D intensity as a proportion of GDP were: innovation and R&D expenditure, economic growth and government expenditure in developing countries; innovation and R&D subsidies; Innovation in the energy sector and R&D expenditure; research policies and the link between academia and industry; and R&D and productivity, public spending and economic growth. It was also seen from the results that the second cluster (green cluster) had the greatest number of citations, topping the list with 392 citations and 56 cites per publication. This means that articles on innovation and R&D subsidies, especially Gorg and Strobl (2007), have clouded the research space of innovation performance and efficiency of R&D intensity as a proportion of GDP since 2007 till date. However, there are some theoretical gaps with the determination of innovation performance and efficiency of R&D intensity as a proportion of GDP. Henceforth, more work is required in this area to tease out the most substantial driving factors of innovation performance for the efficiency of R&D intensity as a ratio of GDP. Future studies should focus on the determination of the most significant driving factors for innovation performance and efficiency of R&D intensity as a fraction of public expenditure.

References

Aghion, P., Howitt, P.: A model of growth through creative destruction. Econometrica Econom. Soc. **60**(2), 323–351 (1992)

Alston, J.M., Norton, G.W., Pardey, P.G.: Science under scarcity: principles and practice for agricultural research evaluation and priority setting. Cab International, Netherlands (1998)

Anderson, N., Potocnik, K., Zhou, J.: Innovation and creativity in organizations: a state-of-the-science review, prospective commentary, and guiding framework. J. Manag. **40**, 1297–1333 (2014)

Antons, D., Kleer, R., Salge, T.O.: Mapping the topic landscape of JPIM, 1984–2013: in search of hidden structures and development trajectories. J. Prod. Innov. Manag. **33**, 726–749 (2016)

Awang, H.: Human Capital and Technology Development in Malaysia, 5. Researchgate (2004)

Bointner, R.: Innovation in the energy sector: lessons learnt from R&D expenditures and patents in selected IEA countries. Energy Policy **73**(C), 733–747 (2014)

Bravo-Ortega, C., García Marin, A.: R&D and productivity: a two way avenue? World Dev. **39**(7), 1090–1107 (2011)

Carboni, O.A.: R&D subsidies and private R&D expenditures: evidence from Italian manufacturing data. Int. Rev. Appl. Econ. **25**, 419–439 (2011)

Cruz-Castro, L., Sanz-Menéndez, L.: The effects of the economic crisis on public research: Spanish budgetary policies and research organizations. Technol. Forecast. Soc. Chang. **113**, 157–216 (2016)

Dangelo, V., Magnusson, M.: A bibliometric map of intellectual communities in frugal innovation literature. IEEE Trans. Eng. Manag. **68**, 653–666 (2021)

Dao, M.Q.: Government expenditure and growth in developing countries. Prog. Dev. Stud. **12**, 77–82 (2012)

Ding, Y., Chowdhury, G.G., Foo, S.: Bibliometric cartography of information retrieval research by using co-work analysis. Inf. Process. Manag. **37**(6), 817–842 (2001)

Durisin, B., Calabretta, G., Parmeggiani, V.: The intellectual structure of product innovation research: a bibliometric study of the journal of product innovation management, 1984–2004. J. Prod. Innov. Manag. **27**, 437–451 (2010)

Fagerberg, J.: Technology and international differences in growth rates. J. Econ. Lit. **32**, 1147–1175 (1994)

Garrone, P., Grilli, L.: Is there a relationship between public expenditures in energy R&D and carbon emissions per GDP? An empirical investigation. Energy Policy **38**(10), 5600–5613 (2010)

Fernandes, A.C., Campello de Souza, B., Stamford da Silva, A., Suzigan, W., Chaves, C.V., Albuquerque, E.: Academy—industry links in Brazil: evidence about channels and benefits for firms and researchers. Sci. Public Policy **37**(7), 485–498 (2010)

Gorg, H., Strobl, E.: The effect of R&D subsidies on private R&D. Economica **74**(294), 215–234 (2007)

Ho, S.-Y.: Bibliometric analysis of biosorption technology in water treatment research from 1991 to 2004. Int. J. Environ. Pollut. **34**, 1–13 (2008)

Independent Evaluation Group: World Bank Group Support for Innovation and Entrepreneurship: An Independent Evaluation. World Bank Independent Evaluation Group. World Bank, Washington, DC, September 2013

Jones, C.I.: R&D-based models of economic growth. J. Polit. Econ **103**, 759–784 (1995)

Kessler, M.M.: Bibliographic coupling between scientific articles. Am. Doc. **14**(1), 10–131 (1963)

Lall, S., Pietrobelli, C.: National technology systems in sub-Saharan Africa. Int. J. Technol. Glob. **1**(3), 311–342 (2005)

Leydesdorff, L.: The Challenge of Scientometrics: The Development, Measurement and Selforganization of Scientific Communications. DSWO Press, Leiden (1995)

Li, X., Wu, P., Shen, G.Q., Wang, X., Teng, Y.: Mapping the knowledge domains of building information modeling (BIM): a bibliometric approach. Autom. Constr. **84**, 195–206 (2017)

Li, Y., Rong, Y., Ahmad, U.M., Wang, X., Zuo, J., Mao, G.: A comprehensive review on green buildings research: bibliometric analysis during 1998–2018. Environ. Sci. Pollut. Res. **28**(34), 46196–46214 (2021). https://doi.org/10.1007/s11356-021-12739-7

Lucas, R.E.J: On the mechanics of economic development. J. Monet. Econ. **22**, 3–42 (1988)

Marzi, G., Ciampi, F., Dalli, D., Dabic, M.: New product development during the last ten years: the ongoing debate and future avenues. IEEE Trans. Eng. Manag. **68**, 330–344 (2021)

Mortazavi, S., Eslami, M.H., Hajikhani, A., Väätänen, J.: Mapping inclusive innovation: a bibliometric study and literature review. J. Bus. Res. **122**, 736–750 (2021)

Nicholas, D., Ritchie, M.: Literature and Bibliometrics. Clive Bingley, London (1978)

Norris, M., Oppenheim, C.: Comparing alternatives to the web of science for coverage of the social sciences' literature. J. Informetr. **1**(2), 161–169 (2007)

OECD: Why modernise official development assistance? In: Third International Conference on Financing for Development, Addis Ababa (2015)

OECD: Main Science and Technology Indicators (MSTI) database (2021)

Peng, R., Chen, J., Wu, W.: Mapping innovation research in organizations: a bibliometric analysis. Front. Psychol. **12**, 750960 (2021)

Randhawa, K., Wilden, R., Hohberger, J.: A bibliometric review of open innovation: setting a research agenda. J. Prod. Innov. Manag. **33**, 750–772 (2016)

Romer, P.M.: Increasing returns and long-run growth. J. Polit. Econ. **94**, 1002–1037 (1986)

Romer, P.M.: Endogenous technological change. J. Polit. Econ. **98**, S71–S102 (1990)

Sarin, S., Haon, C., Belkhouja, M.: A twenty-year citation analysis of the knowledge outflow and inflow patterns from the journal of product innovation management. J. Prod. Innov. Manag. **35**, 854–863 (2018)

Small, H.: Co-citation in the scientific literature: a new measure of the relationship between two documents. J. Am. Soc. Inf. Sci. **24**(4), 265–269 (1973)

Strozzi, F., Colicchia, C., Creazza, A., Noe, C.: Literature review on the "Smart Factory" concept using bibliometric tools. Int. J. Prod. Res. **55**(22), 6572–6591 (2017)

Walker, J.L.: The diffusion of innovations among the American states. Am. Polit. Sci. Rev. **63**, 880–899 (1969)

Weinberg, B.H.: Bibliographic coupling: a review. Inf. Storage Retr. **10**(5–6), 189–196 (1974)

Zaltman, G., Duncan, R., Holbek, J.: Innovations and Organizations. Wiley, New York (1973)

Zhang, D., Mohsin, M., Rasheed, A.K., Chang, Y., Taghizadeh-Hesary, F.: Public spending and green economic growth in BRI region: mediating role of green finance. Energy Policy **153**, 112256 (2021)

Determinants of Small and Medium-Sized Enterprises Access to Financial Services in Ghana

S. N. Dorhetso[✉], L. Y. Boakye, and K. Amofa-Sarpong

Accra Institute of Technology (AIT), Accra, Ghana
samueldorhetso@gmail.com, dorhetso@ait.edu.gh

Abstract. Purpose: The objective of this study is to estimate the relative degree of significance of the determinants of small and medium-sized enterprises (SMEs) access to financial services in Ghana, using a macroeconomic and institutional framework.

Design/Methodology/Approach: A theoretical framework was developed from a mix of the macroeconomic and institutional conceptual dimensions of financial inclusion, and the best-worst method was used to analyse and grade the identified determinants according to their weighted averages.

Findings: The findings of this study indicate that cost of borrowing; collateral requirements/risk perception of borrower; firms' financial characteristics; and, strong banking competition and low information asymmetries are the highest ranked determinants for SMEs access to financial services in Ghana.

Implications/Research Limitation: It is projected that the results of this study would significantly contribute to literature on financial inclusion of small start-ups and medium-sized enterprises, particularly in developing countries, and increase their access to financial services, including affordable credit, and their integration into value chains and markets as postulated in the sustainable development goals (SDG 9.3). However, the study had limitations since it was partially based on the opinions of SMEs and bank/financial institution managers in Ghana, which may be characterised with prejudiced judgment and uncertainty.

Practical Implication: The outcome of the research would also provide guidance for policy makers on significant determinants of small and medium-sized firms' access to funding.

Originality/Value: This distinct research builds an estimation of the most significant determinants of SMEs access to financial services in Ghana, and validates studies on the determinants of SMEs access to financial services. Nevertheless, the method and context of this study differ from hitherto published papers on the matter.

Keyword: Best-worst method · Determinants · Financial · Services · Affordable credit · SMEs

---

© The Author(s), under exclusive license to Springer Nature Switzerland AG 2023
C. Aigbavboa et al. (Eds.): ARCA 2022, *Sustainable Education and Development – Sustainable Industrialization and Innovation*, pp. 278–292, 2023.
https://doi.org/10.1007/978-3-031-25998-2_21

1 Introduction

The significance of financial inclusion is progressively realized by policymakers all over the world. Small and medium-sized enterprises (SMEs) access to financial services, especially, is in the midst of the economic diversification and growth constraints plaguing many countries. Underdeveloped countries still have to face the lack of fair access to funding, even as financial risk assessment has improved in the last few decades in both developed and developing countries, with the need of a stricter regulatory system (Lentner et al. 2020). Refining and escalating access to credit remains an important policy challenge in many countries, with much for policymakers to do. Over the years, policymakers have shifted their emphasis from financial growth to financial inclusion to grasp the unbanked and low-income groups (Johnson and Arnold 2012). Many academics across the world claim that access to funding enables low-income groups to save, borrow and invest in SMEs to take advantage of economic opportunity (Demirguc-Kunt et al. 2014). Nevertheless, notwithstanding the importance of financial inclusion to an economy, about 1.7 billion, a principal share of the world's adult population does not yet own a bank account (Demirguc-Kunt et al. 2018). In Africa, less than twenty-five percent of adults have access to accounts at formal financial institutions (Triki and Faye 2013). In sub-Saharan Africa, especially Ghana, SMEs constitute an important share of businesses. However, these SME businesses delay in economic growth due to constraints with access to financial services. Improving SME access to financial services can help augment economic growth, employment, and the value of fiscal and financial policy. Also, SME access to financial services contributes to financial stability. It is evident from existing literature that there is a surfeit of determinants of SMEs access to financial services (Allen et al. 2012; Dabla-Norris et al. 2015a; Rojas-Suárez 2016; Love and Martinez Peria 2015). These can be categorized as macroeconomic and institutional determinants of financial inclusion (IMF 2019). However, it is also palpably apparent from existing literature that there has been no or negligible studies to estimate the relative importance of the determinants of SMEs access to financial services in Ghana. The rationale of this study is to estimate the relative significance of the determinants of SMEs access to financial services in Ghana, using a macroeconomic and institutional framework, and the best-worst method (Rezaei 2016; Wang et al. 2019). It is expected that the results of this study would significantly contribute to literature on financial inclusion of SMEs, particularly in developing countries, and increase their access to financial services, including affordable credit, and their integration into value chains and markets as stipulated in the sustainable development goals (SDG 9.3). The outcome of the research would also provide guidance for policy makers on the most significant determinants of SMEs access to funding.

2 Literature Review

Financial inclusion is explained by the World Bank (2018) as the process by which entire households and businesses, irrespective of their income levels, have access to and can successfully use the suitable financial services they require in order to improve their livelihood. These financial services have to be provided in an accountable and safe manner to consumers and sustainably given in a suitably regulated setting (Demirguc-Kunt et al. 2015). At a macroeconomic level, financial inclusion can generate a diversified base of deposits that builds a resilient financial system and augmented stability (Garcia 2016). According to the IMF, within a nation's level, financial inclusion is influenced by the limitations oozing out of from numerous macroeconomic outcomes such as equality, stability, and economic growth (Sahay et al. 2015). SMEs are at the core of economic growth, and are key sources of employment (Aga et al. 2015). Most SMEs operating in Sub-Saharan Africa, and in Ghana specifically, are faced with numerous challenges that affect their operations and long-term existence. Many studies (Pandula 2011; Osei-Assibey 2014; Allen et al. 2012; Dabla-Norris et al. 2015a; Rojas-Suárez 2016; Love and Martinez Peria 2015) have teased out the determinants of SMEs access to credit. However, none has been found from existing literature that deals with the relative level of importance of these determinants. The other works were inaudible on the relative significance of the determinants of SMEs access to financial services. This distinctive work seeks to fill the gap by developing a theoretical framework to study the determinants of SMEs access to financial facilities, with regards to the relative importance of these determinants. The purpose of this study is to determine the relative degree of importance of the determinants of SMEs access to financial services. The relevant determinants would be identified from an extensive and scrupulous literature review, and complemented by experts' opinion. The decision framework to be developed and used in this research originates from a blend of macroeconomic and institutional dimensions of the determinants of financial inclusion. This paper concentrates on the relative importance of relevant determinants derived from literature and experts' opinion, and is not skewed towards the discovery of another determinant from literature.

2.1 Identification of the Determinants of SMEs Access to Financial Services

The decision framework developed in this research emanates from the macroeconomic and institutional dimensions of determinants of financial inclusion (IMF 2019). The link between the macroeconomic and institutional contexts characterizes the financial inclusion of SMEs. This distinctive framework was used to explore the weights and rankings of the determinant factors that significantly impact the access of SMEs to financial services (see Table 1). Within each of the two dimensions, there are determinants that were recognized by a blend of literature from preceding studies and opinions of SME and banking industry experts.

2.1.1 Macroeconomic Dimension

Financial inclusion is mainly determined by robust and efficient economic fundamentals. It is evident from preceding studies on the determinants of SMEs access to credit that supply-side market failures may instigate a lack of SMEs access to credit due to refutation from the banks for lack of sustainability of the proposal or high risk and associated costs. Demand-side market failures owing to inadequate information in the project proposal, high rate of bank credit et cetera may also instigate a lack of SMEs access to credit (Pandula, 2011). The availability of infrastructure to provide credit, as well as the lending organizations' risk perception of the borrower are also key determinants of SMEs access to credit (Ahmed and Hamid 2011). Collateral indicates the extent to which

Table 1. A theoretical framework on the dimensions and determinants of SMEs access to financial services.

Dimensions	Factors	References
Macroeconomic (ME)	Supply side factors (ME 1) Demand-side factors (ME 2) Market opportunity/availability of credit infrastructure (ME 3) Cost of borrowing (ME 4) Collateral requirements/risk perception of borrower (ME 5) Better governance, efficiency and productivity (ME 6) Education (ME 7) Macro-financial instability (ME 8)	Adil and Jalil (2020); Sanderson et al. (2018); Sethi and Sethy (2019); Ambarkhane et al. (2016), Dev (2006); Yangdol and Sarma (2019); Ghatak (2013); Kumar et al. (2019); Wolday and Gebrehiwot (2004), Bose et al. (2017); Delis et al. (2014); Aregbeshola (2016), Dabla-Norris et al. (2015b); Karpowicz (2014), Aduda and Kalunda (2012); Mamman et al. (2019) Ahmed and Hamid (2011), Pandula (2011) Dabla-Norris et al. (2015a) Allen et al. (2012) Rojas-Suárez and Amado (2014) Rojas-Suárez (2016)
Institutional (IN)	Firm-specific characteristics (IN 1) Owner/manager's characteristics (IN 2) Firms' financial characteristics (IN 3) Strong banking competition and low information asymmetries (IN 4)	Ajide (2017); Wolday and Gebrehiwot (2004), Alhassan and Sakara (2014); Osei-Assibey (2014). Love and Martinez Peria (2015)

Source: Authors' own construct.

assets are entrusted by borrowers to lenders as security for liability settlement (Gitman, 2003). These security assets would be used to retrieve the principal in case of default by borrowers. Notably, banks evaluate SMEs based on their current financial situation as mirrored by their accounting records. A better physical infrastructure and higher wages increases savings and the loch of available funds, as well as access to funding (Dabla-Norris et al. 2015a). According to Rojas-Suárez and Amado (2014). Better governance can assist the enactment of financial contracts for SMEs, which expedites their access to funding. Also, education is an important determinant of household ownership and the usage of accounts in the formal financial system Allen et al (2012). Furthermore, macro-financial instability can radically influence access to credit and other financial services to SMEs, as banks reinstate their capital ratios by limiting credit, particularly to dicier borrowers (Rojas-Suárez 2016).

2.1.2 Institutional Dimension

Financial institutions deliberate on the creditworthiness of SMEs before granting them credit facilities. According to Pandula (2011), SMEs access to credit depends on firm-specific characteristics; owner/manager's characteristics; and financial characteristics of the institution. Osei-Assibey (2014) found that access to credit is affected by observable socio-economic characteristics of the proprietor of the firm as well as firm's characteristics. Firm-specific characteristics include firm's performance, firm's innovation, firm's size, firm's location, industrial sector of the firm, firm's age, firm's asset structure and the firm-bank relation. Also, owner/manager's characteristics entail owner's gender, level of education, managerial experience and skills, and affiliation or networking with business associations. Financial access of SMEs is enabled by strong banking competition and low information asymmetries. According to Love and Martinez Peria (2015) proliferated banking competition has a positive impact that depends on the coverage of credit bureaus on firms' access to credit.

3 Research Methodology

The conceptual framework proposed in this research entails the trails to prioritising the determinants of SMEs access to financial services., using the BWM. The BWM is used to determine the relative importance of each dimension and determinant of SMEs access to financial facilities. This is done by comparing the best dimension (most important) to the worst (least important) initially and then comparing the other dimensions to the worst afterwards using a linguistic scale for the pair wise comparison.

3.1 Best-Worst Method (BWM)

The BWM is a multi- criteria decision-making model which estimates the weights of criteria by employing two vectors of pair wise comparisons between the most important and the least important criteria (Wang et al. 2019). According to Rezaei (2016), the stages below are involved in determining the weights of criteria using the BWM:

Stage 1: Finalisation of decision criteria.

A set of decision criteria are identified and extracted from an intensive search of literature, and experts' opinions and recorded as $\{C1, C2... Cn\}$ for n main criteria. In this study, the decision criteria are the determinants of SMEs access to financial services.

Stage 2: The best (most important) and worst (least important) criteria are selected.

At this stage, the expert selects the most important and least important criteria from the pool of identified decision criteria in Step 1 based on his/ her opinion.

Stage 3: A matrix is developed by determining the pair wise comparison between the most important criterion and the other decision criteria. The objective of this step is to determine the preference of the most important criterion to the other decision criteria by using a linguistic scale for the BWM having scores from 1 to 9. The outcome of the pair wise comparison of the best criterion and other decision criteria is expressed by a 'Best-to-Others' vector as follows:

$$DB = (dB1, dB2, \ldots, dBn)$$

where dBj represents the preference of the most important criterion B over a criterion j amongst the decision criteria, and $dBB = 1$

Stage 4: The 'Others –to-Worst' matrix is developed by conducting a pair- wise comparison of the other decision criteria against the least important criterion using the linguistic scale for the BWM. The outcome of comparison of the other decision criteria to the worst criterion is expressed as follows:

$$DW = (dW1, dW2, ..., dWn)^q$$

where dWj represents the preference of the criterion j amongst the decision criteria in Step 1 above the least important criterion W, and $dWW = 1$.

Stage 5: Computing the optimal weights $(p1^*, p2^*... pn^*)$.

Weights of criteria are determined such that the maximum absolute differences for all criterion j are minimised over the following set $\{|pB - dBjpj|, |pj - djW\,pW|\}$.

A minimax model can be formulated as:

$$\min \max j\{|pB - dBjwj|, |pj - djWwW|\} \\ \text{Subject to:} \\ \sum_j pj = 1 \quad (1)$$

$pj \geq 0$, for all criterion j

Model (1) can be solved by converting it into the following linear programming problem model:

$$\text{Min } R^L \\ \text{Subject to:} \\ |pB - dpj| \leq R^L, \text{ for all criterion } j \\ |pB - dpW| \leq R^L, \text{ for all criterion } j \\ \sum_j pj = 1 \quad (2)$$

$pj \geq 0$, for all criterion j

Solving the linear model (2), will result in optimal weights $(p1^*, p2^*... pn^*)$ and optimal value R^L.. Consistency (R^L) of comparisons also needs to be estimated. A value nearer to zero is more desired for consistency (Rezaei 2016; Wang et al. 2019) (Fig. 1).

3.2 Data Collection

For this research, questionnaires were designed and used to collect data from participants with a minimum of five years of professional management and decision-making experience in the Ghanaian SME and banking sector. This was done to ensure accuracy of data garnered since the experts were deemed to be sufficiently knowledgeable to effectively complete the survey. The experts were purposely selected from rural banks, commercial banks, microfinance institutions, urban SMEs, and rural SMEs in Ghana. They were assured of the privacy of their reports in order to allow for effective model-structuring and in- depth observation (Nilashi et al. 2016). Also, the participants were

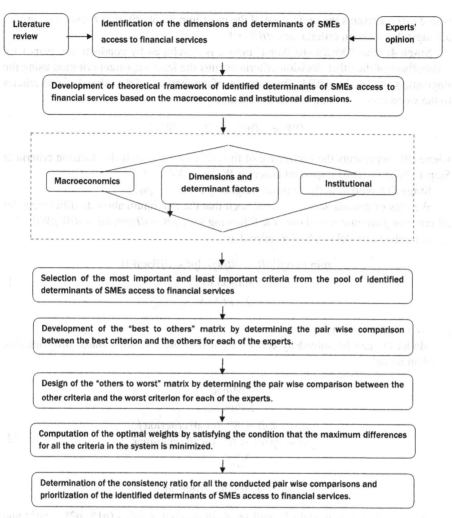

Fig. 1. Research modelling framework.

selected mid-level and above ranking executives, hence their responses satisfactorily represent the SME sector of the economy (Fu et al., 2006).

To conduct the survey, several measures were undertaken to maximise the rate of response and minimise prejudice of response amongst the experts from the selected SME and banking institutions. Initially, a pilot study was done by sending the Google form questionnaires designed for this study to three researchers through emails and interviewing four participants face- to - face to criticise and provide feedbacks. The three researchers that participated in the pilot study were a female and two males who hold PhD degrees and have at least ten years of research experience in SME management, banking, and financial management. The four experts who participated in the pilot study have managerial experience of at least five years in the Ghanaian SME and banking

Table 2. Demographic summary of respondents, and linguistic scale for pair wise comparison in BWM.

Demographic Summary of Respondents		
Characteristic	Number of respondents	Percentage of sample (%)
Gender		
Male	10	55.6%
Female	8	44.4%
Education		
Bachelor/Master degree	11	61.1%
Doctorate degree	7	38.9%
Years of experience		
5–10	6	33.3%
	12	
Above 10	12	66.7%
Roles		
Rural bank manager	3	16.7%
Commercial bank manager	3	16.7%
Microfinance firm manager	3	16.67%
Urban/rural SME manager	4	22.2%
Academic/University lecturer	5	27.7%
Linguistic Scale for Pair Wise Comparison in BWM		
Linguistic attributes	Scores	
Equally important	1	
Equal to moderately more important	2	
Moderately more important	3	
Moderately to strongly more important	4	
Strongly more important	5	
Strongly to very strongly more important	6	
Very strongly more important	7	
Very strongly to extremely more important	8	
Extremely more important	9	

Source: Authors'.

sector. As a result of the feedbacks from the pilot study, the questionnaires were modified and emailed to twenty-five experts. Five experts each were selected to represent the five different roles of the participants as illustrated in Table 2. A follow-up on the respondents was done by phone conversations and personal visits (Yang et al. 2018). Eventually, eighteen completed questionnaires were received out of the twenty-five that were emailed to the experts, a response rate of 72%. This response rate is considered suitable for effective analysis and to yield consistent outcomes, according to the BWM used in this study that does not require a bulky sample size to provide accurate and reliable results (Wang et al. 2019).

4 Results and Discussion

Starting from the first stage of the BWM, the dimensions and determinants of SMEs access to financial services that have been identified from extant literature were evaluated by the decision-makers using questionnaires. The questionnaires were designed to require a 'YES' or 'NO' answer which indicated that a dimension or determinant is 'relevant' or 'not relevant' for SMEs financial inclusion. A simple mean method was used to select the variables that are above the arithmetic mean and the analysis of findings at this phase indicated that all the identified dimensions or determinants were recognised with no ancillary inclusions. Therefore, inclusiveness of relevant data was ensured and content validity was confirmed.

4.1 Calculation of the Weights of Determinants Using BWM

After the finalisation of the determinants of SMEs access to financial services, the weights of the determinants were calculated using the BWM. For this study, eighteen experts performed the identification of the best and worst criteria for the main dimensions as well as subgroup criteria. After obtaining the best and worst criteria all the respondents were bade to give preference ratings of the best criteria to other criteria, and other criteria to worst criteria, for the main dimensions' criteria determinants as well as subcategory criteria determinants. The preference ratings of Expert 1, the first expert, for the main category criteria determinants, as well as subcategory criteria determinants are illustrated in Table 3.

A similar process of the BWM survey, as described in the paragraph above was performed by all the experts in this study to assess the performance ratings of the main category and subcategory determinants of SMEs access to financial services. The entire weights of the determinants for both the main category and subcategory, measured in this study, were obtained using Eq. (1). All the aggregated weights were computed by applying the data sourced from the eighteen experts of this enquiry to Eq. (2) and estimating the mean using the simple average technique. The entire results of the estimation process, enabled by the BWM, are shown in Table 3. The degree of importance of a determinant is revealed by its ranked position in the table. The global ranks of the recognised determinants shown in the table were calculated by multiplying the preference weights of the respective determinant's dimension with the individual weight of the determinant. The rankings of both the main category dimensions and sub category determinants are discussed in detail in the next section of this study.

4.2 Ranking of the Dimensions of the Determinants of SMEs Access to Financial Services

As shown in Table 3, the results point to the macroeconomic dimension as the most significant dimension for SMEs access to financial services. However, the importance

Table 3. Pairwise comparison of main category and subcategory determinants by Expert 1, and aggregate weights of main and subcategory determinants for all the experts.

Main Category Determinants

Best to Others	Macroeconomic (ME)	Institutional (IN)
Best criteria: Macroeconomic (ME)	1	2
Others to Worst	Worst criteria: Institutional (IN)	
Macroeconomic (ME)	3	
Institutional (IN)	1	

Macroeconomic (ME) Subcategory Determinants

Best to Others ME 7 ME 8	ME 1	ME 2	ME 3	ME 4	ME 5	ME 6
Best criteria: ME 4 4 3	5	3	2	1	3	5
Others to Worst	Worst criteria: ME 1					
ME 1	1					
ME 2	5					
ME 3	3					
ME 4	3					
ME 5	2					
ME 6	4					
ME 7	5					
ME 8	3					

Institutional (IN) Subcategory Determinants

Best to Others IN 4	IN 1	IN 2	IN 3
Best criteria: IN 3 5	2	3	1
Others to Worst	Worst criteria: IN 2		
IN 1	3		
IN 2	1		
IN 3	2		
IN 4	3		

Aggregate Weights of Main and Subcategory Determinants for all the Experts.

Main category DFs	Weights of main category DFs determinants	Subcategory determinants	Weights of subcategory determinants	Global weights	Ranking
Macroeconomic (ME)	0.691	ME 1	0..057	0.039	11
		ME 2	0.135	0.093	5
		ME 3	0.067	0.046	9
		ME 4	0.304	0.210	1
		ME 5	0.203	0.140	2
		ME 6	0.102	0.071	6
		ME 7	0.051	0.035	12
		ME 8	0.081	0.056	7
Institutional (IN)	0.309	IN 1	0.181	0.055	8
		IN 2	0.136	0.042	10
		IN 3	0.375	0.117	3
		IN 4	0.308	0.096	4

Source: authors' own construct

of the institutional dimension cannot be discounted. It is learnt from the table that the macroeconomic and institutional dimensions contribute sixty-nine percent and thirty-one percent, respectively, in order of importance. From the results, it can be inferred that the determinants that are associated to the macroeconomic context are remarkably important and should be passably assured for financial inclusion. The managers of SMEs and financial institutions must be invigorated to be in conformity with these determinants to ensure the financial inclusion and access to financial facilities.

4.3 Global Ranks of the Determinants

Table 3 shows the global ranking of the determinants of SMEs access to financial services. The top four determinants under the global ranks, which represent the top 33% of determinants, are part of both dimensions considered in this research. These top determinants are cost of borrowing; collateral requirements/risk perception of borrower; firms' financial characteristics; and, strong banking competition and low information asymmetries. The findings of this study suggest that cost of borrowing is the highest ranked determinant for SMEs access to financial services. Cost of borrowing or high rate of bank credit is a demand-side market failure that may instigate a lack of SMEs access to credit (Pandula 2011). The second most important determinant in the global ranking hierarchy is collateral requirements/risk perception of borrower. Collateral refers to the extent to which assets are assigned by borrowers to lenders as security for liability reimbursement (Gitman 2003). Assets used as collateral would be used to retrieve the principal in case of default by borrowers. The risk perception of the borrower is also a key determinant of SMEs access to credit (Ahmed and Hamid 2011). Firms' financial characteristics is also a highly ranked determinant in this study. SMEs access to credit depends on the financial characteristics of the firm (Pandula 2011; Osei-Assibey 2014). The fourth most important determinant, strong banking competition and low information asymmetries, signifies that increased banking competition has a positive impact on SMEs access to credit, and that depends on the handling of credit bureaus (Love and Martinez Peria 2015).

4.4 Ranking of the Determinants Within Each Dimension

4.4.1 Macroeconomic Determinants

The results of this study show that cost of borrowing has the highest rank in this dimension. As discussed in the earlier paragraphs, cost of borrowing is the rate of bank credit, and an unbearable interest rate on bank credits may impede SMEs access to credit. This implies that managers and other actors in the banking and microfinance sector should endeavour to improve the access of SMEs to financial services, by providing credit at a low cost, in order to enhance sustainability and development. The next ranked in this dimension is the collateral requirements/risk perception of borrower, which has also been previously discussed in preceding sections. The next six ranked determinants in this dimension, in order of significance, are demand-side factors; better governance, efficiency and productivity; macro-financial instability; market opportunity/availability of credit infrastructure; supply side factors; and education respectively.

4.4.2 Institutional Determinants

Firms' financial characteristics is the highest ranked determinant in this dimension. The next most important determinant in this context is strong banking competition and low information asymmetries. The third and fourth ranked determinants in this dimension, in order of their level of significance to the rationale of the study, are firm-specific characteristics and owner/manager's characteristics respectively.

4.5 Theoretical and Practical Implications

Significantly, this study contributes to the theory and practice of the financial inclusion of SMEs to improve their access to credit as stipulated in the third section of the ninth sustainable development goal (SDG 9.3). From a theoretical perspective, the theoretical outline based on the macroeconomic and institutional dimensions of financial inclusion helps in understanding the narrow areas that, when focused on, can facilitate SMEs access to financial services. From the theoretical framework of the study, the financial inclusion of SMEs is influenced by determinants categorized into macroeconomic and institutional dimensions. Theoretically, by virtue of this framework, this study enacts an increase in the level of variance explained on the determinants of SMEs access to financial services in a distinct manner by using the BWM.

The results of this study suggest that cost of borrowing; collateral requirements/risk perception of borrower; firms' financial characteristics; and, strong banking competition and low information asymmetries have significant influence on SMEs access to financial services. These verdicts are in line with the study by Pandula (2011), with regards to how a high cost of borrowing or high rate of bank credit is a demand-side market failure that may initiate a lack of SMEs access to credit. Consequently, managers and other actors in the banking and microfinance sector should put in genuine efforts to augment SMEs access to financial services, by providing credit at an affordable cost. However, contrary to studies by Allen et al. (2012), the findings of this research indicate that education is not highly crucial to the financial inclusion of SMEs in Ghana. This may be attributable to the fact that most SMEs owners usually have some form of formal education. Also, this research revealed that the entirely uneducated proprietors usually have either paid educated assistants or family and friends to provide the needed bookkeeping assistance.

Laconically, this study validates studies on the determinants of SMEs access to financial services. Nevertheless, this method and context differ from hitherto published papers on the matter. For this distinctive study, a theoretical model based on the macroeconomic and institutional dimensions of financial inclusion was used to study the determinants that affect SMEs access to financial facilities. The developed theoretical framework for this study can be used by any banking/microfinance institution and start-up firms to classify the determinants of access to financial services according to their significance levels.

4.6 Managerial Implications

The findings of this study offer an ample and in-depth understanding to SMEs and bank/ microfinance institution managers on effective measures for sustainable financial inclusion as postulated in the ninth sustainable development goal. The study is especially obliging to managers of SMEs and bank/microfinance organisations in developing nations like Ghana, Nigeria, Ivory Coast, and other countries in the west African sub region. Managers may adopt the modelling structure of this research, and focus more on scrutinising the determinants of SMEs access to financial services. Significantly, this work would help managers to underscore the most important determinants of financial inclusion, and concentrate on them by paying attention to details of respective cases of firms for resource distribution.

5 Conclusion

The application of the macroeconomic and institutional model structure for this study precisely exuded the level of variance explained on the determinants of SMEs access to financial services. Using the macroeconomic and institutional concept, this study proposed a comprehensive research framework that is pertinent to the context of SMEs access to financial services in Ghana. It is envisaged that this would give a better understanding of productive measures that must be in place to enhance SMEs access to funding. The results of this study show that cost of credit is still a topical issue in Ghana, and it is characterized by an overdependence on owners' equity. Although owners' personal funds may be more reliable and easily accessible, it may not be sufficient for the sustainable development of the firm.

This study has revealed the value of the determinants of SMEs access to financial services and significantly contribute to literature on financial inclusion of small startups and medium-sized enterprises, particularly in developing countries. The results of the study have underscored the relevance of affordable credit as stipulated in the third section of the ninth sustainable development goal (SDG 9.3). Nevertheless, the study is partially based on the opinions of SMEs and bank/financial institution managers in Ghana, which may be characterised with prejudiced judgment and uncertainty. In future works, fuzzy logic may be used to reduce ambiguity in experts' opinions (Orji and Wei 2016). Also, in future, the theoretical framework developed in this work may be modified to fit other multi-criteria decision methods such as the analytical hierarchy process (AHP) and the technique for order preference by similarity to ideal solution (TOPSIS). A broader angle of this current work may be conducted by garnering data from a bigger pool of experts in the SMEs and banking/financial industries of other countries. Furthermore, a comparative analysis may be carried out by either comparing diverse modelling frameworks on the research subject or comparing results of SMEs and banking/financial institutions of different countries or comparing findings from different SMEs and banking/financial organisations in the country.

References

Adil, F., Jalil, A.: Determining the financial inclusion output of banking sector of pakistan—supply-side analysis. Economies **8**, 42 (2020)

Aduda, Josiah, Kalunda, E.: Financial Inclusion and financial sector stability with reference to Kenya: a review of literature. J. Appl. Finan. Bank. **2**, 95–120 (2012)

Aga, G.A., Francis, D., Rodriguez-Meza, J.: SMEs, age and jobs: a review of the literature, metrics and evidence. World Bank Policy Research Working Paper, 7493 (2015)

Ahmed, H., Hamid, N.: Financing constraints: determinants and implications for firm growth in Pakistan. Lahore J. Econ. **16**, 317–346 (2011)

Ajide, K.B.: Determinants of financial inclusion in Sub-Saharan Africa countries: does institutional infrastructure matter? CBN J. Appl. Stat. **8**, 69–89 (2017)

Allen, F., Demirguc-Kunt, A., Klapper, L., Soledad, M.: The Foundations of Financial Inclusion: Understanding Ownership and Use of Formal Accounts. World Bank Policy Research Working Paper, 6290, Washington, DC. (2012)

Alhassan, F., Sakara, A.: Socio-economic determinants of small and medium enterprises' (SMEs) access to credit from the barclays bank in Tamale-Ghana. Int. J. Human. Social Sci. Stud. **1**(2), 26–36 (2014)

Ambarkhane, D., Ardhendu, S.S., Bhama, V.: Measuring financial inclusion of Indian states. Int. J. Rural Manag. **12**, 72–100 (2016)

Aminu, M., Bawole, J., Agbebi, M., Alhassan, A.R.: SME policy formulation and implementation in Africa: unpacking assumptions as opportunity for research direction. J. Bus. Res. **97**, 304–315 (2019)

Aregbeshola, R.A.: The role of local financial market on economic growth—a sample of three African economic groupings. Afr. J. Econ. Manag. Stud. **7**, 225–240 (2016)

Bose, S., Amitav, S., Habib, Z.K., Shajul, I.: Non-financial disclosure and market-based firm performance: the initiation of financial inclusion. J. Contemp. Account. Econ. **13**, 263–281 (2017)

Dabla-Norris, E., Deng, A., Ivanova, I., Karpowicz, F., Unsal, E., VanLeemput, E., Wong, J.: Financial Inclusion: Zooming in on Latin America. IMF Working Paper 15/206, International Monetary Fund, Washington, DC (2015a)

Dabla-Norris, E., Yan, J, Townsend, R., Unsal, F.: Identifying Constraints to Financial Inclusion and Their Impact on GDP and Inequality: A Structural Framework for Policy. International Monetary Fund: IMF, pp. 15–22 (2015b)

Dev, S.M.: Financial inclusions: issues and challenges. Econ. Polit. Weekly, 4310–4313 (2006)

Demirguc-Kunt, A., Klapper, L., Randall, D.: Islamic finance and financial inclusion: measuring use of and demand for formal financial services among Muslim adults. Rev. Middle East Econ. Finan. **10**, 177–218 (2014)

Demirguc-Kunt, A., Klapper, L., Singer, D., Oudheusden, P.V.: The Global Findex Database 2014: measuring financial inclusion around the world. World Bank Group Policy Research Working Paper, 7255 (2015)

Demirguc-Kunt, A., Klapper, L., Singer, D., Ansar, S., Hess, J.: The Global Findex Database 2017: Measuring Financial Inclusion and the Fintech Revolution. Washington, DC: World Bank (2018)

Delis, M.D., Iftekhar, H., Pantelis, K.: Bank regulations and income inequality: empirical evidence. Rev. Finan. **18**, 1811–1846 (2014)

Fu, J.R., Farn, C.K., Chao, W.P.: Acceptance of electronic tax filing: a study of taxpayer intentions. Inf. Manag. **43**, 109–126 (2006)

Garcia, M.: Can financial inclusion and financial stability go hand in hand? Econ. Issues **21**(2), 81–103 (2016)

Ghatak, A.: Demand side factors affecting financial inclusion. Int. J. Res. J. Social Sci. Manag. **3**, 176–185 (2013)

Gitman, L.J.: The Principles of Managerial Finance, 7th edn. Pearson Education Inc., New York (2003)

IMF. Inclusion of Small and Medium-Sized Enterprises in the Middle East and Central Asia. IMF library (2019). www.elibrary.imf.org.

Kumar, Abhishek, Rama, P., Rupayan, P.: Usage of formal financial services in India: demand barriers or supply constraints? Econ. Model. **80**, 244–259 (2019)

Johnson, Susan, Steven, A.: Inclusive financial markets: is transformation under way in Kenya? Dev. Policy Rev. **30**, 719–748 (2012)

Karpowicz, I.: Financial inclusion, growth and inequality: a model application to Colombia. In: IMF, pp. 14–166 (2014)

Lentner, C., Laszlo, V., Szilard, H.: The Assessment of financial risks of municipally owned public utility companies in Hungary between 2009 and 2018. Montenegrin J. Econ. **16**, 29–41 (2020)

Love, I., Martinez Peria, M.: How bank competition affects firms' access to finance. World Bank Econ. Rev. **29**(3), 413–448 (2015)

Mamman, A., Bawole, J., Agbebi, M., Alhassan, A.: SME policy formulation and implementation in Africa: unpacking assumptions as opportunity for research direction. J. Bus. Res. **97**, 304–315 (2019)

Nilashi, M., Ahmadi, H., Ahani, A., Ravangard, R., Ibrahim, O.B.: Determining the importance of hospital information system adoption factors using fuzzy analytic network process (ANP). Technol. Forecast. Soc. Change **111**, 244–264 (2016)

Osei-Assibey, E.: The rural financial system in ghana: what determines access and sources of finance for rural non-farm enterprises? In: Readings on Key Economic Issues in Ghana, Social Science Series, 4, Department of Economics, University of Ghana, Legon (2014)

Pandula, G.: An Empirical Investigation of Small and Medium Enterprises' Access to Bank Finance: The Case of an Emerging Economy. La Trobe University, Australia (2011)

Rezaei, J.: Best- worst multi- criteria decision making method: some properties and a linear model. Omega **64**, 126–130 (2016)

Rojas-Suárez, L., Amado, M.: Understanding Latin America's Financial Inclusion Gap. Center for Global Development Working Paper 367, Washington, DC. (2014)

Rojas-Suárez, L.: Financial Inclusion in Latin America: Facts, Obsta¬cles and Central Banks' Policy Issues. Discussion Paper IDB-DP-464, Inter-American Development Bank, Washington, DC (2016)

Sanderson, A., Learnmore, M., Pierre, L.R.: A review of determinants of financial inclusion. Int. J. Econ. Finan. Issues **8**, 1–8 (2018)

Sahay, R., et al.: Financial inclusion: Can it meet multiple macroeconomic goals? IMF Staff Discussion Note 15/17 (2015)

Sethi, D., Susanta, K.S.: Financial inclusion matters for economic growth in India: some evidence from cointegration analysis. Int. J. Social Econ. **46**, 132–151 (2019)

Triki, T., Issa, F.: Financial inclusion in Africa. African Development Bank, Tunis (2013)

The World Bank: Research and development expenditure (% of GDP) (2018). https://data.worldbank.org/indicator/GB.XPD.RSDV.GD.ZS. Accessed 10 Feb 2018

Wang, Z.G., Xu, R., Lin, H., Wang, R.J.: Energy performance contracting, risk factors, and policy implications: identification and analysis of risks based on the best- worst network method. Energy **170**, 1–13 (2019)

Wolday, A., Ageba, G.: MSEs Development in Ethiopia: Survey Report. EDRI, Addis Ababa (2004)

Yang, Y., Lau, A.K.W., Lee, P.K.C., Yeung, A.C.L., Cheng, T.C.E.: Efficacy of China's strategic environmental management in its institutional environment. Int. J. Oper. Prod. Manag. **39**(1), 138–163 (2018)

Yangdol, R., Mandira, S.: Demand-side factors for financial inclusion: a cross-country empirical analysis. Int. Stud. **56**, 163–185 (2019)

The Effects of Electronic Taxes on Small and Medium-Sized Enterprises' Access to Financial Services

S. N. Dorhetso([✉]), K. Amofa-Sarpong, and E. Osafoh

Accra Institute of Technology, Accra, Ghana
samueldorhetso@gmail.com, dorhetso@ait.edu.gh

Abstract. Purpose: The purpose of this paper is to estimate the effects of electronic transaction taxes on small and medium-sized enterprises (SMEs) access to financial services.

Design/Methodology/Approach: This quantitative study used cluster sampling to collect primary data via copies of a 5-point Likert scale questionnaire. Correlation analysis was initially used to measure the strength and direction of association that exists between the variables. A linear regression was then run to predict SMEs access to financial services from electronic taxes.

Findings: The findings of this paper revealed that there was a weak, negative correlation between electronic taxes and SMEs access to financial services, which was statistically significant. Also, the results of the linear regression analysis indicated that electronic transaction taxes statistically significantly predicted SMEs access to mobile financial services.

Implications/Research Limitation: It is envisaged that the findings of this study would significantly contribute to literature on the general effects of taxes, especially the effects of electronic transaction taxes on SMEs in developing economies, and increase their access to financial services, including affordable credit, and their integration into value chains and markets as recommended in the third section of the ninth sustainable development goal (SDG 9.3). However, the study is limited by its precincts as the survey was done only in rural and urban suburbs of the foremost city. The opinions of the respondents, which may be characterised with unfair judgment and uncertainty, also posed an acute limitation to the study.

Practical Implication: Practically, the outcome of the research would provide guidance for policy makers on optimal taxation that would prevent a deadweight loss, and promote SMEs access to financial services.

Originality/Value: This distinct study investigated the effects of electronic transaction taxes on SMEs access to financial services and corroborates a critical research area that is currently understudied.

Keyword: Deadweight loss · Electronic taxes · Financial services · SMEs · Taxation

© The Author(s), under exclusive license to Springer Nature Switzerland AG 2023
C. Aigbavboa et al. (Eds.): ARCA 2022, *Sustainable Education and Development – Sustainable Industrialization and Innovation*, pp. 293–302, 2023.
https://doi.org/10.1007/978-3-031-25998-2_22

1 Introduction

Effective financing is a crucial element for the growth of small and medium-sized enterprises (SMEs). Widening access to economic and business opportunities for SMEs can enliven social welfare and improve national productivity. The quantity of SMEs proliferates gradually as national economies grow and capital accessibility needs also rises. Thus, there arises the need to build up innovative financing models that go beyond conventional bank lending to provide timely financing opportunities for SMEs according to their requirements and levels of business development (ADB 2015). There is a plethora of constraints of financial access being faced by SMEs. The constraints may worsen in developing countries where many firms lack financial records, making creditors reluctant to lend due to the increased risk. Nevertheless, these challenges of accessing financial services in developing countries can be surmounted by the use of mobile credit made accessible through mobile phones. A development success story for many developing countries over the past decade has been the growth of mobile money, which has garnered more than a billion registered accounts (GSMA 2020). Mobile money services have financially included underserved SMEs who hitherto had neither the mandatory identity documents nor enough funds to have an official bank account. As a result, and to facilitate the attainment of the sustainable development goals (SDGs) that seeks to support economic development and human well-being, tax policies to keep taxation of mobile money within a threshold that would not instigate a welfare loss should be instituted. Although taxes on mobile phone transactions generate additional revenue for governments, there is a risk they may negatively affect the underserved groups who typically use the service, hypothetically unwinding the gains made in financial inclusion to date, increasing inequity, and undermining the SDGs. It is within this problem setting that this study seeks to estimate the effects of electronic transaction taxes on SMEs access to mobile financial services. The study employs a quantitative approach, and uses a structured questionnaire to collect the information on the effects of electronic taxes on access to financial services from SMEs in both urban and rural communities in Ghana. Regression analysis was used to predict the effects of electronic taxes on SMEs access to mobile financial services in Ghana. It is anticipated that the results of this study would significantly contribute to literature on the effects of taxes on SMEs, particularly the effects of electronic transaction taxes on SMEs in developing countries, and increase their access to financial services, including affordable credit, and their integration into value chains and markets as suggested in the sustainable development goals (SDG 9.3). The results from the study would also provide a guiding light for policy makers on optimal taxation that would prevent a deadweight loss.

2 Literature Review

Mobile money has unsettled traditional financial services and rectified market failures in the formal delivery of financial services during the past decade. According to Kipkemboi and Bahia (2019), one in ten adults in sub-Saharan Africa depend exclusively on mobile money for their access to financial services. Currently, mobile money plays a vital financial intermediation role by allowing savings to be capitalised into the indigenous economy, increasing business efficiency, inspiring job creation and proliferating growth in the economy (Lopez 2019). Mobile money has in excess of one billion registered accounts, and it assists in the attainment of sustainable development goals, contributing to the economic fortification of individuals and communities, including disregarded groups and businesses. In 2019, 372 million mobile money accounts, 35.8% of all registered accounts, were active on a 90-day basis. (GSMA 2020). Mobile money has enabled a long jump in financial infrastructure in many developing economies, by circumventing obsolete payments systems and placing financial services into the hands of those formerly omitted it. It links consumers with suppliers and, together with other mobile phone services generally, resolves information asymmetries that have conventionally challenged marginalised groups' involvement in the formal economy. To facilitate the digital transformation agendas of developing economies, mobile money is poised to offer the payments support system to a wide array of public services, including education, healthcare, and social security. This helps those economies to work towards the 2030 SDGs by: reducing the cost of international remittances (Naghavi and Scharwatt 2018; De 2015); improving resilience in the face of poverty (Ky et al. 2018; Riley 2018; Aron and Muellbauer 2019); strengthening the formal economy (Gosavi 2018); facilitating economic growth (WEF 2015); and improving domestic revenue mobilisation (Aker et al. 2012). The achievements of digital financial services in general, and mobile money in particular, have attracted the attention of governments and revenue collection authorities of developing countries, as a direct source of tax revenue. The current amplified focus on domestic revenue mobilisation by developing countries is prevalent at a period when the tax-to-GDP ratios of these economies ominously trail those of the developed economies. Developing economies are facing both external and internal pressures to widen their tax base, to enable them to finance sustainable development goals and meet their public expenditure commitments. However, this situation risks undermining the development gains in financial inclusion and SMEs access to mobile financial services seen to date. It is for this reason that this study seeks to predict the effects of electronic transaction taxes on SMEs access to mobile financial services in Ghana, using regression analysis. Kodom (2020) studied the drivers and role of regulation for financial inclusion via mobile money services in Ghana and found that the main drivers of mobile money adoption were perceived usefulness, social influence and cost of transaction. He found that whilst perceived usefulness and social influence had a positive effect on adoption, cost of transaction reduced the likelihood of adoption. Ndung'u (2017) contended that poorly designed tax policy will lead to poor outcomes. He argued that, once the optimal tax rate is reached, an additional increase in the excise tax rate produces less tax revenue. He further argued that taxes on retail electronic transactions and bank transactions has the tendency to reverse the advances that technology has contributed to financial inclusion and revert the economy back to the fondness of cash, leading to financial exclusion of low-income earners. Also, several empirical studies, for example Karingi et al. (2001), that have so far been done on the optimal tax rate of excise tax prove that, beyond a certain threshold, additional proliferation of tax rates will generate lower tax revenue.

2.1 Conceptual Framework and Hypothesis

The conceptual framework in Fig. 1, based on the empirical review and theoretical assumptions, was developed for this study. The conceptual framework schematically depicts the effects of electronic transaction taxes on SMEs access to mobile financial services. It is predicted that electronic transaction taxes would have a negative and significant effect on SMEs access to mobile financial services in Ghana.

Fig. 1. Conceptual framework

Since the aim of this study is to estimate the effects of electronic transaction taxes on SMEs access to mobile financial services, linear regression analysis, which is used to predict the value of a variable based on the value of another variable, was used. The assumptions underlying the use of the linear regression analysis are: the two variables should be either interval or ratio variables; there should be a linear relationship between the two variables; there should be no significant outliers; there should an independence of observations, which can easily be checked using the Durbin-Watson statistic; the data should show homoscedasticity, which occurs where the variances along the line of best fit remain similar as you move along the line; and it should be ensured that the residuals (errors) of the regression line are approximately normally distributed. Henceforth, in an attempt to test the hypothesis which states that electronic transaction taxes have no significant effect on SMEs assess to mobile financial services, the study states the following hypothesis for testing:

Hypothesis 1 (H1). *Electronic transaction taxes have a negative and significant effect on SMEs access to mobile financial services.*

Correlation analysis, which is a measure of the strength and direction of association that exists between two variables measured on at least an interval scale, was initially used to detect the existing relationship between the independent and dependent variables. The assumptions underlying the use of the Pearson product-moment correlation coefficient are: the two variables should be measured at the interval or ratio level, that is, they should be continuous variables; there should be a linear relationship between the two variables; there should be no significant outliers; and the variables should be approximately normally distributed.

3 Research Methodology

A post positivist research approach which involves a quantitative survey was adopted for this study. The target population of the study includes all existing SMEs in Accra, Ghana. The sampling frame for this research includes SMEs operating in both urban and rural suburbs of Accra. As a result of this cluster sampling, a probability sampling method, was used to collect data for this study. Cluster sampling is characterised by the division of the whole population into clusters or groups. Consequently, a random sample is taken from these clusters, all of which are included in the final sample (Wilson 2010). Cluster sampling is beneficial to researchers whose subjects are split over large geographical areas as it saves time and money (Davis 2005). The steps for cluster sampling can be summarized as follows: choosing cluster grouping for sampling frame, such as type of company or geographical region; numbering each of the clusters; and selecting sample using random sampling. For this study SMEs operating in Accra are divided into two clusters based on their geographical location, hence SMEs in the suburbs of Accra are sub-grouped as urban and rural SMEs. The cluster sampling method gives each element in the population an equal probability of getting into the sample, and all the choices are independent of one another. Determination of sample size is quite complex, and it depends on factors such as margins of error, degree of certainty, and statistical method. According to Corbetta (2003), sample size is directly proportional to the desired confidence level of the estimates and the variability of the phenomenon being investigated, and inversely proportional to the error that authors are willing to accept.

The sample size is calculated for the favourable case p = q = 0.5 when the size of the population is large and previous studies are unavailable to estimate the variability of an estimate over all possible samples. This study followed the recommendation by Corbetta (2003) in determining the standard deviation, a 95% confidence interval and a 5% sampling error for the calculation of the sample size. The sample size for this study was estimated by employing the Topman formula (Dillon 1993) as follows:

$$n = \frac{z^2 pq}{e^2}$$

n = required sample size
z = degree of confidence (1.96)
p = probability of positive response (0.5)
q = probability of negative response (0.5)
e = tolerable error (0.05)
Therefore, n = $\frac{(1.96)^2 \times 0.5 \times 0.5}{(0.05)^2}$ = 384.16 = 384

3.1 Model Specification and Description of Variables

To examine the effects of electronic taxes on SMEs access to financial services, the following equation, where access to finance is reflected as a function of electronic transaction taxes, is used:

$$AFS = f(ET) \qquad (1)$$

AFS = Access to Financial Services
ET = Electronic Taxes or Electronic Transaction Taxes

3.2 Model Equation of the Study

The above Eq. (1) can be rewritten as the following econometric model with its functional forms.

$$\text{AFS} = \beta 0 + \beta 1 \text{ETt} + C \tag{2}$$

where: β0 is the intercept, β1 represents the coefficient for the independent variable, Electronic Transaction Taxes (measured by proxy questions ranked by Likert scale), and C is the constant of the regression.

The primary data used for this study were collected through questionnaires. The survey was conducted between February and March 2022. The questionnaire involved close-ended types of questions that are germane to the subject of the research and measured by using a 5-point Likert scale, designed in a simple manner to make it easily fillable by respondents. The data garnered through the survey questionnaires were evaluated and discussed via statistic tools, such as bivariate correlation and regression analysis. The data were analysed by using the Statistical Package for the Social Sciences (SPSS) software, version 20. Additionally, secondary data was sourced through review of selected materials, such as SME records, to know the effects of electronic transaction taxes on SMEs access to mobile financial services in Ghana. Lastly, relevant data on current surveys regarding the subject of research was also sourced from the internet.

4 Results and Discussion

A probability sampling method, cluster sampling, was used to collect data for this study. The questionnaire was designed and copies were distributed to the respondents. Although the required sample size of the study was pegged at 384 participants, 400 copies of questionnaires were distributed. This was done in anticipation of the possibility of failure of some respondents to complete the questionnaire appropriately. Subsequently, 325 copies of the questionnaire were retrieved from the participants. Eventually, 304 fully completed questionnaires were retained and used for the study. The variables were represented by proxy questions, and the collected data were analysed using SPSS 20. Correlation analysis was used to check the existing relationship between independent and dependent variables.

4.1 Econometric Analysis

The Pearson's Product-Moment Correlation analysis was initially carried out with SPSS 20 to measure the strength and direction of association that exists between electronic transaction taxes and SMEs access to mobile financial services. The results, as shown in Table 1, indicated that there was a weak, negative correlation between electronic transaction taxes and SMEs access to mobile financial services, which was statistically significant ($r = -.362, n = 304, p = 0.00$).

Based on the significant yield of the correlation analysis, the study proceeded to perform the linear regression analysis. A linear regression was run to predict SMEs access to mobile financial services from electronic transaction taxes. The results indicated that

The Effects of Electronic Taxes on Small and Medium-Sized Enterprises 299

Table 1. Results of correlation analysis

		AFS	ET
AFS	Pearson Correlation	1	−.3.62**
	Sig. (2-tailed)		.000
	N	304	304
ET	Pearson Correlation	−.3.62**	1
	Sig. (2-tailed)	.000	
	N	304	304

**. Correlation is significant at the 0.01 level (2-tailed).

electronic transaction taxes statistically significantly predicted SMEs access to mobile financial services, $F(1, 302) = 119.481$, $p < .0005$, $R^2 = .131$. The results of the linear regression analysis performed using SPSS 20 are recorded in Tables 2, 3 and 4, and each table is comprehensively discussed.

Table 2. Model summary of regression analysis

Model	R	R Square	Adjusted R. square	Std. Error of the estimate	Durbin-Watson
1	.362[a]	.131	.130	3867.772	1.723

a. Predictors: (Constant), Electronic transaction tax.
b. Dependent Variable: Access to financial services.

R^2 is a measure of the extent of the variation of the dependent variable that is explained by the independent variable in the population. Adjusted R^2 is a modification to R^2 that is made with respect to the loss of degrees of freedom associated with adding extra variables (Brooks 2008). For this study, as exuded in the model summary in Table 2, the R^2 and adjusted R^2 have values of 13.1% and 13% respectively. This means that 13.1% of the variation in access to financial services is explained by electronic transaction taxes.

Table 3. Analysis of variance (ANOVA) of regression

	Model	Sum of Squares	df	Mean Square	F	Sig.
1	Regression	1.787E9	1	1.787E9	119.481	.000[b]
	Residual	11.877E9	302	1.4995E7		
	Total	13.665E9	3.0			

a. Dependent Variable: Access to financial services.
b. Predictors: (Constant), Electronic transaction tax.

The remaining 86.9% of the variation in access to financial services may be explained by other predictor variables which are not included in the model for this study.

The p-value of the F-statistic, as exuded in Table 3, indicates the total significance level of the model used for the study. In the F-statistic, the null hypothesis is that the independent variable, electronic transaction taxes, do not have an effect on the dependent variable, access to financial services. If the p-value of the F-statistic is less than 0.05, the null hypothesis is rejected, otherwise we fail to reject the null hypothesis. Rejection of the null hypothesis means that the test is significant, and failure to reject the null hypothesis means that the test is insignificant. For this study, the F-statistic of 119.481, with a p-value of 0.0000, is even significant at 1%. Henceforth, the null hypothesis is rejected at a 1% significance level. Thus, it can be concluded that electronic transaction taxes have a significant effect on SMEs access to mobile financial services.

Table 4. Analysis of coefficients of regression

Model	Unstandardized coefficients		Standardized coefficients	t	Sig.
	B	Std. Error	Beta		
(Constant)	8540.072	406.865		20.990	.000
ET	− 1217.343	111.369	− .362	− 10.931	.000

a. Dependent Variable: Access to financial services.

The model equation of the study is given as:

$$AFS = \beta 0 + \beta 1 ET t + C$$
$$AFS = 8540.072 - 1217.343(ET)$$

The results of the linear regression that was run indicated that electronic transaction taxes have a negative relationship with SMEs access to financial services. Also, the results statistically significantly predicted SMEs access to mobile financial services, $F\,(1, 302) = 119.481, p < .0005, R^2 = .131$. The coefficient of the independent variable indicates that a unit increase in electronic transaction taxes causes SMEs access to mobile financial services to decrease by 1217.343 units, and is statically significant at a 1% significance level. This implies that when existing electronic transaction taxes are either increased or new ones are introduced, SMEs access to mobile financial services will decrease. This finding is consistent with the study by Kodom (2020), with regards to the effects of the cost of transaction on adoption of mobile money financial services. It is also in conformity with the GSMA (2020) report, with respect to the unintended consequences and effects of mobile money transaction taxes on development and financial inclusion. Furthermore, the findings of this study confirm the studies by Ndung'u (2017) and Karingi et al. (2001), regarding the optimal tax rate beyond which lower tax revenue would be generated as a result of a deadweight loss, which culminates into financial exclusion of even entities that were previously financially included.

5 Conclusion

The primary objective of this paper was to estimate the effects of electronic transaction taxes on SMEs access to financial services in Ghana. The results of the Pearson's Product-Moment Correlation analysis initially run with SPSS 20 to measure the strength and direction of association that exists between electronic transaction taxes and SMEs access to mobile financial services indicated that there was a weak, negative correlation between the independent and dependent variables. Subsequently, a linear regression was run to predict SMEs access to mobile financial services from electronic transaction taxes. The results indicated that electronic transaction taxes statistically significantly predicted SMEs access to mobile financial services, and it validates the stated hypothesis of the study. Hence, to sustain a rapid and inclusive economic growth of SMEs, policymakers' must strategize on optimal taxation of mobile financial services that would prevent a welfare loss in the economy. The tax administration methods must further be improved to broaden the tax net, whilst keeping the burden on the common people as low as possible, to achieve real economic growth. Theories and research in economics have confirmed that low taxes, spread over a wider group of payers, pays higher in an economy than high taxes. Also, all stakeholders need to put in efforts to support the expansion of microfinance institutions, and marshal consumers' use of mobile money and mobile banking services to encourage financial inclusion of the underserved, and facilitate achievement of the ninth sustainable development goal (SDG 9.3). This study affords an enhanced and wider insight into SMEs access to financial services, and offers a momentary introspection into the present state and significance of SMEs in Ghana. Nevertheless, it has some limitations. These includes kerbs from the opinions of the respondents, which may be characterised with unfair judgment and uncertainty. Also, there are limitations set by the precincts of the study, since it principally focuses on the effects of electronic transaction taxes on SMEs access to financial services in Ghana. As a result, the survey was limited to only SMEs in rural and urban suburbs of Accra, the capital city of the country. Subsequently, the empirical study was grounded on the results derived from this survey. However, the cogency of deductions of this research need not be restricted only to Ghana, because theoretical and empirical findings are equally relevant in an international setting, particularly in other developing economies.

References

ADB. Asia SME Finance Monitor 2014 (2015). Manila. https://www.adb.org/sites/default/files/publication/173205/asia-sme-finance-monitor2014.pdf

Aker, J.C., Boumnijel, R., McClelland, A., Tierney, N.: Zap it to Me: The Impacts of a Mobile Cash Transfer Program. Tufts University (2012)

Aron, J., Muellbauer, J.: The Economics of Mobile Money: harnessing the transformative power of technology to benefit the global poor. Oxford Martin School (2019)

Brooks, C.: RATS Handbook to Accompany Introductory Econometrics for Finance. Cambridge Books (2008). http://ideas.repec.org/

Corbetta, P.: The qualitative interview. In Social Research: Theory. Methods and Techniques, pp. 264–286. SAGE Publications, Ltd., London (2003)

Davis, D.: Business Research for Decision Making, Australia, Thomson South-Western (2005)

De, S.: Reducing remittance costs and the financing for development strategy. World Bank (2015)
Dillon, J.: Alcinous: *The Handbook of Platonism.* Oxford University Press, Oxford (1993)
Gosavi, A.:. Can mobile money help firms mitigate the problem of access to finance in Eastern sub-Saharan Africa? J. Afr. Bus. (2018)
GSMA. State of the Industry Report on Mobile Money 2019 (2020)
Karingi, S.N., Kimenyi, S.M., Ndung'u, S.N.: Beer Taxation in Kenya: An Assessment. KIPPRA Discussion Paper No. 6. Kenya Institute for Public Policy Research and Analysis, Nairobi (2001)
Kipkemboi, K., Bahia, K.: The impact of mobile money on monetary and financial stability in Sub-Saharan Africa. GSMA (2019)
Kodom, M.: Financial inclusion via mobile money services in Ghana: drivers and the role of regulation. [Unpublished doctoral dissertation]. University of Ghana, Ghana (2020)
Ky, S., Rugemintwari, C., Sauviat, A.: does mobile money affect saving behaviour? evidence from a developing economy. J. Afr. Econ. (2018)
Riley, E.: Mobile money and risk sharing against village shocks. J. Dev. Econ. (2018)
Lopez, M.: Harnessing the Power of Mobile Money to Achieve the Sustainable Development Goals. GSMA (2019)
Naghavi, N., Scharwatt, C.: Mobile money competing with informal channels to accelerate the digitisation of remittances. GSMA (2018)
Ndung'u, N.: Digitization in Kenya: revolutionizing tax design and revenue administration. In: Gupta, S., Keen, M., Shah, A., Geneviève Verdier, G. (eds.) Digital Revolutions in Public Finance. International Monetary Fund, Fiscal Affairs Department (2017). http://www.elibrary.imf.org/doc/IMF073
Wilson, J.: Essentials of business research: a guide to doing your research project. SAGE Publication (2010)
World Economic Forum. How mobile money is driving economic growth (2015)

An Assessment of Practices on Disposal of Solar E-Waste in Lusaka, Zambia

S. Chisumbe[1,3], E. Mwanaumo[2,3(✉)], K. Mwape[1], W. D. Thwala[2], and A. Chilimunda[1]

[1] Department of Environmental Health, Faculty of Health Sciences, Lusaka Apex Medical University, 31909 Lusaka, Zambia
[2] Department of Civil Engineering, College of Science, Engineering and Technology, University of South Africa, Pretoria 0003, South Africa
erastus.mwanaumo@unza.zm, thwaladw@unisa.ac.za
[3] Department of Civil and Environmental Engineering, School of Engineering, University of Zambia, Lusaka, Zambia

Abstract. Purpose: End-of-life (EoL) management of solar products is an emerging issue, primarily because most industrial designs have not considered the impacts of the products when they reach their end-of-life stage. The aim of the study was to assess the practices on the disposal of solar e-waste among the residents of Lusaka, Zambia.

Design/Methodology/Approach: This study employed cross sectional descriptive design, with quantitative methods used. Data was collected using a structured questionnaire from 104 respondents who were purposively sampled and drawn from four zones of Lusaka west namely Hill view, Kapapa, Gomorra and Malcom. These included households using solar as a source of energy. The collected data was analysed using descriptive statistics.

Findings: The study revealed lack of knowledge regarding how solar products must be managed after EoL. Furthermore, that some of practices used in the disposal of solar e-waste were not the best practices, these practices included: throwing in bins or pits, this practice was confirmed with 71.2% participants, sell to informal recyclers 2.0% participants confirmed this and storing confirmed by 79.8% participants. With a mean score of 3.33 the study revealed that financial incentive was a factor contributing to how solar e-waste was being handled by the informal recycling sector. Results further revealed that disposal of Solar E-waste was poorly regulated and lacked sufficiently safe infrastructure. Majority of the respondents with a mean score 4.59 strongly agreed that there is lack of regulation and policy on solar e-waste handling.

Implications/Research Limitations: The present study did not capture other types of electronic-waste during data collection, but rather focused on EoL solar products. Similar study aimed at assessing the management of other types of electronic-waste would provide further information necessary in developing even more comprehensive policies.

Practical Implications: The findings of the current study are essential in informing policy and necessary legal framework on sustainable solar waste management.

© The Author(s), under exclusive license to Springer Nature Switzerland AG 2023
C. Aigbavboa et al. (Eds.): ARCA 2022, *Sustainable Education and Development – Sustainable Industrialization and Innovation*, pp. 303–312, 2023.
https://doi.org/10.1007/978-3-031-25998-2_23

Originality/Value: With the advent of climate change and the need for climate resilient green buildings, the use of solar products as a strategy in reducing energy consumption is being promoted, however, few studies have been done focusing at management of solar e-waste which if not properly managed could be harmful to the environment. The current study is among the first to be carried out in Zambia, and thus provides relevant information necessary in managing solar e-waste.

Keywords: Disposal · Environmental · e-waste · Practices · Solar

1 Introduction

Solar energy is an important renewable energy alternative that reduces carbon dioxide (CO2) emission by generating energy through photovoltaic (PV) thus it also presents a clean energy source (GOGLA 2018). However, due to the fact that equipment used for electricity and electronic devices powered by solar energy become waste in the long run, solar energy presents challenges related to waste management. Firstly, solar products generating power have an average life span of about 25 to 30 years, this implies that at the products' EoL they become e-waste, thus adding to the million tonnes of waste generated world over (Mathur et al. 2020). The generated waste in certain instances has hazardous properties which has an implication for their treatment and disposal practices (GOGLA 2016).

Solar is a clean, affordable and reliable source of electricity. About 30% of the Zambian population has access to solar and only 4% of the rural population has access to it. Solar home system (SHS) kit (solar portable lights (SPL), solar panel and charging battery), are becoming widely used across the country. Most common use is the provision of lighting and charging pots for electric devices since the products consist photovoltaic modules (PV) cells that convert sunlight to electricity (Barnes 2017). Despite of all the benefits associated with solar products, there are several negative effects particularly at the end of product life. The possible negative effects of solar products on the environment and people, should they not be properly managed at the end of the product life, have received little attention. Most consumers of these products just do away with them through burning or dumping which negatively impact on the environment (Batteiger 2015). Typically, the disposal of solar panels by landfilling creates environmental hazards such as leachate of heavy metals like lead (Bill 2015).

Once solar products have reached their end of life (EoL), scrap dealers take them for free. What they do is get valuable and easy to recycle aluminium frames then the remains will then end up at the landfill or temporary disposal forming a large amount of solar electronic waste (solar e-waste) which is an emerging issue. Solar products contain hazardous material such as lead, tin, lithium and cadmium which when not properly managed after EoL will bring environmentally and health related impacts and shortage of critical materials to meet future resource demands (Hamuyuni and Tesfaye 2019).

The solar e-waste management system in rural and urban areas is a challenge because the waste is not recycled locally or reused for different purposes. It is recommended that disposal of used solar be done in plastic bags even though the batteries may no longer be useful in powering devices as it may still have little power left inside which may cause them leak or explode when jammed together. But in the actual sense they are just thrown away end of life (Mathur et al. 2020).

2 Challenges Associated with E-Waste Management in Africa

There are several challenges associated with solar e-waste management such as high collection and logistics costs. This has been attributed to the distance and terrain in most cases (James 2016). Patrick (2016) identified battery diversity to be another challenge. Arguing that there's no lithium battery recycling facility in Africa even though lithium based batteries are dominant. The lack of regulations, legal framework, and infrastructure required to manage E-waste in an environmentally sound manner is another issue that has been brought up (ITU 2018; Robinson 2019). The legal and infrastructural framework for achieving sound management is still quite low in most African countries. Worse still even where the legislation is in place, enforcement of the same is another challenge requiring attention (Widmer 2019). Osibanjo (2018) pointed out that only very few countries have put in place policies and facilities specific for E-waste management and recycling thus what exists is to a large extent in the informal sector. This view is shared by Muntanga et al. (2018) who posited that Zambia lacks E-waste policy as well as capacity and infrastructure for recycling thereby contributing to poor management of E-waste. Lack of information about treating e-waste and inadequate funding for hazardous waste management initiatives are two more challenges in the management of e-waste. Table 1 summarizes the elements cited in the literature that have influenced how solar e-waste is disposed of.

Table 1. Factors identified as being influential is solar e-waste

Factors	Author
Length of use	(James 2016)
Lack of regulation	(ITU (2018; Widmer 2019; Muntanga et al. 2018)
Lack of policy on disposal of solar e-waste	(ITU 2018; Osibanjo 2018; Muntanga et al. 2018)
Community's attitude	(Song et al. 2016; Pasiecznik et al. 2017; Attia et al. 2021)
Lack of financial incentive	(Abila and Kantola 2019; Balasubramanian et al. 2020; Hansen et al. 2022)
Lack of awareness	(Avis 2021; Almulhim 2022)

3 E-Waste Handling and Disposal in Zambia

Generally, disposal of waste in most developing countries including Zambia is characterised by crude dumping of waste, burning of waste as well as burying of waste which may have implications for the ground water and soil (Victor 2018).

The majority of those who handle e-waste in Zambia are involved in the maintenance and repair of electronic devices (Victor 2018). Zambia like most African countries is characterised by inappropriate methods of disposing e-waste. This has largely been attributed to the absence or the inadequacy of regulatory framework for dealing with e-waste (US EPA 2018) as well as lack of systems that adequately address disposal of e-waste and limited recycling infrastructure (Widmer 2019).

According to the Zambia Environmental Management Agency (ZEMA), Zambia only recycles a minimal proportion of e-waste and in most cases the waste taken for recycling falls below the expert criteria for recycling e-waste. Most of the e-waste is buried underground with little consideration for the environment and health of the people (Widmer 2019). Other times, Zambia sends its discarded electronics to South Africa (Mwansa 2021). The remaining material is typically burned or disposed, either in improvised incinerators or landfills. These inadequate techniques of electronic waste disposal don't manage the harmful elements safely or recover valuable materials (Khaliq and Li 2017).

There is need to prioritize treatment and recycling of e-waste in Lusaka and Copperbelt towns with high rate of EEE usage (Victor 2018). Furthermore, there is need to improve on technical competency, environmental awareness and, most of all, training in order to increase the sense of awareness (Victor 2018).

4 Existing Law on Waste Management in Zambia

The Environmental Management Act (EMA) of 2011 is an important piece of the legal framework which provides for how issues of waste management among other things should be conducted and executed in Zambia. According to the act's recommendations for environmental management in connection to waste management, waste generation should be minimized whenever possible, and garbage should then be reused, recycled, recovered, and disposed of properly in a way that doesn't have any negative consequences.

Further, the legislation points out the need for involvement of several stakeholders in management of waste, some of the stakeholders include the local authorities, the community members, and the private sector among others. Through the Act, the local authority is expected to provide for waste management services giving priority to recovery, re-use as well as recycling of waste. The Act also stipulates the need to classify waste into respective categories (Environmental Management Act 2011; Banda et al. 2021).

Through the Act, the producer is of a product has the responsibility of the product to the post-consumer stage. This entails that they are responsible for waste minimisation programmes such as recovery, reuse or re-cycling of waste including measures to reduce the potential impacts of the product on human health and the environment.

However, overall review of this legislation exposes some gaps in that it does not clearly address how e-waste and to be specific solar e-waste should be managed. Considering the dangers associated with wrong disposal of end of life solar products this aspect may need to be addressed with appropriate measures outlined in the corresponding regulations (Ibid).

5 Materials and Methods

A descriptive cross sectional survey design was adopted for this study, with data collected using a structured questionnaire containing closed ended questions. The development of the instrument used for data collection was informed by reviewed literature. The study was conducted in the western region of Lusaka district in Zambia. Four zones in Lusaka west were considered namely Hill view, Kapapa, Gomorra and Malcom. Lusaka west was selected for the study; this is because most of the household in the aforementioned areas use solar to power their homes.

A total of 104 respondents mostly household heads were purposively sampled and participated in the study. The respondents were chosen based on how long they had lived in the research location and whether or not solar energy had been their primary source of household energy. Data was analysed using descriptive statistics.

6 Results

Only 18.2% of the respondents who took part in the survey had lived in the study area for less than two years. According to Fig. 1, the majority of respondents had resided in the region for at least three years. Some of the interviewees had lived in the area under study for more than 15 years.

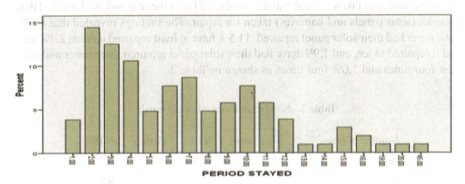

Fig. 1. Respondents number of years stayed in study area. Author (2022)

6.1 Practices on the Disposal of Solar E-Waste

An assessment was carried out to establish the common methods used in the disposal of solar e-waste. With regard to the how solar panels are treated after end of life, 36.5% of the respondents indicated that after product end of life the solar panels are just stock piled. 21.2% of the respondents revealed that they throw the panels in garbage bins or back yard pits together with others solid waste. 11.5% said they sell to informal recyclers, only 1.0% take to formal recyclers as shown in Fig. 2.

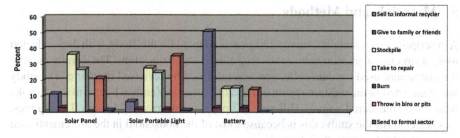

Fig. 2. Practices on the disposal of solar e-waste. Source: (Author, 2022).

Response on how solar portable lights are treated after end of life, 35.6% indicated they throw them in the garbage bins or back yard pits, 27.9% stockpile, whereas 6.7% of the respondent said they sell to informal recyclers. Likewise, on the batteries 51% of the respondents indicated that when batteries reach their end of life, they usually sell to informal recyclers. 15.4% indicated that they stockpile, 15.4% take to repair, while as 14% the batteries in solid waste bins together with other solid waste or back yard garbage pits.

6.2 Number of Repairs

An assessment was taken to establish the number of times the respondents have had their solar kit (solar panels and batteries) taken for repair. The findings revealed that 80.8% have never had their solar panel repaired, 11.5% have at least repaired it once, 2.9% have had it repaired twice, and 1.9% have had their solar panel repaired three times and more than four times and 1.0% four times as shown in Table 2.

Table 2. Number of times taken for repair

Number of times	Solar panel	N (%)	Battery	N (%)
Once	12	11.5	34	32.7
Twice	3	2.9	9	8.7
Three times	2	1.9	6	5.8
Four times	1	1.0	1	1.0
More than four times	2	1.9	0	0
Never	84	80.8	54	51.9
Total	104		104	

Likewise, on how many times respondents had their battery repaired, 51.9% have never had their solar portable light repaired, 32.7% have at least repaired it once, 8.7 have had it repaired twice, and 5.8% have had their battery repaired three times and 1.0% four times.

6.3 Factors Influencing the Disposal of Solar E-Waste

In determining factors influencing the disposal of solar e-waste in Lusaka. The findings revealed that main factors influencing the disposal of solar e-waste were length of use as well as lack of regulations, having scored the mean of 4.65 and 4.59 respectively. The level of agreement on these factors was strong as denoted by the standard deviation of around 1 as shown in Table 3.

Table 3. Factors influencing the disposal of solar e-waste

Factors	Mean	Std	Rank
Length of use	4.65	1.04	1
Lack of regulation	4.59	0.96	2
Lack of policy on disposal of solar e-waste	4.52	1.27	3
Community's attitude	3.65	1.21	4
Lack of financial incentive	3.33	1.25	5
Lack of awareness	1.43	1.03	6

Source: (Author, 2022).

Surprisingly, among the criteria that were determined to have an impact on how solar e-waste was disposed of in Lusaka, lack of awareness came in last. A mean of 1.43 and a standard deviation of 1.03 denoted this. However, the results on disposal practices revealed otherwise considering that others throw away the end of life used products; solar panels (21.2%), portable bulbs (35.6%) and batteries (14%).

7 Discussion of Results

7.1 Practices on Disposal of Solar E-Waste

The findings revealed that throwing off of solar e-waste in garbage bins or backyard pits is a common practice among respondents in the study area. These findings agree with Baden-Baden (2019) that after end of life of solar products, wastes are disposed-off in bins by the customer after which the local authorities or other waste collection company may sometimes collect and take the end of life products as part of waste, then dispose at the landfill.

Further, among other practices the findings revealed that some respondents opt for selling the end of life solar products to informal recyclers this was more common with batteries. Again this aligns with literature, James (2016), established that once a product reaches its producer determined end of life, the customers attempt to use informal expertise to help extend the life of the product. Likewise, Batteiger (2015) posited that consumers find it more cost effective to sell the unwanted end of life solar products. Baden-Baden (2019) pointed out that some of the solar e-waste is been handled by the informal recycling sector, which is poorly regulated and lacks sufficiently safe infrastructure. And has contributed to the improper management of solar e-waste leading to more open dumping and more solar e-waste at the landfill.

The study also revealed that solar panel, solar portable lights and batteries are just stored after end of life. This is similar with the study done by Batteiger (2015), which established that the other option is not to get rid of them, but simply store them. A smaller proportion of respondents opted to burning of solar products after end of life. These findings where similar to a study done by Rwanda (2021) which established that there is improper disposal of end of life solar products especially batteries, through burning. The results implied that some of the practised used in the disposal of solar e-waste where not the best practices.

7.2 Factors Influencing the Disposal of Solar E-Waste

According to the means of 4.52 and 4.59, respectively, the findings indicated that there is a lack of policy and regulation on the disposal of solar e-waste. These findings corroborate the claim stated by Muntanga et al. (2018) that environmental regulations and a legislative framework that are tailored to e-waste are lacking in Zambia. The implication of this being the lack of measures for proper disposal of solar e-waste, which in turn creates negative environmental impacts like land and water pollution due to improper disposal of solar e-waste. This is the view point shared by ITU (2018) that least developed countries often lack the policies, legal instruments, regulatory and infrastructure needed for environmentally sound management e-waste.

7.3 Implications of the Findings

The findings of this research suggest that in order for a developing country like Zambia to achieve sustainable management of e-waste the focus should be on having the necessary policy and legal framework in place. The policy need to stimulate the participation of the private sectors as well as increased knowledge on e-waste management. Therefore, these findings are essential in informing policy and development of relevant legal framework on solar waste management.

8 Conclusion

The study highlights the methods used to manage solar e-waste and shows that there are large gaps in the disposal methods used. More environmentally responsible and secure methods are required. This paper urges Zambia to create a legal framework that allows for the environmentally responsible treatment of solar e-waste. Additionally, the government and companies that make solar-related items must collaborate to ensure that consumers are informed about the best ways to dispose of solar e-waste. The best practices to be used in disposal of solar e-waste should be aimed at promoting recycling and participation of many actors in e-waste management.

References

Abila, B., Kantola, J.: The perceived role of financial incentives in promoting waste recycling—empirical evidence from Finland. Recycling **4**(1), 4 (2019)

Almulhim. A.I.: Household's awareness and participation in sustainable electronic waste management practices in Saudi Arabia. Ain Shams Eng. J. **13**(4), 101729 (2022). https://doi.org/10.1016/j.asej.2022.101729

Attia, Y., Soori, P.K., Ghaith, F.: Analysis of households' e-waste awareness, disposal behavior, and estimation of potential waste mobile phones towards an effective e-waste management system in Dubai. Toxics **9**(10), 236 (2021)

Avis, W.: Drivers, barriers and opportunities of e-waste management in Africa (2021)

Baden-Baden. The Global LEAP Solar E-waste challenge Market Scoping Report., s.l.: Global LEAP Award (2019)

Balasubramanian, S., Clare, D., Ko, S.: Off-Grid Solar E-Waste: Impacts & Solutions in East Africa. Duke University Master of Environmental Management, Durham (2020)

Banda Kachikoti, W., Mwanza, B.G., Mwanaumo, E.M., Banda, I.N.: Governance mechanisms for managing municipal solid waste: a review. In: Proceedings of the 11th Annual International Conference on Industrial Engineering and Operations Management, Singapore, 7–11 March 2021, ISSN/E-ISSN 2169–8767 ©IEOM Societyieomsociety.org. (Virtual Conference) (2021)

Batteiger, A.: Towards a waste management system for solar home systems in Bangladesh. In: Groh, Sebastian, Straeten, Jonas, Lasch, Brian Edlefsen, Gershenson, Dimitry, Filho, Walter Leal, Kammen, Daniel M. (eds.) Decentralized Solutions for Developing Economies. SPE, pp. 133–140. Springer, Cham (2015). https://doi.org/10.1007/978-3-319-15964-5_12

Bill. Conservation and Recovery Act 1976, Senate Bill No. 489- Hazardous Waste: Photovoiltaic Modules: IRENA (2015)

Environmental Management Act. a basis for the growth of an environmental ethos and good environmental governance in Zambia." Environmental law in Africa. Nomos Publishers (2011)

GOGLA. Off-Grid Solar Market Trends Report 2016. Bloomberg New Energy Finance, World Bank, IFC and Global Off-grid Lighting Association (GOGLA) (2016)

GOGLA. Off-Grid Solar Electronic Waste Management. Defining Challenges & identifying solution. Nairobi: E-Waste workshop summary. Mkapa Campus Nairobi (2018)

Hamuyuni, J., Tesfaye, F.: Advances in lithium-ion battery electrolytes: prospects and challenges in recycling. REWAS **2019**, 265–270 (2019)

Hansen, U.E., Reinauer, T., Kamau, P., Wamalwa, H.N.: Managing e-waste from off-grid solar systems in Kenya: Do investors have a role to play? Energy Sustain. Dev. **69**, 31–40 (2022)

ITU. Handbook for the development of a policy framework on ICT/e-waste (2018). https://www.itu.int/en/ITU-D/Climate-Change/Documents/2018/Handbook-Policy-framework-on-ICT-Ewaste.pdf ISBN: 978-92-61-27321-7

James. Key success Factors for Solar Home System After sales services and maintenance in Uganda. Master's thesis. Rotterdam school of management, Rotterdam (2016)

Khaliq, A., Li, J.: Management of electrical and electronic waste: a comparative evaluation of China and India. Renew. Sustain. Energy Rev **76**, 434–444 (2017)

Mathur, D., Gregory, R., Simons, T.: End-of-life management of solar PV panels (2020). https://ris.cdu.edu.au/ws/portalfiles/portal/57428020/EOLManagementSolarPV_Final_e_version.pdf

Muntanga, E., Banda, E., Nabbili, K., Kabuya, S., Hamusuku, G.: An Investigation of how Electronic Media Houses in Lusaka Dispose Off their Obsolete ICT Equipment. The University of Zambia (2018)

Mwansa, J.: Zambia Struggling to introduce E-waste legislation, Lusaka: Techwatch News (2021)

Osibanjo, O.: The challenge of electronic waste mangement in developing countries. In: Waste Mangement Research, pp. 489–501 (2018)

Pasiecznik, I., Banaszkiewicz, K., Syska, Ł.: Local community e-waste awareness and behavior: Polish case study. Environ. Protect. Eng. **43**(3) (2017)

Patrick. Will solar PV create a wave of Toxic Battery Waste in Rural Africa? University of Edinburgh, Uganda (2016)

Robinson, B.: E-waste: an assessment of global production and environmental impacts. Sci. Total Environ. **408**(2), 183–191 (2019)

Rwanda. Innovation and lessons in solar E- waste management. s.l.: Sunny Money e-waste sculpture (2021)

Song, Q., Wang, Z., Li, J.: Residents' attitudes and willingness to pay for solid waste management in Macau. Procedia Environ. Sci. **31**, 635–643 (2016)

US EPA, E.: Sustainable Management of Electronics (2018). http://www.epa.gov/smm-electronics. Accessed 7 Nov 2021

Victor, P.: Electronic Waste reprocessing or processing: an alternative practice for production and extraction of metals in Zambia, pp. 283–293 (2018)

Widmer, R.: Global Perspectives on e-waste. In: Environmental Impacts Assessment Review, pp. 436–458 (2019)

Lean Supply Chain Practices in the Zambian Construction Industry

E. Manda[1], E. Mwanaumo[2,3](✉), W. D. Thwala[2], R. Kasongo[1], and S. Chisumbe[2]

[1] Graduate School of Business, University of Zambia, Lusaka, Zambia

[2] Department of Civil and Environmental Engineering, School of Engineering, University of Zambia, Lusaka, Zambia

erastus.mwanaumo@unza.zm, thwaladw@unisa.ac.za

[3] Department of Civil Engineering, College of Science, Engineering and Technology, University of South Africa, Pretoria 0003, South Africa

Abstract. Purpose: The construction process involves the assembly of various materials and components in order to come up with a final product. This entails that all parties involved in the construction supply chain need to be effective and efficient. However, the Zambian building construction industry like many other developing countries is associated with numerous challenges. This study sought to explore lean practices being implemented in the Zambian construction supply chain.

Design/Methodology/Approach: A quantitative methodological approach was adopted with data collected through a structured questionnaires containing closed-end questions. A total of 151 Building contractors registered with National Council for Construction (NCC) in grades one, two and three were sampled, out of which 102 were responsive. The data collected was analysed using descriptive statistics.

Findings: The results revealed that the most common lean practices being implemented focus on prevention of wastes and defects by; establishing customer needs, collection of multiple quotations, ensuring that firms have updated inventory that ensure flow of construction and ensuring product tests are done for adherence to quality and specification among others.

Implications/Research Limitations: This study was only limited to contractors in grade one to three, a similar study would be good to be conducted among contractors registered in grades four, five and six so as to determine lean practices being implemented among small scale contractors.

Practical Implications: The finding of this paper will help contractors reduce the overall product cost of building construction through elimination of wastes as a result of implementing lean practices.

Originality/Value: This is among the first studies conducted in Zambia with the focus being on establishing the lean practices being implemented on large scale building construction worksites.

Keywords: Construction · Lean · Materials · Practices · Supply chain

© The Author(s), under exclusive license to Springer Nature Switzerland AG 2023
C. Aigbavboa et al. (Eds.): ARCA 2022, *Sustainable Education and Development – Sustainable Industrialization and Innovation*, pp. 313–326, 2023.
https://doi.org/10.1007/978-3-031-25998-2_24

1 The Construction Industry

The complexity of the construction sector makes it more susceptible to numerous challenges, some of which include uncertain construction conditions, specific product demands, as well as fluctuations in demand cycles (Bal et al. 2013). Production in construction is more uncertain and complex than in manufacturing (Lasse 2020). Aziz (2013) mentioned three aspects that add to the complexity of construction and these are site production, temporary and unique products. With regard to uniqueness of products, Bal et al. (2013) posits that the end product of a construction project is usually unique and tailored to meet the client's needs. In addition, even if the outputs of different construction projects have similar characteristics, they can still be understood as unique. Hedman and Hedberg (2015) pointed out that uniqueness of the output is caused by differing preferences of clients, different local prerequisites, and according to the design solutions from the architect. Prasad et al. (2019) shared that the (unique) characteristic of projects entail that the methods used in constructing a building and the inputs used are never the same from one project to another.

It being a site based product, Nikinosheri and Staxäng (2016) posited that construction production is a combination of two characteristics which are fixed position manufacturing and rootedness-in-place. Fixed position manufacturing means that labour resources move around the product, which is quite different from manufacturing where the products move through labour resources. Dubois et al. (2019) highlighted that one of the downsides of fixed position manufacturing is that it leads to having a congested worksite among other things due to the fact that a number of connected processes need to be implemented at the same time and place. Whereas, Rootedness-in-place means that construction projects need to abide by the local laws and regulations. Muhwezi et al. (2014), wrote that there are legal and physical restrictions that come with building, for example building codes and zoning regulations. The physical constraints that could be there include; geographical situations, space constraints, weather, and ground movement will all have an impact on production and organization of project.

Likewise, the construction industry is a multi-faceted industry consisting a number of parties such as contractors, sub- contractors, architects, surveyors, and suppliers. Constructions is the production of unique one-off products through the assembly of various materials. The materials involved can be purchased internally (within the country) or externally (outside the country). Managing a construction project means managing all relationships important to achieving the desired goal. In construction procurement, downstream-upstream integration is key to achieving less cost and high value (Broft et al. 2016). Therefore, applying supply chain management in construction brings with it improved construction processes.

2 Supply Chain

A supply chain is a system that connects a company with many suppliers in order to produce and deliver a good to the customer. The system consists of various individuals, groups, resources and information, and these have to move from upstream to downstream from supplier to consumer. These aspects can all be one firm but all needed to work together to produce value (Felea and Albăstroiu 2013). It refers to the actual steps involved in transforming and delivering a product or service to the final consumer. The steps are purchasing, moving raw materials, transforming them into final products and distributing them to the customer. A supply chain is established and fashioned carefully by organisations in order to cut costs and maintain competitiveness.

A supply chain can also be looked at as a set of companies that work together to bring a product or service to the market. Working in collaboration with the company being supplied to or bought from is essential for business. One company can be part of a long supply chain. In construction for example, one company can simply be involved in making bricks, another simply finishes, another wiring. The different firms are involved in supply chain include among others; manufacturers, sellers, warehouses, transporters as well as distribution-centers. To provide higher customer value at a lower cost in the entire system, supply chain management involves managing relationships with customers and suppliers upstream and downstream (Kain and Verma 2018). The final goal of supply chain management is to manage and arrange the flow of money, flow of information and the flow of physical production in order to meet the client's needs. Management of supply chain involves a group of connected decisions and actions aimed at bringing together all stakeholders in production these include; manufacturers, suppliers, retailers, transporters, and clients in order to ensure timely delivery of materials in right locations, and in correct quantities, so as to minimize costs whilst ensuring customer satisfaction (Mandl 2013).

Today, for businesses to be successful, much attention need to be paid in how operations run and this means being interested and engaged with suppliers and customers. With the expansion of global markets and threatening competition, getting desired products now means paying close attention to the source of raw materials, how they are produced, designed and transported, how your raw materials are eventually transformed into final products and how they are finally transported to the customer (end user). There is need to really find out what customers want, acknowledge and deliver as needed Wisner et al. (2012).

These relationships with producers, suppliers, and customers help businesses remain successful, and they are essential to the practice of supply chain management. Supply chain management is described as the planning and administration of all logistics activities from sourcing, procurement, to conversion, by the Council of Supply Chain Management Professionals (CSCMP). Collaboration among suppliers, customers, and third-party service providers is a component of supply chain management. Supply chain management is defined by the Institute for Supply Administration (ISM) as the design and management of value-added, efficiently operating processes that span organizational boundaries to meet customer needs.

Supply Chain Management (SCM) is a philosophy which defines how organisations ought to be managed in order to attain strategic advantage. The goal is that all activities undertaken should aim at giving the client value for their money as well as achieving end-user satisfaction whilst lowering the cost of production. SCM then can be referred to as the harmonization of the participants' activities in the supply chain, in order to come up with a product or service that satisfies the client and at the same time reduce the cost of the organization (Musonda et al. 2018; Papadopoulos et al. 2016; Newman et al. 2020). Therefore, firms must learn to work together and share information to have a successful supply chain management because competitiveness is no longer solely on one singular organisation but has shifted to supply chains. Information to do with production plans, demand forecasts, marketing strategies, product development, new technologies delivery dates among parties need to be made shared (Wisner et al. 2012).

3 Nature of the Construction Supply Chain

In the construction industry supply chain is characterised by assembling of different kind of materials and component from various sources in order to come up with the final product (Papadopoulos et al. 2016; Nazir et al. 2020). The "construction factory" is organized around a single product, as opposed to production systems where a variety of items circulate through the factory and are distributed to a large number of clients. In addition, through frequent reconfiguration of the project organisation, it produces a unique product by careful management of its transient supply chain. However, due to its transient nature in many cases it suffers from fragmentation, as well as instabilities in managing multiple players. Even more so, every project results in the creation of a fresh product or prototype, making it a conventional make-to-order supply chain. With a few minor exceptions, there is hardly much duplication.

Large projects have very complex supply chains. The major complexity comes from the fact that there are a variety of material involved in construction and different parties like suppliers and subcontractors working together (Mwanaumo et al. 2014). The complexity of a project is also attached to the scope of the project, therefore, it requires careful planning, organisation and coordination amongst actors. For instance, a construction company interacts with a lot of sub-contractors and suppliers in delivering a large scale project. The nature of construction production process is in such a way that final-product is for the client/s who owns the project; each product is different, in terms of place, equipment used and methods of productions. Likewise, there is a high rotation-index among the contractors workforce, and sometimes construction material is even stored off-site.

3.1 Zambian Construction Supply Chain

The Zambian construction supply chain suffers from numerous challenges which negatively affect it. The construction supply chain like many other developing countries is known for having a lot of material wastages, delays in material delivery, and poor supplier-contractor coordination resulting from poor planning as well as due to the fact that construction supply chain management is too traditional and underdeveloped (Samuel et al. 2015; Mostafa and Dumrak 2020). The Zambian Construction supply chain is characterized by ineffective planning and scheduling of projects, short-term, poorly managed contactor–supplier relationships leading to loss of competitiveness, delays in material delivery, inappropriate selection criteria, and frequent changes in the specification. This ultimately brings about increased production costs (Turkyilmaz et al. 2013). Traditional supply chain management is responsible for the majority of the issues in Zambia (Kumar 2016). Only a few studies on the application of lean construction principles to building construction projects have been done in Zambia.

Therefore, the purpose of this study was to determine which lean practices are being used in the Zambian construction industry (ZCI). Recognizing the many challenges attributed to traditional supply chain management and the subsequent importance of adopting lean principles in the supply chain. This paper discusses the challenges in the construction supply chain, different types of wastes evident in the Zambia construction industry and lean practices being implemented with the potential to bring about a reduction in supply chain-related delays and the associated costs (Turkyilmaz et al. 2013).

4 Lean

Lean was first introduced by Toyota motors. It was used as a secret weapon for winning competitive advantage over its competitors. Thereafter, the Japanese manufacturing industry followed suit and began applying the principles in their operations. The term 'lean' was coined by an American business man Krafcik in his 1988 article, "Triumph of the lean production system" (Kumar 2014). Lean is simply tools that help to identify and eliminate waste. Elimination of waste is key because as waste is eliminated, simultaneously, quality is improved and time and costs are reduced. The types of waste highlighted at Toyota motors company included unnecessary movements, underutilization of employees creativity, idle time (waiting), defects, a lot of inventory, over producing as well as over-processing among others (Nimeh et al. 2018). Martínez-Jurado and Moyano-Fuentes (2014) adds that the focus of lean is mainly aimed at eliminating operations which do not add any value and wastages in the production processes.

A lean organization is an organisation that understands customer value and ensures that all its activities are centered around-it. The process of value delivery needs to be seamless. To make lean a success, Arif-Uz-Zaman and Ahsan (2014) emphasized that the level of thinking needs to changed and focus on management. Elimination of waste has to be along the entire value streams, instead of isolated points. Babalola et al. (2019) defined lean as a methodical style aimed at identifying and elimination of wastages by embracing the practice of continuous improvement aimed achieving value for the client as well end-user satisfaction.

4.1 Lean in Supply Chain

Adoption of lean principles in supply chain has contributed to the reduction in material delivery time, material cost as well as improved efficiency and efficiency (Ugochukwu et al. 2012). Furthermore, implementation of lean as a strategy aimed at managing supply chain has increased and is been viewed as a philosophy meant to guide organisations in achieving overall-business performance. Alkhoraif et al. (2019) contends that research on lean supply chain has focused more on lean application within single organisation or shop floor, and not the entire supply chain. Adding that development need to be extended throughout the whole supply-chain. Having considered the merits of lean application in management approach (Emuze 2012) opined that adoption of lean principles contributes to achieving effectiveness in the management of the construction supply chain. Overall when it comes to project implementation, lean brings about improved productivity, effectiveness and efficiency in management of projects finite resources (Barbosa et al. 2013). Siyam et al. (2015), describing lean refers to it as a close alignment from raw material to buyer through collaboration. Hence, organisations wanting to have an integrated supply chain need to adopt lean practices. Lean supply chain therefore, refers application of lean principles in supply chain management.

4.2 Lean Practices in Construction Supply Chain

Implementation of lean practices in the construction supply chain is aimed at ensuring among other things efficiency and effectiveness throughout the process by eliminating wastages in time and material inventory. Literature evidence reveals a number of practices as been essential in achieving an efficient and effective supply chain management in construction these include; ensuring that customer needs are established, getting multiple quotations from suppliers, integration of firm's systems with suppliers, ensuring product tests are done for adherence to quality and specification, as well as firms giving feedback to suppliers on quality and delivery among other factors as shown in Table 1.

Table 1. Identified lean practices in construction supply chain management

Lean practices	Author
Establishing customer needs	Yala (2016)
Collection of multiple quotations	Locatelli et al. (2013)
Ensuring flow of construction through having updated inventory	Aziz (2013), Kaynak (2013)
Testing products to ensure adherence to specifications and quality	Kosky (2021)
Delivering only what satisfies the customer	Wisner et al. (2012), Modrak (2014)
Efficient utilisation of space and machines by the firm	Aziz (2013)
Firms storing only what is needed	Ansah et al. (2016), Gnich (2012)
Contractor-supplier collaborations	Kain and Verma (2018)
Feedback from the firm to supplier material quality and deliveries	Bento et al. (2020)
Firm ensures there are effective transportation systems	Arif-Uz-Zaman and Ahsan (2014), Kelly (2002)
Ensuring that firms have integrated their systems with the suppliers	Emuze and Smallwood (2013), Aigbavboa et al. (2014)
Monitoring of supplier performance	Chiponde et al. (2014)

5 Materials and Methods

This study adopted a cross sectional descriptive design with quantitative approach. Data was collected using a structured questionnaire containing closed-ended questions from a total of 151 respondents drawn from contractors registered in grades one to three of the national council for construction (NCC). These included project managers, quantity surveyors, site engineers, as well as procurement managers. The development of the questionnaire (instrument) used for data collection was informed by the extensive literature review from which lean practices were identified and included on the questionnaire. Descriptive statistics was used for the analysis of collected data. Table 2 shows a summary of population size as well as successful response rate.

Table 2. Summary of population sizes, sample sizes and response rates

Population category	Population size	Sample size	Successful responses	Response rate
Grade 1	67	57	36	63%
Grade 2	64	55	39	71%
Grade 3	43	39	27	69%
Total	**174**	**151**	**102**	**67.5%**

Of the successive responses 37% were project managers, 26% were quantity surveyors, 11% site engineers, whereas procurement managers accounted for a total 6% as shown in Fig. 1.

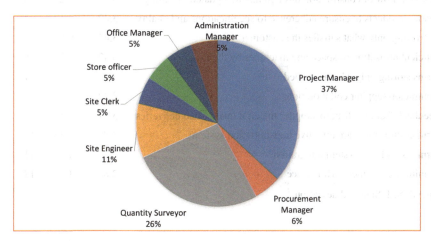

Fig. 1. Position of the respondent in the firm.

6 Results

6.1 Lean Supply Chain Management Practices

With regards to the lean practices which are been implemented in the Zambian construction supply chain, the findings revealed that; establishment of customer needs, and Collection of multiple quotations were the most common lean supply chain practices, with the means of 3.67 and 3.65 respectively. These were followed by Firm ensures that they have updated inventory that ensure flow of construction with a mean score of 3.53, Product tests are done for adherence to quality and specification with a mean score of 3.47, delivering only what satisfies the customer with a mean score of 3.47. The results showed that Firm efficiently utilizes its space and machines had a mean score of 3.42, firms storing only what is needed had a mean score of 3.42, There is close collaborations with supplier had a mean score of 3.22 and feedback from the firm to supplier material quality and deliveries had a mean score of 3.11. Lastly Firm ensures there are effective transportation systems had a mean of 3.05, firm – suppliers' system integration had a mean of 2.37 and Monitoring supplier performance had a mean of 2.84. For all the results, the values for the standard deviations were in the range of 0.84 to 1.50 as shown in Table 3.

Table 3. Application of lean practices in supply chain management

Application of lean	Mean	Std.	Rank
Customer needs are established	3.67	1.14	1
Multiple quotations are collected	3.65	1.27	2
Ensuring flow of construction through having updated inventory	3.53	1.22	3
Testing products to ensure adherence to specifications and quality	3.47	1.17	4
Delivering only what satisfies the customer	3.47	0.84	5
Efficient utilisation of space and machines by the firm	3.42	1.22	6
Firms storing only what is needed	3.42	1.07	7
Contractor-supplier collaborations	3.22	1.26	8
Feedback from the firm to supplier material quality and deliveries	3.11	1.10	9
Firm ensures there are effective transportation systems	3.05	1.18	10
Firm – suppliers' system integration	2.37	1.46	11
Monitoring supplier performance	2.84	1.50	12

Legend: **Std**: Standard deviation.

6.2 Wastes Associated with Construction Supply Chain Management

With regards to the types of wastes evident in the construction industry, the results revealed that that time and cost overruns were the most common wastes identified as denoted by the group means of 3.84 and 3.42 respectively. These were followed by Movement of equipment, material and people to a job site before they are needed with mean score of 2.68; too much inventory (too many materials on site) with mean score of 2.42; adding features or activities that do not add value to the client with mean of 2.42; and reworks with mean score of 2.32. For all the results, the value for the standard deviations were in the range of 1.06 to 1.30 as shown in Table 4.

Table 4. Types of wastes

Types of wastes	Mean	Std.	Rank
Cost overrun	3.84	1.30	1
Time overrun	3.42	1.22	2
Waiting for material from Suppliers	3.21	1.13	3
Movement of equipment, material and people to a job site before they are needed	2.68	1.11	4
Too much inventory (too many materials on site)	2.42	1.07	5
Adding features or activities that do not add value to the client	2.42	1.22	6
Reworks	2.32	1.06	7

Legend: **Std**: Standard deviation.

6.3 Challenges in Construction Supply Chain Management

With regards to the barriers associated with implementing supply chain management practices, the findings revealed that the top four challenges were Limited resources, Lack of supply chain integration, Lack of stakeholder interaction at design Phase and Lack of contractor/Subcontractor inclusion in planning, with mean scores of 4.00, 3.89, 3.89 and 3.84 respectively. These were followed by Lack of top management commitment with a mean score of 3.84 and Misunderstanding of lean with a mean score of 3.84. The other challenges with lower scores were Lack of communication & information sharing among parties which had a mean score of 3.68 and Lack of long-term relationships with suppliers with a mean score of 3.63. The results revealed that the values for the standard deviation were in the range of 0.77 to 1.32 as shown in Table 5.

Table 5. Challenges of supply chain management practices

Challenges of supply chain management practices	Mean	Std.	Rank
Limited resources	4.00	0.77	1
Lack of supply chain integration	3.89	1.05	2
Lack of stakeholder interaction at design Phase	3.89	1.13	3

(*continued*)

Table 5. (*continued*)

Challenges of supply chain management practices	Mean	Std.	Rank
Lack of contractor/subcontractor inclusion in planning	3.84	0.96	4
Lack of top management commitment	3.84	1.12	5
Misunderstanding of lean	3.84	1.01	6
Lack of communication & information sharing among parties	3.68	1.29	7
Lack of long-term relationships with suppliers	3.63	1.16	8
Lack of employees training and incentives	3.63	1.01	9
Resistance to change	3.56	1.20	10
Lack of proper planning	3.47	1.07	11
Lack of supplier reliability	3.42	1.22	12
Firms limited control and monitoring of suppliers	3.32	1.06	13
Negative attitude from employees	3.28	1.18	14
Incomplete designs	3.21	1.27	15
Inaccurate material delivery schedule	3.16	1.12	16
Conflicts with other initiatives of the company	3.11	1.10	17
Lack understanding on benefits of lean	2.79	1.32	18

Legend: **Std**: Standard deviation.

7 Discussion of Results

7.1 Lean Supply Chain Management Practices

This study found that the most common lean practices been implement on Zambian construction jobsites include; establishing customer needs, getting multiple quotations from suppliers, testing products to ensure adherence to specifications and quality, as well as ensuring flow of construction by having updated inventory. The four aspects were ranked as the most implemented lean supply chain practices in the building construction sector. A study by Papadopoulos et al. (2016) however, indicated that in order to achieve the implementation of supply chain management four critical factors should be considered these include; integration of information, collaboration and integration of operations, client consideration as well as development of management strategies for the entire supply chain.

7.2 Challenges of Supply Chain Management Practices

In establishing the challenges associated with the implementation of lean practices in the Zambian construction supply chain management. The findings revealed that the main challenges include: limited resources; lack of supply chain integration; lack of stakeholder interaction at design Phase; lack of contractor/Subcontractor inclusion in planning; lack of top management commitment; as well as misunderstanding of lean.

These findings agree with literature, according Heravi et al. (2015) failure to combine expertise and knowledge from various participants at an early stage leads to failure in identifying and solving problems beforehand. On contractor's inclusion in planning, Rahman and Alhassan (2012) suggests an early involvement of the contractors in the earlier phases of the building in order to detect defects earlier and make changes needed. Long-standing contractor-supplier relationships is one of the approaches which should be enhanced through supply chain integration. Another result which agrees with literature is that the construction industry is characterized with poor information sharing among stakeholders, competitive pricing resulting from increased number of suppliers on the market, as well as presence of fearful atmosphere and corruption (Fulford and Standing 2014).

Lack of top management commitment to lean implementation is another challenge. Kundu and Manohar (2012) confirmed that management commitment which can be demonstrated through allocation of sufficient resources, planning as well as promoting continuous improvement is necessary in ensuring successful implementation of lean practices. With regards to understanding of lean, Remon (2013) posits that it is very difficult for contractors to apply lean principles in the supply chain, if they lack basic knowledge on lean construction.

7.3 Implications of the Findings

The finding of this paper will help contractors reduce the overall product cost of building construction through elimination of wastes as a result of implementing lean practices. It is also important to note this study was only limited to contractors in grade one to three, a similar study would be good to be conducted among contractors registered in grades four, five and six so as to determine lean practices being implemented among small scale contractors.

8 Conclusion

Lean and Supply Chain Management in Construction are still practices being developed. Lean has shown to be adaptable to construction and the lean practices that are being implemented were reflected in this study. Some practices within Lean have showed more adaptability than others, which are still not fully embraced. In order to have a successful implementation of lean in construction, there has to be an adequate level of awareness and commitment among construction stakeholders. The study recommends enshrining of lean principles in the project life-cycle right from the inception phase of the project in order to ensure quality, efficiency and timely deliver on material.

References

Aigbavboa, C.O., Thwala, W.D., Mukuka, M.J.: Construction project delays in Lusaka, Zambia: causes and effects. J. Econ. Behav. Stud. **6**(11), 848–857 (2014)

Alkhoraif, A., Rashid, H., McLaughlin, P.: Lean implementation in small and medium enterprises: literature review. Oper. Res. Perspect. **6**, 100089 (2019)

Ansah, R.H., Sorooshian, S., Mustafa, S.B.: Lean construction: an effective approach for project management. ARPN J. Eng. Appl. Sci. **11**(3), 1607–1612 (2016)

Arif-Uz-Zaman, K., Ahsan, A.N.: Lean supply chain performance measurement. Int. J. Prod. Perform. Manag. **63**, 588–612 (2014)

Aziz, R.F.: Applying lean thinking in construction and performance improvement. Alex. Eng. J. **52**(4), 679–695 (2013)

Babalola, O., Ibem, E.O., Ezema, I.C.: Implementation of lean practices in the construction industry: a systematic review. Build. Environ. **148**, 34–43 (2019)

Bal, M., Bryde, D., Fearon, D., Ochieng, E.: Stakeholder engagement: achieving sustainability in the construction sector. Sustainability **5**(2), 695–710 (2013)

Barbosa, G., Andrade, F., Biotto, C., Mota, B.: Implementing lean construction effectively in a year in a construction project. In: Proceedings of the 21st Annual Conference of the International Group for Lean Construction (IGLC) (2013)

Bento, G.S., Schuldt, K.S., Carvalho, L.C.: The influence of supplier integration and lean practices adoption on operational performance (2020)

Broft, R.D., Badi, S., Pryke, S.: Towards supply chain maturity in construction. Built Enviro. Proj. Asset Manag. **6**(2) (2016). https://doi.org/10.1108/BEPAM-09-2014-0050

Chiponde, B.D., Tembo, C.S., Mutale, L., Chisumbe, S.: Information and Communication Technology Supported Inventory Management: A Tool for Improving Cost Control in the Zambian Construction Industry (2014). https://www.google.com/url?sa=t&rct=j&q=&esrc=s&source=web&cd=&ved=2ahUKEwiovfCTodX5AhWznVwKHa9UAvAQFnoECAQQAQ&url=https%3A%2F%2Fwww.diiconference.org%2Fassets%2Ffiles%2FDII-2014-Full-Proceedings.pdf&usg=AOvVaw0vSYdB5e6WphUT4XBLDpLN

Dubois, A., Hulthén, K., Sundquist, V.: Organising logistics and transport activities in construction. Int. J. Logist. Manag. **30**, 620–640 (2019)

Emuze, F.: Qualitative content analysis from the lean construction perspective: a focus on supply chain management. Acta Structilia **19**(1), 1–18 (2012)

Emuze, F., Smallwood, J.: Lean supply chain decisions implications for construction in a developing economy. In: 21th Annual Conference of the International Group for Lean Construction, Fortaleza, Brazil (2013)

Felea, M., Albăstroiu, I.: Defining the concept of supply chain management and its relevance to Romanian academics and practitioners. Amfiteatru Econ. J. **15**(33), 74–88 (2013)

Fulford, R., Standing, C.: Construction industry productivity and the potential for collaborative practice. Int. J. Proj. Manag. **32**(2), 315–326 (2014)

Gnich, S.: Lean transportation: applying lean thinking basics to transportation. Unpublished master thesis, Copenhagen Business School (2012)

Hedman, M., Hedberg, N.: Construction logistics from a subcontractor perspective. Master's thesis (2015)

Heravi, A., Coffey, V., Trigunarsyah, B.: Evaluating the level of stakeholder involvement during the project planning processes of building projects. Int. J. Proj. Manag. **33**(5), 985–997 (2015)

Mandl, J.S.: Implementation of critical risk factors in supply chain management. Int. J. Manag. Res. Bus. J. **2**(1), 104–120 (2013)

Kain, R., Verma, A.: Logistics management in supply chain–an overview. Mater. Today Proc. **5**(2), 3811–3816 (2018)

Kaynak, H.: Total Quality Management and Just-in-Time Purchasing: Their Effects on Performance of Firms Operating in the US. Routledge, New York (2013)

Kelly, C.: Clearer need. In: Supply Chain Yearbook 2002, pp. 76–77 (2002)

Kosky, P.: Manufacturing engineering. In: Wise, G. (ed.) Exploring Engineering, 5th edn. (2021)

Kumar, A.: Supply chain quality management-an empirical study. Doctoral dissertation (2016)

Kumar, S.: Lean manufacturing and its implementation. Int. J. Adv. Mech. Eng. **4**(2), 231–238 (2014)

Kundu, G., Manohar, B.M.: Critical success factors for implementing lean practices in it support services. Int. J. Qual. Res. **6**(4), 301–312 (2012)

Lasse, H.: The Just in Time concept in construction (2020). http://apppm.man.dtu.dk/index.php/The_Just_In_Time_concept_in_construction

Locatelli, G., Mancini, M., Gastaldo, G., Mazza, F.: Improving projects performance with lean construction: State of the art, applicability and impacts. Organ. Technol. Manag. Constr. Int. J. **5**(Special), 775–783 (2013)

Martínez-Jurado, P.J., Moyano-Fuentes, J.: Lean management, supply chain management and sustainability: a literature review. J. Clean. Prod. **85**, 134–150 (2014)

Modrak, V. (ed.): Handbook of Research on Design and Management of Lean Production Systems. IGI Global, Hershey (2014)

Mostafa, S., Dumrak, J.: A waste elimination process: an approach for lean and sustainable manufacturing systems. In: Sustainable Business: Concepts, Methodologies, Tools, and Applications, pp. 567–598. IGI Global (2020)

Muhwezi, L., Acai, J., Otim, G.: An assessment of the factors causing delays on building construction projects in Uganda. Int. J. Constr. Eng. Manag. **3**(1), 13–23 (2014)

Musonda, M.M.J., Mwanaumo, E.M., Thwala, D.W.: Risk management in the supply chain of essential medicines. In: Chau, K.W., Chan, I.Y.S., Lu, W., Webster, C. (eds.) Proceedings of the 21st International Symposium on Advancement of Construction Management and Real Estate, pp. 1275–1287. Springer, Singapore (2018). https://doi.org/10.1007/978-981-10-6190-5_112

Mwanaumo, E., Thwala, W.D., Pretorius, J.H.: Assessing health and safety requirements in construction contracts in Botswana. J. Econ. Behav. Stud. **6**(1), 37–43 (2014). https://doi.org/10.22610/jebs.v6i1.468

Nazir, F., Edwards, D.J., Shelbourn, M., Martek, I., Thwala, W.D., El-Gohary, H.: Comparison of modular and traditional UK housing construction: a bibliometric analysis. J. Eng. Des. Technol. **19**, 164–186 (2020)

Newman, C., Edwards, D.J., Martek, I., Lai, J., Thwala, W.D., Rillie, I.: Industry 4.0 deployment in the construction industry: a bibliometric literature review and UK-based case study. Smart Sustain. Built Environ. **10**(4), 557–580 (2020)

Nikinosheri, R., Staxäng, F.: Contractor-supplier relationships in the construction industry-a case study. Master's thesis (2016)

Nimeh, H.A., Abdallah, A.B., Sweis, R.: Lean supply chain management practices and performance: empirical evidence from manufacturing companies. Int. J. Supply Chain Manag. **7**(1), 1–15 (2018)

Papadopoulos, G.A., Zamer, N., Gayialis, S.P., Tatsiopoulos, I.P.: Supply chain improvement in construction industry. Univ. J. Manag. **4**(10), 528–534 (2016)

Prasad, K.V., Vasugi, V., Venkatesan, R., Bhat, N.S.: Critical causes of time overrun in Indian construction projects and mitigation measures. Int. J. Constr. Educ. Res. **15**(3), 216–238 (2019)

Rahman, M., Alhassan, A.: A contractor's perception on early contractor involvement. Built Environ. Proj. Asset Manag. **2**(2), 217–233 (2012). https://doi.org/10.1108/20441241211280855

Remon, F.A.: Applying lean thinking in construction and performance improvement. AEJ Alex. Eng. J. **52**(4), 679–695 (2013)

Samuel, D., Found, P., Williams, S.J.: How did the publication of the book The Machine That Changed The World change management thinking? Exploring 25 years of lean literature. Int. J. Oper. Prod. Manag. **35**(10), 1386–1407 (2015)

Siyam, G.I., Wynn, D.C., Clarkson, P.J.: Review of value and lean in complex product development. Syst. Eng. **18**(2), 192–207 (2015)

Turkyilmaz, A., Gorener, A., Baser, H.: Value stream mapping: case study in a water heater manufacturer. Int. J. Supply Chain Manag. **2**(2), 32–39 (2013)

Ugochukwu, P., Engström, J., Langstrand, J.: Lean in the supply chain: a literature review. Manag. Prod. Eng. Rev. **3**, 87–96 (2012)

Wisner, J., Tan, K., Leong, G.: Supply Chain Management: A Balanced Approach, 3rd edn. South-western Cengage Learning, Boston (2012)

Yala, J.O.: Lean supply chain management practices and operational performance of the manufacturing firms in Kenya. Unpublished MBA Project, University of Nairobi (2016)

The Performance Assessment of Zambia Railways Transport Service Quality

E. Mwanaumo[1,2(✉)], C. Bwalya[2], W. D. Thwala[2], and S. Chisumbe[2,3]

[1] The Department of Civil and Environmental Engineering, School of Engineering, University of Zambia, Lusaka, Zambia
erastus.mwanaumo@unza.zm

[2] Department of Civil Engineering, College of Science, Engineering and Technology, University of South Africa, Pretoria 0003, South Africa
thwaladw@unisa.ac.za

[3] Department of Environmental Health, Faculty of Health Sciences, Lusaka Apex Medical University, 31909 Lusaka, Zambia

Abstract. Purpose: This study examined the railway transport service quality using Zambia railways limited as a case study.

Design/Methodology/Approach: The methodological approach adopted was quantitative with a case study strategy. A 138 respondents contributed to the study, these included Zambia railways limited (ZRL) employees and rail freight customers. Non-probabilistic sampling techniques namely purposive and snowball were used in arriving at the respondents.

Findings: The study revealed that the revealed that there is low level of service quality at ZRL, more so, that the service which customers were more satisfied with was goods security management having scored a mean score of 3.04 on a scale of 5. This was followed by access to accurate information with the mean score of 2.65, as well as customer changing Transportation specifications with the mean score of 2.51. Furthermore, evaluating service quality impact on the organizational profitability by means of the RAILQUAL model factors and statistical evidence revealed that the service quality factor of Staff Service Skills and Knowledge at ZRL had significant effect on profitability indicated by the significance level of 0.000 with $P < 0.05$. Further, the service quality factor of goods security management employed by ZRL had statistically significant effect on organizational profitability indicated by importance level of 0.001, with $p < 0.05$.

Implications/Research Limitations: A comparative analysis of the two major railway companies operating in Zambia would enhance further even the findings of the current study.

Practical Implications: This study adds new knowledge and identifies areas for improvement in the quality of railway transport service that should be addressed by rail transport service providers. More importantly, it educates managers of organizations that offer rail services on mechanisms to boost customer happiness through enhancing the quality of service that manages transport period and flexibility. It implies that in order to gain an advantage in the liberalized markets, managers must comprehend the significance of service quality.

Originality/Value: Studies on service quality have been done in other regions especially in developed countries. However, this is the first study to be carried out in Zambia using this methodological approach. Thus it provides useful information even to other countries in the sub-Saharan African countries.

Keywords: Customer · Satisfaction · Service · Quality · Rail freight

1 Introduction

Transportation involves the movement of persons, goods, services and information through varies explicit methods of roads, rail, air, and water vessels (Fadare and Omole 1991). Public transport service is an essential part of human development everywhere. This is because it enables mobility of people from one point to another as they undertake different trading and transactional activities. According to Marinov et al. (2013), one of the pillars of globalization and economic progress is the ability to transport goods rapidly, safely, and affordably, and this is dependent on a strong freight transport system.

However, while providing this service to the public at an affordable price, any transport system must be able to meet its administrative and operational costs in the process. Martin (2019) alludes that in any sector of the economy, rail transport inclusive, the main idea is to eliminate inefficiencies in the production processes, reduce wastes and minimize costs while at the same time maximize revenue and profits for the Organization. Polat (2012), adds that the fundamental challenge for public transportation decision-makers is how to provide the best possible service while reducing operational, environmental, and other (social) costs. On the one hand, this means increasing customer satisfaction and drawing in as many customers as necessary to meet predetermined goals. This study sought to assess Zambia Railways Transport Service Quality performance.

2 Railways Transport

The origin of railway transport can be traced from as far back as 1820 with the invention of Mechanized rail transport systems in England. Later, the prototype of the modern-day railway' emerged, combining a specialized track, public traffic, passenger and freight conveyance Coulls (1999). Ever since then, railway transport has continued as the main transport system on land in the world. Dolinayova et al. (2019) contended that railway transport is among the most contributors to increase in the value of life of citizens as it allows the optimum use of public funds. Martin (2019) posited that the lean production management of rail transport optimizes businesses and makes it an effective mode of transport. Adding that rail transport is environmentally responsive, with nearly zero overcrowding and has a superior energy adeptness compared to other methods of transportation.

There is no single indicator used in measuring the effectiveness of a rail transport, however, Credo (2013) pointed out that a dynamic network, will increase customer usage, demonstrate robust progress, and improve the market stake of all freights within the region. Thus, an effective network would guarantee a client approval, the ability to fulfill demand, a solid service infrastructure, a suitable, current, and accessible fleet.

In developing countries, rail transport plays a very crucial role as it has proved to be the most convenient means of transport. This is so because of the cost of transporting both people and sometimes goods, as well as the high poverty levels associated with developing countries. Given these factors, rail transport has the advantage of being cheaper, safer, as well as its ability to carry bulk goods and passengers. Thus in African countries, a poorly utilized rail system negatively affects entire transportation network and decreases the contribution to the national economy (Mwila and Mwanaumo 2016).

2.1 Zambia Railways Transport

Situated in the Southern Africa, Zambia's economy is categorized as a lower-middle-income whose projected population is just over 19 million as of 2020. Of this total population, 41.7% of them live in urban areas, while the rest comprises of rural dwellers (World Factbook 2022). For the period 2022–2025, growth in economic activities have been projected to pick up to an average of 3.8%. However, at the same time, national poverty remains high, with an average growth in population of 2.8% per year (World Bank 2022). Given this growth in the population, the demand for services, transport services inclusive, is expected to increase proportionally over time. Further, the growth in the economy is expected to improve the capacity of all economic players to move from one place to another, as they trade. In Zambia, there are two main railway lines namely Tanzania-Zambia Railway (TAZARA) and the Zambia Railways Limited (ZRL). The TAZARA Railways operates between Zambia and Tanzania, and co-owned by the two Governments, while Zambia Railways operates between Zambia and Zimbabwe, and is wholly owned by the Zambian Government.

Zambia Railways was originally formed as North Western Region of Rhodesia Railways in 1903 during the colonial rule. The town of Kabwe in Zambia was the regional head office, although the company head offices were in Bulawayo, Zimbabwe (Zambia Daily Mail 2018). The line was until 1967 jointly owned by the present day Zambia and Zimbabwe respectively. In April 1967, Zambia enacted the Railways Act of 1967, which recommended a formation of the Board for Zambia Railways. Later, in 1982, the Government of the Republic of Zambia vested all assets and liabilities in Zambia Railways limited.

With the new policy of privatization that Government engaged in 1991, the company was privatized and was then called the Railways Systems of Zambia by December 2003, through a concession agreement. This new concession, was now responsible for operating both the cargo and commuter trains (ZRL 2019). However, due to the deteriorating infrastructure in the company, Government cancelled the concession agreement and repossessed the operational rights. The company was again reverted to the earlier name of Zambia Railways System (ZRL 2013). Zambia Railways transport has so far proved critical in the provision of convenient transport especially to the lower class in the country who cannot afford other forms of transport such as road and air transport, which are comparatively more expensive. The transport system has been providing bulk carriers services for cargo ranging from different sectors of the economy, ensuring a safer, friendlier and less congested environment for its customers. With its vision of becoming the prominent bulk and heavy freight transportation company in Zambia and the most dependable linkage in Southern regional rail network functioning in a cost effective, proficient and safe manner (ZRL 2013), Zambia Railways, has the potential to improve on its quality of services.

However, the transport system has also faced a number of challenges in the provision of its services. However, the absence of reinvestment over the years resulted in the entire rail industry failing to restock and replenish their rolling stock (Zambia Daily Mail 2018). A study conducted by Mwila and Mwanaumo (2020) determine that rail transport is underutilized in Zambia, with 73% of respondents engaged indicating that they preferred road transport to any other mode of transport in Zambia. The study also established that the road, transported over 83% of freight for exports, imports, transit, and local consumption, while more than 50% of bulk and heavy transportation, traditionally handled by rail, remained transported by road. The study concluded that poor accessibility to railways was caused by, among other things, poor rail infrastructure, unfavorable government policies, poor railway management, limited routes, and very slow speeds.

2.2 Railway Transport Service Quality

Several studies have looked at how the quality of service, business image, client happiness, loyalty, and behavioural intents are related. While, most of such studies are marketing, there is very little that separates the relevance of the variety of service features to customer satisfaction and identifying improvement objectives for the high speed rail under study from a transportation planning standpoint. This knowledge is essential for improving high speed rail design and service. Shainesh and Mathur (2000) Shainesh and Mathur (2000) identified the features considered by customers when assessing the quality of railroad freight services and created a tool called RAILQUAL that the railroads can use to get customer input. From the study, RAILQUAL proved to be an ideal tool to monitor, control, and improve railway freight services and competitiveness.

In order to assess the level of service provided for railway freight, Palaitis and Ponomariovas (2012) conducted a study. The study concluded that there are a wide variety of criteria that might be presented for the evaluation of rail service quality. The investigation revealed that reliability, punctuality, information, and rolling stock are the most often used factors, highlighting how dependability, adaptability, punctuality, information management, and average traffic speed are the primary issues with rail-based freight transportation.

Passenger satisfaction is another factor used to describe service quality. Passenger satisfaction is a measure of how satisfied consumers are with their transportation service (TCRP 2003). Customer satisfaction valuation aids service providers in setting strategic development goals and determining service development priorities within restricted budgets, resulting in improved service provider performance (Irfan et al. 2012; Cao and Cao 2017; de Oña and de Oña 2015; Zhang et al. 2017; Ellis et al. 2021, Newman et al. 2020).

Literature revealed that a number of indicators are used is measuring service quality in railways transport. Among the common indicators used include: freight charges and rates, staff service skills and knowledge, efficiency of service, service reliability, access to accurate information, products/services range, goods security management, settlement of issues/complaints, as well as customer changing transportation Specifications as shown in Table 1.

Table 1. Identified service quality indicators in the railway transport

SN	Variables	Source/author
1	Freight charges and rates	Parasuraman et al. (1988), Shainesh and Mathur (2000)
2	Staff service skills and knowledge	Grönroos (2000), Kotler (2003), Lu (2017)
3	Efficiency of service	Zeithaml et al. (1990), Kotler (2003), Eboli and Mazzulla (2007)
4	Service reliability	Zeithaml et al. (1990), Eboli and Mazzulla (2007), Anderson et al. (2013), Haron et al. (2016), Perera and Bandara (2016)
5	Access to accurate information	Grönroos (2000), Zeithaml et al. (1990), Pašaitis and Ponomariovas (2012)
6	Products/services range	Haron et al. (2016)
7	Goods security management	Zeithaml et al. (1990), Eboli and Mazzulla (2007), Saputra (2010), Anderson et al. (2013), Haron et al. (2016), Perera and Bandara (2016)
8	Settlement of issues/complaints	Grönroos (2000), Anderson et al. (2013)
9	Customer changing transportation specifications	Yilmaz et al. (2021)

3 Materials and Methods

This research employed a quantitative cross-sectional descriptive design, where data was collected using a structured questionnaire designed with closed-ended questions. Variables considered for questionnaire development in measuring of quality were informed by comprehensive literature review. A total of 138 respondents were sampled using purposive and snowball sampling from Zambia railways (ZRL) as well as railway freight customers. The data collected was analysed using analysis of variance (ANOVA) to establish the statistical significance whether the level of service quality at ZRL affected the performance of the organization. Respondents were asked to rate the effectiveness of the service quality of ZRL by analysing statements which they responded to across a range of possible responses from strongly disagree to strongly agree on a 5-point Likert scale. They were also asked to rate the performance of the organization from very effective to very ineffective on a 5-point Likert scale.

4 Results

Out of 138 respondents who participated in the study, Zambia railways employees accounted for 27.5%, 31.9% were customers from the mining sector, 11.6% were Energy sector customers, 13.8% were customers from the agriculture sector, 8.7% were clients in the construction and manufacturing sector, whereas, 6.5% were customers from trade and other industries as shown in Table 2.

Table 2. Respondents categorization

Category	Responses	% rate
Zambia railways employees	38	27.5%
Customers from the mining sector	44	31.9%
Energy sector customers	16	11.6%
Customers from the agriculture sector	19	13.8%
Construction and manufacturing	13	8.7%
Customers from trade and industry	9	6.5%
Total	**138**	**100.0%**

From the findings, the age analysis of the respondents revealed that 40.6% had been with ZRL for 3 to 8 years, 34.4% of the customers had ZRL provide services between 1 year and 3 years, while 25% of the respondents for less than 1 year as shown in Fig. 1.

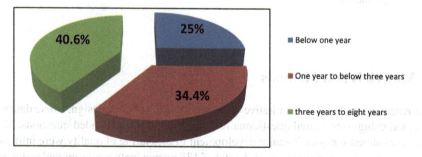

Fig. 1. Customer age analysis

With regard to the customer overall quality service level of satisfaction with ZRL, the findings revealed that out of 100 respondents, 6.2% indicated that the service was good, 37.5% indicated that the service was poor while 56.2% felt that the overall service level was moderate as shown in Fig. 2.

4.1 Extent of Railway Freight Transport Service Quality at ZRL

Nine (9) factors of the services were selected to test the extent to which customers are satisfied with the service quality of freight services provided by ZRL and consequently

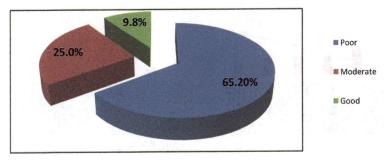

Fig. 2. Customer overall quality service level satisfaction with ZRL

the impact on performance. For three (3) of them were found to be satisfying to the respondents on a scale of 1–5 whereby a mean score above 2.5 implies satisfaction. The service which customers were satisfied most was Goods Security Management with the mean score of 3.04 on the scale of 5. This was followed by Access to Accurate Information with the mean score of 2.65 and then followed by Customer Changing Transportation Specifications with the mean score of 2.51. The rest of the services were found to be at low levels as shown in Table 3. The average of all the 9 mean scores was found to be 1.86 suggesting a less than moderate customer satisfaction rating for the services quality offered by the ZRL amounting to a ratio of 37.2% as the extent of low level service quality at ZRL.

Table 3. Statistics on the selected quality services metrics at ZRL

	Freight charges and rates	Staff service skills and knowledge	Efficiency of service	Service reliability	Access to accurate information	Products/services range	Goods security management	Settlement of issues/complaints	Customer changing transportation specifications
Valid	100.00	100.00	100.00	100.00	100.00	100.00	100.00	100.00	100.00
Missing	0.0	0.0	0.0	0.0	0.0	0.0	0.0	0.0	0.0
Mean	1.90	1.57	1.17	1.4	2.65	1.09	3.04	1.42	2.51
Std. dev.	.718	.498	1.143	.402	0.730	.456	.636	.643	.814

4.2 Reasons for Low Levels of Railway Freight Transport Service Quality at ZRL

The reasons for low level of service quality at ZRL were also analysed by applying the Pareto analysis as depicted in Fig. 3. The findings were that 5 factors of the nine (9) factors formulated >75% of the reasons magnitude. These were: Staff Service Skills and knowledge (SSS)-23%, Efficiency of Service (ES)-16%, Settlement of Issues/Complaints (SIC)-15%, Access to Accurate Information (AAI)-14% and Service Reliability (SRL)-9%. Other are the Goods Security Management (GSM)-8%, Customer Changing Transportation Specifications (CCTS)-7%, Products/Services Range (PSR) 4%and Freight Charges and rates (FCR)-4%.

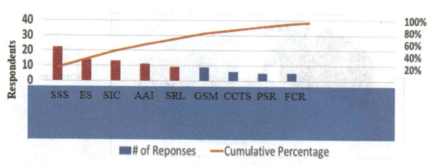

Fig. 3. Pareto chart of reasons for low levels of service quality at ZRL

4.3 Relationship Between Railway Freight Transport Service Quality and Profitability at Zambia Railways Limited

In assessing the relationship between railway freight transport service quality and Profitability One-way Analysis Of Variance (ANOVA) was used. It was established that service quality factor of Staff Service Skills and Knowledge at ZRL had significant effect on profitability. Further established that the service quality factor of Goods Security Management employed by ZRL had statistically significant effect on Zambia Railways Limited profitability as shown in Table 4.

Likewise, there was statistical evidence that the service quality factor of Service Reliability, with which the ZRL provided services to customers, influenced profitability, with Significance level = 0.000, where $P < 0.05$. Table 4 also shows that there is statistically significant evidence that the service quality factor of Freight Charges and Rates at ZRL influenced the profitability, Significance level = 0.002, where $P < 0.05$. However, it was found out that there was no significant statistical evidence that the service quality factor of Settlement of Issues/Complaints employed by ZRL influenced profitability of the organisation significantly, with significance level = 0.217, where $p > 0.05$ as shown in Table 4.

5 Discussion of Results

5.1 Railway Freight Transport Service Quality at ZRL

The service which customers were satisfied most was goods security management with the mean score of 3.04 on the scale of 5. This finding agrees with literature on the importance of goods security management in achieving quality customer service (Anderson et al. 2013; Haron et al. 2016; Lubinda et al. 2020). This was followed by access to accurate information with the mean score of 2.65, likewise, this result agrees with Anderson et al. (2013)'s assertion that when it comes to service quality, it's important to have dynamic passenger information available in stations and on trains, access to real-time travel information during service disruptions, and interactions that provide accurate ticketing and route information. Thirdly, customer changing transportation specifications followed with a mean score of 2.51, equally this agrees with literature that customer changing specification (Nassazi 2013). The findings revealed that the rest of the services were found to be at low levels.

Table 4. One-way analysis of variance – ANOVA

Service quality factors		Summation of Sqs	df	Av. Square	F	Sig.
Freight charges and rates	Btn Grps	6.32	1.0	6.32	9.99	0.002
	In Grps	62.049	98.0	0.63	–	–
	Total	68.36	99.0	–	–	–
Settlement of issues/complaints	Btn Grps	0.42	1.0	0.42	1.55	0.217
	In Grps	26.33	98.0	0.27	–	–
	Total	26.75	99.0	–	–	–
Service reliability	Btn Grps	22.76	1.0	22.76	64.83	0.000
	In Grps	34.40	98.0	0.35	–	–
	Total	57.16	99.0	–	–	–
Goods security management	Btn Grps	1.50	1.0	1.50	11.09	0.001
	In Grps	13.26	98.0	0.145	–	–
	Total	14.76	99.0	–	–	–
Staff service skills and knowledge	Btn Grps	3.56	1.0	3.56	15.78	0.000
	In Grps	22.08	98.0	0.23	–	–
	Total	25.64	99.0	–	–	–

Btn – Between, Grps- groups, av- average/mean

5.2 Relationship Between Railway Freight Transport Service Quality and Profitability at ZRL

This research established statistically that service quality factor of Staff Service Skills and Knowledge at ZRL had significant effect on profitability. This result agrees with Nassazi (2013) who pointed out that employee's skill levels significantly influence performance and ultimately profitability in business organisations. The study also found that the service quality factor of Goods Security Management adopted by ZRL affects profitability of the railway company. Likewise, statistically, there is indication that the service quality factor of Service Reliability that was adopted to provide the services by the ZRL to the clienteles affected the profitability. This agrees with Abate et al. (2013) who established that reliability has a positive impact on productivity of railway business, which can lead profitability. There is statistically significant evidence that the service quality factor of Freight Charges and Rates at ZRL influenced the profitability. Again this agrees with Nag (2011) who established that freight charges and rates have significant influence on profitability in the railway transport business. Conversely, evidence from statistics indicate that the service quality factor of Settlement of Issues/Complaints employed by ZRL influenced profitability of the organization significantly.

5.3 Implications of the Findings

This study adds new knowledge and identifies areas for improvement in the quality of rail shipping services that should be addressed by the rail service providers. More importantly, it educates those who provide rail services on best ways of boosting client happiness through assuring service quality enhancements. It implies that for managers to have a competitive edge in the liberalized markets, they must comprehend the significance of service quality.

6 Conclusion

From the study, it is evident that service quality or excellence is an underlying part of an effective and viable railway system, sector and operation. This excellence suggests balancing client's expectations and the technical capacities of ZRL as a service-provider and thus plays a pivotal role in gaining and maintaining market share which yields profitability. Consequently, this service quality would have a noteworthy influence on the organizational performance and hence profitability at ZRL. Therefore, to improve on service quality, the study recommends that ZRL establish a regular monitoring and evaluation system that will monitor consumer satisfaction by conducting market surveys on a regular basis, to determine freight customer requirements, because consumer needs and wants change over time and across sectors, and should integrate total quality management when providing services, so that all departments within organizations are synchronized with one another. Furthermore, the approach for the processes regarding up scaling service quality must be based on a critical analysis of factors such as; Goods Security Management; Service Reliability through time-to-delivery of service, automation, capacity building; integrating user experience; as well as improved access to information through application of technology.

References

Abate, M., Lijesen, M., Pels, E., Roelevelt, A.: The impact of reliability on the productivity of railroad companies. Transp. Res. Part E Logist. Transp. Rev. **51**, 41–49 (2013)

Anderson, R., Condry, B., Findlay, N., Brage-Ardao, R., Li, H.: Measuring and valuing convenience and service quality: a review of global practices and challenges from mass transit operators and railway industries (2013)

Cao, J., Cao, X.: Comparing importance-performance analysis and three-factor theory in assessing rider satisfaction with transit. J. Transp. Land Use **10**, 65–82 (2017)

Coulls, A.: As World Heritage Sites. International Council on Monuments and Sites (ICOMOS) (1999)

Credo: The effectiveness of the rail network across Great Britain a comparative analysis. White Paper (2013). https://kipdf.com/the-effectiveness-of-the-rail-network-across-great-britain-a-comparative-analysi_5ab079f61723dd329c635aae.html

Dolinayova, A., Danis, J., Cerna, L.: Regional railways transport—effectiveness of the regional railway line. In: Fraszczyk, A., Marinov, M. (eds.) Sustainable Rail Transport, pp. 181–200. Springer, Cham (2019). https://doi.org/10.1007/978-3-319-78544-8_10

De Oña, J., de Oña, R.: Quality of service in public transport based on customer satisfaction surveys: a review and assessment of methodological approaches. Transp. Sci. **49**, 605–622 (2015)

Eboli, L., Mazzulla, G.: Service quality attributes affecting customer satisfaction for bus transit. J. Public Transp. **10**(3), 21–34 (2007)

Ellis, J., Edwards, D.J., Thwala, W.D., Ejohwomu, O., Ameyaw, E.E., Shelbourn, M.: A case study of a negotiated tender within a small-to-medium construction contractor: modelling project cost variance. Buildings **11**, 260 (2021). https://doi.org/10.3390/buildings11060260

Fadare, S.O., Omole, T.: Assessing the quality of rail passengers services. The case of Nigerian Railway Corporation. Ilorin J. Bus. Soc. Sci., 36–42 (1991)

Mwila, F., Mwanaumo, E.M.: Framework for attracting traffic back to the railways in Zambia. In: Popkova, E.G., Sergi, B.S., Haabazoka, L., Ragulina, J.V. (eds.) Supporting Inclusive Growth and Sustainable Development in Africa - Volume II, pp. 185–202. Springer, Cham (2020). https://doi.org/10.1007/978-3-030-41983-7_14

Grönroos, C.: Service management and marketing: a customer relationship management approach (2000)

Haron, S., Nasir, M.S., Mohamad, S.S.: Rail transport service performance indicators in Klang Valley. In: AIP Conference Proceedings, vol. 1774, no. 1, p. 030022. AIP Publishing LLC, October 2016

Irfan, S.M., Kee, D.M.H., Shahbaz, S.: Service quality and rail transport in Pakistan: a passenger perspective. World Appl. Sci. J **18**, 361–369 (2012)

Kotler, P.: Management Marketing, New Jersey, USA, p. 415 (2003)

Lu, M.: Evaluation of railway performance through quality of service. Doctoral dissertation, University of Birmingham (2017)

Sakanga, L.M., Mwanaumo, E., Thwala, W.D.: Identification of variables proposed for inclusion into a regional railway corridor transportation economic regulatory framework: a case of the Southern African Development Community North-South Corridor. J. Transp. Supply Chain Manag. **14**, a504 (2020). https://doi.org/10.4102/jtscm.v14i0.504

Marinov, M., et al.: Urban freight movement by rail. J. Transp. Lit. **7**, 87–116 (2013)

Martin, R.L.: The High Price of Efficiency (2019). https://hbr.org/2019/01/the-high-price-of-efficiency

Mwila, F., Mwanaumo, E.M.: A situational study of rail transport usage in Zambia. J. Bus. Adm. Manag. Sci. Res. **5**(6), 77–88 (2016)

Nag, B.: Determinants of profitability of US Class I freight railroads. (2011). SSRN 2342470

Nassazi, A.: Effects of training on employee performance: evidence from Uganda (2013)

Parasuraman, A., Zeithaml, V., Berry, L.: SERVQUAL: a multiple item scale for measuring consumer perceptions of service quality. J. Retail. **64**(1), 12–40 (1988)

Newman, C., Edwards, D.J., Martek, I., Lai, J., Thwala, W.D., Rillie, I.: Industry 4.0 deployment in the construction industry: a bibliometric literature review and UK-based case study. Smart Sustain. Built Environ. **10**(4), 557–580 (2020)

Pašaitis, R., Ponomariovas, A.: Assessment of rail freight transport service quality. Transp. Telecommun. J. **13**(3), 188–192 (2012)

Perera, R.A.S.A., Bandara, A.B.D.M.: Foreign travelers' perspective towards service quality of railway service in Sri Lanka; a study based on Kandy railway travelers (2016)

Polat, C.: The demand determinants for urban public transport services: a review of the literature. J. Appl. Sci. **12**, 1211–1231 (2012)

Saputra, A.D.: Analysis of train passenger responses on provided services (case study Pt Kereta Api Indonesia and Statens Jarnvargar (SJ) AB, Sweden) (2010)

Shainesh, G., Mathur, M.: Service quality measurement: the case of railway freight services. Vikalpa **25**(3), 15–22 (2000)

TCRP: Transit capacity and quality of service manual, TCRP Report 100, 2nd edn. TRCP, Cooperative Research Program, Transit Washington, DC (2003)

The World Factbook: Zambia (2022).. https://www.cia.gov/the-world-factbook/countries/zambia/

World Bank: Zambia (2022). https://www.worldbank.org/en/country/zambia/overview

Yilmaz, V., Ari, E., Oğuz, Y.E.: Measuring service quality of the light rail public transportation: a case study on Eskisehir in Turkey. Case Stud. Transp. Policy **9**(2), 974–982 (2021)

Zambia Daily Mail (2018). http://epaper.daily-mail.co.zm/epaper_1_1_71_2018-03-12_Zambia%20Daily%20Mail.html

ZRL: The History of Zambia Railways Limited (2019). http://www.zrl.com.zm/history/

ZRL: Strategic Business Plan for the years 2014–2018, Zambia Railways Limited, Lusaka (2013)

Zeithaml, V.A., Parasuraman, A., Berry, L.L.: Delivering Quality Service, New York (1990)

Zhang, C., Cao, X., Nagpure, A., Agarwal, S.: Exploring rider satisfaction with transit service in Indore, India: an application of the three-factor theory. Transp. Lett. **11**, 469–477 (2017)

Effective Cost Management Practices for Enabling Sustainable Success Rate of Emerging Contractors in the Eastern Cape Province of South Africa

A. Sogaxa[1(✉)] and E. K. Simpeh[2]

[1] Department of Built Environment, Walter Sisulu University, Mthatha, South Africa
`asogaxa@wsu.ac.za`
[2] Centre for Settlements Studies, Kwame Nkrumah University of Science and Technology, Kumasi, Ghana
`eric.simpeh@knust.edu.gh`

Abstract. Purpose: The paper investigated the cost management practices adopted by emerging contractors to ensure long-term success in the South African construction industry.

Design/Methodology/Approach: The data was collected using survey questionnaires, and the data was analyzed using the Statistical Package for Social Sciences (SPSS) version 25. Both descriptive and inferential statistics were used for the data analysis. The Cronbach's Alpha test was also used to assess the constructs' internal consistency.

Findings: The findings revealed the following factors as most crucial for emerging contractors, effective cost control during the project delivery, managing cost through work production, subcontracting work to transfer financial responsibility, training employees on how to effectively manage the cost of project and effective administration of variation and contract instruction.

Implications/Research Limitations: Given the fact that the study is limited to the Eastern Cape Province, further studies may be conducted across the remaining eight provinces of South Africa to establish whether similar findings would emerge.

Practical Implications: In effect, effective cost management practices could help emerging contractors' sustainability and may subsequently have a positive impact on construction project delivery.

Originality/Value: The success rate of emerging contractors in developing nations in terms of project delivery has been significantly declining due to ineffective cost management strategies. The findings from this study contributes to the body of knowledge towards enhancing emerging contractors' sustainability through effective cost management practices.

Keyword: Cost management · Cost control · Cost monitoring · Emerging contractors · Sustainable success rate

© The Author(s), under exclusive license to Springer Nature Switzerland AG 2023
C. Aigbavboa et al. (Eds.): ARCA 2022, *Sustainable Education and Development – Sustainable Industrialization and Innovation*, pp. 339–357, 2023.
https://doi.org/10.1007/978-3-031-25998-2_26

1 Introduction

Construction infrastructure projects are the key drivers of socio-economic development in South Africa (Juanzon and Mhuhi 2017). According to Offei et al. (2019), the government is spending more money no construction infrastructure development. This is achieved through empowering local emerging construction contractors referred to as SMEs in this study (Offei et al. 2019). Emerging contractors are defined as contractors that is owned, managed and controlled by historical disadvantaged persons (Bikitsha and Amoah 2020). In South Africa, emerging contractors are considered as catalyst for enhancing the economic growth and development of the country (Paek et al. 2016; Bikitsha and Amoah 2020). Also, Seeletse and Ladzani (2012) noted that the advantage of emerging contractors is the fact that well establish need SMMEs needed to secure construction project. However, emerging contractors' success is hindered by inadequate cash flow planning, lack of financial planning, lack of knowledge, and poor level of management competency (Bikitsha and Amoah 2020). Furthermore, these challenges affecting emerging contractors constitute a significant problem to the growth and survival of emerging contractors in South Africa (Olabisi et al. 2012). In addition, emerging contractors adopt an unreasonable limit with outdated costing method which hinders the accuracy of the tender values (Seeletse and Ladzani 2012). Cost management practices of emerging contractors are regarded as the most significant management skills which enhance emerging contractors' competitive advantage in the industry and enable emerging contractor's sustainable success rate in terms of construction project delivery (Hussin et al. 2013).

Previous studies have identified causes with possible solutions and recommendations concerning SME contractors cost management practices in terms of how to improve SMEs' poor performance (Deros et al. 2012; Jindrichovska 2013). Furthermore, the government have developed policies to assist SME contractors to survive in the construction industry (Pu et al. 2021). However, emerging contractors are still not sustainable and this is partly as a result of poor cost management practices adopted by emerging contractors. Further study by Olawale and Garwe (2010) focused on obstacles affecting SME contractor's growth in South Africa. However, this study dwelt on business environment that includes both internal and external environment within the construction business sector. This study however acknowledges that internal environment such as access to finance, management skills, location and networking cost of production and external environment such as economic variables, crime, and market contributes to emerging contractors' failure. Nonetheless, it appears that there is limited literature focusing on the most effective cost management practices that enables emerging contractors' sustainability. Hence, the problem investigated in this study is stated as follows: emerging contractors are often confronted by inadequate cost management practices during construction project delivery and subsequently fail to complete the project due to inadequate cost management practices adopted. As a result, the research question for this study is as follows:

i. What are the most significant cost management practices that could be adopted by emerging contractors to enhance sustainable business performance in South Africa?

This paper seeks to develop effective cost management practices for emerging contractors to achieve sustainable success rate in terms of project delivery. Fortunately, cost management practices enhances emerging contractors effective cost control based on inflow and outflow cost which improves project operations by reducing expenses (Deros et al. 2012). The following sections consist of the literature review related to the study, the research methodologies adopted to achieve the study objectives, the analysis and discussions of results and conclusions and recommendations.

2 Literation Review

2.1 Theories of Cost Management Practices

This study is underpinned by cost estimating theory and capabilities and competencies theories since there is a link between competencies and general construction management. According to Hsu (2013), cost estimating theories allows the estimator to uncover an early warning of the project during the planning stage. On the other side, Gunduz et al. (2011) argued that the cost estimation model includes historic information that is currently adopted in the construction industry as well as the new data related to the proposed project. The model developed by Hsu can be applied by emerging contractors to make a case if the project would be profitable or not. The Hsu model, which is noteworthy, evaluates the unit costs assigned to the cost elements of capital as well as operations and maintenance costs. On the other hand, Wilson (1994) described the competencies model as the skills, abilities, and personal attributes required by the construction manager. In addition, Swanson et al. (2020) summarized that the management competencies model has emerged as the dominant model in the construction industry.

2.2 Business Sustainability

Worldwide sustainability has been an emerging challenge. According to Gross-Gołacka et al. (2020), the term sustainability requires contractors to search for and implement approaches that allow the contractor to adopt the integration of economic, social, and environmental development goals. In the context of emerging contractors, sustainability consists of contractors' actions to enhance the environment and social well-being and remain competitive and profitable (Tur-Porcar et al. 2018). However, emerging contractors are mainly managed by business owners, and most believe that they do not have an environmental and social environment (Prabawani 2013). In addition, Patma et al. (2021) allude that there are costs associated with the sustainable production process which need to be discussed in planning and project production. Hence, the sustainability of emerging contractors is very significant and assists emerging contractors to understand the environment they operate, sustainability approach is crucial in the construction industry, hence, the purpose of this study is to develop cost management practices that will enhance the sustainability of emerging contractors. This is an effort to enhance social, economic, environmental, and management-related causes which lead to emerging contractor's failure within the first five years of existence (Prasanna et al. 2019). Furthermore, Prasanna et al. (2019) summarized that emerging contractors' sustainability is affected by numerous challenges such as lack of capital, inadequate technical know-how, basic utilities, and low training and skill development all known as the factors hindering the growth performance of emerging contractors.

2.3 Emerging Contractors Cost Management

Emerging contractors' cost management practices are associated with an accurate estimate, cost control planning, financial knowledge, and cash flow management (Hongyi and Huanxue 2011; Toosi and Chamikarpour 2021). Jindrichovska (2013) outlined that cost management practices in the construction industry are crucial for managing and running a business within management practices that allow the contractor to run the business successfully and sustainably. However, Muneer et al. (2017) posited that emerging contractors are confronted by problems linked to accounting information, record keeping, and cost monitoring, and as the result, cost management practices have a positive effect on emerging contractors' business performance. In addition, Alvarez et al. (2021) postulated that emerging contractors are challenged by poor management skills, unqualified human resources to manage the project, poor financial resources, and inadequate accounting management skills adopted by emerging contractors.

3 Functions of Cost Management Practices

3.1 Resource Planning

Emerging contractor resource planning requires a systematic approach to resolve a resource allocation challenges and a correlation between time analysis and cost (Pocebneva et al. 2018). Also, resource planning in construction industry is considered to be the base of running a business. Emerging contractors are affected by ineffective resource planning as compared to well-established contractors because of the lack of technology use and inadequate integration of information systems to remain competitive in the construction industry (Shehab et al. 2004). According to Chowdhury et al. (2021), resource planning improves the following: task completion; communication barriers, and cost of resources. In addition, resource planning enables the contractor to optimize the internal resources by providing effective use of information and making it easily accessible by multiple functional areas within the organization (Amade et al. 2022).

3.2 Cost Estimation

Cost estimation is an important aspect in emerging contractors tendering and securing work by providing the grounds for identifying the cost of resources for a construction project (Enshassi et al. 2007; Matel et al. 2022). Cost estimation enables the contractor to evaluate the cost feasibility of a construction project and enhances strategies for effective cost control during project delivery (Matel et al. 2022). The existing literature reveals factors leading to inadequate cost estimations as follows: market competition; project time frame; drawings with lack of specification; inadequate project scope; inflation; errors in judgment; lack of historic data related to the project, and lack of similar project experience (Enshassi et al. 2005). Furthermore, cost estimation is affected by material price, the bills of quantities, construction approaches, and the assumptions used in preparing the project estimate are among the factors affecting cost estimation (Ibrahim and Elshwadfy 2021). However, Matel et al. (2022) uncovered that cost estimation is achieved through intensity, number of project team members, project duration, collaborating disciplines, contract type, project phases, and scale of work.

3.3 Budgeting

Budgeting is defined as a contractor tool representing resource planning and costs to enhance contractors' project priorities and objectives (Jowah and Mkuhlana 2021). Budgeting in the construction industry is very important from the very early stages of the construction project (Xenidis and Stavrakas 2013). Also, traditionally budgeting strategies focus on expenses that are included in the previous project, as a result, at the start of each project the contractor prepares a budget to minimize expenses (Kazar et al. 2022). Notably, budgeting plays an important role in guiding the decision-making process of the construction management team (Smith 2005). In addition, budgeting planning is a fundamental technique in managing construction project costs (Ali and Kamaruzzaman 2010). Furthermore, Chinedum (2019) opines that cost budgeting impacts cost control and the project quality of decision-making through information sharing. Jowah and Mkuhlana (2021) added that extensive budget training enables the contractor to be knowledgeable in terms of construction project cost planning.

3.4 Monitoring and Controlling

According to Al-Jibouri (2003), a construction project is likely to have challenges during the implementation stage of a project, especially when a project has been expressed in some much-needed detail. Cheng (2014) added that for a project without an effective cost control mechanism, emerging contractors will likely fail at the project level, and end up increasing the total project cost and lowering the intended profit. Toosi and Chamikarpour (2021) acknowledged that cost control and monitoring in the PMBOK guide includes planning, budgeting, estimating, budgeting, financing, fundraising, managing, and controlling towards reducing project costs. Hence, in developing countries, construction project cost overrun has become a major problem (Azhar et al. 2008).

3.5 Cost Reporting

According to Rogošić (2021), for emerging contractors to improve sustainably, the contractor has to take into account the impact of construction project cost management. In addition, Adler and Smith (2009) believe that construction firms need sound and effective methods for cost accumulation reporting to achieve project goals. Also, the research reveals that construction project cost reporting is found to be significant in enhancing the sustainability of a construction contractor and assisting the manager in decision-making and planning (Taniguchi and Onosato 2018). Hence, LaBonte et al. (2013) opined that contractors need to adopt a cost breakdown structure (CBS) to track project costs that accrue during construction project delivery. Adedipe and Shafiee (2020) added that CBS identifies the major cost drivers during construction project delivery.

3.6 Tracking Earned Value

Tracking earned value is known as a systematic model adopted to integrate and measure the cost, schedule, and scope of each project activity (Vyas and Birajdar 2016). Additionally, Patil et al. (2012) revealed that earned value originated from the planned cost of a construction project and the rate of production during project delivery.

4 Methodology

4.1 Research Approach

This study adopted a quantitative research method using a questionnaire survey to determine the effective cost management practices adopted by emerging contractors to enhance sustainability in terms of construction project delivery.

4.2 Population, Sampling and Response Rate

The targeted population in this study comprises emerging contractors registered under cidb Grade 1–4 who have previously completed construction projects. In total, a combined list of 2721 emerging contractors registered on cidb Grade 1–4. Table 1 presents the list of registered emerging contractors.

Table 1. Cidb registered contractors in the Eastern Cape

Cidb grade	No. of emerging contractors in General Building (GB)
1	2554
2	71
3	40
4	56
Total	2721

Quantity surveyors, site agents, construction managers, and business owners/directors involved in the sphere of emerging contractors make up the population used in this study. In this study four (4) metropoles in the Eastern Cape Province were used namely: East London, Port Elizabeth, Mthatha, and Butterworth. Cluster sampling was adopted to group emerging contractors registered under the cidb Grade 1–4 into groups of thirty-two (32) in which respondents were purposively selected. Out of the four metropoles, emerging contractors were grouped into a cluster of 32 to make a total of 128 emerging contractors used in this study with at least one participant from each contractor. In addition, of 128 emerging contractors selected, 57 returned complete responses, and thus, a response rate of 46% was achieved. Moyo and Crafford (2010) recommend that the built environment survey response rate in a broader perspective varies between 7% and 40%. This recommendation validates the sample size used in this study to be suitable.

4.3 Data Collection

To gather the quantitative data, closed-ended survey questionnaires were formulated to obtain data. Closed-ended survey questionnaires were formed in two sections namely: section A was demographic information of the respondents and section B was effective cost management practices adopted by emerging contractors. This study adopted a 5-point Likert scale, to determine effective cost management practices adopted by emerging contractors to enhance sustainable business performance. The Likert scale was formed in this order: SD (1) = Strongly disagree, D (2) = Disagree, N (3) = Neutral, A (4) = Agree, SA (5) = Strongly agree. In this research, some statements used in the survey questionnaires were extracted from the existing literature, and other statements were developed by the researchers to determine effective cost management practices adopted by emerging contractors to enhance sustainable construction project success.

4.4 Data Analysis and Interpretation

In this study, the data were analysed using SPSS version 25 and presented the results in the form of tables and diagrams. Notably, descriptive statistics were used to analyze, categorize, and narrow down vast quantities of data (Creswell and Creswell 2018). According to Conner and Johnson (2017), descriptive statistics enables the researcher to analyse a large volume of data gathered using frequencies, averages, and patterns. In addition, descriptive statistics allow the researcher to generalize the research findings from the selected population (Creswell and Creswell 2018; Petscher et al. 2013; Gabor 2010; Pallant 2011). The results in this study were organised using mean ranking and Standard deviation. Furthermore, Leavy (2017) stated that after completing the variable rankings, these ratings indicate the degree to which the variables are affected, and the ranking displays the hierarchy. Furthermore, this study used Cronbach's Alpha to measure the internal reliability of the variables. Maree (2007) revealed that the acceptable Cronbach's alpha is between 0.70 to 0.95. Thus, the results of Cronbach's alpha in this study was 0.87 which was acceptable. Table 2 presents the reliability test.

Table 2. Reliability statistics

Cronbach's alpha	Cronbach's alpha based on standardized items	N of items
.872	.873	18

Also, this study adopted principal component analysis (PCA) to analyse emerging contractors' effective cost management practices and reduce the measured cost management variables into fewer elements that collectively explain the variance. In this study factor analysis was conducted to determine the most effective cost management practices adopted by emerging contractors using the Meyer-Olkin (KMO) and Bartlett's Test. This assessment was also performed to validate the consistency of the quantitative analysis.

5 Analysis and Discussion of Results

5.1 Profile of the Respondents

Figure 1 below present the profile of the respondents. A total of 59 participants from different emerging contractors in South Africa were used. Figure 1 indicates the age group of the respondents with 54% falling in age group between 26 and 39. In addition, 24% fall into age group of 40 and 49. A minimal 12% of the respondents under in age group of 50 and 59 years. Notable there were few participants in age group of 18 to 25 with only 10%.

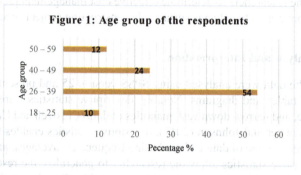

Fig. 1. Age group of the respondents

Figure 2 also indicates that for the total of 59 respondents 37% have industry experience ranging from 1 to 5 years, this is followed by 26% of respondents with relevant

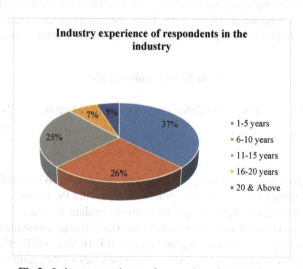

Fig.2. Industry experience of respondents in the industry

experience of 6 to 10 years. In addition, an overwhelming 25% have work related experience of 11 to15 years. Notable a combination of 7% and 5% of experience representing 16 and 20 years and over 20 years respectively.

Figure 3 also present educational qualification of the respondents with 47% hold a qualification of National Diploma, 25% holding a Degree qualification, while 14% of the respondents had other qualifications related to the industry. There were 12% of respondents with matric certificate and only 2% were lower than matric.

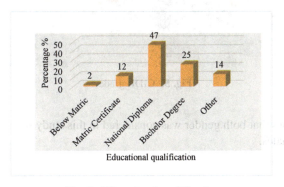

Fig. 3. Educational qualification

Figure 4 present the role of the respondents with 38% of the respondents were site agent, 32% of respondents were having other role in a construction site. Also, 20% of respondents were construction managers, lastly 10% were Quantity Surveyors.

Fig. 4. Role of the respondents

Figure 5 indicates the cidb Grading categories of the emerging contractors, with 36% of the emerging contractors falling under cidb category Grade 3, while 32% under grade 4. Also, 22% under grade 2 and lastly 10% of contractors in grade 1.

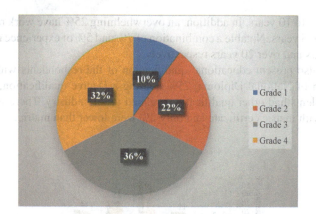

Fig. 5. CIDB grade

Figure 6 shows that both gender was considered in this study with 63% were males and 37% are females.

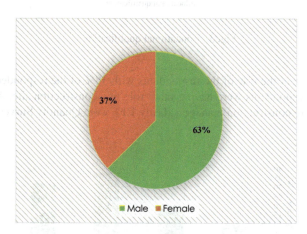

Fig. 6. Gender of the respondents

5.2 Emerging Contractors Effective Cost Management Practices

Table 3 presents the results related to effective cost management practices for emerging contractors to enhance sustainable business performance in South Africa. A 5 point Likert scale was used with Strongly Disagree = SD (1); Disagree = D (2); Neutral = N (3); Agree = A (4); and Strongly Agree = SA (5). The results from Table 3 reveals that emerging contactors adopts effective circulation of drawings and specifications as the strategy to reduce construction cost is ranked first with a mean score of 4.08. Furthermore, managing cost through work production is ranked second, with a mean score of 4.07. The results from Table 3 also indicates that the appointment of experienced estimators with construction related experience with a mean score of 4.02 is ranked third. However, emerging contractors ranked precise cost estimate during procurement of the project as the lease recognized cost management practice with mean score of 3.66. It should be noted that the average mean score is 3.9, which is above the mid-point threshold of 3.0.

Table 3. Emerging contractors effective cost management practices

Cost management practices 1 = Strongly Disagree … 5 = Strongly Agree	Cronbach's alpha 0.87 Descriptive statistics			
	No	Mean	Std.	Rank
Effective circulation of drawings and specification among management team to identify abortive works	59	4.08	.65	1
Managing cost through work production	59	4.07	.69	2
Appointment of experienced estimators who are conversant with the industry	59	4.02	.84	3
Subcontracting work to transfer financial responsibility	59	4.00	.74	4
Effective system for monitoring cash flow of business/project	59	3.98	.82	5
Resource allocation on a project is well defined	59	3.97	.74	6
Managing cost through effective allocation of budget to each activity	59	3.95	.78	7
Managing cost through project cost reporting	59	3.93	.81	8
Procurement of materials based on comparative market analysis	59	3.93	.72	8
Effective administration of variations and contract instructions	59	3.92	.75	9
Employees are trained on how to effectively manage the cost of projects	59	3.92	.68	10
Effective cost control during project delivery	59	3.88	.87	11
Access to financial institutions for project funding	59	3.80	.91	12
Ability to manage capital raised by owners of SMEs	59	3.76	1.01	13
Money due to suppliers is paid on time to prevent any interest cost on projects	59	3.76	.88	13
Scope creep is well managed to avoid over-budget/escalation	59	3.76	.70	13
Timely progress payment by client	59	3.75	1.18	14
Precise cost estimating during the procurement stage of a project	59	3.66	.76	15
Average	**59**	**3.90**		

5.3 Factor Analysis

5.3.1 Identifying the Most Significant Cost Management Practices for Emerging Contractors

Factor analysis was used to reduce a large number of factors into smaller groups by identifying the most significant cost management practices adopted by emerging contractors to enable sustainable project delivery. In total, 18 cost management practices were assessed. This assessment was also performed to validate the consistency of the quantitative analysis. According to Pallant (2012), factor analysis can be performed in three main steps to test the significance of the study. Kaiser-Meyer-Olkin (KMO) and Bartlett's test of sphericity was performed to determine the factorability of the dataset relating to the variables influencing the emerging contractors' sustainable project delivery in South Africa. Table 4 presents the results of the KMO and Bartlett's test sphericity. Pallant (2012) recommends that for significant factor analysis, the value of KMO should range between 0 and 1, and the minimum value is suggested to be 0.60. In addition, Field (2013) reveals that the Bartlett test is the indicator of the relationship among the variables, and for this research, Bartlett test requirements are considered. For factor analysis to be considered significant and appropriate, the Bartlett test associated with significance level should be $p < 0.005$. Table 4 indicates the KMO value as 0.765, which is more than the minimum value of KMO of 0.60. In addition, the Bartlett's test sphericity significance level was $p = 0.000$, which is less than the minimum value of $p < 0.005$. These results prove that the results meet the minimum requirements and factor analysis can be performed.

Table 4. KMO and Bartlett's test of SMEs cost management practices

KMO and Bartlett's test		
Kaiser-Meyer-Olkin measure of sampling adequacy		.765
Bartlett's test of sphericity	Approx. Chi-Square	455.865
	Df	153
	Sig.	.000

5.3.2 Principal Components of Cost Management Practices

The step which has been followed after checking the significance of the variables is factor extraction. With regard to factor extraction, Pallant (2012) noted the most commonly-used extracting factor techniques are: Kaiser-Meyer-Olkin criterion; where eigenvalues greater than 1 are considered most significant and Cattell's scree test; retaining all factors above the elbow in the structure. The principal components analysis was adopted to determine the most significant cost management practices adopted by emerging contractors to enhance sustainable business performance. Table 5 presents five (5) factors that have their eigenvalues greater than one, which is adopted by emerging contractors as a tool to achieve sustainable project success. Table 5 indicates the eigenvalues of the five (5) extracted components as follows: 5.905, 2.258, 1.616, 1.235, and 1.118. Furthermore, Table 5 presents the most significant factor extracted with factor one capable of explaining 32.804% of the variance, the second factor is 12.545% of the variance, the third factor extracted is 8.976% of the variance, while the fourth and the fifth factors are 6.859% and 6.209% of the variance respectively. Nonetheless, the combined components extracted constitute 67.393% of the variance, and these components are most significant for emerging contractors' sustainability in construction project delivery.

Component loading was performed using a component matrix on five components. As indicated in Table 5, the values are all greater than 0.30, and all the variables less than 0.30 were suppressed. Furthermore, components that are most important for emerging contractors, the variable that coverages on component 1 is "effective cost control during the project delivery", and consist of five factors (ECMCMP2, ECMCMP1, ECMCMP4, ECMCMP5, and ECMCMP8). Secondly, component 2 was characterized as "managing cost through work production" and comprised five factors (ECMCMP7, ECMCMP12, ECMCMP10, ECMCMP9, and ECMCMP11). Thirdly, component 3 was classified as "Managing cost to complete work within budgeted cost" and consisted of four factors (ECMCMP14, ECMCMP3, ECMCMP6, and ECMCMP13), while component 4 was on "Adequate training to effectively manage the cost of the project" and comprised two factors (ECMCMP16 and ECMCMP18). Fourthly, component 5 was on "Effective administration of suppliers' payment and variation" and consist of two factors (ECMCMP15 and ECMCMP17).

Table 5. Rotated component matrix

Code		Component				
		1	2	3	4	5
	Component 1: Effective cost control during the project delivery					
ECMCMP2	Effective system for monitoring cash flow of business/ project	0.853				
ECMCMP1	Ability to manage capital raised by owners of SMEs	0.754				
ECMCMP4	Access to financial institutions for project funding	0.654				
ECMCMP5	Timely progress payment by client	0.605				
ECMCMP8	Resource allocation on a project is well defined	0.359				

(*continued*)

Table 5. (*continued*)

Code		Component				
		1	2	3	4	5
	Component 2: Managing cost through work production					
ECMCMP7	Precise cost estimating during the procurement stage of a project		0.793			
ECMCMP12	Procurement of materials based on comparative market analysis		0.733			
ECMCMP10	Managing cost through work production		0.720			
ECMCMP9	Managing cost through project cost reporting		0.633			
ECMCMP11	Managing cost through effective allocation of budget to each activity		0.592			
	Component 3: Managing cost to complete work within budgeted cost					
ECMCMP14	Employees are trained on how to effectively manage the cost of projects			0.741		
ECMCMP3	Scope creep is well managed to avoid over-budget/ escalation			0.683		
ECMCMP6	Effective cost control during the project delivery			0.623		
ECMCMP13	Subcontracting work to transfer financial responsibility			0.457		
	Component 4: Adequate training to effectively manage the cost of project					
ECMCMP16	Appointment of experienced estimators who are conversant with the industry				0.770	
ECMCMP18	Effective circulation of drawings and specification among management team to identify abortive works				0.650	
	Component 5: Effective administration of suppliers payment and variation					
ECMCMP15	Money due to suppliers is paid on time to prevent any interest cost on projects					0.826

(*continued*)

Table 5. (continued)

Code		Component				
		1	2	3	4	5
ECMCMP17	Effective administration of variations and contract instructions					0.524
	Eigenvalue	5.905	2.258	1.616	1.235	1.118
	Variance (%)	32.804	12.545	8.976	6.859	6.209
	Cumulative variance (%)	32.804	45.350	54.326	61.184	67.393

5.4 Discussion of Results

The results from the survey indicate that emerging contractors adopt effective circulation of drawings and specifications to enhance sustainable business performance in South Africa. According to Magzoub, Salehi, Hussein, and Nasser (2020), effective circulation of drawings and specifications enables the contractor and the project team to identify changes in a construction worksite. In addition, Zhang and Cai (2021) outlined that effective circulation of drawings and specifications speeds up the project time and improves the understanding among the project stakeholders. In addition, emerging contractors appoint experienced estimators to ensure sustainable business performance in South Africa. According to Monyane and Okumbe (2012), experience enables the estimator to identify important factors to estimate construction costs effectively. The findings also reveal that emerging contractors manage costs through work production to maximize profit. Also, in the management of construction project costs, increasing work production is viewed as one of the most important costs management indicators for emerging contractors to enhance sustainable success rate in terms of construction project delivery (Cooray et al. 2018).

The findings from the factor analysis reveal that emerging contractors adopt effective cost control during construction project delivery to enhance sustainable construction project delivery in South Africa. This finding aligns with Cheng (2014) who stated that emerging contractors adopt various financial control approaches to manage cost which subsequently influence total project cost and ultimately affect profit if cost is not properly controlled. Additionally, emerging contractors subcontract part of their work to transfer financial responsibility and fast-track the construction project. Subcontracting is found to be crucial in the construction industry for the implementation of the construction project (Yoke-Lian et al. 2012). Furthermore, emerging contractors' employees are trained on how to effectively manage the cost of the project. This finding aligns with that of Martin (2010) who emphasized that employees' training enhances their knowledge through effective managerial support, peer encouragement, adequate resources, opportunities to apply learned skills, technical support, and consequences for using training on the job. The findings also uncovered that emerging contractors adopt effective project administration to avoid project disputes (Zakaria et al. 2013).

6 Conclusions and Recommendations

This study investigates emerging contractors' effective cost management practices for enabling sustainable construction business performance in South Africa. Literature was reviewed to back up the problem faced by emerging contractors in South Africa. This study adopted a quantitative research approach to determine effective cost management practices adopted by emerging contractors.

The study's findings revealed that emerging contractors adopt the following practices to enhance sustainable business performance in the South African construction industry: "effective cost control during the project delivery"; "managing cost through work production"; "managing cost to complete work within budgeted cost"; "adequate training to effectively manage the cost of the project", and "Effective administration of suppliers' payment and variation". These management practices enable emerging contractors to enhance sustainable business performance in South Africa. Based on these findings and as depicted in Fig. 7, it can be concluded that effective cost management practices for emerging contractors would enable them to survive in a competitive environment and increase the sustainability and profitability of emerging contractors.

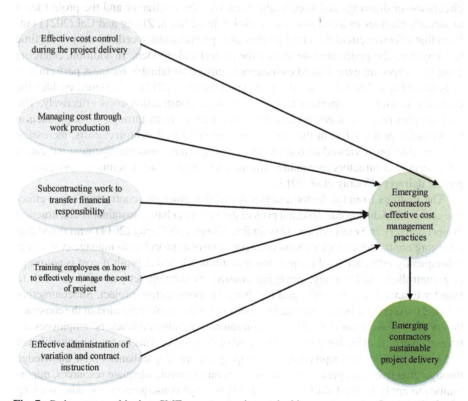

Fig. 7. Pathway to achieving SMEs contractors' sustainable success rate. Source: Author's construct

Hence, it is necessary to carry out a robust investigation on emerging contractors effective cost management practices to enable emerging contractors not to only focus on factors leading to effective cost management at project level but also categorise such cost in a way that lead to business profitability and sustainable performance. The rational here is if emerging contractors can effectively manage cost, then they can be sustainable in the construction industry.

7 Limitations

The study focuses on effective cost management practices adopted by emerging contractors to enhance sustainable construction project delivery in South Africa. The main focus of this article is to know the emerging contractor's costs management practices. Subsequently this study will address emerging contractor's sustainability issues. The results of this study is only based on quantitative study.

References

Adedipe, T., Shafiee, M.: An economic assessment framework for decommissioning of offshore wind farms using a cost breakdown structure. Int. J. Life Cycle Assess. **26**(2), 344–370 (2020). https://doi.org/10.1007/s11367-020-01793-x

Adler, T.R., Smith, W.L.: How organisational cost reporting practices affect project management: the issues of project review and evaluation. Int. J. Proj. Organ. Manag. **1**(3), 309–320 (2009)

Ali, A.S., Kamaruzzaman, S.N.: Cost performance for building construction projects in Klang Valley. J. Build. Perform. **1**(1), 28–36 (2010)

Al-Jibouri, H.: Monitoring systems and their effectiveness for project cost control in construction. Int. J. Proj. Manag. **21**(2), 145–154 (2003)

Alvarez, P., Sensini, L., Bello, C., Vazquez, M.: Management accounting practices and performance of SMEs in the hotel industry: evidence from an emerging economy. Int. J. Bus. Soc. Sci. **12**(2), 24–35 (2021)

Amade, B., Ogbonna, A.C., Nkeleme, E.I.: An investigation of the factors affecting successful enterprise resource planning (ERP) implementation in Nigeria. J. Constr. Dev. Ctries. **27**(1), 41–63 (2022)

Azhar, N., Farooqui, R.U., Ahmed, S.M.: Cost overrun factors in construction industry of Pakistan. In: First International Conference on Construction in Developing Countries (ICCIDC–I), Advancing and Integrating Construction Education, Research and Practice, Karachi, Pakistan, 4–5 August 2008 (2008)

Ben Chinedum, N.: Implications of budgeting and budgetary control on construction project delivery in Nigeria. Civ. Environ. Res. **11**(8), 83–87 (2019)

Bikitsha, L., Amoah, C.: Assessment of challenges and risk factors influencing the operation of emerging contractors in the Gauteng Province, South Africa. Int. J. Constr. Manag. (2020). https://doi.org/10.1080/15623599.2020.1763050

Chowdhury, M.S.A., Rahman, M.T., Shahabuddin, A.M., Hassan, M.R., Chowdhury, M.S.R.: Implementation of enterprise resource planning (ERP) in Bangladesh - opportunities and challenges. Int. J. Bus. Manag. **16**(11), 1–11 (2021)

Cooray, N.H.K., Somathilake, H.M.D.N., Wickramasighe, D.M.J., Dissanayke, T.D.S.H., Dissanayake, D.M.M.I.: Analysis of cost control techniques used on building construction projects in Sri Lanka. Int. J. Res. **5**(23), 909–923 (2018)

Creswell, J.W., Creswell, J.D.: Research Design, Qualitative, Quantitative, and Mixed Methods Approaches, 5th edn. (2018)

Deros, B.M., Rahman, N., Zainal, N.H., Rahman, M.N.A., Ismail, A.R.: Development of an effective cost management method for Malaysian SMEs. Aijstpme 5(2), 27–32 (2012)

Enshassi, A., Mohamed, S., Madi, I.: Factors affecting accuracy of cost estimation of building contracts in the Gaza Strip. J. Financ. Manag. Prop. Constr. 10(2), 115–124 (2005)

Enshassi, A., Mohamed, S., Madi, I.: Contractors' perspectives towards factors affecting cost estimation in Palestine. Jordan J. Civ. Eng. 1(2), 186–193 (2007)

Gross-Gołacka, E., Kusterka-Jefmanska, M., Jefmanski, B.: Can elements of intellectual capital improve business sustainability?-the perspective of managers of SMEs in Poland. Sustainability 12, 1–23 (2020)

Hsu, L.R.: Cost estimating model for mode choice between light rail and bus rapid transit systems. J. Transp. Eng. 139, 20–29 (2013)

Hussin, M.R.A., Alias, R.A., Ismail, K.: An action research approach for the development of cost management skills training programme among the owners of small and medium enterprises (SMEs) in Malaysia. Procedia Soc. Behav. Sci. 91, 515–521 (2013)

Ibrahim, A.H., Elshwadfy, L.M.: Factors affecting the accuracy of construction project cost estimation in Egypt. Jordan J. Civ. Eng. 15(3), 329–344 (2021)

Jindrichovska, I.: Financial management in SMEs. Eur. Res. Stud. 16(4), 79–95 (2013)

Jowah, L., Mkuhlana, X.: Budgeting systems and project execution at a selected government department of the Western Cape Province, South Africa. J. Public Adm. Dev. Altern. 2, 16–31 (2021)

Kazar, G., Mutlu, U., Tokdemir, O.B.: Development of zero-based budgeting approach for multinational construction contractors. Eng. Constr. Archit. Manag. (2022). https://www.emerald.com/insight/0969-9988.htm

LaBonte, A., O'Connor, P., Fitzpatrick, C., Hallett, K., Li, Y.: Standardized cost and performance reporting for marine and hydrokinetic technologies. In: Proceedings of the 1st Marine Energy Technology Symposium, METS 2013, Washington, 10–11 April 2013 (2013)

Leavy, P.: Research design, quantitative, qualitative, mixed methods, art-based, and community-based participatory research approaches (2017)

Magzoub, M.I., Salehi, S., Hussein, I.A., Nasser, M.S.: Loss circulation in drilling and well construction: the significance of applications of crosslinked polymers in wellbore strengthening: a review. J. Petrol. Sci. Eng. 185, 1–13 (2020)

Monyane, T.G., Okumbe, J.O.: An evaluation of cost performance of public projects in the free state province of South Africa. In: Second NMMU Construction Management Conference, Protea Marine Hotel, Port Elizabeth, South Africa, 25–27 November 2012 (2012)

Muneer, S., Ahmad, R.A., Ali, A.: Impact of financial management practices on SMEs profitability with moderating role of agency cost. Inf. Manag. Bus. Rev. 9(1), 23–30 (2017)

Offei, I., Kissi, E., Nani, G.: Factors affecting the capacity of small to medium enterprises (SME) building construction firms in Ghana. J. Constr. Dev. Ctries. 24(1), 49–63 (2019)

Olabisi, J., Sokefun, A.O., Oginni, B.O.: Kaizen cost management technique and profitability of small and medium scale enterprises (SMEs) in Ogun State, Nigeria. Res. J. Financ. Account. 3(5), 103–112 (2012)

Olawale, F., Garwe, D.: Obstacles to the growth of new SMEs in South Africa: a principal component analysis approach. Afr. J. Bus. Manag. 4(5), 729–738 (2010)

Paek, H., Oh, S., Hove, T.: How fear-arousing news messages affect risk perceptions and intention to talk about risk. Health Commun. 31(9), 1051–1062 (2016)

Pallant, J.: SPSS Survival Manual – A Step by Step Guide to Data Analysis Using the SPSS Program, 4th edn. McGraw Hill, Berkshire (2011)

Patil, S., Patil, A., Chavan, P.: Earned value management for tracking project progress. Int. J. Eng. Res. Appl. 2(3), 1026–1029 (2012)

Patma, T.S., Wardana, L.W., Wibowo, A., Narmaditya, B.S., Akbarina, F.: The impact of social media marketing for Indonesian SMEs sustainability: lesson from Covid-19 pandemic. Cogent Bus. Manag. **8**(1), 1–16 (2021)

Petscher, Y., Schatschneider C., Compton, D.L.: Applied Quantitative Analysis in Education and the Social Sciences, 1st edn. (2013)

Pocebneva, I., Belousov, V., Fateeva, I.: Models of resource planning during formation of calendar construction plans for erection of high-rise buildings. E3S Web Conf. **33** (2018). https://doi.org/10.1051/e3sconf/20183303032

Prabawani, B.: Measuring SMEs' sustainability: a literature review and agenda for research. Int. J. Manag. Sustain. **2**(12), 193–207 (2013)

Prasanna, R.P.I.R., Jayasundara, J.M.S.B., Gamage, S.K.N., Ekanayake, E.M.S., Rajapakshe, P.S.K., Abeyrathne, G.A.K.N.J.: Sustainability of SMEs in the competition: a systemic review on technological challenges and SME performance. J. Open Innov. **5**, 1–18 (2019)

Pu, G., Qamruzzaman, M., Mehta, A.M., Naqvi, F.N., Karim, S.: Innovative finance, technological adaptation and SMEs sustainability: the mediating role of government support during COVID-19 pandemic. Sustainability **13**, 1–27 (2021)

Rogošić, A.: Quality cost reporting as a determinant of quality costing maturity. Int. J. Qual. Res. **15**(4), 1233–1244 (2021)

Seeletse, S., Ladzani, W.: Project cost estimation techniques used by most emerging building contractors of South Africa. Acta Structilia **19**(1), 106–125 (2012)

Shehab, E.M., Sharp, M.W., Supramaniam, L., Spedding, T.A.: Enterprise resource planning an integrative review. Bus. Process. Manag. J. **10**(4), 359–386 (2004)

Smith, J.: Cost budgeting in conservation management plans for heritage buildings. Struct. Surv. **23**(2), 101–110 (2005)

Taniguchi, A., Onosato, M.: Effect of continuous improvement on the reporting quality of project management information system for project management success. Int. J. Inf. Technol. Comput. Sci. **1**, 1–15 (2018)

Toosi, H., Chamikarpour, A.: Developing a cost control system to increase competitiveness in construction projects based on the integration of the Performance Focused Activity Based Costing and target costing. Revista De Contabilidad Span. Account. Rev. **24**(1), 31–47 (2021)

Tur-Porcar, A., Roig-Tierno, N., Mestre, A.: Factors affecting entrepreneurship and business sustainability. Sustainability **10**, 1–12 (2018)

Vyas, A.B., Birajdar, B.V.: Tracking of construction projects by earned value management. Int. J. Eng. Res. Technol. **5**(3), 829–831 (2016)

Xenidis, Y., Stavrakas, E.: Risk based budgeting of infrastructure projects. Procedia Soc. Behav. Sci. **74**, 478–487 (2013)

Yoke-Lian, L., Hassim, S., Muniandy, R., Teik-Hua, L.: Review of subcontracting practice in construction industry. Int. J. Eng. Technol. **4**(4), 442–445 (2012)

Zakaria, Z., Ismail, S.B., Yusof, A.B.M.: An overview of comparison between construction contracts in Malaysia: the roles and responsibilities of contract administrator in achieving final account closing success. In: Proceedings of the 2013 International Conference on Education and Educational Technologies (2013)

Reflections on Real Options Valuation Approach to Sustainable Capital Budgeting Practice

S. Aro-Gordon[1(✉)], M. Al-Salmi[2], G. Chinnasamy[3], and G. Soundararajan[1]

[1] Department of Business and Accounting, Muscat College, Bousher Street, Muscat, Oman
{stephen,soundararajan}@muscatcollege.edu.om
[2] Department of Budget and Projects Engineering, Haya Diam, Muscat, Oman
[3] CMS Business School, Bangalore, Karnataka, India
dr.gopalakrishnan_c@cms.ac.in

Abstract. Purpose: Business exists to create value for society. One factor that is often overlooked in business valuation is the real options value that a project may have. This paper aims to revisit the real options valuation approach as an advanced capital budgeting technique, based on new data obtained between December 2020 and May 2022 and to reflect on how the results could be integrated with previous research.

Design/Methodology/Approach: Structured questionnaire was distributed to managers working mainly in the large manufacturing firms in the emerging market of Oman. Two hundred and two managers responded to the survey. A panel of experts and Cronbach's Alpha coefficient confirmed the instrument's validity and reliability. The perspectives gained from two virtual focus group capital budgeting expert discussions held in November 2020 and May 2022 were integrated with the survey results.

Findings: The results show that the ROV method is yet to gain traction as a veritable capital investment appraisal technique. A significant challenge against using ROV revolves around top management's willingness to explore its benefits, its perceived complexity, the minimal level of understanding of how it could be used and the depth of data-driven corporate culture. The nature of the industry and the firm's size are also influential factors.

Research Limitations/Implications: A notable limitation of this study is that choice of the population may lead to a biased result toward large firms.

Practical Implications: The paper reflects on opportunities for managers to improve investment decision-making quality in an increasingly unpredictable global business environment. Real options offer robustness and flexibility to deal with capital investment uncertainty. Given the heightened risks and uncertainties in the global business environment, managers need to start using the real options reasoning (ROR) strategy. Adequate training and capacity building at the highest level of management is imperative.

Originality/Value: The paper is probably the first real options study in Oman and thus joins the flow of recent country-specific real options empirical research, notably from the Nordic economies and Nigeria. The paper's main argument is that integrating ROR into traditional capital budgeting is imperative for achieving the steady flow of a future stream of shared well-being. The paper adds to the literature on the relatively more advanced capital budgeting approach.

Keyword: Black-Scholes-Merton · Capital-budgeting · Flexibility · Investment · Real-options

1 Introduction

This paper revisits the real options valuation approach as an advanced capital budgeting technique, based on new empirical data obtained in 2020–2021 and reflects on how the results could be integrated with previous studies. Negative net present value (NNPV) projects are often rejected as a default requirement of theoretical discounted cash flow (DCF) decision rules. In contrast, the realistic volatility and timing elements surrounding NNPVs may give organisations immense opportunities to realise long-term, strategic benefits; thus, identifying potentially rewarding NPVs becomes a critical research interest. The option pricing approach is thought to be superior to the DCF techniques because they clearly reflect the value of flexibility, to the extent that project value represents the sum of DCF and options value (McKinsey & Company 2000; Ardalan 2022). Thus, real options reasoning (ROR) incorporates the value of retaining the right to make future decisions under risky and uncertain situations as a conceptual framework for sustainable strategic investment. A real option conveys to the firm the right but not the obligation to explore specific projects or investments. Thus, real options can include a wide range of choices, such as expanding, contracting, deferring, waiting, switching, or abandoning a project entirely. Interestingly, research has examined the value of the real options valuation approach in healthcare management (Williams amd Hammes 2007).

The significance of the real options valuation approach to capital budgeting lies in the value of managerial flexibility as the management tries to examine all possible options while evaluating a strategic investment project (Jensen and Kristensen 2022). Real options reasoning (ROR) (also called contingency claims analysis) gives management the flexibility to postpone an investment as a veritable option to take advantage of potential prospects. Critically evaluating option possibilities tends to inherently reduce financial risks, thereby reducing the downside risk if conditions turn unfavourably. In effect, real options minimise downside risk by enabling firms to flexibly manage the uncertainties in the business environment (McGrath 1999). In the traditional financial options context, the underlying asset's volatility is thought to positively correlate with the option's value. Similarly, real options have derivatives that derive value from the variability of future decisions, thus providing the capital budgeting framework that allows managers to manage the downside risk of long-term projects more sustainably (Farsani 2012; Collins and Hansen 2011). Uncertainty accelerates the option's value because it provides managers with a framework to take advantage of future contingent opportunities (McGrath 1999; Ipsmiller et al. 2019; Jensen and Kristensen 2022).

While ROR has managerial benefits of enhancing the quality or flexibility of capital investment decision-making, the approach comes with some implementation challenges. Chance and Peterson (2002) remind us that real options value could be overvalued. Moreover, the technique is often thought to come with intricate mathematical structures that may not lend themselves to simple analytical formulas. Nonetheless, research has raised the prospects for readily operable, and computationally resourceful frameworks that can make real options work well (Kamrad 1995; van Putten and MacMillian 2004). Research has shown the plethora of opportunities available to firms for integrating the DCF analysis with the binomial option pricing model in a spreadsheet, as pedagogically illustrated by McKinsey and Company (2000) in its chapter twenty on "Using Option Pricing Methods to Value Flexibility".

Despite the increasing research on capital budgeting, there are a couple of unanswered questions (Lambrecht 2017; Bengtsson and Olhager 2002): To what extent are managers familiar with the benefits of real options in capital budgeting? To what extent have managers used the real options reasoning to complement the traditional methods of the likes of net present value (NPV), payback period (PB), return on investment (ROI), and internal rate of return (IRR)? What are the barriers to using real options, especially in emerging market economies like Oman? How relevant are the profiles (e.g., age groups and education, experience) of managers to their adoption of ROR in capital budgeting policy and practice? Is there any intent to use a real options approach in future? Although its potential benefits are large, real options analysis is hardly used in today's financial risk management practice, hence the need to reflect further on the real options science, techniques, principles, theories, and concepts from an empirical perspective, using an emerging market economy of Oman as the study area. Thus, the present study was designed to investigate (i) the degree to which managers have used the real options reasoning to complement the traditional capital budgeting methods such as NPV and PB, (ii) to examine the extent to which managers are familiar with the benefits of real options in capital budgeting, to find out if there is intent to use real options approach in future, and (iii) to uncover the barriers to using the real options valuation approach.

This paper is organised into five sections, including this as introduction Sect. 1. Section 2 highlights key aspects of the extant literature on real options relative to capital budgeting, while Sect. 3 outlines the research methodology. The study's key findings are presented and discussed in Sect. 4 while Sect. 5 concludes the paper including suggestions for further research.

2 Literature Review

2.1 The General Concept of Real Options Valuation – Value of Flexibility

Value-based management is grounded in universal economic principles in which the fundamental purpose of business is value creation. The real options theory's central claim is that there is value in flexibility. Conventionally, value is created by investing in projects at rates of return that surpass the costs of capital (McKinsey & Company 2000; Damodaran 2013; Palebu et al. 2016; Desai 2019; Hillier et al. 2021). As noted earlier, real options theory suggests the value of a capital budgeting project is the sum of the values of its DCF and ROR value of the project (Ardalan 2022). External factors, dynamics of technology changes, government policies, volatility of oil prices, pandemic occurrences, and other global developments such as the war in Ukraine create strategic challenges and opportunities for companies to generate cash flows. The business systematic and unsystematic risks are measurable and typically reflected in the popular capital asset pricing model (CAPM) in estimating the cost of equity capital. In the end, an option's value increases as the variability of the underlying asset's value increases because the project's value using option pricing will always the exceed project's value using NPV (McKinsey & Company 2000; CFA Institute 2022a,b).

Option valuation models use two basic approaches: the binomial model based on the discrete-time and the Black-Scholes-Merton (BSM) model based on continuous time (CFA Institute, 2022). The weighted average cost of capital (WACC), DCF values of the project, Monte Carlo analysis, event trees, decision trees, and sensitivity analysis, characterise the overall approach to options pricing application, as shown in the four-step process displayed in Appendix 3. A principal assumption of the BSM option valuation approach is that the return of the underlying asset follows the geometric Brownian motion, suggesting that the asset's return follows a lognormal distribution. Thus, to the extent that the main goal of the firm is to maximise shareholder value, the options pricing approach complements the traditional DCF forecasting processes in business analysis and valuation (Lambrecht 2017; Chance and Peter 2002; Hillier et al. 2021). Uncertainty and risk characterise business decision-making such that valuation essentially reflects various scenarios and ranges of value that reflect variability prospects – a key purpose of the options valuation models.

2.2 The Real Options Approach Offers a Wide Scope for Innovative Business Applications

Using traditional appraisal methods (as in NPV, PB, PI, and IRR, among others) to evaluate innovative business ideas in the emerging digital world presents obvious challenges to investors, entrepreneurs, and managers. Happily, the research to date has shown a wide range of opportunities for applying the ROR in the evaluation of capital investments across several fields of endeavours. Appendices 1 and 2 at the end of this paper display respectively, the wide range of scope for real options in capital budgeting and the overall approach to the options valuation process. The evidence provided in extant literature should encourage managers and investors to enhance investment decision-making processes with real options valuation analysis. This presents opportunities for managers to determine the level of investment to be accomplished in a phased manner thus enabling better monetization of customer futures on e-commerce sites. Hence, as Collins and Hansen (2011) argued, with the disciplined, strategic outlook offered by the options pricing technique, companies can thrive despite today's increasingly uncertain, unpredictable, tumultuous, fast-moving, unstable global business environment. Some of the notable applications emerging from a review of the literature include digital technology marketing and business applications as in algorithms/ML-AI-driven e-commerce apps (Bhattacharyya 2022; Kandinskaia and Lusbian 2021; Adkins and Paxson 2014), aviation (Miller and Bertus 2005), energy exploration and production (E&P) (Schiozer et al. 2008), and abandonment options (Pfeiffer and Schneider 2010). The value of the real options' application to healthcare management has earlier been noted in the wake of the covid-19 pandemic (Wang et al. 2022). Further options application to other capital investment endeavours in multinational enterprises (MNEs), multi-stage commercial real estate developments, entrepreneurship, and equity valuation are equally notable

(Jensen and Kristensen 2022; Ayodele and Olaleye 2021; Guthrie 2013; Vahdatmanesh et al. 2021; Ardalan 2022).

2.3 The Adoption Level of Real Options Reasoning In Capital Budgeting Practice

Against the backdrop of the strengths of ROR, as highlighted in the preceding paragraphs, one of the motivations for the present paper was to investigate the extent to which the real options valuation approach is used to complement traditional capital budgeting methods. In this regard, survey-based research has shown mixed results regarding the adoption level of ROR in capital budgeting practice. Some researchers have documented relatively low levels of adoption of the real options approach across various markets; 14.3% among the US Fortune 1000 companies (Block 2007); 6% among the Scandinavian largest companies (Horn et al. 2015), and 8% among large firms in Canada (Bennouna et al. 2010). A similar observation was made about the Spanish companies (de Andres et al. 2012). Interestingly, a much higher level of adoption (up to 50%) was found among the non-financial Bombay Stock Exchange (BSE) 200 index companies. Many reasons could be adduced for the relatively low level of options technique adoption among capital project appraisers across the globe. The advanced analytics thought to be involved in options valuation was earlier noted. Besides, the use of real options is often associated with high capital and research and development budgets typical of the energy and biotech sectors. Additionally, where the chief financial officers (CFOs) were familiar with the ROA, the complexity of real options was cited as a major barrier thus pointing to higher education and executive ROR strategy training needs. In this context, it is noteworthy that the CFA Institute (2022) reviewed sixteen topic areas for its professional investment management training programme and mapped its innovations over the past decades since the 1960s. The present contribution believes that five of the CFA topics have a bearing on real options as an evolving innovation of the ongoing 2020s, especially concerning financial analysis and valuation which borders on capital budgeting. This includes financial analysis, valuation, wealth/risk management, data analytics/visualisation and ethical, social, and governance (ESG) considerations in investing.

In sum, the review of the literature in the preceding sections has pointed to the need for further real options capital budgeting research in several dimensions. First, the relatively low level of adoption of real options in notable economies like the US and Canada compared to the higher level of usage in the developing markets of India, among others, raises some questions for further scientific enquiry. Second, there is a need for real options research methodology to be more inclusive in terms of exploring alternative data gathering tools beyond the survey and experimentation realms to strengthen the data quality, a motivation for including the focus group decision method in the present study. Third, previous research in the Omani context (notably, Al-Awaid, 2014; Al-Ani 2015) have focused on the traditional capital budgeting methods in the oil and gas industry. Thus, leveraging the resultant conceptual framework depicted in Fig. 1, the present contribution was designed to attempt to fill the gaps by exploring the utility of relatively more advanced capital budgeting methods in real options, and in the more diversified sectors as economies shift away from carbon-based sectors.

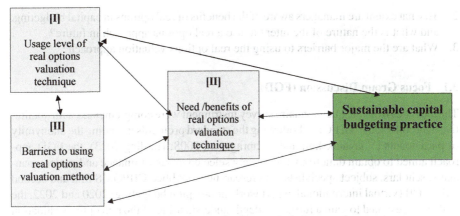

Fig. 1. Real options valuation research framework

3 Methodology

This study was designed to revisit the real options valuation approach as an advanced capital budgeting technique, based on new data obtained between December 2020 and May 2022 and to reflect on how the results could be integrated with similar previous studies. While there has been a couple of experimental analyses in the previous research, it is noteworthy that most of the country-specific studies have adopted the survey/questionnaire research design. The present study in the Omani context has followed suit, but, in addition, the research strategy has been completed by expert focus group discussion groups (FGDs). The FGDs were in the form of international research workshops in November 2020 and May 2022. The combined survey and FGD approaches adopted for the present study were justified by the need for in-depth analysis to generate further insights from the managers and other stakeholders regarding the adoption level of real options valuation and the nature of barriers to using the more advanced approach. There was also the need for a more representative sample in the non-oil sector of an emerging market like Oman; this is to aid the drawing of generalization of study findings (Harding 2013; Vijayalakshmi and Sivapragasam 2008)

A structured questionnaire was distributed to managers working mainly in publicly quoted manufacturing firms in the emerging market of Oman. Two hundred and two managers responded to the survey. The survey includes enquiry regarding why there is a low level of adoption of advanced capital budgeting decision techniques. A panel of experts and Cronbach's Alpha coefficient confirmed the instrument's validity and reliability. The perspectives gained from two virtual focus group capital budgeting expert discussions held in 2020 and 2022 respectively were integrated with the survey results. As the percentage of missing data is fewer, missing data has been filled with the median value (neutral) response. Aligned with the earlier-stated objectives, the present contribution is concerned with three main research questions, namely:

1. To what degree have managers used the real options reasoning to complement the traditional capital budgeting methods such as NPV, PB, ROI, PI, and IRR?

2. To what extent are managers aware of the benefits of real options in capital budgeting, and what is the nature of the intent to use a real options approach in future?
3. What are the major barriers to using the real options valuation approach?

3.1 Focus Group Discussion (FGD)

The analytics from the questionnaire survey instrument were complemented with qualitative data collected from FGDs. Following the standard protocols including the anonymity of participants (Vijayalakshmi and Sivapragasam 2008; Harding 2013), the FGD approach aimed to obtain data from a purposely selected group of finance and project managers, scholars, subject specialists, and researchers. Dubbed CBPOMS first and second VIEW-CB (virtual international expert workshop-capital budgeting) 2020 and 2022, the FGD was designed to gain a further in-depth understanding of the real options valuation approach to capital budgeting.

4 Results and Discussion

4.1 Respondents Profile

Figure 2 shows (a) the distribution of the respondents/managers in terms of a and (b) educational qualifications. The respondents' working experience (a) and sector of employment (b) are presented in Fig. 3. It is shown that most (66%) of the respondents were relatively young managers; this is representative of the country's general demographics (62.16%) as of September 2021 (Index-Mundi 2022). Nearly half (46%) are aged between 31–40 years.

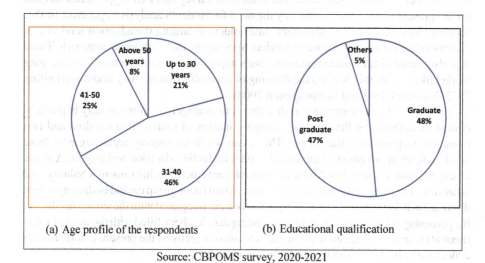

Source: CBPOMS survey, 2020-2021

Fig. 2. The distribution of the respondents/managers in terms of age and educational qualifications

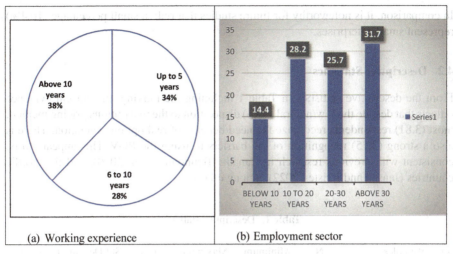

Fig. 3. The distribution of working experience and employment sector

It is shown in Fig. 3(a) that the respondents are generally mature professionals, well-educated, and experienced in capital appraisal and related activities. Further, it is reflected in Fig. 3(b) that most (58.9%) of the respondents are in the target industrial sectors, while some work in finance (13.9%), and other sectors (27.2%). It can be deduced that, generally, the sampled managers have the foundational knowledge and experience conducive to further capacity development in the manufacturing industries. Figure 4(a) shows the distribution of the company age and (b) the perceived size of the business based on total assets and sales turnover. Figure 4(a) shows that about three out of every five sampled managers work in firms that have been in business for between 10 and 30 years, while one out of every three have operated for more than thirty years. In Fig. 4(b), it is reflected that most (90%) respondent organisations are classified as medium-large.

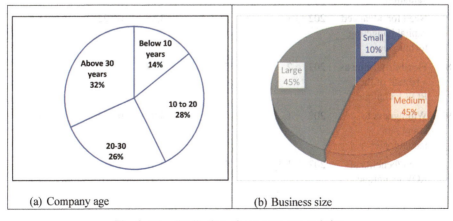

Fig. 4. The distribution of company age and size

In comparison, it is noteworthy for future studies that only a small percentage (10.4%) represent small enterprises.

4.2 Descriptive Statistics

From the descriptive statistics in Table 1 including the ranking, the current evidence reflects that despite the low/neutral (3.43) disposition to the real options pricing method, most (3.81) respondents recognize the need/benefits of real options valuation. There is also a strong (3.75) recognition of the barriers to using the ROV. This appears to be consistent with previous research in Canada (Bennouna et al. 2010), and the Nordic countries (Jensen and Kristen 2022), among others.

Table 1. Descriptive statistics

Q	Variables	N	Minimum	Maximum	Mean	Std Deviation	Rank
1	Usage level of real options valuation (ROV) technique	202	1	5	3.43	1.096	9
2	Training on ROV	202	1	5	3.71	0.890	8
3	Technique's complexity	202	1	5	3.59	0.872	8
4	Top management commitment	202	1	5	3.85	0.882	3
5	Data-driven culture	202	2	5	3.86	0.788	2
6	Costly and time-consuming	202	1	5	3.75	0.876	7
	Barriers to using real options valuation		1.2	5	3.75	0.862	7
7	Need for advanced capital budgeting techniques	202	1	5	3.88	0.822	1
8	Flexibility value as a benefit of the ROV technique	202	2	5	3.74	0.750	6
9	Willingness to adopt new technologies	202	1	5	3.82	0.876	4
	Need /benefits of ROV technique	202	1.3	5	3.81	0.815	5

4.3 On Usage Level of Real Options Valuation Technique

Figure 5 reflects the degree of capital budgeting techniques adoption. Similar to the findings in Table 1, Fig. 5 shows the widespread use of the traditional methods of ARR, PB, and NPV, but with a lower level of disposition towards the real options valuation method. This finding aligns with similar findings in Block (2007); and Horn et al. (2015), among others.

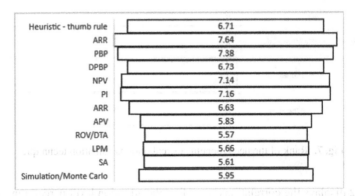

Fig. 5. Rank of usage of capital budgeting techniques

4.4 Barriers to Using Real Options, Especially in Emerging Market Economies Like Oman

The ranking of responses concerning the five main barriers is highlighted in Fig. 6. It can be inferred from Fig. 6 that the topmost issues border on the recognition of the real options valuation (ROV) in its capital budgeting practices, the imperatives for the company's data-driven culture encourage the use of advanced capital budgeting techniques, and the top management's willingness to implement advanced capital budgeting methods including the emergent new technologies such as ML and AI, to enhance its capital budgeting practices. Interestingly, like the observations in previous research (Horn et al. 2015; Bennounal et al., 2010), the (mathematical) complexity typically associated with real options discourages its usage among managers.

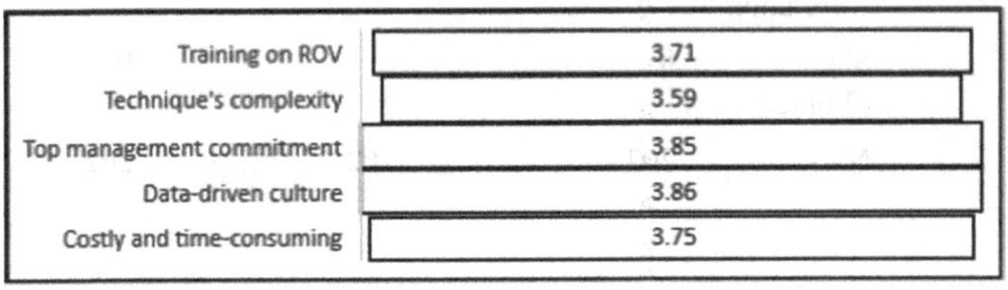

Barrier	Score
Training on ROV	3.71
Technique's complexity	3.59
Top management commitment	3.85
Data-driven culture	3.86
Costly and time-consuming	3.75

Fig. 6. Rank of the barriers to using the real options valuation method

4.5 Need/benefits of Real Options Valuation Technique

The ranking of responses with respect to three aspects of the need/benefits is highlighted in Fig. 7. The current evidence suggests that managers are generally aware (3.88) of the benefits and the need for advanced capital budgeting techniques such as ROV. There is also a favourable ranking (3.82) of intent to adopt relatively advanced capital budgeting techniques.

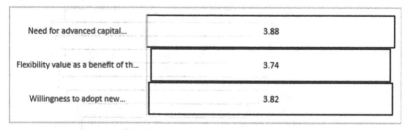

Fig. 7. Rank of the need /benefits of real options valuation technique

Table 2 displays the correlations regarding the adoption level perception of capital budgeting methods and barriers to using the advanced techniques, notably, the real options valuation method, relative to sustainable capital budgeting practice. The Pearson correlation was .420 for usage perception and .504 with respect to the challenges or barriers to the adoption of the real options valuation approach.

Table 2. Correlations

		Usage Perception	Barriers to ROV technique	Sustainable capital budgeting practice
Usage perception	Pearson Correlation	1	.491**	.420**
	Sig. (2-tailed)		0	0
	N	202	202	202
Challenges	Pearson correlation	.491**	1	.504**
	Sig. (2-tailed)	0		0
	N	202	202	202

4.6 Reliability Analysis

The Cronbach's alpha coefficients indicate that the level of internal consistency of the survey instrument is good and acceptable as none of the coefficients including the usage level of real options valuation technique is less than .700.

4.7 Findings from Focus Group Discussions (FGDs)

The invited experts and investment analysts indicated that the more familiar capital budgeting approaches, PB, NPV, IRR, ARR, and PI, remain frequently used by managers across the globe. The three most used methods in the Omani context were PB, NPV, and IRR, with PB, being reported as the most applied in the country. Thus, as far as the traditional capital project appraisal methods are concerned, there is little or no theory-practice gap. The practitioners suggest managers to adopt multiple methods in gauging the viability of project proposals. Contrary-wise, the theory-practice gap seems to exist when it comes to using more advanced capital budgeting methods, especially the real options pricing techniques which one participant termed "calculated gambles". Nevertheless, discussants noted the role of options in enhancing capital investment decision-making quality. For example, managers should exercise care before rejecting negative NPV proposals. Rather, managers should consider asset options, and key drivers of the option value, notably, volatility and time horizon, among other critical determinants.

In the end, the utility of real options in capital budgeting, an expert reflected, may hinge on the size and nature of the industry; smaller firms may not find enough motivation to engage in options pricing techniques. The barriers to adopting real options border on lack of awareness of its benefits, availability of experts, measuring non-financial factors, and the blurring of the dividing line between operating expenses (OPEX) and capital expenses (CAPEX) in a range of projects. For managers to use the real options valuation approach, making the method relatively simple, transparent, realistic, and easy to understand, is imperative.

5 Conclusion

This paper attempted to revisit the real options reasoning strategy to capital budgeting technique, based on new empirical data obtained in 2020–2021 from an emerging market economy, Oman. The paper also reflected on how the results could be integrated with similar previous studies. Three major findings emanated from the present study. First, consistent with previous research, there is a low level of adoption of the real options valuation approach. Second, while managers are generally familiar with the need for real options for capital investment appraisal, particularly for capital-intensive projects, there is inadequate knowledge of how to recognise the value of flexibility in the options appraisal. Third, the current results have shown the imperatives for the company's data-driven culture to encourage the use of advanced capital budgeting techniques, and the low level of top management's willingness to implement a real options approach. Thus, the results from this study may have very important implications for capital budgeting policy and practice; three management development dimensions are particularly apposite:

i. Managers should be trained to recognize real options in project proposals. The corollary is for stakeholders, government agencies, business organisations, and higher education institutions, to provide project funding for further research in real options valuation technique, with emphasis on prospects for application of new data-driven technologies such as ML and AI.
ii. Real options capacity-building programmes should expose the methodological advances and archetypes that have made options pricing techniques easier to understand and apply. Real options technique should be part of the undergraduate business and project management curricula.
iii. Greater top management commitment to adopting real options valuation technique as a critical element in capital budgeting policy and practice.

Real options should not be seen as superior but complementary to the traditional capital budgeting approaches – payback period, profitability index, benefit-cost ratio, break-even analysis, NPV, ARR, IRR, etc. – in that it encourages exploration, innovation, experimentation, and greater strategic flexibility for sustainable capital budgeting. The future direction of project appraisal is pointing towards more integrated, innovative, agile, relatively more advanced approaches represented by the real options method. To this end, real options complement, not a substitute for, discounted cash flow analysis; managers should integrate the two approaches in selecting the best growth projects,

5.1 Limitations and Scope for Future Advanced Capital Budgeting Research

The present analysis and reflections have not made a distinction between enterprise DCF and equity DCF, and, thus, excluded the possibility of adopting the APV (adjusted present value) model to adjust the cost of capital for the tax benefit of interest expense. Besides real options, there is a research opportunity to examine other related capital budgeting techniques such as linear programming model (LPM), integer linear programming model (ILPM), and sensitivity analysis /scenario analysis, among others The present study reflected that most (90%) of the respondent organisations came from medium-large firms; future studies may want to look at the subject matter from the perspectives of small businesses.

Traditional options	Equipment capital budgeting	Real options techniques	License valuation options (Aerospace industry)	Bhattacharyya's (2022)'s six spaces for testing real options
- Invest now/ exploit - Expand - Abandon/ withdraw - Switch/ redeploy - Defer - Wait/delay - Call - Put - Compound exchange - Rainbow options - Learning options - Liability options - Compound options (multiphase investment)	• Unexpected technology progress • Anticipated technology progress • Uncertain (volatile) technology progress • Deferred replacement of incumbent equipment.	• European call, • Dual asset, • Exchange, • Perpetual option. • Appropriate sensitivity analyses	• traditional delay licenses, • contingent investment licenses, • licenses with cost uncertainties, • indefinite delay licenses.	- Never invest - Immediately invest - Present-day investment position - Possibly invest later - Invest probably later - Possibly never invest.

Sources: Miller and Bertus (2005); Pfeiffer and Schneider (2010); ACCA (2015), Adkins and Paxson (2014), Bhattacharyya (2022)

Steps	[1] Compute base case present value (PV) without flexibility using DCF	[2]. Model the uncertainty using event trees	[3] Identify and incorporate managerial flexibilities creating a decision tree	[4] Calculate option value
Objectives	Compute base case PV without flexibility at t=0	Understand how the PV develops concerning the changing uncertainty. Choose multiplicative or additive stochastic process.	Analysing the event tree to identify and incorporate managerial flexibility to respond to new information.	Value the total project using a simple algebraic methodology and spreadsheet.
Comments	Traditional PV without flexibility.	Still no flexibility; this value should equal the value from Step 1. Explicitly estimate uncertainty	Flexibility is incorporated into event trees, which transforms them into decision trees The flexibility has altered the risk characteristics of the project, therefore the cost of capital has changed.	The option value method will include the base case PV without flexibility plus the option (flexibility) value. Under high uncertainty and managerial flexibility option value will be substantial

Source: McKinsey and Company (2000, p. 418)

Acknowledgement. The research leading to these results has received funding from the Research Council (TRC) of the Sultanate of Oman, under the Block Funding Programme. TRC Block Funding Agreement No. TRC/BFP/MC/O1/2019.

References

Adkins, R., Paxson, D.: Stochastic equipment capital budgeting with technological progress. Eur. Financ. Manag. **20**(5), 1031–1049 (2014)

Al-Ani, M.K.: A strategic framework to use the payback period in evaluating the capital budgeting in Oman's energy and oil and gas sectors. Int. J. Econ. Financ. Issues **5**(2), 469–475 (2015)

Al-Awaid, M.: Oman oil and gas: estimating versus the GAO's best practices in capital budgeting: a benchmark study. PM World J. **3**(9), 1–22 (2014)

Ardalan, K: Underdiversification puzzle, volatility puzzle and equity premium puzzle: a common solution. Stud. Econ. Financ. 1086–7376 (2022, ahead-of-print). Emerald Publishing Limited. https://doi.org/10.1108/SEF-01-2022-0005

Ayodele, T.O., Olaleye, A.: Flexibility decision pathways in managing uncertainty in property development: experience from an emerging market. J. Financ. Manag. Prop. Constr. **26**(3), 408–432 (2021). https://doi-org.ezproxy-s2.stir.ac.uk/https://doi.org/10.1108/JFMPC-05-2020-0037

Bengtsson, J., Olhager, J.: Valuation of product-mix flexibility using real options. Int. J. Prod. Econ. **78**(1), 13–28 (2002)

Bennouna, K., Meredith, G.G., Marchant, T.: Improved capital budgeting decision making: evidence from Canada. Manag. Decis. **48**(1–2), 225–247 (2010)

Bhattacharyya, S.S.: Monetization of customer futures through machine learning and artificial intelligence-based persuasive technologies. J. Sci. Technol. Policy Manag. 2053–4620 (2022). https://doi.org/10.1108/JSTPM-09-2021-0136

Block, S.: Are "real options" actually used in the real world? Eng. Econ. **52**(3), 255–267 (2007). https://doi.org/10.1080/00137910701503910

CFA Institute (2022a). The future of work in investment management: Skills and learning. CFA Institute (2022a). https://www.cfainstitute.org/en/research/survey-reports/future-of-work-content/?s_cid=dsp_FOSLConsult_EnterprisingInvestor. Accessed 9 July 2022a

CFA Institute Refresher reading: Valuation of contingent claims (2022b). https://www.cfainstitute.org/en/membership/professional-development/refresher-readings/valuation-contingent-claims. Accessed 9 July 2022

Chance, D.M., Peter, P.P.: Real Options and Investment Valuation. The CFA Institute Research Foundation (2022)

Collins, J., Hansen, M.T.: Great by Choice: Uncertainty, Chaos, and Luck – Why Some Thrive Despite All. HarperCollins Publishers (2011)

Desai, M.A.: How Finance Works: The HBR Guide to Thinking Smart About the Numbers. Harvard Business Review Press (2019)

de Andres, P., de la Fuente, G., San Martin, P.: The CEO and capital budgeting practices in Spanish firms. Universia Business Review **36**, 14–31 (2012)

Farsani, F.A.: Investigation of the relationship between real option method and escalation of commitment in capital budgeting. Life Sci. J. **9**(3), 2094–2099 (2012)

Guthrie, G.: Real options analysis as a practical tool for capital budgeting. Pac. Account. Rev. **25**(3), 259–277 (2013)

Harding, J.: Qualitative Data Analysis from Start to Finish. SAGE Publishers, Thousand Oaks (2013)

Hillier, D., Ross, S., Westerfield, R., Jaffe, J., Jordan, B.: Corporate Finance, 4th European edn. McGraw-Hill Education, London (2021)

Horn, A., Kiaerland, F., Steen, B.W.: The use of real option theory in Scandinavia's largest companies. Int. J. Fin. Anal. **41**, 74–81 (2015)

Indexmundi: Oman age structure (2022). https://www.indexmundi.com/oman/age_structure.html. Accessed 30 Sept 2022]

Ipsmiller, E., Brouthers, K.D., Dikova, D.: 25 Years of real option empirical research in management. Eur. Manag. Rev. **16**(1), 55–68 (2019)

Jensen, C.H., Kristensen, T.B.: Relative exploration orientation and real options reasoning: survey evidence from Denmark. Eur. Bus. Rev. **34**(2), 191–223 (2022). https://doi.org/10.1108/EBR-07-2020-0172

Kamrad, B.: A lattice claims model for capital budgeting. IEEE Trans. Eng. Manage. **42**(2), 140–149 (1995)

Kandinskaia, O., Lopez-Lubian, F.: Assessing value of a digital company: Uber's IPO 2019. The Case J. **17**(4), 588–624 (2021). https://doi-org.ezproxy-s2.stir.ac.uk/https://doi.org/10.1108/TCJ-08-2020-0111

Lambrecht, B.M.: Real options in finance. J. Bank. Finance **81**, 166–171 (2017)

McGrath, R.G.: Falling forward: real options reasoning and entrepreneurial failure. Acad. Manag. Rev. **24**(1), 13–30 (1999)

McKinsey & Company: Valuation: Measuring and Managing the Value of Companies, 3rd edn.. John Wiley & Sons, Inc., New York (2000).

Miller, L., Bertus, M.: License valuation in the aerospace industry: a real option approach. Rev. Financ. Econ. **14**(3–4), 225–239 (2005)

Mittendorf, B.: Information revelation, incentives, and the value of a real option. Home Manag. Sci. **50**(12), 1615–1761 (2004)

Palebu, K.G., Healy, P.M., Peek, E.: Business Analysis and Valuation, IFRS Cengage Learning, Fourth Edition (2016)

Pfeiffer, T., Schneider, G.: Capital budgeting, information timing, and the value of abandonment options. Manag. Account. Res. **21**(4), 238–250 (2010)

Schiozer, R.F., Lima, G.A.C., Suslick, S.B.: The pitfalls of capital budgeting when costs correlate to the oil price. J. Can. Pet. Technol. **47**(8), 57–61 (2008)

Singh, S., Jain, P.K., Yadav, S.S.: Capital budgeting decisions: evidence from India. J. Adv. Manag. Res. **9**(1), 96–112 (2012)

Song, N., Xie, Y., Siu, T.K.: A real option approach for investment opportunity valuation. J. Ind. Manag. Optim. **13**(3), 1213–1235 (2017)

Vahdatmanesh, M., Firouzi, A., Rotimi, J.O.B.: Real options analysis of revenue risk sharing in post-disaster housing reconstruction. J. Financ. Manag. Prop. Constr. (2021). ahead-of-print https://doi-org.ezproxy-s2.stir.ac.uk/https://doi.org/10.1108/JFMPC-02-2021-0018

van Putten, A.B., MacMillian, I.C.: Making real options really work. Harvard Bus. Rev. **82**(12), 134 (2004)

Vijayalakshmi, G., Sivapragasam, C.: Research Methods: Tips and Techniques. MJP Publishers (2008)

Wang, Z., Arvind Upadhyay, A., Kumar, A.: A real options approach to growth opportunities and resilience aftermath of the COVID-19 pandemic. J. Model. Manag. 1746–5664 (2022). https://doi.org/10.1108/JM2-12-2021-030w

Williams, D.R., Hammes, P.H.: Real options reasoning in healthcare: an integrative approach and synopsis. J. Healthc. Manag. **52**(3), 170–186 (2007)

A Reconstructionist Approach to Communalism and the Idea of Sustainable Development in Africa

J. O. Thomas[✉]

Department of Religion and Philosophy, Faculty of Arts, University of Jos, Jos, Nigeria
thomasj@unijos.edu.ng

Abstract. Purpose: Communalism, either in its unrestricted or restricted versions, is a social principle that involves the commitment of individuals in a community to shared goals, demanding of these individuals the obligation to promote communal values and other social practices. Communalism, on this score, appears to be incapable of providing credible development for Africa in the twenty-first century and so needs to be reconstructed.

Methodology: The reconstructionist and critical methods of philosophy were employed to show that communalism, in its present forms, stifles Africa's capacity for meaningful development. This is done by showing its ignoring of a very crucial element needed for development namely, the ontology of the individual.

Findings: Communalism therefore in this present form seems to be committed to the idea of 'undifferentiated holism' involving an idea of collectivism that defines the African communal universe with an ontological primacy over individuals, presupposing that an African conception of reality is to be located in the community.

Research Limitations: The present study, though discussed within the scope of Africa's development can further be expanded to a discourse relating to a credible participation of Africa in the emerging world order.

Practical Implications: This paper argues that the recognition of the condition of social differentiation, which describes an individual laced with a rationalist or critical attitude, provides a viable option for Africa toward a credible developmental strategy in the twenty-first century. This is the hallmark of a reconstructionist approach to communalism.

Originality/Value: Existing studies on development in Africa have dwelt on its socio-political, economic and industrial dimensions with minimal attention paid to the ontology of the individuals whose summation constitutes the society.

Keyword: African · Collectivism · Communalism · Development · Reconstructionist

1 Introduction

In accordance with the United Nations resolution, A/RES/70/1 of September 2015, (Ekanola and Lawal 2016) specifying global initiative referred to as Sustainable Development Goals (SDGs), the theme of this conference is perceived to be considerably apt in addressing the problems associated with sustainable development in Africa with particular attention to the infrastructural sustainability. Most often, infrastructure is considered in terms of the material structure that underlies our reality conducing to a materialist interpretation of an African conception of reality. However, literature on infrastructure has often ignored the other and most deserving interpretation which ought to infuse our minds in the search toward sustainable development in Africa. This is what I refer to as an idealist interpretation of the discourse of infrastructural reality that serves as a credible base for any theory of sustainable development. The description of infrastructure this paper adopts is that which considers the foundation required for the functioning of a community. It is upon the unveiling of this foundation that we seek to establish the values that undergird the functioning of African community. The problem of infrastructural deficit in Africa cannot be exhaustively addressed without a recourse to the substratum that forms the desideratum or essential substance of the discourse of development in Africa. In the words of Nkrumah, there may be no practice without thought for it will lead to blindness, whereas thought without practice will be a candidate for emptiness. It is thought that directs our action and practices and so a materialist conception of reality in Africa must be antedated by an idealist provocation. In this paper therefore, I shall show that a discussion on the infrastructural reality as a framework for addressing sustainable development in Africa must utilize the resources contained in the belief system that underlies Africa's existence. This resource is found in the sort of principle of social ordering which describes the relationship that may exist amongst individuals whose beings constitute the summation of the society, and between the individuals and the community in Africa. This idealist construct I shall take as communalism.

In Africa, communalism is believed to be a distinctive mode of social living involving a commitment to communal life. It is a practice and principle of communal living involving the affirmation of the ontological priority of the community over that of the individual. As a principle of social ordering, particularly in terms of the consideration of the interests relating to the individual and the community, the community is held superior. Perhaps it is important to mention that communitarian social life is not to be seen as a challenge given the fact that every modern democracy, either civic or ethnic, is to a certain degree communitarian in nature. Often the challenge appears in form of the debate concerning the space of the individual in the community and the ultimate realization of social order. This current conversation then requires us to demonstrate that communalism, in the first place, has a substantial bearing on the discourse of sustainable development in Africa and that its various renditions seem to have failed to sufficiently address the African condition. The paper further argues that communalism that admits of the critical or rationalist attitude best provides the platform for Africa in the consideration of the individual ontology as a paradigm for development. Consequently, a reconstructionist approach, that sees beyond ordinary materialist interpretation, now becomes imperative.

2 Theories of Development: Some Shortfalls

Various theories of development which were advocated to address issues relating to development of the nations of the third world have been perceived as materialist in approach and therefore taken as candidates for imperialism and renewed attitude toward subjugation and hegemony. For instance, modernization theory draws heavily from the discourse of naturalism and rationalism to argue that "natural environments create societies and people and that these have different potentials for development" (Peet and Hartwick 2009). Drawing from the positions of Max Weber, Auguste Comte, and the enlightenment scholarship, modernization theory, which seems to idealize science by restricting meaningfulness of any discourse within the boundary of science, is perceived as imperialistic and so does not reflect the circumstances of the people for whom such theory is advanced. The interpretation, which is derived from other theories such as dependency, neo-liberalist, globalization, etc., equally suggest efforts that seem to have ignored the condition of the people for which these theories were developed in the first place. The condition of underdevelopment supposedly brought upon Africa bears out a theory of underdevelopment, which describes this condition as a direct outcome of dependency on the core capitalist economy and so stifles the efforts of the periphery toward development. The notion of dependence then emerges within the context of the relationship that exists between the center and the periphery in which the former dominates most resources available for development leaving the dominated nations dependent. As a counterintuition to these theories, the position of post-developmental theorists reveals that the idea of development propagated by the West is at best Eurocentric and merely describes norms and ethos of Northern oligarchy. An idea of development, therefore, perceived from this perspective, is an attempt to classify the global community into developed and undeveloped regions dominated by Western economic paradigms. The result of this is a gradual erosion of indigenous values system, a threat to the environment and creation of corresponding feelings of inferiority in the South. In addition, a consideration of the principle which defines the conditions of Africans aptly represented in communalism equally shows that theories of development couched in terms of the principle of social ordering in Africa seems to have failed to sufficiently address the African condition. It is perhaps important to briefly demonstrate some variants of this principle and show their ignoring of what I consider a fundamental basis for African development.

2.1 Radical Communalism

Perhaps, an official position in literature may have been recorded prior to J. S. Mbiti's epochal study. It would not be inappropriate to maintain that Mbiti's work on African Religion and Philosophy, written at the end of the sixth decade of the twentieth century, marked a reference point for Africa as a communal or communitarian state. Having examined African religious beliefs beginning with God through the Spirit and arriving at men, Mbiti presents the place of man, as an individual being within the corpus of the existential reality of Africa. According to him, the individual is a corporate being without which he ceases to be an individual. Mbiti writes.

In traditional life, the individual does not and cannot exist alone except corporately. He owes his existence to other people, including those of past generations and his contemporaries. He is simply part of the whole. The community must therefore make, create or produce the individual; for the individual depends on the corporate group. Physical birth is not enough: the child must go through rites of incorporation so that it becomes fully integrated into the entire society (Mbiti1969)

He continues to assert that

Only in terms of other people does the individual become conscious of his own being, his own duties, his privileges and responsibilities towards himself and towards other people. When he suffers, he does not suffer alone but with the corporate group; when he rejoices, he rejoices not alone but with his kinsmen, his neighbors and his relatives whether dead or living. When he gets married, he is not alone, neither does the wife 'belong' to him alone. So also, the children belong to the corporate body of kinsmen, even if they bear only their father's name. Whatever happens to the individual happens to the whole group and whatever happens to the whole group happens to the individual (Mbiti1969)

From the above, there may be no individual in traditional African conception according to Mbiti, independent of the social environment which evidently produced him. This is communalism of a preponderant magnitude. Mbiti sums up this position in a cardinal point namely, 'I am, because we are, and since we are, therefore I am' (Mbiti1969).

An affirmation of radical communalism is equally expressed by Jomo Kenyatta when he asserts that

According to Gikuyu ways of thinking, nobody is an isolated individual. Or rather, his uniqueness is a secondary fact about him; first and foremost, he is several people's relative and several people's contemporary…individualism and self-seeking were ruled out…. The personal pronoun 'I' was used very rarely in public assemblies. The spirit of collectivism was [so] much ingrained in the mind of the people (Kenyatta 1965)

Although Kenyatta did not express his communal concern in the famous cliche, his position substantially pre-empted Mbiti's four years later. Menkiti, years later, re-echoed this idea when he argues that the African communal universe is ontologically prior to the individual pointing to the ontological derivativeness of individuals. He maintains that,

as far as Africans are concerned, the reality of the communal world takes precedence over the reality of the individual life histories, whatever these may be. [For him], it is the community which defines the person as person, not some isolated static quality of rationality, will, or memory…[this understanding implies that] the notion of personhood [is] acquired…[and as] far as African societies are concerned, personhood is something at which individuals could fail (Menkiti 1984)

Gyekye agrees with the theses of Mbiti and Menkiti and unequivocally perceives them as candidates for providing the ideological groundwork for African socialism represented in the works of Leopold Senghor, Kwame Nkrumah, and Julius Nyerere. All these are a strong affirmation of the African communitarian social arrangement which conceives of the self as embedded or constitutive of its ends involving the conception of the common good and shared meaning. This implies that the idea of self-determination of an individual, independent of the conception of the intrinsic good, and the autonomy of the individual to potentially and rationally revise his projects are considered as a social absurdity. This temperament towards communalism Gyekye considers misleading and attempts to proffer an alternative which he calls 'restricted or moderate' communalism.

2.2 Moderate Communalism

The idea of natural sociality or relationality, Gyekye argues, whittles down or depletes the self-determination and self-sufficiency of an individual in a way the self cannot realize him/herself without cooperating with other human beings in the society. It is then maintained that an individual is so described to the degree he endorses the communal ethos and practices. From the foregoing, the primacy of the community over the individual is immediately suggested. However, it might equally be argued that since individuals, who are distinct in their autonomy and freedom, embody the community without which a community may not be so defined, a recognition of these individuals as necessarily prior to the community might not be an inappropriate inference. This is to re-echo the liberals' conception of the self as prior to its ends, and so the ontological derivativeness of the community is forthwith implied (Gyekye 1997).

Implied in Gyekye's argument is the attempt to present a version of communalism which acknowledges the intrinsic worth and dignity of human beings, recognizing individuality and its responsibility to the attainment of his goals and tasks. Gyekye attempts to construct a model of communalism which elevates the condition of the individual within the conception of the African communal universe. Nonetheless, he prescribes an interface between claims of the individual and the community in the recognition of 'status of equal moral standing' (Gyekye 1997). In the pursuance of this, Gyekye invokes certain African proverbs particularly in the Akan language, which clearly suggest that, though Africa is communal, an individual's personal responsibility and initiatives needed to attain his goals and ends are not completely ruled out. According to him, proverbial such as,

i) 'Life is as you make it yourself'.
ii) 'It is by individual effort that we can struggle for our heads'.
iii) 'The lizard does not eat pepper for the frog to sweat' (Gyekye 1997)

By these statements, the idea of individual's efforts in the exercise of autonomy is forthwith revealed. This, in a way, presents an account of causality as a prevalent condition of an individual's self-determination particularly in relation to the proposition "as you lay your bed, so you lie on it". The fulfilments of our needs and goals in life largely depend on the responsibility an individual takes towards his condition. Our lives are often defined by certain conditions of continuous struggles involving success, failures, frustration, expectations, which require individual's responsibility and accountability for reasonable attainment of our well-being. Gyekye summarizes his moderate version of communitarianism roughly in the following lines.

> The view seems to represent a clear attempt to come to terms with the natural sociality as well as the individuality of the human person. It requires recognizing the claims of both communality and individuality and integrating individual desires and social ideals and demands…But, in view of the fact that neither can the individual develop outside the framework of the community nor can the welfare of the community as a whole dispense with the talents and initiative of its individual members, I think that the most satisfactory way to recognize the claims of both communality and individuality is to ascribe to them the status of an equal moral standing (Gyekye 1997)

The above is a crude representation of the model of communitarian restricted-minded scholars wish to project. Underlying this assumption is an individual who is responsible for his ends particularly in ways continuous struggles are defined by a commitment to, and fulfillment of these goals and tasks. However, there may not be a fulfillment of an individual's goals or self-determination outside the enabling environment provided by the community. This is in two ways: either the community provides a credible environment for an individual's autonomy to thrive, or an individual's autonomy will not thrive unless and until it is brought under state perfectionism or recognition of the intrinsic goods. Both ways appear to stifle the idea of autonomy. How the tenets of moderate communitarianism would achieve its mandate remains a dubious claim if examined within the position which, according to Oladipo, denies limit to self-sacrifices and the extent an individual is willing to take full moral responsibilities (Oladipo 2009). Gyekye, in a discourse of supererogationism and nature of moral conduct, writes as follows.

> Which form of self-sacrifice can or should be required of the moral agent, and how do we determine that? For some people, providing the slightest assistance of any kind to someone in distress will be a self-sacrifice; others, however, will not consider such acts as sending huge amounts of money to help people in famine-stricken areas within their nation or outside it, or helping to get someone out of real danger, as self-sacrificial or heroic or saintly. What all this means surely is that the field of our moral acts should be left open: the scope of our moral responsibilities should not be circumscribed. The moral life, which essentially involves paying regard to the needs, interests, and well-being of *others,* already implies self-sacrifice and loss, that is, in my view, no need, therefore, to place limits on the form of the self-sacrifice and, hence the extent of our moral responsibilities (Gyekye 1997).

It would seem then, from the above, that there is no limit to the call of moral duty and responsibility one may have toward the other. From this, it can be deduced that the moderate version is not different from its radical model particularly in the idea of unrestricted self-sacrifice and moral responsibilities as prescribed in the last part of the above quote. Moreover, it might be argued that if the communitarian principle, both in its unrestricted and restricted versions, discourages an individual's right to rational revisability of its projects, then ascribing status of equal moral standing between the individual and the community seems inappropriate. This is a conspiracy theory, for moderate communitarians cannot be committed to individuality in a sense relating to their conclusion. In the absence of a liberalist outlook to this debate, an individual cannot be recognized as having any moral standing in the first place.

2.3 Some Other Versions of Communalism/Communitarianism

I shall consider the discussion under this section as referring to some other variants of what has been discussed above. I begin with African socialism which, according to Gyekye, provides a strong affirmation for the African communitarian social arrangement that conceives the self as embedded or constitutive of its ends involving the conception of the common good and shared meaning. The intellectual resources eminently presented in African socialism are clearly represented in the works of Leopold Senghor, Kwame Nkrumah, and Julius Nyerere.

Nyerere, for example, developed a developmental template referred to as UJAMAA, to advocate first for the aspirations and togetherness of the people of Tanzania but was ultimately intended to provide a basis for the collective desires and progress of Africa. As a form of political democracy, it is a commitment to people in the preservation of public participation and ownership (Nyerere 1968). It is an attitude of the mind which recognizes the need for people to care for one another (Nyerere 1968). Explicitly demonstrated in Nyerere's version of socialism is an affirmation of a socialist consciousness that is unadulterated by colonialism, suggesting the need for a rediscovery of autochthonous values which defined the traditional Africa. This according to Matolino Bernard, suggests a commitment to certain communitarian ethos which fail to recognize the contemporary African condition characterized by incessant conflicts between the community interests and those of the individuals who constitute it. However, it is important to commend Nyerere's formulation represented in the mental attitude from which the care of the other is derived and deeply impressed on our minds. This is the kernel of the socialist disposition firmly expressed in terms such as 'equality', 'classlessness', 'solidarity', 'brotherhood', 'familyhood', respect for others, and other related values. These dispositions can be summed up in Nyerere's rejection of exploitation in all its ramifications which are embedded in capitalism (Okeregbe 2012).

Kwame Nkrumah's consciencism is an attempt to highlight the importance of social circumstances in the analysis of facts and events and how philosophy helps this analysis to further the enrichment of the society, particularly in terms of human experience. It is the thought process that arises from the African intellectual consciousness capable of initiating a practice geared toward the resolution of the African existential condition. According to him, "practice without thought is blind; thought without practice is empty" (Nkrumah 1970). Toward a social revolution involving philosophy and ideology for decolonization, Nkrumah identifies three important segments of the African society namely, the traditional African, the Euro-Christian, and the Islamic values all of which co-exist in a certain manner. These three components of the African society must be accommodated as experiences of the African conscience to reflect the egalitarian structure of the African universe whose emancipation must be founded on certain intellectual consciousness derived from the social environment and condition of the African people. At the core of this social emancipation is a philosophical consciencism, though deeply rooted in materialism, that involves a certain disposition that creates a synergy or harmony between the traditional African Society, the Western and Islamic orientations thereby providing a platform for the emergence of the African personality (Nkrumah 1970)). Nkrumah presents a socio-political ideology that unveils a philosophical standpoint which is grounded on African conscience, forming the intellectual contents of the African practice and paving a way for a consideration of socialism as the most eminent journey to the envisaged African revolution. However, Nkrumah seems to have ignored the very sensitive interplay between the individual and the society; a discourse which has inundated socio-political literature in recent times.

Leopold Sedar Senghor's idea of 'negritude' was initially coined by Aime Cesaire to describe a whole range of ideologies of black civilization represented in a revolt against French domination. It is "the ensemble of values of the civilization of the black world…a black manner of living…a way of living as black" (Gary 2015). It is an embodiment of black consciousness which, according to Senghor, was extant in pre-colonial Africa. Negritude, as a way of being, describes certain ontological categorizations. As a way of perception, it is epistemic. In its ontological consideration, it is predicated on the idea of 'vital force' which animates all things and predates the existence of all things. Though the 'vital force' is anterior to all beings, it seeks an expansion and an increase in the activity of producing beings whether humans, animals, or plants. To be is to be a force that can be strengthened or attenuated. Years earlier, Placide Tempels, a Belgian priest, had made an inquiry into the Bantu worldview for which he was required to examine the ontological foundation of their belief system to ultimately provide credible explanations in other areas of their existence, particularly in the manner the gospel can become meaningful to them. He identified the principle of the 'vital force' which was found to be responsible for the existence and graduation of beings in their various categorization, making his work a prelude to Senghor's formulation. Senghor understands the 'vital force' and its categorization according to their strength beginning from the gods with a descent to humans, animals, and plants, while the ancestors and the living dead are placed at a level higher than the humans within this categorization. The epistemic justification of 'negritude' draws heavily on the recognition of the black consciousness as distinct from the western civilization. As a response to the position of some western anthropologists such as Lucien Levy-Bruhl, which perceives the western societies as imbued with rationality and thus superior to other non-western, Senghor conversely attributed emotions to the black consciousness by which African societies have been adjudged as primitive and inferior compared to their western counterparts. Although Senghor's position, suggesting that Africans are people defined by emotions, has suffered severe criticisms in literature, the reading of 'negritude' provides the unveiling of African socialism predicated upon notions of solidarity and brotherhood on the African continent and beyond- a position carefully expressed in his idea of 'humanistic universalism' (Walter 1965).

Bernard Matolino, in his attempt to evolve a political theory which could address Africa's condition, argues that African socialism in all its variants as well as Wiredu's idea of consensual democracy have been insufficient to serve as socio-political theories needed for the emancipation of Africa on the grounds that they appeal to certain communitarian ethos which constitute the structure of socio-political and economic realities in Africa (Matolino 2019). While the former is deemed to be committed to a one party politics, consensual democracy is said to exemplify a non-party template. For him, such a theory of politics relevant in addressing Africa's problem must not be committed to certain ideological proclivities involving a conception of Africa's essentialism (Matolino 2019). Such a political theory, which Matolino refers to as Afro-communitarianism, must appreciate as priority the relationships between the individual and the institutions which regulate their existence. Very suitable for Africa for him therefore, in terms of a socio-political template, is a thin and normative version of communitarianism which is neither essentialist nor traditional but one that projects an understanding of the community as representing an organic dimension of people's lives with the readiness and vision to respond to these aspirations through strong institutions which are designed to regulate people's existence. Matolino seems to suggest a strong proposal that recognizes voluntary cooperation amongst the people undergirded by values of mutual recognition and benefication (Matolino 2019).

Closely related to the discourse of communalism is a conception that considers an Afro-relational ontological dimension of the individual involving the relational properties of the self through which the essence of an object is determined. This metaphysical conception is further strengthened by the determination of the moral status of an individual involving the recognition of a modal-relational approach understood in terms of the causal or intentional connection with the other and ultimately the community. This for Metz, offers a credible alternative to holism and individualism. Metz further argues that the possible conciliation of the rupture and gaps concerning the conception of difference in Western-liberalist orientation and Afro-communalism consists in the Afro-communal ethic (Metz 2020). However, the position of these communitarians concerning the space of the individual in the community, in terms of the ontological commitment of Africa communal universe, is particularly not too clear.

In the consideration of various renderings of communalism, there seems to be undue emphasis on a retrieval of Africa's past in terms of the rediscovery of its autochthonous values thereby suggesting certain ideological proclivities that seem to give credence to the idea of Africa's essentialism. Moreover, the different positions on communitarianism in scholarship, though conscious of the individual, seem to ignore the nature of the ontology of these individuals whose summation constitutes the being of the community, presupposing that an African conception of reality is to be located in the community. A position in this direction forecloses a fair discourse of development in Africa because there seems to be the ignoring of a certain underlying metaphysics of the individual; a metaphysics that defines the very being of the individuals who constitute members of the society.

The degree of impartation derived from communalism in Africa can be accessed from the interpretation of the politics of the common good. By common good, it is meant a condition that makes a community fulfill its end defined by shared meanings and values. The spirit of communitarianism involves shared purposes, interests, and the understanding of the good which demands of all citizens the willingness and commitment to promote communal values and practices and to advance its course. However, that individuals are equally deemed to have the right to self-determination and to pursue their ends according to their worth is not to be trivialized. Common good then is justified only to the degree in which the state can deploy its resources for the fulfillment of individuals' conceptions of the good within a social context. Recognizing the ontology of the individuals in addition to that of the community as constituting the basis for Africa's metaphysical conception of reality must be a very sensitive task worthy of attention. I now turn to a reconstructionist approach to communalism.

3 Result and Discussions

In other to reconstruct communalism for the purpose of providing viable template for African development, this paper finds useful resources in the works of Karl Popper and Joseph Cardinal Ratzinger. Popper's treatise exemplifies moral and intellectual temperaments represented in the rationalist attitude. It is an attitude that recognizes and holds as sacrosanct the willingness to listen to critical argumentations with a view to realize one's errors and to learn from them. Popper first alludes to this axiom as early as in his 1932 publication of *Open Society* under the rubrics of 'the revolt against reason', with this theme recurring in his later works such as *Myth of the Framework, Conjectures and Refutations, Logic of Scientific Discovery*, etc. This rationalist attitude admits that "*I may be wrong, and you may be right, and by an effort, we may get nearer to the truth*". The degree of profundity of Popper's conviction, involving an idea that rationality ought to be without foundation, forms the core of 'critical rationalism' which is an attempt to provide a common path between two opposing forms of authoritarianism that have defined rationalist discourse such as in science and society. Here, Popper speaks of some sort of dogmatism derived from the community practice and the relativism of the individualist orientation.

It is worthy to know that within the discourse of politics, scholarship has focused on the sort of relationship that may exist between individuals and the community with minimal success. A major aspect of this concern refers to the rights the individual has to self-realization and rational revisability of ends without being sucked in by communal ethos and principles on the one hand (an individual unencumbered by social circumstances), and the quest to a sustainable social order on the other. Either of these extremes is capable of engendering a form of authoritarianism. A major concern of Popper's philosophy then is to unveil a thesis which seeks to curb the excesses of science and workings of the society in a formulation of rationalist attitude or critical attitude required of every rational being involving the interrogation of the world as an individual on the one hand, and for the society to cede her values and practices to further strictures on the other. The method of rationalist attitude which describes all rational discussions is that which utilizes the ideals of criticism toward resolutions of our problems. This is, according to Popper to argue that growth of knowledge consists in conjectures and refutations of our hypothesis, theories, and traditions involving subjecting every attempt at solving problems to severe criticism. With such a bold attempt at solving problems often checked by a series of criticism, no authority remains an authority anymore both in science and socio-political environment. An individual can no longer see himself as an atomized, impenetrable windowless self nor should the community extol its feature of perfectionism; for either of these can at best lead to a state of dysfunctional bureaucracy. Contrarily, both must relinquish their authorities in a rationalist attitude. Human fallibilism is a given but the need for a critical attitude is a requirement of truth for it is in the awareness of our fallibility and deliberate criticism of solutions to our problem that we move nearer the truth. Although the pursuit of truth is a commitment, achieving it is a myth. No wonder Popper maintains that clarity and distinctiveness are not clear criteria of truth, but obscurity and confusion are indicators of error (Stefano 2009).

How then should this attitude impart on our practices, attitudes, values, and traditions? How can an individual, though embedded in social values, distance himself from them? Popper invites us to consider the tradition of the Greeks, which over the years, can be said to inform the rationalist tradition. In the consideration of tradition as myths according to Popper, the ancient Greek philosophers did not just replicate the thoughts and myths of the primitive and pre-scientific mythmakers by accepting their religious traditions uncritically. On this score, Popper argues for analysis of tradition. He identifies the attitude of discussing and challenging the matter in order to accept or reject the myth which is in contention. This is to suggest that the understanding and explanation of nature is usually formulated as myths and the rationalist tradition requires that a myth is challenged and discussed not necessarily to supplant the tradition of mythmaking; for the outcome of this is represented by another myth, but to eliminate the taboos embedded in them. Eliminating taboos embedded in our myths then requires the adoption of the tradition of a critical attitude toward them involving telling the myth first, understanding them, then discussing and challenging them with the view to improving upon them. This then presupposes theories in science as well as traditions, are mere guesses and hypotheses involving myths that must be constantly discussed, challenged, and revised. Just as in the natural sciences, social sciences ought to entrench a tradition of critically discussing the myth involving a process of interrogating a story or narrative handed on but accompanied by a second-order tradition of criticism which is, in itself, a myth. Often, this rationalist tradition involves that the man to whom a myth is handed on, reserves the right to criticize it in order to improve on what has been handed on to him. He, in turn, hands-on the myth upon the same clear mandate. A continuous and sustaining attitude of discussing and challenging a myth in this manner over time draws us nearer to the truth though it may, by an inch, elude us. This then is the basis for the rationalist tradition (Popper 1963).

The critical attitude described above is an integral part of the ontology of the individual dictated by the rational nature of man. Rationality, though essential to the being of the individual, does not only define the constitution of this ontology, for a community of rational people can still be oblivious of the conditions applicable for sociality. Sociality does not just emerge because there are rational people in a community, nor is it manifested solely on the account that an individual is born into a society; a process over which he possesses no control. These conditions are not sufficient, though necessary. Such a community constituted under the conditions described above can at best produce a community of people who are rationally insensitive to and intolerant of others. Another crucial element is by all means, necessary for the proper unveiling of a community whose sensibilities would provide a win-win situation between it and its members in such a manner as to make common good less controversial. A reconstituted order for Africa requires more than such an attitude described above. It must equally utilize the resources of dialogue that is capable of engendering credible sociality culminating in the enthronement of critical or rational communalism for Africa. It is already noted that a reasonable process of sociality is not just a product of a mere natural sociality of the individual but in a consideration of a robust metaphysical conception of reality in Africa particularly in terms of the ontology of the individual engaging that of the community. A very crucial element of this has been discussed above. This paper now turns to the other element as we conclude.

The critical attitude described above is very important in the reconstruction of communalism. But a mere veneration of this element only makes a community a congregation of the deaf for individuals in a community can develop such a scientific attitude and yet be unable to maintain a credible sociality. What is intended here is for the individual in a community, having imbibed a spirit of criticism, to recognize his limitations and thus the need to encounter others in a credible interaction. This is the content of the principle of rational dialogue according to Ratzinger. Prior to any meaningful process of sociality, there has to be rational dialogue in the same way Plato conceives it as important for intellectual interaction within the context of his "Academy". What then constitutes this dialogue that is needed for meaningful interaction? Prior to any dialogue is the recognition of the individual who, in his 'individualness', possesses being that is distinctive from the other. This is to affirm the ontological precedence of the individual in such a way as to deduce a mind free from all constraints and encumbrances external to it. It is then upon this clear understanding of an individual with ontological independence and separateness that meaningful dialogue can emerge amongst two interlocutors who are persuaded by the interior master namely, "truth" in order to reach consensus. In dialogue, Ratzinger avers that the art of listening is fundamental. To listen implies to be open to the reality of other things or people, knowing, acknowledging, and allowing them to step into one's own realm of being. The idea of mutuality and reciprocity here potentially leads to the unity of beings of interlocutors in a process in which the "I" of the individual is now concealed and suppressed. Cardinal to a process of aggregating individuals in the society in their bid to reach consensus is the place of truth. Obedience to this interior master, and being guided and purified by it, unify the society and it is by this that credible interactions and sociality are achieved, Ratzinger concludes. The absence of it forecloses such relationality and makes a community a congregation of the deaf, and so, efforts toward social cohesion become intractable (Ratzinger.1995).

4 Conclusion

The present paper has attempted a description of the African condition particularly in terms of what is present in it which continues to make theories of development an optical illusion. In this consideration, this paper notes that a materialist conception of the infrastructural reality in Africa is warped and continues to make genuine efforts toward sustainable development a mirage. Consequently, the current discussion has proposed the recognition of the ontology of the individual as the basis of credible theory of development in the African communal universe through which I believe, a credible effort toward sustainability can be achieved. Embedded in this proposal is a reconstructionist approach to communalism from its restricted form to a more rational or critical stance which utilizes the thesis 'critical rationalism' of Karl Popper and the discourse of rational dialogue of Joseph Ratzinger. Very crucial to the process of reconstructionism is a re-examination of the condition of impoverishment represented in the conceptual/mental colonization through which Africans continue to grope in darkness, thereby engendering loss of identity. It would then not be inappropriate to infer that at the core of the African condition is the mental impoverishment manifested in the inability of Africa to interrogate its condition consisting in worldviews, institutions, orientations, dispositions, traditions, cultures, and policies. An ignoring of this condition can only engender a state of hopelessness thereby presenting vivid images of Africans as people uncritical of their environment and tradition, and ultimately preventing the embracement of the rationalist attitude. Conversely, a credible theory of development, this paper notes, would emerge in the recognition of the individual who, though possesses an ontological independence, but oriented toward commonality in a rational dialogue. This is then the basis for an idealist conception of infrastructural sustainability required in the process of development in Africa.

References

Barry, H.: More than the sum of its parts: holism in the philosophy of Emmanuel Oyecherere Osigwe Anyiam- Osigwe. In: Olu-Owolabi, K.A., Ekanola Ibadan, A.B. (eds.) Holistic Approach to Human Existence and Development, pp. 3–14. Hope Publications. Nigeria (2013)

Ekanola, B.A., Lawal, L.A.: Godwin Sogolo and metaphysics of development. In: Oyeshile, O.A., Offor, P. (eds.) Ethics, Governance and Social Order in Africa; Essays in Honour of Godwin, S. Sogolo. Zenith Book House Ltd., Ibadan (2016)

Gary, W.: Freedom Time: Negritude, Decolonization and the Future of the World. Duke University Press, London (2015)

Gyekye, K.: Traditions and Modernity: Philosophical Reflection on the African Experience. Oxford University Press, Oxford (1997)

Kenyatta, J.: Facing Mount Kenya. Vintage books, New York (1965)

Lee, M.B.: Understanding and Ontology in Traditional African Thought. African Philosophy: New and Traditional. Oxford University Press, Oxford (2004)

Matolino, B.: Afro-Communitarian Democracy. Lexington Books, London (2019)

Mbiti, J.S.: African Religions and Philosophy. Heinemann, London (1969)

Menkiti, I.: Person and community in African traditional thought. In: Wright, R.A. (ed.) African Philosophy: An Introduction, 3rd edn., pp. 171–180. University Press of America, Lanhan (1984)

Metz, T.: African communalism and difference. In: Imafidon, E. (ed.) Handbook of African Philosophy of Difference. Switzerland, pp 31–51. Springer, Cham (2020). https://doi.org/10.1007/978-3-030-04941-6_2-1

Nkrumah, K.: Consciencism: Philosophy and Ideology for Decolonization. Panaf books Ltd, London (1970)

Nyerere, K.J.: UJAMAA: Essays on Socialism. Oxford University Press, vii, London (1968)

Okeregbe, A.: A critical evaluation of the concept of Ujamaa in Julius Nyerere's Political Philosophy. In: Okeregbe, A., Jegede, S., Ogunkeye, D. (eds.). A Study in African Socio-Political Philosophy, pp. 45–67. University of Lagos Press, Lagos: (2012)

Oladipo, O.: Introduction: Africa's challenges. In: Remaking Africa: Challenges of Twenty First Century. Hope Publication, Ibadan (1998)

Oladipo, O.: Philosophy and Social Reconstruction in Africa. Hope Publication, Ibadan (2009)

Olu-Owolabi, A.K.: African philosophy: retrospect and prospect. In: Kolawole, O.A. (ed.) Issues and Problems in Philosophy. Grovas Network, Ibadan (2007)

Olufemi, T.: Africa Must be Modern: A Manifesto. Indiana University, Bloomington (2014)
Peet, R., Hartwick, E.: Theories of Development: Contentions, Arguments, Alternative, 2nd edn. The Guilford Press, New York (2009)
Popper, R.K.: Open society and its enemies. In: Popper, R.K. (ed.) Logic of Scientific Discovery, 1st edn. Routledge, London (1945)
Popper, R.K.: Conjectures and Refutations: The Growth of Scientific Knowledge. Harper Touchbooks Publisher, New York (1965)
Popper, R.K.: Objective Knowledge: An Evolutionary Approach. Clarendon press, Oxford (1972)
Popper, R.K.: The myth of the framework. In: Notturno, M.A. (ed.) Defense of Science and Rationality. Routledge, London (1994)
Ratzinger, C.J.: The Nature and Mission of Theology: Approaches to Understanding its Role in the light of Present Controversy. Ignatius Press, San Francisco (1995)
Walter, A. E. S.: Leopold Sedar Senghor and African socialism. J. Mod. Afr. Stud. 3(3), 349–369 (1965)
Wiredu, K.: Democracy by consensus: some conceptual consideration. Social. Democr. 21(3), 155–170 (2007). https://doi.org/10.1080/05568640109485087
Stefano, G.: Karl Popper's Philosophy of Science: Rationality without Foundations. Routledge, London (2009)

Provision of Digital Library, a Catalyst for Scholastic Creativity Among Undergraduates in South-West, Nigeria

V. O. Amatari[✉] and I. U. Berezi

Faculty of Education, Niger Delta University, Amassoma, Bayelsa State, Nigeria
dramatariodiri@gmail.com

Abstract. Purpose: Contemporary society craves creativity that juxtaposes with advanced development emanating from an enabling environment in the citadel of higher education. This study, therefore, investigated the provision of the digital library as a catalyst for scholastic creativity among undergraduates in South-West, Nigeria.

Design/Methodology/Approach: A correlational survey design was adopted. A randomly selected sample of 89 undergraduates from private universities in the South-West, Nigeria was utilized for data collection. The instrument of data collection was a structured questionnaire and was validated before administration. A Cronbach Alpha value of 0.75 estimated the internal consistency of items. Data were analyzed using descriptive statistics and the Pearson moment correlation coefficient.

Findings: Results show that a greater percentage of undergraduates access and utilize the services rendered by digital libraries in their institutions. Further, there is significant positive correlation between undergraduates' utilization of services rendered by digital library and their knowledge of creativity ($r = .206$, $N = 89$; $p < 0.05$); attitude to creativity ($r = .292$, $N = 89$; $p < 0.01$).

Research Limitation: The justification for the provision of digital libraries in the institutions of university education as a vital tool to enable a creative atmosphere is underscored in this study. Creativity is an agent of change for sustainable development. This study would have been more remarkable if it was carried out among undergraduates in public universities. But the public universities are closed down as a result of an ongoing union strike.

Practical Implications: The paucity of infrastructure in Nigerian institutions of higher learning, especially in the public domain is a major concern to stakeholders. The incessant strikes by university unions are pinned on inadequate and obsolete infrastructures within. If scholastic creativity is dependent on the provision of technologically driven library facilities in schools, then the findings of this study extrapolate the need to adequately fund the university system.

Originality/Value: Previous studies had been carried out on scholastic creativity and several on infrastructural degeneracy in higher institutions. However, this study is angled at exploring the relationship between scholastic creativity and the provision of resilient infrastructure such as a school digital library for sustainable development in society.

Keywords: Access · Creativity · Digital library · Provision · Utilization

1 Introduction

An essential facility, that every educational institution should be able to provide is the school library. This is a facility that enables students, teachers, and entire staff of the institution to access variety of resources for learning, teaching, information, entertainment, and professional development. As a major component of school infrastructure, the school library provides resources to motivate students to read, teachers to upgrade knowledge and others to be adequately informed in all the spectra of society. International Federation of Library Associations and Institutions (IFLA) (2015) defined school library as the school's physical and digital and learning space where reading, inquiry, thinking, imagination and creativity are central to students' information, to knowledge and to their personal, social and cultural growth. White (2012) described the role of libraries as gateways to knowledge and culture as they play a fundamental role in society. He went further to report that libraries offer informational resources and deliver services that give opportunities to learn, enhance scholarship and help to influence fresh ideas and perspective that seem to be pivotal to unique and productive culture. Accordingly, libraries help to confirm an incredible history of information developed and acquired by prior creation. An entire globe without repositories (libraries), it would be hard to accelerate studies in research, human knowledge or safeguard world's literature and inheritance for intergenerational consumption.

The conventional school library is organized and operated in a smooth manually coordinated system. However, with contemporary advancement in science and technology, a conventional library system can no longer meet the societal needs of a creative and innovative world. Hence, the digitalization of library services to accommodate the insatiable search for knowledge, science, creativity and innovation to fit into the present technological driven society. The ubiquity of technology has changed the way learners access information and interact with others (Organization for Economic Co-operation & Development (OECD), 2014).

Fakir and Shilpa (2019) posited that digital repositories are becoming a core part of digital learning era. Their extremely importance is growingly becoming a scale of library's contribution to the revolution and advancement of a nation and the society. Simply put, digital library applies information communication technology (ICT) to the routine of library services. It is also known as online or internet or digital repository that uses an online database of digital objects, documents, media that is accessible through the internet. Trivedi (2010) defined digital library as a library in which collections are stored in digital formats as opposed to print, microform or other media and these collections are accessible by computer. Accordingly, Oluwaseye and Abraham (2013) reported that digital library is not a singular entity. It needs the fusion of technology with collection of many resources.

Fakir and Shilpa (2019) listed characteristics of digital library as:

- Resources of digital library could be used and interpreted by many users as regards to their personally identifiable information.
- Digital library has remote access to variety of online informational sources that are massive and global and fairly to everyone possessing website access.
- Digital library may encapsulate numerous online information ranging from content to photographs and multichannel audio.
- Digital library is used to inhibit useless information.
- There is largely a decrease on the need for the amount of space compelled for construction and maintenance of the physical structure
- Time, space and language difficulties are broken with digital library.
- Digital library provides enhanced searching and collection innovation.
- Users may develop their own private library

The concept of creativity is central to students' development in the three domains of learning (i.e., knowledge, values and skills). In other words, the acquisition of head skills, soft skills and go-get-skills is what every educational institution should sell to its product. Creativity as a phenomenon implies forming something new and valuable. The created object may be physical, imaginable or intangible. The concept of creativity alludes novelty, originating and appropriateness. Boden (2004) defined creativity as the ability to come up with ideas or artefacts that are new, surprising and valuable. Actually, we are all wired to create.

Kaufman and Gregorie (2015) reported that creativity is multifaceted. Creativity is diversified on the standard of the mind, personal style and the design process and can also be demonstrated in numerous direction; from intensely emotional joy of unmasking a new notion or experience that enables one to express self through words, pictures, dressing style and other day after day designs to highly regarded designers that surpass ages. From the theoretical perspective, Kronfeldner (2009), explored the two diversified views in explaining the concept of creativity. The first view is the neuropsychology and behavioural sciences that look into creativity through the telephoto of a materialist worldview. These natural sciences attempt to articulate how the higher mental tries to be innovative and make the assumption that a theory realism interpretation is possible. The second is from some philosophers' view that retain that creativity to them is truly remarkable, unexplainable, unknowable and as such unforeseen in theory. According to Kronfeldner, though the views may deviate, but both imagine an implicit and explicit concept of creativity.

In his own conceptual analysis, Kronfeldner (2009) emphasized the polyphonous construct of creativity. This is the anthropological, the psychological, and the metaphysical concepts of creativity. While the anthropological concept regards humans as creators of culture and therefore, we are all creative; the psychological concept focuses on originality and spontaneity as key elements of creativity. The historical concept talks about creating something that is culturally new. The item created can either be unique to the inventor, innovative for a group of individuals who already are bound together by tradition or innovative in the perception of being its first appearance in the universe. The metaphysical concept of creativity sees creativity as beyond the horizon of a naturalistic science of mind. It is a mystery which will always lie beyond the reach of science. It is unexplainable and will always remain so.

The psychological concept of creativity focuses on originating and spontaneity as key elements of creativity. And it is a fact that the teeming youths of the population of a nation occupy the higher institutions. These youths are inherently built for new adventures, discoveries and zeal for self-actualization. They are wired for originality and spontaneity. However, they are inhibited in several ways. One major way is the unfriendly and unenabling environment that is offered to them in terms of the provision of infrastructures in school.

The infrastructure theory is a turning point of underpinning equal importance to the practices of school library with the other infrastructures found in the school location and not merely supporting the activities that take place in the classroom. (Centerwall and Nolin 2019; Schatzki 2010). The general view in research in education is that the classroom is the most crucial and visible facility at the premises of a school compound. Other engaged infrastructures and services are tailored towards assisting the acts and sayings of the classroom (Ott 2017; Stevens et al. 2016). However, infrastructure studies view all the infrastructure in school as equally important. The invisibility of other infrastructures except the classroom is deemphasized. The infrastructural approach is useful in making school libraries observable and the current discovery that also be produced within standpoint.

Originality is a core value in creativity studies. An enabling environment that promotes creativity is essential in our citadel of higher education. An environment that embraces creativity as part of learning, makes use of the most effective strategies that make open variety of opportunities to explore students' skills and talents while in schools and having opportunity to participate in or create program to develop creative skills, encourage and enhance scholastic creativity.

Today's world craves for creativity and therefore the need to overhaul the curriculum, policy and implementation, pedagogy and infrastructures provided in the educational system. A curriculum given that do not open students to think outside the box is no longer tenable. Policy formulation that is not garnished with adequate funding for implementation is a misplaced priority. Pedagogy that is built on traditional methods of passing information and ideas to students with little or no opportunities to train their minds to think can no longer stand the test of time. Inadequate, redundant and ill-equipped infrastructures in schools is a significant inhibitor. This is the current state of the educational system in the nation. A state that restrains creative awakening and inventiveness of the learners. A state where the outcomes of higher education can't see and think beyond their certificates but these certificates prepare them for a world that no longer exist. This is why the public school system in an underdeveloped or developing could not stay competition with that of the advanced nations where policies and structures are sloped towards inventiveness, psyche-exploration and development by the student.

The traditional view that the classroom is the most central and visible infrastructure at the site of the school is gradually fading out. A school library which in the past is backgrounded and viewed as just a supporting facility can no longer stand the test of time. Many of the universities in Nigeria have prioritized the provision of digital library in the citadel of learning. While the first-generation universities had keyed into digitalization for more than a decade running, the relative new universities have bought into the digitalization of services given by school library. Actually, one of the main attractions of private university is the provision of digital school library and this probably accounts for the exorbitant school fees charged by this category of university institutions.

What percentage of the undergraduates access and utilize the services rendered by digital library in their institutions and does utilization of such services relate to or enhance their knowledge of creativity as well as their attitude to creativity? To give answers to these questions, this study is hinged on investigating the relationship between utilization of digital library and undergraduates' knowledge and attitude to creativity.

This study is guided by these set of research questions.

1. What percentage of undergraduates have access to the type of digitalized library service in their school?
2. What percentage of undergraduates utilize the digitalized library services provided?
3. What percentage of undergraduates have knowledge of creativity?
4. What is the attitude of undergraduates to creativity?
5. What relationship exists between utilization of digital library and undergraduates' knowledge of creativity?
6. What relationship exists between utilization of digital library and undergraduates' attitude to creativity?
7. What relationship exists between undergraduates' knowledge of creativity and their attitude to creativity?

2 Methods

The correlational survey is adopted for this study. The population of undergraduates in the two popular and purposively selected private universities in the South-West of Nigeria was eight thousand, nine hundred and ninety-nine. The choice of only private university is because at the time this study was carried out, the academic staff of public universities were on strike and therefore accessing the students was infeasible. The Taro Yamane method was adopted to calculate the sample size. However, only 23% of the estimated sample size (i.e., eighty-nine) undergraduates were randomly selected as the sample for the study. A self-developed and structured questionnaire tagged 'Digital library and creativity Questionnaire' was the instrument of data collection. A type of Likert response scale format was adapted. A reliability coefficient of 0.75 was attained for internal consistency of items using Cronbach Alpha. Data collected were analyzed with percentages, mean and standard deviation and Pearson Product Moment correlation coefficient. However, the discrepancies in the total sample size in the analysis is because insignificant few of the respondents did not attempt some items in the instrument.

3 Results

Research question 1. What percentage of undergraduates has access to a type of digitalized library service in their school?

Table 1. Students' accessibility to digital library

Type of digital library available	No of students	Accessibility
1. Fully digitalized library	66	93%
2. Partially digitalized library (i.e. subscription to resource base data)	17	

Finding shows that 93% of the students has ease at accessing digitalized services provided by their institutional library whether partially or fully digitalized library service, while 6(7%) of the students selected either of the options.

Table 2. Undergraduates' utilization of digitalized library services

Statement Items	Yes	No
1. I utilize my school library for learning 24/7 h	32(36%)	57(64%)
2. I utilize my school library for variety of materials 24/7 h	37(42%)	51(58%)
3. I utilize the library facility at school hours only	60(68%)	28(32%)
4. I don't have any business doing in my school library	25(28%)	68(72%)
5. My smartphone/Laptop/Tablet adequately meets all my ICT	67(77%)	20(23%)
	Mean = 7.52 SD = 1.08	

Findings shows that greater percentage of the students (68%) utilized the services of digital library only during school hours. 77% of the students meet their ICT needs using phone, laptop and tablet.

Research Question 2. What percentage of undergraduates utilize the digitalized library services provided?

Research Question 3. What percentage of undergraduates have knowledge of creativity?

Table 3. Descriptive statistics on knowledge of creativity.

Item statements	SA	A	D	SD	m-score	$\bar{x}$	SD
1. Only science and Technology students should think of creativity	12	30	40	55	1.54		3.81
2. Creativity is solving problems in new ways	180	114	10	1	3.43		
3. Creativity is producing solutions, which have value	188	114	4	–	3.44		
4. Making the best of difficult situation is creativity	156	111	20	2	3.25		39.62

(*continued*)

4. Research Question 4: What is the attitude of undergraduates to creativity.

Table 3. (*continued*)

Item statements	SA	A	D	SD	m-score $\bar{x}$	SD
5. The promotion of creativity among undergraduates will lead to development in Nigeria	216	87	4	4	3.49	
6. Doing old things in new ways is creativity	152	126	14	4	3.33	
7. There is creativity in the arts discipline	148	141	6	2	3.34	
8. The social sciences talk about creativity as well	128	138	14	2	3.17	
9. There is creativity in all areas of life	208	99	4	1	3.51	
10. Adapting to change in life is creativity	100	135	22	7	2.97	
11. It is impossible for creativity to eliminate some problems	60	78	66	13	2.44	
12. Creativity is only for the intelligent	44	24	56	41	1.85	
13. Everybody has the ability to be creative	232	24	2	3	2.93	

From Table 3, with a total mean of 39.62; SD = 3.81, it is estimated that 83% of the undergraduates have knowledge of creativity. A low coefficient of variation of 0.1 (i.e., mean = 39.62; SD = 3.81) buttressed that respondents have about the same knowledge of creativity.

Research Question 5: What relationship exists between utilization of digital library and undergraduates' knowledge of creativity?

Research Question 6: What relationship exists between utilization of digital library and undergraduates' attitude to creativity?

Table 4. Descriptive statistics on attitude to creativity

Item statements	SA	A	D	SD	M-score	$\bar{x}$	SD
1. It doesn't matter whether one is creative or not.	60	105	64	5	2.63		4.31
2. Talking to undergraduates about creativity is a waste of time	16	33	86	31	1.87		
3. People should stop bothering themselves about creativity	24	39	78	31	1.93	39.10	
4. Creativity is an enjoyable enterprise	132	156	8	–	3.33		
5. Everybody should take creativity more seriously	132	141	10	3	3.21		
6. Those who speak about creativity should worry about it alone	32	30	86	27	1.97		
7. Those who talk about creativity put a strain on the	24	66	94	12	2.20		

(*continued*)

Table 4. (*continued*)

Item statements	SA	A	D	SD	M-score	$\bar{x}$	SD
8. Success in life is possible whether one is creative or not	96	144	28	2	3.03		
9. There is no need to lose sleep over creativity	32	117	64	8	2.48		
10. Creativity will help everyone to be relevant in competitive world	148	135	14	–	3.34		
11. Creativity should be the least priority in the University	12	51	94	22	2.01		
12. Passing exams is more important than thinking about creativity	32	75	90	10	2.32		
13. People would rather listen to political discussion than talk of creativity	28	90	64	20	2.27		
14. Creativity will enhance better performance	176	117	8	2	3.40		
15. More priority is given to ICT based crimes than genuine creativity	76	126	36	8	2.76		

From Table 4, 74% of undergraduates are favorably disposed to creativity (mean = 39.10; SD = 4.31) and a low coefficient of variation of 0.1 implies that responses tilted to the same range of attitude among undergraduates. Further, with a criterion mean of 2.50, items 4, 5, 8, 10 and 14 had scores far above the criterion as their statements are tilted towards creativity.

Table 5. Utilization and knowledge of creativity

Variables		Utilization	Knowledge
Utilization of digital library	Pearson correlation	1	.206
	sig (2-tailed)		.05*
	N	89	89
Knowledge of creativity	Pearson correlation	.206	1
	sig (2-tailed)	.05*	
	N	89	89

* Statistically significant at 0.05 alpha level.
There is a significant positive relationship between the utilization of services provided by digital library and undergraduates knowledge of creativity (r = .206; N = 89; p < 0.05).

Research Question 7: What relationship exists between undergraduates' knowledge of creativity and their attitude to creativity?

Table 6. Utilization and attitude to creativity

Variables		Utilization	Knowledge
Utilization of digital library	Pearson correlation	1	.292
	sig (2-tailed)		.01**
	N	89	89
Attitude to creativity	Pearson correlation	.292	
	sig (2-tailed)	.01**	
	N	89	89

**Correlation is significant at .01 level.
There is a significant relationship between undergraduates' utilization of digital library services and their attitude to creativity (r = .292; N = 89; p < 0.01).

Table 7. Knowledge and attitude to creativity

Variables		Knowledge	Attitude
Knowledge of creativity	Pearson correlation	1	.207
	sig (2-tailed)		.05*
	N	89	89
Attitude to creativity	Pearson correlation	.207	
	sig (2-tailed)	.05*	
	N	89	89

*Significant at p < 0.05.
A significant and positive association is observed between undergraduates' knowledge of creativity and their attitude to creativity (r = .207; N = 89; p < 0.05).

4 Discussions of Findings

The sampled institutions in this study provide digital library as one of the essential infrastructures that is vital to the present digital learning age. Finding in Table 1 shows that 93% of undergraduates affirmed that they access the services provided by digital library with ease. This implies that the undergraduates have access to advance searching and retrieval technology provided by digital library. The visibility of the school library as an equally important infrastructure in a given school set up is underscored when ease of accessibility is attained.

Accessibility facilitates utilization. Findings in Table 2 show a greater percentage of 80% (i.e. mean = 7.52; SD = 1.08) of undergraduates utilized services provided by their school digital libraries. However, a significant percentage of 61% of undergraduates agreed that they do not utilize the digitalized services twenty-four hours daily and seven days in a week. This is likely due to the fact that there is limitation to service accessibility outside the regulated restrictions. This is buttressed by the response of 68% of undergraduates who utilize the library services only at school hours. This contradicts the main features and advantage of digital library over the conventional library; unrestricted open access and retrieval of digital resources without boundaries. In a previous study Matusiak (2012), investigated the use of digital library (DL) resources in two undergraduate classes and explored faculty and students' perception of educational digital library. Finding showed that students and faculty use academic DL primarily for textual resources. However, they preferred the open web for visual and multimedia resource. They used search engines when searching for visual resources.

Table 3 findings show undergraduates' knowledge of creativity as measured in this study. 83% of undergraduates agree to have a basic knowledge of creativity. Statements such as 'creativity is solving problems in new way'; 'there is creativity in all areas of life'; 'doing old things in new ways is creativity'; and everybody has the ability to be creative' had remarkable scores above the criterion mean (3.43; 3.51; 3.33; and 2.93 respectively). However, Matusiak's (2012) study that found the perception of digital library by undergraduates to be limited to a place of primarily textual resources is likely to inhibit the scope and depth of knowledge that digital library is designed to impart to users.

Understanding knowledge to an extent determines an individual's disposition or tendency to respond either favorably or unfavorably toward a particular task and its consequent effect on behaviour. Adequate knowledge influences positive attitude. Findings in Table 4 show that 74% of undergraduates are favourably disposed to creativity. This alludes a relationship between knowledge of creativity and attitude to creativity. Result from Table 7 aligns with this stance. A significant and positive correlation ($r = .207$; $N = 89$; $p < 0.05$) was observed between undergraduates' knowledge of creativity and their attitude towards creativity. A positive correlation connotes that an increasing knowledge in creativity produces an increasing attitude to creativity.

Inference to knowledge-attitude-behaviour approach which seeks to measure not only knowledge but the heightening of attitude and the impact of knowledge and attitude on behavioural change, formed the baseline for the finding of this study. Knowledge is indeed a factor of attitudinal behaviour to creativity. However, in this study, the continuum observed is that provision of digital library enables undergraduates to access and utilize multi-dimensional knowledge which broaden their knowledge base; enhance improved collaboration and interaction among student. Subsequently, an intrinsic motive is being built that helps them to think and do things in a new way. Information is vital to promotion of creativity among undergraduates. A creative enabling end enriching environment fosters innovation and sustainable development that developing countries are yearning for.

Justifiably, findings in Tables 5 and 6 indicate a significant positive correlation between undergraduates' utilization of services rendered by digital library and their knowledge of creativity ($r = .206$, $N = 89$; $p < 0.05$); attitude to creativity ($r = .292$, $N = 89$; $p < 0.01$).

Implication of findings is anchored on the need to build functional and resilient school infrastructures, a focus on digital school library that will foster creativity for sustainable development in the institutions of higher learning. However, there are challenges confronting the provision of digital library in the university system. Some of the challenges include paucity of funds, poor internet connectivity, under use of subscription of resource base data before its expiration, technophobic nature of staff, licensing issues and security.

Igboechesi and Dang (2019) enumerated the challenges that the development and sustenance of digital libraries have been facing in the case of University of Jos. These included traditional bureaucratic walls, expertise challenge, librarian's in-depth training challenge, change challenge, inadequate infrastructure/financial challenge and the challenge of network and connectivity. On their part Oluwaseye and Abraham (2013) in their study that investigated the development of academic digital in Oyo state, Nigeria, found that lack of access to internet, inadequate web-search skills, computer illiteracy and power instability posed a great deal of challenge to most of the institutions that have digital libraries.

Findings of this study have underscored the relationships between undergraduates' utilization of digital library and their creative knowledge and attitude. However, these findings are premised on the provision of digital library in the institutions. Based on findings, the following recommendations are made:

- Services rendered by digital library should be upgraded to enhance its ubiquitousness wherein accessibility by undergraduates is not restricted and can be accessed in any part of the campus and outside the campus.
- Provision of resilient digital library in all institutions of higher education is advocated. This is a requisition that will enable undergraduates to acclimate to the present digital learning age.
- There is need to begin to foster the creative mindset of undergraduates, therefore deliberate and planned curriculum, pedagogical strategies and accessibility to vast digital information should be operational. This will motivate students to generate, create or discover new ideas, solutions and possibilities is underscored.
- Adequate funding and deliberate implementation of policy on Information Communication Technology (ICT) will mitigate the challenges encountered in operating digital library in the universities.

5 Conclusion

This study highlighted the provision of digital library as a catalyst for scholastic creativity among undergraduates in South-West, Nigeria. Findings established significant positive relationships between undergraduates' utilization of services rendered by digital library and their knowledge and attitude to creativity. Originality is a core value of creativity. Creativity heightens innovations that are currently needed for sustainable development. Everyone has the ability to be creative. What is needed is an enabling and enriching environment that will awaken this ability. An essential component of this environment is the provision of resilient digital school library infrastructure. The accessibility and utilization of digital library by undergraduates will accentuate and capture a creative-innovative 'atmosphere' in our institutions of tertiary education. Therefore, provision of school digital library as a needs assessment in the university can no longer be dealt with lightly.

References

Oluwaseye, A.J., Abraham, A.O.: The challenges in the development of academic digital library in Nigeria. Int. J. Educ. Res. Dev. **2**(6), 152–157 (2013)
Boden, M.A.: The Creative Mind, 2nd edn., Routledge, London (2014)
Centerwall, U., Nolin, J.: Using an infrastructural perspective to conceptualize the visibility of school libraries in Sweden. Inf. Res. **24**(3) paper 831 (2019). https://InformationR.net/ir/243/paper831
Fakir, A.S., Shilpa, S.W.: Digital library: services and its applications in the information age. Int. J. Adv. Innov. Res. **6**(1), 51–55 (2019)
Igboechesi, G.P., Dang, T.L.: Challenges of digital library in Nigeria: an overview of the University of Jos. In: Banerjee, S.N. (ed.) Pioneering Concepts in Strategic Management & Entrepreneurship. Exceled Open (2019)
International Federation of Library Associations and Institutions (IFLA): IFLA School Library Guidelines (2015). www.ifla.org
Kaufman, S.B., Gregoire, C.: Wired to Create: Unravelling the Mysteries of the Creative Mind. Perigee Books, New York (2015)
Kronfeldner, M.E.: Creativity naturalized. Philos. Q. **59**(237), 577–592 (2009). https://doi.org/10.1111/j.1467-9213.2009.637x
Matusiak, K.K.: Perception of usability and usefulness of digital libraries. Int. J. Hum. Arts Comput. **6**(1–2), 133–147 (2012). https://doi.org/10.3366/ijhac.2012.0044
Organization for Economic Co-operation & Development (OECD): Education at Glance: OECD Indicators. OECD Publishing (2014)
Ott, T.: Mobile phones in schools: from disturbing objects to infrastructure for learning. Unpublished doctoral dissertation. Gothenburg University (2017). https://gupes.ut.gu.se/bitstream
Shatzki, T.: Materially and social life. Nat. Cult. **5**(2), 123–149 (2010)
Stevens, R., Jona, K., Ramey, K.E Hilppo, J., Penuel, W.: FUSE: an alternative infrastructure for empowering learners in schools. In: International Conference of the Learning Sciences (ICLS), Singapore. International Society of the Learning Sciences (2016)
Trivedi, M.: Digital Libraries: Functionality, Usability and Accessibility. Library Philosophy and Practices (2010). ISSN:1522-022
White, B.: Guaranteeing access to knowledge: the role of library. WIPO Mag. (14 August 2012). Article. 6004. https://www.wipo.int/

Evaluation of the Factors Influencing the Intention-To-Use Bim Among Construction Professionals in Abuja, Nigeria

S. Isa(✉) and M. O. Anifowose(✉)

Department of Quantity Surveying, School of Environmental Technology,
Federal University of Technology, Minna, Nigeria
suleimanqs@gmail.com

ABSTRACT. Purpose: This paper evaluated the factors influencing the intention to use BIM among construction professionals in Abuja with the view to understanding the variables that influence BIM adoption in Nigeria.

Design/Methodology/Approach: A questionnaire was used as part of the study's quantitative research strategy. The population of the study included construction professionals registered with Federal Inland Revenue Service (FIRS) which is 4,195. To determine the sample size for the investigation, the Kothari sample size calculation was used (352 professionals). Cronbach's Alpha reliability testing was done prior to analysis, and the resultant coefficient (0.784) shows that the data gathering tool was reliable. The mean item score was used to examine the obtained data. Using the UTAUT constructs and the one-sample t-test, the mean score was used to rank the factors impacting the intention to utilise BIM.

Findings: The study assessed the factors influencing the intention of using BIM among construction professionals in Nigeria using the UTAUT constructs as a yardstick. Mean item score was used to rank the identified variables from the literature review that influence the intention-to-use BIM. The highly ranked variables in each of the constructs include; Facilitating Conditions ("My clients have an interest in the use of BIM"), Performance Expectancy ("Using BIM is of benefit to me"), Effort Expectancy ("I do not have difficulty in explaining why using BIM may be beneficial"), Social Influence ("People who are important to me (e.g. family, friends) think that I should use BIM"). Additionally, using the one-sample t-test value of 3.5, the results showed that respondents believed all of the proposed solutions to be statistically significant ($p < 0.05$). This shows that the constructs in the research model have a huge impact on BIM adoption which is the main dependent construct.

Research Limitation/ Implications: Inability to reach all the thirty-six (36) states in Nigeria would affect the generalization of the study and it is also anticipated that some knowledgeable construction professionals would be reluctant to fill the questionnaire due to to work engagement and other personal reasons.

Practical Implication: Complete adoption of BIM by few number of professionals may practically increase the number of unemployment in the built environment/construction sector.

Originality/Value: This study is the first in Nigeria to explore the behavioural intention of construction professionals using the UTAUT model and include critical variables that influence BIM adoption. This study would serve as a theoretical foundation for future studies and as well assist construction industry stakeholders to develop appropriate policies to improve BIM adoption in Nigeria.

Keywords: BIM · Construction professionals · Evaluation · Usage · UTAUT

1 Introduction

Construction is one of the sectors that significantly contributes to economic growth in Nigeria. (Olanrewaju et al. 2018). It is evident from previous studies that developing countries like Nigeria are hotspots for infrastructural investors. Nonetheless, some critical issues such as time and cost overrun (Bin Seddeeq et al. 2019; Idrees and Shafiq 2021), poor productivity (Loosemore et al. 2021), labour shortage (Pradhananga et al. 2021), material wastage (Fayisa and Wayessa 2021) and slow adoption of emerging technologies (Okpala et al. 2021; McNamara and Sepasgozar 2021) is attributable to the construction sector. Hence, the application of emerging technologies in a meaningful way solves some of the issues in the construction industry (Akdag and Maqsood 2019).

The construction industry has witnessed significant changes in the previous several decades, including a rise in interest in information technology with a view to enhancing construction productivity and cost control. The information technology applied in the construction industry includes 3D printing (Pessoa et al. 2021), big data analytics (Aghimien et al. 2021), building information modelling (BIM) (Olanrewaju et al. 2021), digital twin (Opoku et al. 2021), blockchain (Scott et al. 2021) among others. BIM is one of the most researched emerging technologies in the construction industry with diverse applications.

The construction industry started using BIM in mid-2000 (Zhao 2017) and it has gained momentum among researchers in many developed and developing countries of the world including New Zealand (Doan et al. 2021), China (Cui et al. 2021), United States (Mutis and Mehraj 2022), United Kingdom (Dalu et al. 2021), and Seychelles (Adam et al. 2021).

Some researches have also evaluated the BIM capabilities of built environment experts and the condition of BIM in Nigeria (Olorunfemi et al. 2021). This shows that BIM is a crucial technology that may improve Nigeria's construction sector by lowering costs, speeding up the completion of projects, and enhancing project quality and productivity. Due to the numerous advantages it offers in construction projects, including increased productivity, decreased rework, decreased conflict among building experts, and cost savings, it is necessary to examine the behavioural intention to utilise BIM by construction professionals in Nigeria.

Studies have recently begun examining BIM adoption from the standpoint of construction professionals' behavioural intentions in both developed and developing nations. For instance, Cui et al. 2021) studied the use intention of architectural designers towards BIM in China. Similar to this, Wu et al. (2021) examined the theory of planned behaviour to examine the adoption of BIM in China (TPB). By using the Unified Theory of Acceptance and Use of Technology, Abubakar and Oyewobi (2019) investigated the BIM preparedness of construction professionals (UTAUT).

However, few or no studies have studied the behavioural intention of construction professionals in Nigeria towards the use of BIM. Brown et al. (2002) and Howard et al. (2017) emphasized that the attitude of people towards innovation can affect its implementation negatively. This is because the acceptance of an innovation is hinged on individual acts based on their perception of the innovation. This current study contends that the use intention of construction professionals influences the adoption of BIM.

2 Theories Underpinning the Study

The purposes and capacities of BIM have been attempted to be contextualised by a number of studies. According to Lee et al. (2006), BIM is a modelling tool, communication platform, and a digital virtual model.

According to RICS (2014), no specific definition of BIM has been widely accepted since it is always expanding as new frontiers and sectors encroach into previously established bounds. Nevertheless, each definition of BIM is an attempt to capture the characteristics, functions and abilities inherent in the BIM system which is based on individual use. The definition of BIM exist in many forms as described below:

- According to the US National Building Information Standard [NIBS] (2012), BIM is a digital representation of a facility's structural and functional details. It described BIM in more depth as a shared knowledge resource for information about a facility that provides a trustworthy basis for choices throughout its life cycle, which is defined as beginning with initial conceptualization and ending with destruction.
- BIM is a methodology used to integrate digital descriptions of connections between building digital objects. This enables stakeholders to query, simulate, and estimate actions and their effects on the building process as a lifecycle entity.

Theoretical Framework (UTAUT Theory)
Primarily, there exist eight theories upon which the study of human behaviour towards the adoption or execution of required behaviour is hinged, these theories have undergone progressive development from the 1960s as more factors influencing the behavioural attitude of humans emerges. The innovation diffusion theory (IDT), the theory of reasoned action (TRA), the theory of planned behaviour (TPB), the social cognitive theory, the motivational model, the model of perceived credibility (PC) utilisation, the technology acceptance model (TAM), and a combination of the TAM and TPB are among the eight theories mentioned above. For this study, three of these models (theory of reasoned action, Theory of Planned behaviour and Technology Acceptance Model, as well as a hybrid model Unified Theory of Acceptance and Use of Technology), would be considered.

Vankatesh *et al.* (2003), established the Unified Theory of Acceptance and Use of Technology (UTAUT) model in 2003, this was to eliminate the several disadvantages observed from the utilization of the previously existing eight models especially the technology Adoption Model (TAM), whilst incorporating social factors and human behaviours in the operation of the model. This model refines, integrates and corroborates the existing theories. The UTAUT model sprung up from the painstaking evaluation and comparative analysis carried out on the existing eight models in the longitudinal field of behavioural studies.

UTAUT (Theory) FACTORS
According to the Unified Theory of Acceptance and Use of Technology (UTAUT), the adoption of any new technology by a professional is dependent on four factors: the degree to which the professional believes the new technology will improve his or her performance (performance expectancy), how easy the professional feels the technology will be to use (effort expectancy), other experts' opinions about the importance of using the technology (social influence), and the degree to which the professional believes that the necessary infrastructure exists to support the use of the new technology(facilitating conditions). There are thus five factors that need to be in place before a professional will adopt a new technology: Awareness; Performance expectancy; Effort expectancy; Social influence, and Facilitating Conditions.

Assessing the Level of Readiness to Adopt BIM Amongst Built Environment Professionals in Selected Northern Nigerian States
This study assesses the level of readiness of building professionals in using BIM by exploring the factors guiding the acceptance of new technologies; awareness, performance expectancy, effort expectancy, social influence and facilitating conditions with the view to encourage BIM adoption. The study discovered that the Nigerian building professionals are nowhere near ready to adopting Building Information Modelling (BIM) and there is a dire need for Nigerian buildings professionals to restore their good name and regain their sense of professional pride in providing the highest quality buildings and structures.

3 Research Methodology

In this study, a quantitative research methodology was applied. The goal of quantitative research is to gain a better understanding of society. Olarewaju *et al.* (2020) and Babatunde et al. (2020) are two examples of past BIM-related research that successfully used quantitative approaches. Thus, this supports the use of quantitative techniques in this research. All Abuja-based construction industry experts are included in the study's sample. The Federal Inland Revenue Service's (FIRS) database's names and addresses of Abuja-based construction companies served as the study's population. This database was considered credible because it captures professionals that regularly pay their taxes, suggesting that they are active in the field. A total of 4,195 professionals in Abuja were registered on the database as of November 2019. The sample size of 352 was arrived at using Kothari's formula (2004) is given in Eq. (1):

$$n = \frac{Z^2 pqN}{e^2(N-1) + Z^2 pq} \quad (1)$$

The building industry experts in Abuja, Nigeria, were chosen using a straightforward random sample approach. The problem is that there is not an official publication that lists how many professionals have used BIM for a while. In order to allow everyone in the public the chance to participate in the research if they are informed about BIM, simple random sampling was chosen, as also recommended by Ibrahim et al. (2006). As a research tool for this study, a questionnaire was used since it ensures the viewpoint of the intended responder and is appropriate for evaluating unobservable phenomena (Tharenou et al. 2007).

The entire survey was divided into three sections: questions about respondents' backgrounds and experiences; questions regarding construction professionals' intentions to utilise BIM in Abuja, Nigeria; and questions seeking recommendations and opinions on how to ensure the adoption of BIM projects. On a Likert scale with 1 being "strongly disagree," to 5 being "strongly agree," the respondents were asked to score the identified elements (variables). A total of 156 questionnaires, or a 44% response rate, were collected after a total of 352 questionnaires were distributed and served at random to the public. Idrus and Newman (2002) believed that any query with a response rate of 20% to 30% was sufficient for research in the construction sector. Li et al. (2005) conducted a similar questionnaire study in the UK and achieved an 11% response rate. A mean item score and a t-test were used to analyse the data that was gathered.

4 Results and Discussion

Table 1. Demographics characteristics of respondents

Variables		Frequency	Percentage (%)
Highest academic Qualification	Higher National Diploma	14	8.97
	Bachelor Degree	72	46.15
	Post Graduate Diploma	11	7.05
	Master Degree	58	37.18
	Doctorate Degree	1	0.64
	Total	**156**	**100.00**
Profession	Architect	65	41.67
	Builder	7	4.49
	Engineer	25	16.03
	Estate surveyor	9	5.77

(*continued*)

Table 1. *(continued)*

Variables		Frequency	Percentage (%)
	Quantity surveyor	50	32.05
	Total	156	100.00
Age group	21-30	70	44.87
	31-40	50	32.05
	41-50	17	10.90
	50 above	19	12.18
	Total	156	100.00
Years of experience	10–15 years	4	2.56
	15–20 years	33	21.15
	20 years above	31	19.87
	5–10 years	85	54.49
	Less than 5 years	3	1.92
	Total	156	100.00
Main class	Government		19
	Private	137	87.82
	Total	156	100.00
	Large	19	12.18
	Medium	96	61.54
	Small	41	26.28
	Total	156	100.00

Result and Discussion on Variables Influencing the Intention-To-Use BIM (UTAUT Construct)

The descriptive data for the UTAUT concept are shown in Table 1. It displays the components' relative importance to the mean scores in descending order. The table also demonstrates that, according to the respondents, all of the proposed solutions are statistically significant ($p < 0.05$) using the one-sample t-test value of 3.5 (the same cutoff was used in a prior BIM research by Olanrewju et al. (2020). The "Facilitating circumstances" factors' mean scores vary from 3.40 to 3.96. The highest-ranking of these factors was "My clients have an interest in the usage of BIM" (mean = 3.96; SD = 0.074; t (155) = 7.75; p = 0.00 > 0.05), while the lowest-ranking was "BIM is compatible with other systems that I use" (mean = 3.40; SD = 1.22; t (156) = −1.05; p = 0.29 > 0.05). With reference to the p-value, which should be less than 0.05 for each of the relevant factors, a threshold of 3.5 was chosen in order to get the most significant experiences based on the mean score. Therefore, only three (3) of these facilitating condition variables were considered significant because they were above the set 3.5 thresholds.

Table 2. Variables influencing the Intention-to-use BIM (UTAUT construct)

S/N	UTAUT constructs	MS	SD	t-value ($\mu = 3.5$)	df	Sig. (2-tailed)	R
Facilitating conditions							
FC5	My clients have an interest in the use of BIM	3.96	0.74	7.75	155	0.00	1
FC4	I have the required subscription for BIM packages	3.57	1.04	0.85	155	0.40	2
FC1	I have the resources necessary to use BIM	3.54	1.15	0.49	155	0.63	3
FC3	A specific person or group is available for assistance with difficulties concerning the use of BIM	3.47	1.08	-0.37	155	0.71	4
FC2	BIM is compatible with other systems that I use	3.40	1.22	-1.05	155	0.29	5
Performance expectancy							
PE1	Using BIM is of benefit to me	4.13	0.93	8.52	155	0.00	1
PE4	Using BIM improves my performance in my job	4.03	0.73	9.09	155	0.00	2
PE3	Using BIM will increase my productivity	3.90	1.25	3.99	155	0.00	3
PE5	Using BIM will increase clients satisfaction	3.17	1.10	-3.78	155	0.00	4
PE2	Using BIM will enable me to accomplish my design more quickly	3.11	0.86	-5.67	155	0.00	5
Effort expectancy							
EE4	I do not have difficulty in explaining why using BIM may be beneficial	4.15	0.68	11.90	155	0.00	1
EE3	I clearly understand how to use BIM	4.04	0.79	8.62	155	0.00	2

(*continued*)

Table 2. (*continued*)

S/N	UTAUT constructs	MS	SD	t-value (μ = 3.5)	df	Sig. (2-tailed)	R
EE2	Learning to use BIM will be easy for me	4.00	0.87	7.16	155	0.00	3
EE1	It will be easy for me to become skillful at using BIM	3.99	0.88	6.98	155	0.00	4
EE5	My clients would love a BIM-enabled project	3.35	1.30	-1.42	155	0.16	5
Social influence							
SI2	People who are important to me (e.g. family, friends) think that I should use BIM	4.19	1.06	8.16	155	0.00	1
SI4	My clients think I should use BIM for them	4.18	0.74	11.46	155	0.00	2
SI1	People who influence my behaviour think that I should use BIM	4.13	1.08	7.25	155	0.00	3
SI5	My clients feel BIM would increase their project social status	3.82	1.27	3.16	155	0.00	4
SI3	People who are important to me (e.g. colleagues) think that I should use BIM	3.64	1.07	1.64	155	0.10	5

The mean score of the variables under "***Performance Expectancy***" ranges between 3.11 and 4.13. These variables ranged from "Using BIM is of benefit to me" (mean = 4.13.; $SD = 0.93$; $t(155) = 8.52$; $p = 0.00 < 0.05$) which is the highest-ranked to "Using BIM will enable me to accomplish my design more quickly" (mean = 3.11; $SD = 0.86$; $t(155) = -5.67$; $p = 0.00 < 0.05$) which is the least ranked. To get the most significant experiences based on the mean score, a threshold of 3.5 was set with a reference to the *p*-value which should be less than 0.05 for each of the significant variables. Therefore, only three (3) of these performance expectancy variables were considered significant because they were above the set 3.5 thresholds.

The mean score of the variables under "***Effort Expectancy***" ranges between 3.35 and 4.15. These variables ranged from "I do not have difficulty in explaining why using BIM may be beneficial" (mean = 4.15.; $SD = 0.68$; $t(155) = 11.90$; $p = 0.00 < 0.05$) which is the highest-ranked to "My client would love a BIM-enabled project" (mean = 3.35; $SD = 1.30$; $t(155) = -1.42$; $p = 0.16 > 0.05$) which is the least ranked. To get the most significant experiences based on the mean score, a threshold of 3.5 was set with a reference to the p-value which should be less than 0.05 for each of the significant variables. Therefore, only four (4) of these effort expectancy variables were considered significant because they were above the set 3.5 thresholds.

The mean score of the variables under "*Social Influence*" ranges between 3.64 and 4.19. These variables ranged from "People who are important to me think that I should use BIM" (mean = 4.19.; SD = 1.06; t (155) = 8.16; p = 0.00 < 0.05) which is the highest-ranked to "People who are important to me (e.g. colleague" (mean = 3.64; SD = 1.07; t (155) = 1.64; p = 0.10 > 0.05) which is the least ranked. To get the most significant experiences based on the mean score, a threshold of 3.5 was set with a reference to the p-value which should be less than 0.05 for each of the significant variables. Therefore, all the five (5) variables of social influence were considered significant because they were above the set 3.5 thresholds. The mean values for constructs in the conceptual model based on the UTAUT model including facilitating conditions, effort expectancy, performance expectancy, behaviour intention, actual use, adoption of BIM and social influence indicators are between 3.11 and 4.19 which is above average (around 62–84%). The descriptive statistics also suggest that most of the respondents agree with the statements in the questionnaire as observed in the table.

5 Discussion of Findings

This study's objective was to assess the factors influencing construction professionals' intentions to employ BIM in Nigeria using the UTAUT constructs as a benchmark. The discovered factors from the literature review that impact the intention to utilise BIM were ranked using the mean item score. Facilitating conditions ("My clients have an interest in the use of BIM"), Performance expectancy ("Using BIM is of benefit to me"), Effort expectancy ("I do not have difficulty in explaining why using BIM may be beneficial"), and social influence ("People who are important to me (e.g. family, friends) think that I should use BIM") are the variables that receive the highest rankings in each of the constructs. The constructs are represented by variables that are comparable to those used by (Mahamadu *et al.* 2014).

6 Conclusion and Recommendation

Effective construction project management in terms of cost, time, and quality depends on the usage of BIM. This report offers information to those involved in the building business on how to increase the use of BIM in the Nigerian construction sector. To change the environment for BIM deployment, a full understanding of the factors affecting BIM acceptance is essential. In light of this, the study evaluated, using the UTAUT model, the behavioural intentions of Nigerian construction professionals about the use of BIM. A questionnaire was used to obtain primary information from 156 Abuja-based construction industry experts. In order to draw logical conclusions from the data obtained, descriptive statistics (charts and tables) and inferential statistics (t-test) were used to examine the data. According to the report, enabling conditions, intention to use BIM, actual use of BIM, and crucial success aspects of the intention-to-use BIM are the four key factors impacting BIM adoption in the Nigerian construction sector. The report makes the following suggestions in an effort to accelerate BIM adoption in the Nigerian construction sector. Collaboration between the government and construction industry stakeholders: In order to make BIM implementation for building projects a reality, the government must work with industry stakeholders including professionals, clients, and professional bodies, among others.

References

Abubakar, I.T., Oyewobi, L.O.: Assessing the level of readiness to adopt building information modeling (BIM) amongst built environment professionals in selected Northern Nigerian States. In: Proceedings of 3rd International Engineering Conference (IEC 2019). School of Electrical Engineering and Technology School of Infrastructure, Process Engineering and Technology, Federal University of Technology, Minna (2019)

Adam, V., Manu, P., Mahamadu, A.M., Dziekonski, K., Kissi, E., Emuze, F., Lee, S.: Building information modelling (BIM) readiness of construction professionals: the context of the Seychelles construction industry. J. Eng. Design Technol. ahead-of-print No. ahead-of-print (2021). https://doi.org/10.1108/JEDT-09-2020-0379

Aghimien, D.O., Ikuabe, M., Aigbavboa, C., Oke, A., Shirinda, W. (2021). Unravelling the factors influencing construction organisations' intention to adopt big data analytics in South Africa. Constr. Econ. Build. **21**(3) (2021)

Akdag, S.G., Maqsood, U.: A roadmap for BIM adoption and implementation in developing countries: the Pakistan case. ArchNet-IJAR: Int. J. Archit. Rese. **14**(1), 112–132 (2020)

Anifowose, M., Babarinde, S.A., Olanrewaju, O.I.: Adoption level of building information modelling by selected professionals in Kwara State. Environ. Technol. Sci. J. **9**(2), 35–44 (2018)

Babatunde, S.O., Ekundayo, D., Adekunle, A.O., Bello, W.: Comparative analysis of drivers to BIM adoption among AEC firms in developing countries: a case of Nigeria. J. Eng. Des. Technol. **18**(6), 1425–1447 (2020). https://doi.org/10.1108/JEDT-08-2019-0217

Bin Seddeeq, A., Assaf, S., Abdallah, A., Hassanain, M.A.: Time and cost overrun in the Saudi Arabian oil and gas construction industry. Buildings **9**(2), 41 (2019)

Brown, S.A., Massey, A.P., Montoya-Weiss, M.M., Burkman, J.R.: Do I really have to? User acceptance of mandated technology. Eur. J. Inf. Syst. **11**(4), 283–295 (2002)

Cui, Q., Hu, X., Liu, X., Zhao, L., Wang, G.: Understanding architectural designers' continuous use intention regarding BIM technology: a China case. Buildings **11**(10), 448 (2021)

Dalui, P., Elghaish, F., Brooks, T., McIlwaine, S.: Integrated project delivery with BIM: A methodical approach within the UK consulting sector. J. Inf. **26**, 922–935 (2021)

Doan, D.T., GhaffarianHoseini, A., Naismith, N., GhaffarianHoseini, A., Zhang, T., Tookey, J.: An empirical examination of Green Star certification uptake and its relationship with BIM adoption in New Zealand. Smart Sustain. Built Environ. ahead-of-print No. ahead-of-print (2-021). https://doi.org/10.1108/SASBE-05-2021-0093

Edwards, D.J., Pärn, E.A., Love, P.E.D., El-Gohary, H.: Machinery, manumission and economic machinations. J. Bus. Res. **70**, 391–394 (2017). https://doi.org/10.1016/j.jbusres.2016.08.012

Fargnoli, M., Lombardi, M.: Building information modelling (BIM) to enhance occupational safety in construction activities: research trends emerging from one decade of studies. Buildings **10**(6), 98 (2020)

Fayisa, G.W., Wayessa, S.G.: Cause of construction material wastage on public building project in Western Oromia. Am. J. Civil Eng. **9**(2), 55–62 (2021)

Ganah, A., John, G.A.: Integrating building information modeling and health and safety for onsite construction. Saf. Health Work **6**(1), 39–45 (2015)

Hamma-adama, M., Kouider, T.: What are the barriers and drivers toward BIM adoption in Nigeria? In: Creative Construction Conference 2019, pp. 529–538. Budapest University of Technology and Economics (2019)

Howard, R., Restrepo, L., Chang, C.Y.: Addressing individual perceptions: an application of the unified theory of acceptance and use of technology to building information modelling. Int. J. Project Manage. **35**(2), 107–120 (2017)

Holden, R.J., Karsh, B.T.: The technology acceptance model: its past and its future in health care. J. Biomed. Inf. **43**, 159–172 (2010)

Idrees, S., Shafiq, M.T.: Factors for time and cost overrun in public projects. J. Eng. Project Prod. Manag. **11**(3), 243–254 (2021)

Kothari, C.R.: Research Methodology: Methods and Techniques. New Age International, New Delhi (2004)

Loosemore, M., Alkilani, S.Z., Luperdi, S.: Productivity and industrial relations in the Australian construction industry. In: Proceedings of the Institution of Civil Engineers-Management, Procurement and Law, vol. 40(Ahead of Print), pp. 1–10 (2021)

Li, B., Akintoye, A., Edwards, P.J., Hardcastle, C.: Critical success factors for PPP/PFI projects in the UK construction industry. Constru. Manage. Econ. **23**, 459–471 (2005)

McNamara, A.J., Sepasgozar, S.M.: Intelligent contract adoption in the construction industry: concept development. Autom. Constr. **122**, 103452 (2021)

Mutis, I., Mehraj, I.: Cloud BIM governance framework for implementation in construction firms. Pract. Period. Struct. Des. Constr. **27**(1), 04021074 (2022)

Muhammadu, A.M., Mahdjoubi, L., Booth, C.A.: Determinants of building information modelling (BIM) acceptance for supplier integration: a conceptual model. In: Proceedings 30th Annual ARCOm Conference. Portsmouth, UK (2014)

Okpala, I., Nnaji, C., Awolusi, I., Akanmu, A.: Developing a success model for assessing the impact of wearable sensing devices in the construction industry. J. Constr. Eng. Manag. **147**(7), 04021060 (2021)

Olanrewaju, O.I., Babarinde, S.A., Chileshe, N., Sandanayake, M.: Drivers for implementation of building information modeling (BIM) within the Nigerian construction industry. J. Financ. Manag. Prop. Constr. **26**(3), 366–386 (2021)

Olanrewaju, O.I., Chileshe, N., Babarinde, S.A., Sandanayake, M.: Investigating the barriers to building information modeling (BIM) implementation within the Nigerian construction industry. Eng. Constr. Archit. Manag. **27**(10), 2931–2958 (2020)

Olanrewaju, O.I., Idiake, J.E., Oyewobi, L.O., Akanmu, W.P.: Global economic recession: causes and effects on Nigeria building construction industry. J. Survey. Constr. Prop. **9**(1), 9–18 (2018). https://doi.org/10.22452/jscp.vol9no1.2

Olanrewaju, O.I., Kineber, A.F., Chileshe, N., Edwards, D.J.: Modelling the relationship between Building Information Modelling (BIM) implementation barriers, usage and awareness on building project lifecycle. Build. Environ. **207**, Part B, January 2022, 108556 (20221)

Olanrewaju, O., Babarinde, S.A., Salihu, C.: Current state of building information modelling in the Nigerian construction industry. J. Sustain. Architect. Civil Eng. **27**(2), 63–77 (2020b)

Olorunfemi, E.T., Oyewobi, L.O., Olanrewaju, O.I., Olorunfemi, R.T.: Competencies and the penetration status of building information modelling among built environment professionals in Nigeria. In: Proceedings of International Federation of Surveyors – FIG, FIG e-Working Week 2021 Smart Surveyors for Land and Water Management - Challenges in a New Reality Virtual, 21–25 June 2021

Opoku, D.G.J., Perera, S., Osei-Kyei, R., Rashidi, M.: Digital twin application in the construction industry: a literature review. J. Build. Eng. **40**, 102726 (2021)

Owusu-Manu, D., Edwards, D.J, Mohammed, A., Thwala, W.D., Birch, T.: Short run causal relationship between foreign direct investment (FDI) and infrastructure development. J. Eng. Design Technol. **17**(6), 1202–1221 (2019). https://doi.org/10.1108/JEDT-04-2019-0100

Pärn, E.A., Edwards, D.J., Sing, M.C.P.: Origins and probabilities of MEP and structural design clashes within a federated BIM model. Autom. Constr. **85**, 209–219 (2018). https://doi.org/10.1016/j.autcon.2017.09.010

Pessoa, S., Guimarães, A.S., Lucas, S.S., Simões, N.: 3D printing in the construction industry-a systematic review of the thermal performance in buildings. Renew. Sustain. Energy Rev. **141**, 110794 (2021)

Pradhananga, P., ElZomor, M., Santi Kasabdji, G.: Identifying the challenges to adopting robotics in the US construction industry. J. Constr. Eng. Manag. **147**(5), 05021003 (2021)

Scott, D.J., Broyd, T., Ma, L.: Exploratory literature review of blockchain in the construction industry. Autom. Constr. **132**, 103914 (2021)

Tender, M., Couto, J.P., Fuller, P.: Improving occupational health and safety data integration using building information modelling. In: Arezes, P.M., et al. (eds.) Occupational and Environmental Safety and Health III. SSDC, vol. 406, pp. 75–84. Springer, Cham (2022). https://doi.org/10.1007/978-3-030-89617-1_7

Tharenou, P., Donohue, R., and Cooper, B.: Management Research Methods. Cambridge University Press, Melbourne (2007)

Venkatesh, V., Morris, M.G., Davis, G.B., Davis, F.D.: User acceptance of information technology: toward a unified view. MIS Q. **27**, 425–478 (2003)

Wu, Z., Jiang, M., Li, H., Luo, X., Li, X.: Investigating the critical factors of professionals' BIM adoption behavior based on the theory of planned behavior. Int. J. Environ. Res. Public Health **18**(6), 3022 (2021)

Zhao, X.: A scientometric review of global BIM research: analysis and visualization. Autom. Constr. **80**, 37–47 (2017)

Determinants of Farmers' Satisfaction with Access to Irish Potato Farmer Co-operatives' Services in Northern and Western Provinces, Rwanda

C. Uwaramutse[✉], E. N. Towo, and G. M. Machimu

Moshi Co-Operative University, Moshi, Tanzania
uwacharles3@yahoo.fr

Abstract. Purpose: Satisfaction of members with services offered by co-operatives is key for a co-operative success. However, it remains questionnable whether co-operatives have really achieved their expected objectives. This paper analysed the determinants of farmers' satisfaction with access to services offered by Irish Potato Farmer Co-operatives in Northern and Western Provinces of Rwanda.

Design/Methodology/Approach: The study employed descriptive design in cross-sectional research. Data were analysed descriptively and inferentially. Service accessibility level among Irish potato farmers was measured by developing an index. In assessing the level of farmers' satisfaction, satisfaction index was adapted. Demographic and socio-economic factors influencing farmers' satisfaction with Irish potato farming services were analyzed using multiple linear regression.

Findings: The regression results indicate that only gender, primary occupation, livestock ownership, and co-operative membership significantly affected farmers' satisfaction with co-operative services. Findings reported a low level of farmers' satisfaction with farming services, and co-operatives in the study area failed to resuscitate their activities, forcing some farmers' exit from Irish potato farming activities.

Practical Implications: The findings of this study generate facts to inform IPFCs, community development partners, and policymakers about farmers' satisfaction with co-operative services and how they should be improved. In addition, the paper contributes to the literature by analyzing farmers' accessibility to farming services and satisfaction with co-operative services in developing countries.

Originality/Value: This paper took a holistic perspective to cover all services that members expect from their co-operatives.

Keywords: Co-operative · Co-operative services · Farmers · Satisfaction · Irish potato · Rwanda

1 Introduction

Worldwide, farmer co-operatives are considered to be the backbone of agricultural development (Ma et al. 2021) by offering an extensive range of services to smallholder farmers, including improved access to agricultural inputs, information communication, credit, agro-processing training, and extension (ILO 2021; Lepe 2016; Zheng and Song 2011). Likewise, they serve to organise adequate storage facilities in collection centres, find markets for members' produce, promote improved technologies, and support farmers by strengthening their collective bargaining power (Seneerattanaprayul and Gan 2021; Abebaw and Haile 2013). However, smallholder farmers in developing countries face several challenges that include lack of improved technologies, access to agricultural inputs, improved storage facilities, managerial skills, weak bargaining power (Liu et al. 2021; Zheng et al. 2021; Grashuis and Dary 2021) and poor access to credit services (Ma et al. 2018).

The evolving function of farmer co-operatives has prompted many studies on the members' satisfaction with co-operative services. Morfi et al. (2021) and Morfi et al. (2015) have proved a strong relationship between co-operative membership and satisfaction with farming services. As stated by Grashuis and Cook (2019) & Tarekegn (2017), satisfaction of members is essential for a co-operative to achieve its goals and objectives. Satisfied co-operative farmers actively participate in their co-operatives' activities, hence the improved performance (Prasertsaeng et al. 2020). Co-operatives should thus move beyond maximization of financial performance as their sole criteria of success and give priority to maximizing satisfaction of members' needs through offering a range of services that can improve their social and economic status.

In Rwanda, co-operatives are central to national development (MINICOM 2018). The government of Rwanda (GoR) expects a significant contribution of co-operatives in achieving Vision 2050 (GoR 2020) and the National Transformation Strategy 2018–2024, which aims to accelerate the transformation and economic growth with the private sector (MINAGRI 2018). GoR has established an environment conducive to the development of the co-operative movement. This encompasses law N° 024/2021 governing co-operatives and other regulations for co-operative governance. The Government has also formulated a national policy of 2018 on the promotion of co-operatives to ensure that they are profitable and productive enterprises capable of delivering services and creating surpluses for themselves and their members. In addition, the Government collaborates with co-operatives in activities such as value chain development, research, and extension (MINICOM 2018).

Furthermore, in 2002, the GoR launched a Crop Intensification Programme (CIP) to increase national agricultural productivity and food security. Irish potato was prioritised as one of the priority crops (FAO 2016). Production of Irish potatoes covers 40.6% of the gross agricultural production value and 28.7% of the total cultivated area (NISR 2016). Irish Potato Farmer Co-operatives (IPFCs) were chosen to be the strategic vehicle in improving the production. Given the Government policy to organize Irish potato farming, every farmer has to join IPFCs. The aim is to make co-operatives stronger to manage collection centers (Mbarushimana 2018). Within IPFCs, farmers can easily get subsidies, financial credit, training on best farming practices, and storage facilities in collection centers to reduce exploitation by middlemen (MINAGRI 2018).

Despite the above initiatives, IPFCs failed to improve their services in the face of competition from private investors (FAO 2015). Members of IPFCs in Rwanda are unsatisfied with market for their production due to speculative pricing by unscrupulous buyers. Consequently, they do business with private traders, which strongly affect performance of smallholder farmer co-operatives (Kanamugire 2017).

While in a considerable number of studies (Grashuis and Cook 2019; Singh et al. 2019), performance assessment in co-operatives is dominated by financial ratios, researchers use the satisfaction of members with co-operatives services to measure the success of these organizations. Satisfaction of farmers with services offered by co-operatives as key for co-operative success (Sultana et al. 2020; Marete 2010), is viewed as an important measure of co-operative performance, and target for policy formulation (López-Ridaura et al. 2002). However, there are still limited studies conducted on farmers' satisfaction with co-operative services.

In this perspective, this study intended to fill the gap by analysing determinants of farmers' satisfaction with the services offered by IPFCs in Northern and Western Provinces, Rwanda. This paper specifically measured service accessibility level among co-operative farmers; analysed co-operative and non-co-operative farmers' access to farming services; assessed the level of co-operative and non-cooperative farmers' satisfaction with co-operatives' services, and determined demographic and social-economic factors influencing farmers' satisfaction with access to Irish potato farming services. The rest of the paper is organised into theoretical and empirical framework, methodology, results and discussion, and finally, conclusion and recommendations.

2 Theoretical and Empirical Framework

2.1 Expectancy Disconfirmation Theory

This study was guided by Expectancy Disconfirmation Theory (EDT). The EDT is a theory of customer satisfaction developed by Oliver (1977) and originated from a subject of study for antecedents of satisfaction (Anderson and Sullivan, 1993). Basically, the theory was developed to measure satisfaction of customers based on difference between their expectations and experience in perceived services (Spreng and Page 2003). When the service or product offered to the customer cannot meet his expectations, negative disconfirmation arises and results in dissatisfaction (Oliver 1980). If this happens, most dissatisfied customers decide not to complain; instead, they exit the service (Osarenkhoe and Komunda 2013). The theory was used to assess whether perceived services provided by IPFCs met farmers' expectations, particularly non-co-operative members.

2.2 Empirical Review and Hypothesis Development

Access to farm inputs is one of the significant challenges expressed by both co-operative and non-co-operative farmers (Ajah 2015), which negatively impacts the overall agricultural production (Anglade et al. 2021). Several studies (Sultana et al. 2020; Abate 2018

& Ajah 2015;) report the differences between the two groups of farmers, whereas other studies revealed benefits in favour of co-operative members (Grashuis and Su 2019; Anderson et al. 2014).

A study by Ajah (2015) showed that co-operative members' access level to agricultural inputs is higher than that of non-members. Co-operative membership provides a secured market than non-co-operative farmers (Sultana et al. 2020; Giagnocavo et al. 2018), more access to loan and storage facilities (Ajah 2015), improving bargaining power of smallholder farmers and market information (Serra and Davidson 2021). In Ethiopia, Abebaw and Haile (2013) observed a positive impact of co-operative membership on fertilizer adoption. Compared to farmers who are not in cooperative, co-operative farmers are more likely to access agro-chemicals among smallholder farmers in China (Ma et al. 2018). Morfi et al. (2021) and Morfi et al. (2015) have proved a strong relationship between co-operative membership and satisfaction with farming services. The above discussion leads to the following hypotheses.

H_1: There is a significant difference between co-operative members and non-members' access to farming services.

H_2: There is a significant difference between co-operative members and non-members' satisfaction with co-operative services.

There are different factors influencing farmers' satisfaction (Barham and Chitemi 2009; Hellin et al. 2009). Some are connected with demographic factors of farmers (Ahmed and Mesfin 2017; Ma et al. 2018) and others are related to socio-economic status of farmers (Morfi et al. 2021; Ahmed and Mesfin 2017). Comparing older and younger smallholder farmers, the former are more satisfied with farming services than the latter (Lavis and Blackburn, 1990; Terry and Israel 2004). However, Elias et al. (2015) oppose Lavis and Blackburn's study, stating that older farmers are often reluctant to engage in innovative activities fearing of risk. Education background and farm size were also reported as factors that influence farmers' satisfaction (Higuchi et al. 2020; Ma et al. 2018; Bernard and Spielman 2009).

H_3 There is a relationship between demographic and socio-economic factors with farmers' satisfaction.

3 Methodology

3.1 Research Design and Target Population

The study employed descriptive design in cross-sectional study. A concurrent mixed-method approach was employed as recommended by Creswell (2009). The study was conducted in Rwanda in Northern and Western Provinces. It included four separate Districts of Musanze, Burera, Nyabihu and Rubavu. The targeted population of this study was 76 co-operatives which had 25332 members in the above Districts (NCCR 2018). For comparative purposes, non-co-operative members were also included in the study.

3.2 Sampsling Techniques and Sample Size

A multistage sampling approach was employed to select the co-operatives, their members, and non-members. In the stage one, the above Districts were selected purposively due to their predominance in Irish potatoes farming (NISR 2017). In stage two, in selecting Irish Potato Farmer Co-operatives in the above Districts, 30% were selected as recommended by (Cooper and Schindler 2006). Hence, a sample of 23 co-operatives out of 76 was selected.

A purposive sampling technique was applied to ensure that large and small co-operatives are included in the sample. In this stage, the criterion was based on co-operative share capital, the number of active members and quantity of production. In stage three, the sample size calculation was based on Taro's (1967) formula from a population of 8096 co-operative members across 23 IPFCs (NCCR 2018). Using Taro formula, the sample size of co-operative members was computed as follows:

$$n = \frac{N}{1 + N * e^2} \quad (1)$$

where n is the sample size, N is the population size and e is the margin of error (5%).

$$n = \frac{8096}{1 + 8096(0.05)^2} = 381.17 \simeq 382$$

In stage four, the determined sample size of co-operative members was distributed to each co-operative on the basis of Probability Proportional to Size (PPS) (Appendix Table A1). PPS formula adopted according to (Kothari 2004) as presented below.

$$n_1 = \frac{nN_1}{N} \quad (2)$$

where n = determined sample size, N = target population, N_1 = total number of population in each co-operative, n_1 = number of samples in each co-operative. In selecting member respondents from the sample, a list of members in the selected co-operatives was entered into Microsoft Office Excel to make a random selection.

Concerning non-co-operative members, co-operative and village leaders have facilitated identifying Irish potato farmers who are non-co-operative members. They were selected using convenience and snowball sampling techniques. Convenience selection was used since the information was obtained from the ones readily available during data collection (Etikan 2016). Snowball sampling was also applied because some non-cooperative members have assisted in identifying others. This technique was applied to avoid bias from village leaders in identifying non-co-operative members' respondents (Naderifar et al. 2017). In computing the sample size for non-co-operative members, Cochran formula for unknown population (Cochran, 1977) was employed and obtained 167 respondents as computed below:

$$n_0 = \frac{(z_\alpha)^2 pq}{e^2} \quad (3)$$

where n_0 = sample size, z_α confidence level of 2.58, p estimated population of 0.5, q is 1-p and e is precision which was 0.05. Thus, the sample size is $n_0 = \frac{(2.58)^2 * 0.5 * 0.5}{(0.1)^2} = 167$

3.3 Instruments and Data Collection Techniques

In this study, concurrent mixed-method research was employed (Creswell 2009). Data were collected using a structured questionnaire, Key Informants Interviews (KIIs), and Focus Group Discussion (FGD). A structured questionnaire was designed to collect information from both co-operative and non-co-operative farmers on demographic and socio-economic characteristics, accessibility of farming services and their level of satisfaction with co-operative services. KIIs guide was applied to collect qualitative data from representatives of the National Co-operative Confederation of Rwanda, Irish Potato Federation, and Chairpersons of co-operative unions, Districts Co-operative Officers, Sector Executive Secretaries, and all co-operative managers. Concerning FGDs, four were conducted with Board members & Supervisory committee; two were in large co-operatives and two in small co-operatives. Each FDG was composed of five Board Members of primary co-operatives and three members of supervisory committee. Furthermore, four FGDs were conducted with co-operative members and non-members: There were two with members (one from large co-operative and one from small co-operative) and two with non-members. The ones having more ideas were excluded from individual interviews to avoid monotony and formed part of FGD.

To ensure the quality of scales employed, it was checked whether they meet the criteria of reliability and validity. Cronbach's alpha coefficient was used for that case and the result indicated a good internal consistency of 0.876 which is above the acceptable standard of 0.7.

3.4 Analysis and Model Specification

This section discusses the methodological approaches used to describe the services offered by IPFCs, the level of satisfaction between the two groups of farmers with co-operatives' services, compares co-operative and non-members farmers' access to farming services, and analyses the factors influencing farmers' satisfaction with services provided. Descriptive statistics were used to describe the services offered. Service accessibility level among Irish potato farmers was measured by developing Service Accessibility Index (SAI). The index was derived as follows:

$$SAI = \frac{\sum_{i=1}^{t} p_i}{t * n} * N \qquad (4)$$

where SAI is the Service Accessibility Index, p_i stands for points of a sub-service, t is the number of sub-services, n number of respondents, N is the total number of services. SAI was developed to assess whether Irish Potato farmers were able to improve their accessibility to farming services.

The response weights were yes (1) and no (0). Thereafter, each service was allocated points, and all the points were summed to get the overall scores for service accessibility. The overall scores ranged from 0 to 23. This measure was finally divided into three categories after computing the mean score (5.3), median (5.0), minimum (1.0), and maximum scores (12). In this context, the categories were high service accessibility (5.1

to 23), moderate service accessibility (5.0), and low service accessibility (1.0 to 4.9). It has to be pointed out that the cut-off points were selected using the computed median.

In assessing the level of farmers' satisfaction, the Farmer Satisfaction Index (FSI) was developed using Factor Analysis (FA) with Principal Components Analysis (PCA) method. In developing the index, responses were assigned weights, strongly agree (5), agree (4), undecided (3), disagree (2) and strongly disagree (1). The responses were thereafter subjected to Principal Component Analysis for data reduction. The respective weights from the set of statements were added up and divided by the number of statements that remained after data reduction to develop the index. Orthogonal Varimax (Variable Maximization) rotation was used to identify and group the causes that explain farmers' satisfaction. Variables with communalities greater than 0.5 and components whose Eigenvalue is at least 1 were selected. Finally, variables to merge were found in the Rotated Component Matrix.

$$FSI = \left(\frac{\sum_j x_{ij}}{X_m} \right) (i = 1, 2, \ldots, x; j = 1, 2, \ldots, m) \quad (5)$$

where FSI is the satisfaction index, x_{ij} is the weight by respondent i to statement j on satisfaction, Xm represents the number of statements on each of satisfaction variables after PCA data reduction, and x denotes the total number of respondents.

The level of farmers' satisfaction was determined by calculating the interval size (Adel and Nahed 2016). The interval size $= \frac{5-1}{5} = 0.8$. Levels of satisfaction are presented below.

Strongly Dissatisfied	Dissatisfied	Moderately Satisfied	Satisfied	Strongly Satisfied
[1.00–1.8 [	[1.8–2.6 [	[2.6–3.4 [	[3.4–4.2 [	[4.2–5[

In comparing service accessibility and service satisfaction between the two groups, independent samples t-test was run to check if there is a significant difference between the means in two groups. After that, Eta squared and Cohen's D was applied to determine the magnitude of differences between the two groups of farmers. Eta squared ranges from 0 to 1 and indicates the proportion of variance (Lakens 2013). As proposed by Cohen (1988), this shows how it is interpreted: 0.01 = small; 0.06 = moderate; 0.14 = large magnitude.

$$\text{EtaSquared} = \frac{t^2}{t^2 + (n_1 + n_2 - 2)} \quad (6)$$

t = t-test score, n_1 = sample size of members, n_2 = sample size of non-members. In testing the hypothesis guiding this paper, multiple regression analysis was adopted to determine factors that influence farmers' satisfaction with the services of IPFCs. Before running the model, normality of data was checked using Kolmogorov-Smirnov Test and Shapiro-Wilk Test. The test indicated that the data were not normally distributed. As recommended by Field (2009), data transformation was used to solve the problem.

Therefore, data were transformed to the natural logarithm. Moreover, Tolerance and Variance Inflation Factor (VIF) was checked to explore the presence of multicollinearity and indicated that multicollinearity was not a problem in the model.

The following model was estimated:

$$Y = \beta_0 + \sum_{i=1}^{13} \beta_i X_i + \varepsilon \qquad (7)$$

where Y denotes farmer's satisfaction which is measured in terms of five levels (Strongly Dissatisfied, Dissatisfied, Moderately Satisfied, Satisfied, and Strongly Satisfied), X_i are age, gender, household size, marital status, education qualification, primary occupation, land size, livestock ownership, savings, loan service, training, and non-livestock assets respectively, β_i are regression coefficients, and ε is the error term. Concerning description of variables as specified in the regression analysis (see Appendix Table A2). Qualitative data obtained from KIIs and FGDs were analysed using content analysis. The interview data were transcribed, sorted, and arranged in this case. Subsequently, the information obtained was coded into different themes which were further interpreted into meaningful information.

4 Results and Discussion

4.1 Demographic and Socio-economic Characteristics

Demographic and Socio-economic characteristics of heads of households are summarized in Appendix Table A3. The results indicated significant differences observed between members and non-members, such as age, dependency ratio, and others with p-values less than or equal to 0.05. It is shown that among co-operative members, most of the respondents (69%) were male, whilst 31% were female. With regard to non-members, 77% were male, whereas 23% were female. This result is roughly in accordance with what is revealed in Rwanda Co-operative Agency (2018); 60% of agriculture co-operative members were male, and 40% were female. This is because most women are involved in housework, while men are interested in remunerated work.

Concerning the age of respondents, the current study was conducted to the population with an age group ranging between 16 and 74 years. The youth population (16–30) represents 7%, while the adults (31–74) represent 93% of the total respondents. It was different for non-members, of whom 22% comprised the youth population, while 78% were adults. Many young people are reluctant to engage in agriculture activities (FAO 2018), and most of them do not own land. Co-operative members interviewed (61%) have attended at least primary school; 10% of member respondents have no formal education; only 28% have attended secondary schools, vocational training, and university. In the study area, no significant differences were observed in the level of education between the two groups. This information concurs with what was revealed by Ministry of Agriculture and Animal Resources (2018), which stated that formal education in Rwanda among farmers is still low. The majority of co-operative members (90%) in the study area are married; this is almost similar to non-members (85%). This majority is due to the fact that agriculture is the sector absorbing the biggest part of the Rwandan population, and married people are mostly involved in farming activities, as they are responsible for survival of their families.

Regarding dependency ratio, which describes how much pressure working people face in supporting non-productive group, such as the children and elderly, it was revealed from the study area that the child dependency ratio is 98% or 98 children for every 100 co-operative members and 90.8% for non-members. Conversely, the elderly dependency ratio was 4.2% and 7% for members and non-members, respectively. This indicates that there is a little burden to support older people given that they are very few as the life expectancy is 58 years in Rwanda. It was also reported a total dependency ratio of 102.2% for members and 97.8% for non-members. This percentage still indicates how much pressure working people face in supporting the elderly and the children in the study area. The above percentages are higher than those of the World Bank (2019), which reported the child dependency ratio of 70.3% and 5% for the elderly.

4.2 Service Accessibility Level Among Farmers

As mentioned in the background section, co-operative members are expected to get an extensive range of services above what they can achieve individually at a lower cost than non-members. However, in spite of eminent benefits associated with membership in smallholder farmer co-operatives, not all smallholder farmers join co-operatives. As reported by different researchers, the reasons for not joining co-operative are linked with farmers' previous experience with co-operative mismanagement, high membership fees, which is a major limitation for poor farmers, delayed payment of members' deliveries, lack of trust for the management, meeting obligations and penalty for not showing up and not aware of membership advantages (Kayitesi 2019 & Balgah 2019). This study measured service accessibility level among smallholder farmers by employing Service Accessibility Index as presented in Table 1.

Table 1. Service accessibility level among farmers

Service Accessibility	Co-operative members access level		Non-co-operative members access level		T-test	
	Score Index	Level	Score Index	Level	T	p
Access to agricultural inputs	7.0	High	6.4	High	−5.434	.000
Access to storage facility	1.6	Low	1.4	Low	1.092	0.275
Access to Agri implements	3.4	Low	3.3	Low	−1.756	0.080

(*continued*)

Table 1. (*continued*)

Service Accessibility	Co-operative members access level		Non-co-operative members access level		T-test	
	Score Index	Level	Score Index	Level	T	p
Access to market	13	High	13.2	High	−0.007	0.994
Access to transport	2.6	Low	3.4	Low	−2.439	0.015
Access to finance	2.2	Low	2.5	Low	−0.666	0.506
Access to land	4.8	Low	4.7	Low	−2.663	0.008
Access to market information	10.7	High	9.8	High	−1.438	0.151
Access to extension and training	2.2	Low	1.8	Low	1.436	0.152
Overall access to services					2.123	0.034

Results in Table 1 indicate that the services accessed by farmers interviewed were reported by several studies to be important in farming activities (Lepe 2016; Abebaw and Haile 2013). It is reported from the study that both groups of farmers have highly accessed agricultural inputs such as seeds, fertilizer, and pesticides, with a slight difference in favour of co-operative members. The problem remains the dissatisfaction with cost of inputs, as shown in Table 3. This is explained by small number of co-operatives licensed to sell agricultural inputs in the study area; only three co-operatives out of twenty-three are licensed to sell the inputs to farmers. Co-operative members have scored an index of 7.0, while non-members have a score index of 6.4, which implies that the difference in scores with access to inputs is significant. The results seem to corroborate with a study by Alemayehu (2008), which urged co-operatives to provide credits for agricultural inputs. Hence, members are supposed to have more access to inputs in their farming activities than non-members. However, both groups of farmers complain about the high cost of agricultural inputs compared to the income generated from selling Irish potatoes. One of the farmers in a FGD, elaborated on the issue, saying that: "*agricultural inputs are available to the market, but they are costly; in future, only large farmers will afford them. Our co-operatives fail to help us get the inputs at a reasonable price. As a result, we incur losses, and some farmers have shifted to other crops* (co-operative farmer, 18th September 2019). This caption indicates that even though agricultural inputs are available to farmers, their cost is still higher than the revenue generated for some farmers. Usually, smallholder farmers join co-operatives with the expectation to get inputs at a lower price than other sources. However, as mentioned above, few co-operatives have managed to comply with conditions to be licensed as sellers of agricultural inputs. This has resulted in a market dominated by private traders imposing prices beyond the capacity of a smallholder farmer to afford.

Concerning storage facilities, the accessibility level for co-operative and non-co-operative members is low, 1.6 and 1.4, respectively. This witnessed a challenge for potato farming in Rwanda. Furthermore, none of the IPFCs in the study area owns cold room storage. Consequently, farmers always rush into selling with no storage option even in case of lower prices. This issue was explained by a member of the supervisory committee who said: *"As long as we do not have improved storage facilities to keep our harvests for an extended period, farmers will always be susceptible to exploitation by corrupt traders. We are incurring losses because, during harvest, we rush into selling for any price. We do not have financial capacity to construct improved storage; we need support from Government"* (Member of the supervisory committee, 14th October 2019). In KII with District Co-operative Officer (DCO), he has explained the mechanisms adopted by local Government to mitigate the problem: *"It is our responsibility to bolster co-operative sector; currently we have linked some of the co-operatives with an NGO called Post-Harvest and Agribusiness Support Project (PASP) which has agreed to support in constructing storage facilities, and the activities are in progress"* (DCO, 19th October 2019). The above findings concur with FAO (2018), which reports lack of storage facilities in Rwanda. As a result, farmers sell their production at a low price during harvest to avoid damage.

Observations from the study further show the low level of accessibility to agriculture implements amongst farmers; 3.4 score index for co-operative members and 3.3 for non-members. None of farmers owns tractors or animal traction for cultivation in the study area. In contrast to the above services, co-operative and non-co-operative members in the study area enjoy their market with 13.0 and 13.2 score indices. Extension service and training is also an issue noticed in the study area. Generally, the above findings reveal the low level of service accessibility among co-operative and non-co-operative farmers with a slight difference.

The Independent t-test for overall services in Table 1 provides the p-value of 0.034, which is less than Alpha of 0.05, leading to accept the hypothesis. There is a difference between co-operative members and non-members' access to farming services. To check the magnitude of differences, an eta test was applied and results are presented below:

Eta Squared $= \frac{-2.123^2}{-2.123^2+(394+167-2)} = 0.0080$, indicating small difference in service accessibility between members and non-members.

4.3 Service Accessibility in Co-operatives Compared with Other Sources

Multiple response analysis was used to assess the source of farming services among farmers since farmers can get services from different sources. As presented in Table 2, only 15.3% of members and 11.8% of non-members have obtained agricultural inputs from co-operatives.

Table 2. Service accessibility in co-operatives compared with other sources

Farming services	Co-operative members' access (%)		Non-co-operative members access (%)	
	From co-operative	Others sources	From co-Operative	Other sources
Access to agricultural inputs	15.30	84.70	11.80	88.20
Access to storage facility	15.20	84.80	10.45	89.55
Access to agriculture implements	0.00	100.00	0.00	100.00
Access to market	63.60	36.40	48.60	51.40
Access to transport	7.60	92.40	4.50	95.50
Access to finance	11.93	88.07	0.00	100.00
Access to land	3.70	96.30	0.00	100.00
Access to market information	61.67	38.33	44.90	55.10
Access to extension and training	33.20	66.80	0.85	99.15

This dampens members' enthusiasms from cooperatives that they have joined with the expectation of obtaining services that could not be affordable from other sources. These findings lead to agree with Lepe (2016) who recommends that farmer co-operatives should support smallholder farmers by offering an extensive range of services, including improved access to agricultural inputs.

Despite the ministerial order to sell Irish potatoes through co-operatives, as reported in Table 2, only 63.6% of co-operative members and 48.6% of non-members sell their production through co-operatives. An interviewed co-operative farmer in a FGD has given the reason saying: *"We do not sell to co-operative due to their mode of payment; most of the time they do not have enough cash to pay immediately. Consequently, we prefer selling to private traders when we urgently need money"* (Co-operative farmer" 27th September 2019). This implies limited financial capacity among IPFCs in the study area, which constitutes a serious drawback to satisfaction of members.

In some of the co-operatives, it was observed that even when they have cash at bank, cash withdrawal requires permission from a local government authority, thus delaying co-operative activities. In FGD with board members, one said: *"We are experiencing a big challenge: To withdraw our money from SACCO when we need to carry out any transaction, we are forced to get authorization from Sector Executive Secretary. This delays our activities when he is not in the office to approve. The other issue is that our co-operatives must pay through a bank account; farmers dislike this mode of payment, especially those living far from banks. Consequently, our member farmers and non-members decide to sell through private traders who are ready to pay immediately"*

(Board member, 13th October 2019). This interference of local authorities within the administration of co-operatives is a serious violation of the co-operative principle of autonomy and independence, which is a real indicator of poor management among IPFCs. Comparing both groups in terms of their source of finance, only 11.93% of members have obtained credit through their co-operatives. The results show that none has obtained credit from co-operative among non-members. This is a challenge for members to improve their production since they expect to get credit from their co-operatives at a lower cost than other finance sources. It was also observed in Table 2 that only 33.20 of members and 0.85% of non-members have accessed extension and training through co-operatives. The findings reveal that there is much more yet to be done for farmers to boost their farming practices through provision of due services in accordance with principles and objectives of cooperatives.

4.4 Satisfaction Level Among Irish Potato Smallholder Farmers

In assessing the level of smallholder farmers' satisfaction, the Farmer Satisfaction Index (FSI) was developed. The level of satisfaction was determined by calculating the interval size as mentioned in data analysis and model specification. The satisfaction with agricultural inputs was assessed by acquisition cost, quality and quantity of inputs, and timeliness.

Table 3. Satisfaction level among farmers in Northern and Western Provinces

Service satisfaction	Northern Province		Western Province		T-Test	
	Co-operative members' satisfaction level	Non-Co-operative members' satisfaction level	Co-operative members' satisfaction level	Non-Co-operative members' satisfaction level		
	Index	Index	Index	Index	t	P
Access to agricultural inputs	2.49	2.26	2.56	2.34	0.658	0.511
Access to storage facility	2.25	1.78	1.87	1.70	0.337	0.736
Access to farm infrastructure	1.73	1.18	1.68	1.25	1.524	0.129
Access to market	3.13	2.96	3.65	3.48	2.214	0.028
Access to transport	2.45	1.83	2.36	1.99	-0.374	0.709
Access to finance	1.78	1.70	2.94	2.69	-0.287	0.774
Access to land	1.72	1.64	1.91	1.78	5.529	0.000

(*continued*)

Table 3. (*continued*)

Service satisfaction	Northern Province		Western Province		T-Test	
	Co-operative members' satisfaction level	Non-Co-operative members' satisfaction level	Co-operative members' satisfaction level	Non-Co-operative members' satisfaction level		
	Index	Index	Index	Index	t	P
Market prices	1.74	2.14	1.71	2.27	-3.067	0.002
Access to market information	3.95	3.85	3.47	3.43	3.797	0.000
Extension and training	2.41	1.79	2.64	1.62	2.503	0.013
Overall statistics	**2.36**	**2.11**	**2.48**	**2.25**	**2.657**	**0.008**

As revealed in Table 3, both groups of farmers in both provinces were dissatisfied with agricultural inputs (2.49 and 2.26 score indices for members of co-operative and non-co-operative members respectively in Northern Province compared with 2.56 score index for co-operative members and 2.34 score index for non-co-operative members in Western Province). It was observed that most farmers are dissatisfied with the availability, quality, and cost of inputs (seeds, fertilizer, and pesticides). As shown above, this issue is explained by a small number of co-operatives licensed to sell agricultural inputs in the study area. Similar to non-co-operative members, the members of co-operatives are incurring losses due to high costs and poor quality of inputs. One of the co-operative board members explained why agricultural inputs are costly and suggested the solution: *"The cost of inputs is high compared to revenues from our sales. This is due to lack of competition; only one company in our area is authorized for that business. The authority should remove barriers and allow our co-operatives to enter this business; otherwise, we will continue suffering. We are expected to sell the inputs to our members, but authorities are reluctant to authorize"* (Board member, 23rd September 2019). This implies that few companies in the study area monopolize the sale of agricultural inputs.

It is further noticed in Table 3 that smallholder farmers in both provinces were dissatisfied with storage facilities (2.25 score index for co-operative members and 1.78 score index for non-co-operative members in Northern Province compared with 1.87 score index for co-operative members and 1.70 score index for non-co-operative members in Western Province). As long as there is no intervention to avail improved storage facilities, members will always rush into side-selling to avoid damages. Concerning farm infrastructure, findings also report dissatisfaction among farmers. There are no adequate roads for easy transportation of harvests in some areas. Lack of tractors for cultivation and irrigation facilities constitutes another challenge facing Irish potato farming in Northern and Western Province. Due to the lack of an irrigation system, farmers get losses during heavy rain and drought.

In contrast to the above services, both groups of farmers were satisfied with market for their harvests. However, despite ministerial order requesting all smallholder farmers to sell through co-operatives, some farmers are reluctant, as revealed in Table 2. Several factors explained the reasons, including lack of members' loyalty to their co-operatives. The interviewed respondents said that they were forced to join co-operatives as a condition to sell Irish potatoes, contrary to the co-operative principle of open and voluntary membership (ICA 2006). As a result, most farmers lack co-operative ownership; there is no shared vision, and members are not interested in the growth of their co-operatives. It was also observed that some leaders of co-operatives in the study area sell to private traders; they all blame their co-operative for late payment.

The other factor influencing members' reluctance to sell through co-operative was due to dissatisfaction with the price, as indicated in Table 3. This dissatisfaction was explained by an interview in FGD with one of the board members, saying that: *Farmers are very dissatisfied with the prices of Irish potatoes. MINICOM sets prices, but private traders to whom we sell do not respect that ministerial order. We buy Irish potatoes from members at a price set by MINICOM and we get less than expected when we deliver them to Nzove wholesalers. We thus decide to buy from our members at a lower price to avoid big losses; some members decide to sell to private traders. Furthermore, the price set by MINICOM is low compared to what a farmer expects, considering the cost of inputs. Again, when MINICOM's price is high, private traders abstain, and co-operatives buy from farmers and, subsequently, private traders buy from the co-operatives at a lower price* (Board member, 9th October 2019). Irish potato co-operatives operate in a market like any other business where supply and demand very often dictate the price. During April, October, and November, Irish potato production becomes abundant in the market, resulting in a price decrease, which is sometimes overlooked. Generally, both groups of farmers are dissatisfied with farming services. Mainly, the cost of inputs is very high compared with the revenue earned. Consequently, some farmers in both provinces have decided to exit for other businesses. To be successful, a co-operative is expected to perform its functions and strive to provide services for improved member satisfaction (Liebrand and Ling 2014).

In comparing satisfaction between co-operative and non-cooperative farmers with farming services, the result of independent t-test in Table 3 reports the difference between the two groups. Thereafter, effect size statistics was used to determine the magnitude of differences. The results are presented below:

Eta Squared $= \frac{2.657^2}{2.657+(394+167-2)} = 0.012$. This shows a small difference between the compared groups in terms of satisfaction with co-operative services.

4.5 Regression Results

The main objective of this paper was to determine the demographic and socio-economic factors influencing farmers' satisfaction with co-operatives' services. Multiple linear regression was adopted since all assumptions required were not violated. Appendix Table A4 shows that the independent variables statistically and significantly predict the values of dependent variable, $F(13, 529) = 45.983$, $p(.000) < 0.05$, i.e., the regression model is a good fit of the data.

As revealed by multiple regression output, VIF used to detect multicollinearity among independent variables were less than 10, and all values of tolerance were greater than 0.1, indicating that multicollinearity was not a major problem in the model. Furthermore, results of the regression analysis in Appendix Table A4 indicates that, among demographic and socio-economic factors, only gender of household, livestock ownership, and co-operative membership significantly affected farmers' satisfaction with co-operative services, as their p value < 0.05 and primary occupation of household is significant at 10%.

The results indicate a negative and statistically significant relationship between gender of household and farmers' satisfaction with Irish potato farming services at five percent significant level (p = 0.024). As presented in Appendix Table A3, male and female respondents are 69% and 31%, respectively. Given the small number of female-headed households, the negative relationship shows that females are more effective in managing farming activities than their counterparts in the study area, considering the low level of satisfaction with co-operatives' services observed among farmers. Regarding primary household occupation, it also has a negative and significant relationship with farmers' satisfaction with Irish potato farming services at a 10 percent significant level (p = 0.098). As shown in Appendix Table A3, among heads of households, 99% practice Irish potato farming as their primary occupation. This implies that being restricted to the farming of Irish potatoes negatively affects the access to agricultural inputs since at the time a farmer experience poor production, it limits his/her ability to afford high cost of farming services for the next farming season contrary to the other farmer who adopts crop diversification. According to Elias et al. (2015), practicing off-farm activities to earn additional income helps to afford the expenses of service inputs.

The results also indicated a positive and highly significant relationship between livestock ownership and farmers' satisfaction with co-operative services at a 5 percent significant level (p = 0.010). This implies that households with livestock are more likely to get cash income easily and improve their satisfaction with farming services than non-livestock assets. This is because, apart from manure to improve soil structure and fertility, as well as water retention, farmers can also get money to buy other agricultural inputs for improved farming satisfaction. According to Jabbar (1996), cash income earned in livestock supports purchasing food and farm inputs, such as fertilizers, pesticides, and seeds.

From regression output, co-operative membership has a negative and significant relationship with farmers' satisfaction at a one percent significant level (p = 0.000). The following caption from one of the co-operative members in a FGD explained why this happened: *"Due to lack of financial capacity, our co-operatives do not provide expected services to members; we do not see any benefits from our co-operatives. At least non-members have some choices about where they can sell their harvests. Irish potato co-operatives in our areas fail because they were not formed under the principle of open and voluntary membership; most of us were forced to join these co-operatives"* (Co-operative farmer, 27th September 2019). This is simply because being a cooperative member restricts a farmer from accessing farming services from other sources when they can be obtained from co-operative. The issue especially arises when members want to sell as per ministerial order that restricts their sales to co-operatives as a sole channel, even if the price is lower than prices practiced in the mainstream market.

As shown in Appendix Table A4, loans and savings services among farmers have not significantly affected their satisfaction with co-operative services. This is due to the small number of farmers working with SACCOs and banks. Most of them opt for illegal money lenders, commonly known as Bank Lambert and solidarity tontine, which are informal and unreliable sources of finance, but effective in financing farming activities given their flexibility compared with banks and SACCOs, the latter being mostly faced with liquidity and cash flow problems to provide demand-driven services to farmers. The effect of family size is negatively insignificant, implying that less satisfied farmers have more family members than highly satisfied ones. This is because a large number of family members increases expenses to sustain the family; hence, a hindrance to satisfaction with co-operative services. Age and educational background are not the factors contributing to farmers' satisfaction. This is explained by a large number of older (93%) and a high level of illiteracy among farmers in the study area. According to Elias et al. (2015), older farmers are often reluctant to engage in innovative activities fearing of risk.

4.6 Discussion of the Results

As result of the study, the hypotheses formulated were tested. The independent t-test shows differences between co-operative members and non- members' access to farming services (H$_1$), leading to accept the hypothesis. Surprisingly, an eta test shows a small difference between the two groups when checking the magnitude of differences. This result does not support the previous studies by Abate (2018), Ajah (2015), and Sultana et al. 2020) who found differences between co-operative members and non-members. According to Sultana et al. (2020) and Giagnocavo et al. (2018), co-operative membership provides a more secure market than non-co-operative farmers. Co-operative members have more access to agricultural inputs, loans, storage, and processing equipment than farmers who are not in co-operatives (Ajah 2015). Co-operatives help their members to improve their bargaining power, and market information (Serra and Davidson 2021).

The study also hypothesised that there is a significant difference between co-operative members and non-members' satisfaction with co-operative services (H_2). The result shows differences between the two groups. Furthermore, eta test indicates a small satisfaction difference. The findings of this study do not conform to the study by Morfi et al. (2021) and Morfi et al. (2015) that proved a strong relationship between co-operative membership and satisfaction with farming services. Finally, in determining demographic and socio-economic factors affecting farmers' satisfaction (H_3), results indicate that gender, livestock ownership, co-operative membership, and off-farm income significantly affected farmers' satisfaction with access to co-operatives' services. In contrast, age, household size, marital status, educational qualification, land size, savings, loans, farmers' training, and no-livestock assets do not affect farmers' satisfaction.

The above result concurs with the study by Elias et al. (2015) who reported a positive and significant effect of off-farm income on farmers' satisfaction. Similar to the results of this study, Elias *et al.* further reported that age, education, and training did not significantly affect farmers' satisfaction. However, the findings of this study do not conform to the study by Higuchi et al. (2020), Ma et al. (2018) and Bernard and Spielman (2009) that reported education and farm size as socio-economic characteristics that differentiate satisfied and non-satisfied members, and Elias et al. (2015) who found that family size and credit significantly affect farmers' satisfaction.

In accordance with EDT, when actual performance of products or services does not meet customer's expectation, negative disconfirmation occurs. Findings in this study concur with what is hypothesised by EDT, because the study found that there was farmers' negative disconfirmation, as services offered by IPFCs in the study area did not meet their expectations. Consequently, as noticed, some dissatisfied farmers decided to exit Irish potato co-operatives for other businesses including a shift to other crops.

5 Conclusion and Recommendations

The results of the study show a low level of satisfaction with farming services among farmers in Northern and Western provinces. As observed, nothing can motivate non-co-operative farmers to join IPFCs in the study area since they suffer in the same way as co-operative members in accessing farming services. Nevertheless, Irish potato farmers in Western Province strive to be market-oriented compared to their counterparts in Northern Province, who mostly practice subsistence farming. In general, co-operatives in the area failed to resuscitate their activities, resulting in the exit of Irish potato farming activities for some of the farmers, as reported above. If this problem persists, it will negatively impact the overall production of Irish potatoes in Rwanda.

In the endeavour to improve Irish potato farming and enhance the level of farmers' satisfaction, it is recommended to the IPFCs, on the basis of research findings, to be market-oriented so as to be successful and provide the expected services to members. They should also mobilise their members to work closely with financial institutions to improve their farming activities. Since private traders are the ones enjoying more benefits from Irish potato farming, with government support, co-operatives are finally recommended to change their existing Irish potato market channel by taking control and management of the whole chain of distribution from farm areas through collection centers to wholesale points in the city of Kigali.

It is recommended that the Ministry of Agriculture and Animal Resources provide storage facilities with cold rooms to help IPFCs cope with price fluctuation. Furthermore, Rwanda Agriculture Board is recommended to boost up research on seeds appropriate to a specific area and support Irish potato co-operatives to enjoy the privilege of selling agricultural inputs. On the other hand, Rwanda co-operative Agency is recommended to strengthen IPFCs' capacity building for self-governance to curtail the interference by local authorities within the administration of co-operatives. To deal with inadequate Irish potato seeds, Rwanda Agriculture Board is finally recommended to use the area of *Nyagahinga* in *Butaro* for seed multiplication given its favorable soil.

The findings of this study generate facts to inform IPFCs, community development partners, and policymakers about determinants of the farmers' satisfaction with co-operative services and how they should be improved to attract non-co-operative members instead of being forced to join co-operative as a condition to sell their products. In addition, the paper contributes to the literature by analyzing farmers' accessibility to farming services and satisfaction with co-operative services in developing countries.

Appendix

Table A1. Sampled co-operatives and Probability Proportionate to Size

Province	District	Cooperative	Membership Number	Probability Proportionate to Size (PPS) Members	Non members
Northern	Musanze	BUNYENYERI	412	19	9
		ABASERUKANASUKA	268	13	6
		KABUKA	116	5	2
		KOTEMUSHI	150	7	3
		KOJYAMUGA	95	4	2
	Burera	ISHEMA RY'UMUHINZI	205	10	4
		COAIBGI	71	3	1
		KTMKI	90	4	2
		KOUGIKA	139	7	3
		KOABINYA	65	3	1
		KOAIKAKA	99	5	2
		KOABUTA	833	39	17
		COVMB	1400	66	29
		COOPIGATE	96	5	2
Western	Nyabihu	KOTMUIRU	656	31	14
		KMIRJ	116	5	2
		KOAGIRU	925	44	19
		KOIKAGA	484	23	10
		KOAIGAMU	128	6	3
	Rubavu	IKEREKEZO	961	45	20
		KOKIKA	526	25	11
		KOTUGO	165	8	3
		KOABINYARU	96	5	2
	Total		8096	382	167

Source: Calculated from Secondary data, NCCR (2019)

Table A2. Description of variables as specified in the regression analysis

Variable Category	Variable name (X-covariates)	Variable Description
Demographic and Socio-economic and factors Farmers' Satisfaction with Access to IPFCs Services	Age	Age of respondent (in years)
	Gender	Gender of respondent (1 = male, 0 = female)
	Household size	Household size (in numbers)
	Marital status	Marital status of client (1 = married, 0 = otherwise)
	Educational qualification	Education of respondent (1 = no formal education, 6 = primary education, 12 = secondary education, 13 = vocational training, 15 = tertiary education)
	Primary occupation	Primary occupation of head of household (1 = farming, 0 = others)
	Land size	Land size used for Irish potatoes (in acres)
	Livestock ownership	Livestock ownership (1 = yes, 0 = no)
	Savings	Savings per month (1 = yes, no = 0)
	Loan service	Loan service (1 = yes, no = 0)
	Training	Training (1 = yes, no = 0)
	Membership	Co-operative Membership (1 = yes, no = 0)
	Non-livestock assets (Radio, bicycle, cell phone, TV, motorcycle, hoes, pangas, rakes, spades, axes, slashers, sickles, watering cane, wheelbarrow, ox-ploughs, chemical sprayer, manual irrigation pumps, other agricultural implements.) Access to agricultural inputs Access to storage facility Access to farm infrastructure Access to market Access to transport Access to finance Access to land Market prices Access to market information	Non-livestock assets owned by farmers (1 = yes, 0 = no) Strongly Dissatisfied [1.00–1.8[, Dissatisfied [1.8–2.6[, Moderately Satisfied [2.6–3.4[, Satisfied [3.4–4.2[, Strongly Satisfied [4.2–5[

Table A3. Demographic and socio-economic characteristics of respondents

Variable		Membership				t-test	
		Co-operative members		Non - members		t-value	p-value
		Frequency	%	Frequency	%		
Gender	Male	265	69	128	77	−1.890	0.060
	Female	117	31	39	23		
	Total	382	100	167	100		
Age	16–30	26	7	37	22	11.179	0.000
	31–74	353	93	130	78		
	Total	379	100	167	100		
Education level	No formal education	39	10	14	8	0.779	0.436
	Primary	234	61	115	69		
	Secondary	69	18	28	17		
	Vocation training	22	6	4	2		
	University	18	5	6	4		
	Total	382	100	167	100		
Marital status	Single	37	10	25	15	1.440	0.151
	Married	345	90	142	85		
	Total	382	100	167	100		
Dependency ratio	Child		98		90.8	4.246	0.000
	Aged		4.2		7	2.786	0.000
	Total		102.2		97.8	-1.116	0.095
Primary occupation of head of household	Farming of potatoes	378	99	147	88	4.384	0.000
	Other	4	1	20	12		
	Total	382	100	167	100		
Land size	< 50 acres	112	29	64	38	3.756	0.000
	[50 – 100 acres[	106	28	45	27		
	≥ 100 acres	164	43	58	35		
	Total	382	100	167	100		
Livestock	Yes	312	82	131	78	1.086	0.278
	No	70	18	36	22		

(*continued*)

Table A3. (*continued*)

Variable		Membership				t-test	
		Co-operative members		Non - members		t-value	p-value
		Frequency	%	Frequency	%		
	Total	382	100	167	100		
Savings	Yes	228	59.7	84	50.3	2.030	0.043
	No	154	40.3	83	49.7		
	Total	382	100	167	100		
Loan service	Yes	92	23.4	28	16.7	2.295	0.022
	No	302	76.6	139	83.3		
	Total	382	100	167	100		
Training	Yes	289	75.6	87	52.1	5.300	0.000
	No	93	24.4	80	47.9		
	Total	382	100	167	100		

Source: Survey Data (2019)

Table A4. Demographic and Socio-Economic Factors of Farmers' Satisfaction with access to IPFCs services

Model	Unstandardized Coefficients		Sig	Collinearity Statistics	
	B	Std. Error		Tolerance	VIF
(Constant)	2.698	0.104	0.000		
Age	−0.001	0.001	0.494	0.626	1.598
Gender	−0.052	0.023	0.024**	0.750	1.333
Household size	−0.008	0.005	0.114	0.635	1.574
Marital status	−0.002	0.034	0.961	0.843	1.186
Educational qualification	0.002	0.004	0.559	0.659	1.517
Primary occupation of household	−0.087	0.052	0.098*	0.733	1.364
Land size	0.000	0.000	0.268	0.674	1.484
Livestock ownership	0.072	0.028	0.010**	0.768	1.302
Savings	−0.015	0.019	0.410	0.979	1.022

(*continued*)

Table A4. (*continued*)

Model	Unstandardized Coefficients		Sig	Collinearity Statistics	
	B	Std. Error		Tolerance	VIF
Loan service	−0.017	0.023	0.472	0.910	1.098
Farmers' training	−0.001	0.020	0.976	0.920	1.087
Co-operative Membership	−0.490	0.024	0.000***	0.686	1.458
Non-livestock Assets	0.080	0.109	0.446	0.538	1.860
	The good fit of regression model				
Model	R	R Square	Adjusted R Square	Std. Error of the Estimate	
	0.728[a]	0.531	0.519	0.211	
	Df		F	Sig	
Regression	13		45.983	0.000[b]	

* = Significant at 10%, ** = Significant at 5%, *** = Significant at 1%

References

Abate, G.T.: Drivers of agricultural cooperative formation and farmers' membership and patronage decision in Ethiopia. J. Co-oper. Organ. Manage. **6**(2), 53–63 (2018). https://doi.org/10.1016/j.jcom.2018.06.002

Abebaw, D., Haile, M.G.: The impact of co-operatives on agriculture technology adoption: empirical evidence from Ethiopia. Food Policy **38**(1), 82–91 (2013). https://doi.org/10.1016/j.foodpol.2012.10.003

Adel, L., Nahed, A.: Measuring Farmers' Satisfaction with the Services of Agricultural Service Providers in Minya and BeniSuef Governorates, p. 48. CARE International in Egypt, Cairo (2016)

Ahmed, M.H., Mesfin, H.M.: The impact of agricultural cooperatives membership on the wellbeing of smallholder farmers: empirical evidence from Eastern Ethiopia. Agricult. Food Econ. **5**(6), 1–20 (2017). https://doi.org/10.1186/s40100-017-0075-z

Ajah, J.: Comparative analysis of cooperative and non-cooperatives farmers' access to farm inputs in Abuja, Nigeria. Eur. J. Sustain. Dev. **4**(1), 39–50 (2015). https://doi.org/10.14207/ejsd.2015.v4n1p39

Alemayehu, M.S.: Farmers' Perception on the Effectiveness of Co-operatives in Disseminating Agricultural Technologies in Ethiopia. MSc Dissertation. Sokoine University of Agriculture, Morogoro, Tanzania (2008). 102pp.

Anderson, C., Brushett, L., Gray, T., Renting, H.: Working together to build cooperative food systems. J. Agricult. Food Syst. Commun. Dev. **4**(1), 3–9 (2014). https://doi.org/10.5304/jafscd.2014.043.017

Anderson, E.W., Sullivan, M.W.: The antecedents and consequences of customer satisfaction for firms. Mark. Sci. **12**(2), 125–143 (1993). https://doi.org/10.1287/mksc.12.2.125

Anglade, B., Swisher, M.E., Koenig, R.: The formal agricultural input sector: A missing asset in developing nations. Sustainability **13**(19), 1–19 (2021)

Balgah, R.A.: Factors influencing coffee farmers' decision to join co-operatives. Sustain. Agricult. Res. **8**(1), 42–58 (2019). https://doi.org/10.22004/ag.econ.301852

Barham, J., Chitemi, C.: Collective action initiative to improve marketing performance: lessons from farmer groups in Tanzania. Food Policy **34**(1), 53–59 (2009). https://doi.org/10.1016/j.foodpol.2008.10.002

Bernard, T., Spielman, D.J.: Reaching the rural poor through rural producer organisations? A study of agricultural marketing cooperatives in Ethiopia. Food Policy **34**(1), 60–69 (2009). https://doi.org/10.1016/j.foodpol.2008.08.001

Cochran, W.G.: Sampling Techniques, 3rd edn., p. 448. Wiley, New York (1977)

Cohen, J.: Statistical Power Analysis for the Behavioural Sciences, 2nd edn. Routledge Academic, New York (1988). https://doi.org/10.4324/9780203771587

Cooper, D.R., Schindler, P.S.: Business Research Methods, 9th edn. McGraw-Hill, USA (2006). 744pp.

Creswell, J.W.: Research Design: Qualitative, Quantitative, and Mixed Methods Approaches, 3rd edn. SAGE Publications, London (2009). 295pp. https://doi.org/10.2307/1523157

Elias, A., Makoto, N., Kumi, Y., Akira, I.: Farmers' satisfaction with agricultural extension service and its influencing factors: a case study in North West Ethiopia. J. Agric. Sci. Technol. **18**(1), 39–53 (2015)

Elias, A., Nohmi, M., Yasunobu, K., Ishida, A.: Effect of agricultural extension program on smallholders' farm productivity: evidence from three peasant associations in Highlands of Ethiopia. J. Agricultural Sci. **5**(8), 163–181 (2013). https://doi.org/10.5539/jas.v5n8p163

Etikan, I.: Comparison of convenience sampling and purposive sampling. Am. J. Theor. Appl. Stat. **5**(1), 1–4 (2016)

FAO: Global Forum on Food Security and Nutrition "Youth Employment in Agriculture as a Solid Solution to Ending Hunger and Poverty in Africa: Engaging through Information and Communication Technologies (ICTs) and Entrepreneurship". Food and Agriculture Organization (FAO): Kigali (2018). 9pp.

FAO: Strengthening linkages between small actors and buyers in the Roots and Tubers sector in Africa : Rwanda Work Plan. Food and Agriculture Organization (FAO): Kigali (2015). 15pp.

Franken, J.R.V., Cook, M.L.: Informing measurement of cooperative performance. In: Windsperger, J., Cliquet, G., Ehrmann, T., Hendrikse, G. (eds.) Interfirm Networks, pp. 209–226. Springer, Cham (2015). https://doi.org/10.1007/978-3-319-10184-2_11

Giagnocavo, C., GaldeanoGómez, E., Pérez-Mesa, J.C.: Cooperative longevity and sustainable development in family farming. Sustainability **10**(7), 1–15 (2018)

GoR (2020). Vision 2020. Government of Rwanda (GoR): Kigali. 53pp.

Grashuis, J., Cook, M.L.: A structural equation model of cooperative member satisfaction and long-term commitment. Int. Food Agribus. Manage. Rev. **22**(2), 247–264 (2019). https://doi.org/10.22434/IFAMR2018.0101

Grashuis, J., Cook, M.L.: A structural equation model of cooperative member satisfaction and long-term commitment. Int. Food Agribus. Manage. Rev. **22**(2): 247–263 (2019). https://doi.org/10.22434/IFAMR2018.0101

Grashuis, J., Dary, S.K.: Design principles of common property institutions: the case of farmer cooperatives in the upper West Region of Ghana. Int. J. Commons **15**(1), 60–62 (2021). https://doi.org/10.5334/ijc.1056

Grashuis, J., Su, Y.: A review of empirical literature on farmer cooperatives: performance, ownership and governance, finance, and member attitude. Annals of Public and Cooperative Economics **90**(1), 77–102 (2019). https://doi.org/10.1111/apce.12205

Hansen, M.H., Morrow, J.L., Jr., Batista, J.C.: The impact of trust on co-operative membership retention, performance, and satisfaction: an exploratory study. Int. Food Agribus. Manage. Rev. **5**(1), 41–59 (2002). https://doi.org/10.1016/S1096-7508(02)00069-1

Hellin, J., Lundy, M., Meijer, M.: Farmer organisation, collective action, and market access in Meso-America. Food Policy **34**(1), 16–22 (2009). https://doi.org/10.1016/j.foodpol.2008.10.003

Higuchi, A., Coq-Huelva, D., Arias- Gutiérrez, R., Alfalla-Luque, R..:. Farmer satisfaction and cocoa cooperative performance: evidence from Tocache, Peru. Int. Food Agribus. Manag. Rev. **23**(2), 217–234. https://doi.org/10.22434/IFAMR2019.0166

ILO: Assessment of the potential for cooperative development in selected agriculture sectors in Ethiopia (Amhara and Sidama Regions). International Labour Organization (ILO): Ethiopia (2021). 55pp.

Jabbar, M.: Energy and the evolution of farming system: the potential of mixed farming in the moist Savannah of Sub-Saharan Africa. Outlook Agricult. J. **25**(1), 27–36 (1996). https://doi.org/10.1177/003072709602500106

Kanamugire, J.: Rwanda Irish Potato farmers decry losses as move to curb middlemen fails. The East African of Monday July 31 2017 (2017)

Kayitesi, C.: Determinants of membership and benefits of participation in pyrethrum co-operatives in Musanze District, Rwanda. Master thesis in Agricultural and Applied Economics. University of Nairobi (Kenya) (2019).81pp.

Kothari, C.R.: Research Methodology: Methods and Techniques, 2nd edn., p. 401. New Age International Publishers, New Delhi (2004)

Lakens, D.: Calculating and reporting effect sizes to facilitate cumulative science : a practical primer for t -tests and ANOVAs. Front. Psycol. 1–12 (2013). https://doi.org/10.3389/fpsyg.2013.00863

Lavis, K.R., Blackburn, D.J.: Extension clientele satisfaction. J. Ext. **28**(1), 28–56 (1990)

Lepe, M.: The role of agriculture cooperatives and farmer organizations on the sustainable agricultural practices adoption in Uganda. Master thesis in Rural Development. Ghent University (Belgium) (2016). 96pp.

Liebrand, C.B., Ling, K.C.: Member Satisfaction With Their Cooperatives : Insights From Dairy Farmers. United States Department of Agriculture (USDA) Rural Business Co-operative Programs Research Report. Washington, D.C. (2014). 24pp.

Liu, M., Min, S., Ma, W., Liu, T.: The adoption and impact of e-commerce in rural China: application of and endogenous switching regression model. J. Rural. Stud. **83**, 106–116 (2021). https://doi.org/10.1016/j.jrurstud.2021.02.021

LópezRidaura, S., Masera, O., Astier, M.: Evaluating the sustainability of complex socio-environmental systems. The MESMIS framework. Ecol. Indicat. **2**(2), 135–148 (2002). https://doi.org/10.1016/S1470-160X(02)00043-2

Ma, W., Abdulai, A., Goetz, R.: Agricultural cooperatives and investment in organic soil amendments and chemical fertilizer in China. Am. J. Agr. Econ. **100**(2), 502–520 (2018). https://doi.org/10.1093/ajae/aax079

Ma, W., Zheng, H., Zhu, Y., Qi, J.: Effect of cooperative membership on financial performance of banana farmers in China: a heterogeneous analysis. Annals of Public Cooperative Econ. 1–23 (2021). https://doi.org/10.1111/apce.12326

Marete, M.: The influence of co-operative structure on member commitment, satisfaction and success: the Murang'a nutribusiness co-operative Kenya. Doctoral dissertation. Pennsylvania University (2010). 137pp.

Mbarushimana, J.P.M.: The contribution of Irish potato collection centres in linking potato smallholder farmers to markets. A Case of Musanze District. Master Thesis in Agricultural Production Chain Management. Van Hall Larenstein, University of Applied Sciences (Netherlands) (2018). 50pp.

MINAGRI: Strategic Plan for Agriculture Transformation 2018–24: Planning for Wealth. Ministry of Agriculture and Animal Resources (MINAGRI): Kigali (2018). 235pp.

MINICOM: National Policy on Cooperatives in Rwanda: Toward Private Co-operative Enterprises Business Entities for Socio-Economic Transformation. Ministry of Trade and Industry (MINICOM): Kigali (2018). 53pp.

Morfi, C., Nelson, J., Hakelius, K., Karantininis, K.: Social networks and member participation in cooperative governance. Agribusiness 3(2), 264–265 (2021)

Morfi, C., Ollila, P., Nilsson, J., Feng, L., Karantininis, K.: Motivation behind members' loyalty to agricultural cooperatives. In: Windsperger, J., Cliquet, G., Ehrmann, T., Hendrikse, G. (eds.) Interfirm Networks, pp. 173–190. Springer, Cham (2015). https://doi.org/10.1007/978-3-319-10184-2_9

Mukarugwiza, E.: The hope for rural transformation: A rejuvenating co-operatives movement in Rwanda. International Labor Organization (ILO): Rwanda (2010). 22pp.

Naderifar, M., Goli, H., Ghaljae, F.: Snowball sampling: a purposeful method of sampling in qualitative research. Strides Dev. Med. Educ. 14(3), 1–6 (2017). https://doi.org/10.5812/sdme.67670

NISR: Seasonal Agricultural Survey 2016. National Institute of Statistics of Rwanda (NISR): Kigali (2016). 114pp.

NISR: Seasonal Agriculture survey 2017. National Institute of Statistics of Rwanda (NISR): Kigali (2017). 167pp.

Oliver, R.L.: A cognitive model of the antecedents and consequences of satisfaction decisions. J. Mark. Res. 460–469 (1980). https://doi.org/10.2307/3150499

Oliver, R.L.: Effect of expectation and disconfirmation on postexposure product evaluations: an alternative interpretation. J. Appl. Psychol. 62(4), 480–486 (1977). https://doi.org/10.1037/0021-9010.62.4.480

Osarenkhoe, A., Komunda, M.B.: Redress for customer dissatisfaction and its impact on customer satisfaction and customer loyalty. J. Market Dev. Compet. 7(2), 102–114 (2013)

Pallant, J.: SPSS Survival Manual: A Step by Step Guide to Data Analysis Using the SPSS Program, 4th edn. Allen & Unwin, Berkshire (2011)

Prasertsaeng, P., Routrary, J.K., Ahmad, M.M., Kuwornu, J.K.M.: Factors influencing farmers' satisfaction with the activities of horticultural cooperatives in Thailand. Int. J. Value Chain Manage. 11(1), 42–62 (2020)

RCA: Statistics on Co-operatives in Rwanda. Rwanda Co-operative Agency (RCA): Kigali (2018). pp45

Ruane, M.J.: Essentials of Research Methods: A Guide to Social Science Research. Blackwell Publishing: USA (2006). 312pp.

Seneerattanaprayul, J., Gan, C.: Effects of agricultural co-operative services on rural household welfare in Thailand. Int. Soc. Sci. J. 71(6), 1–18 (2021). https://doi.org/10.1111/issj.12277

Serra, R., Davidson, K.A.: Selling together: the benefits of cooperatives to women honey producers in Ethiopia. J. Agric. Econ. 72(1), 202–223 (2021). https://doi.org/10.1111/1477-9552.12399

Singh, K., Misra, M., Kumar, M., Tiwari, V.: A study on the determinants of financial performance of U.S. agricultural co-operatives. Journal of Business Economics and Management, 20(4), 633–647 (2019). https://doi.org/10.3846/jbem.2019.9858

Spreng, R.A., Page, T.J.: A test of alternative measures of disconfirmation decision sciences. Decis. Sci. 34(1), 31–62 (2003). https://doi.org/10.1111/1540-5915.02214

Sultana, M., Ahmed, J.U., Shiratake, Y.: Sustainable conditions of agriculture cooperative with a case study of dairy cooperative of Sirajgonj District in Bangladesh. J. Co-oper. Organ. Manage. 8(1), 100–105 (2020). https://doi.org/10.1016/j.jcom.2019.100105

Tarekegn, M.: Factors contributing for members' satisfaction with their co-operatives: the case of co-operative in South Wollo Zone, Ethiopia. Eur. J. Bus. Manage. 9(25), 20–28 (2017)

Taro, Y.: Statistics: An Introductory Analysis, 2nd edn., p. 919. Harper and Row, New York (1967)

Terry, B.D., Israel, G.D.: Agent Performance and customer satisfaction. J. Ext. 42(6) (2004)

World Bank. Age dependency ratio, old (% of working age population) (2019). https://data.worldbank.org/indicator/SP.POP.DPND.OL. Accessed 24 Nov 2021

Zheng, H., Ma, W., Wang, F., Li, G.: Does internet use improve technical efficiency of banana production in China? Evidence from a selectivity-corrected analysis. Food Policy, **102**, 1–12 (2021). https://doi.org/10.1016/j.foodpol.2021.102044

Zheng, S., Wang, Z., Song, S.: Farmers' behaviours and performance in co-operatives in Jilin Province of China. Soc. Sci. J. **48**(3), 449–457 (2011). https://doi.org/10.1016/j.soscij.2011.05.003

Determining Factors Influencing Out-of-Pocket Health Care Expenditures in Low- and Middle-Income Countries: A Systematic Review

R. Muremyi[1,4(✉)], D. Haughton[2,3], F. Niragire[4], and I. Kabano[4]

[1] African Centre of Excellence in Data Science, University of Rwanda, Kigali, Rwanda
muremyiroger@gmail.com
[2] Mathematical Sciences and Global Studies, Bentley University, Boston, USA
[3] Université Paris 1 Panthéon Sorbonne (SAMM), Université Toulouse 1(TSE-R), Paris, France
[4] Department of Applied Statistics, University of Rwanda, Kigali, Rwanda

Abstract. Purpose: This study determines the factors influencing the increase in out-of-pocket medical costs in low- and middle-income countries from the reviewed literature. As there is little relevant information available on health expenditure, the findings will help identify research gaps and provide the basis for future studies on rising healthcare costs in developing countries. Failure to address this situation may threaten sustainable development goals.

Methodology: A systematic literature search was used to identify relevant studies on out-of-pocket medical expenditure papers published from 2000 to 2022 were included. Of the 3,933 papers on out-of-pocket medical costs in developing countries found from the University of Gothenburg Library through popular databases such as CINAHL, and GUPEA, JUNO, Mediearkivet, NE.se, PubMed, and Scopus, only 14 papers meeting the criteria were included in this study. Relevant literature has been searched by using the following keywords: out-of-pocket, and medical expenses. In research, Boolean terms are used to separate keywords (AND; OR). In our search for eligible studies, health expenditure has been used, PubMed was selected as the one containing the papers under study, and Rayyan software was used as tool for inclusion and exclusion of the selected papers.

Findings: Several factors have been identified as having an impact on the rise in out-of-pocket medical costs. However, vulnerable groups who do not understand insurance systems, occupational status, supplemental health insurance, rural residents, people with disabilities, people with chronic illnesses, and families with elderly people, and female-headed households, distance to health facilities, transportation are all factors that contributed to the increase of direct payment of medical expenditure.

Research Limitations: The classification of countries is evolving, and the World Bank index is continuously updated. A country that was categorized as low- and middle-income in 2000, for instance, was categorized as a high-income country in 2022. This restriction makes it crucial to continually look at these nations whose economic classifications have changed.

Practical Implications: The results will inform both governments and health policymakers to fight against the factors that are pushing the increase of healthcare

© The Author(s), under exclusive license to Springer Nature Switzerland AG 2023
C. Aigbavboa et al. (Eds.): ARCA 2022, *Sustainable Education and Development – Sustainable Industrialization and Innovation*, pp. 441–450, 2023.
https://doi.org/10.1007/978-3-031-25998-2_32

spending worldwide, especially in developing countries and put in place the strategies of removing those factors in order to achieve the universal health coverage as recommended by world health organization.

Originality/Value: Little research has been conducted to date on the factors that influence the direct payment of health care services for in the developing countries. The findings of this research will contribute to the understanding of the features influencing rising healthcare costs in developing countries.

Keyword: Health care · Health expenditure · Spending · Systematic review · Rwanda

1 Introduction

Out-of-pocket payment (OOPP) refers to the amount a patient pays for health care services not covered by a health insurance scheme (Brinda et al. 2014). The World Health Organization (WHO) defines out-of-pocket medical payments as payments made directly by an individual for medical services that are not covered by health insurance (Garg et al. 2009). However, personal spending may include deductibles, coinsurance, copayments, and costs for non-covered medical services. Disaster Health Expenditure (DHE)represents out-of-pocket spending (OOP) on health care above a household-specific expenditure threshold where a household's income or ability to pay (Adam et al. 2018). There is no consensus on the threshold above which medical costs are considered catastrophic (Bredenkamp et al. 2011) define DHE as Expenses that exceed 10% of the household's monthly earnings.

The World Health Organization (2000) defined disaster of out of pocket health expenditure for direct spending that exceeds 40% of net household income that are used for health services. The creation of a financial protection program aims to pool funds to protect individuals from catastrophic medical costs. However, public health insurance reduces households' vulnerability to high out-of-pocket costs in case of illness by reducing direct health care costs and reducing income due to poor health (Aimable 2008). Direct Payment is generally considered as the least preferred payment method for healthcare. There is no pooling risk and cross-subsidization among people with different healthcare needs. People with greater health care needs bear the heaviest financial burden, regardless of their ability to pay.

Direct payments can also put households at risk of catastrophic spending on health care services. This is a situation where a household spends a large portion of its income on health care, to the detriment of other needs such as children's clothing and education, expenses in food. For intervention purposes, the various stakeholders must be well informed on how OOPP affects different populations, health care systems, localities, and genders because without such information, Sustainable Development Goals cannot be achieved (Koomson et al. 2021).

The achievement of universal health coverage for all people worldwide is one of the Sustainable Development Goals. However, direct payment for health care spending for some households are the most useful source of access to health care, but little is known about their ability to predict the burden associated with health care spending patterns.

Therefore, a healthy population is more productive and the government needs to legislate to reduce health care costs (Yang 2020). In this regard, all governments have tried to develop policies to protect households from the financial crisis that can be caused by the use of money spent on health services (Liu et al. 2019; Woldemichael et al. 2016).

Households pay to get medical care by taking out health insurance and paying directly when using the health care service. Direct payments are generally considered as the least preferred payment method for healthcare. This is the fact that there is no risk of pooling or cross-subsidization among people with different health needs. People with higher health needs bear the heaviest financial burden, regardless of their ability to pay. Therefore, payment of medical expenses is not fair. Direct payments also expose households to disaster risk costs. Households spend most of their income on health services rather than on other needs such as clothing and education for children in this situation. Furthermore, total direct spending as a percentage of private health care costs in 2013 in the East African region was shocking. In Burundi it was 44.7%, Kenya 75.5%, Rwanda 44.6%, Tanzania 52.1% and in Uganda it was 54% (WHO 2019).

According to the above statistics, the less GDP a country spends on health care, the more it spends on direct payments between people. This means that Kenyan households are the most responsible for medical expenses. There is a current tendency to run out of health insurance during and after hospitalization for treatment and before the actual diagnosis of the problem is made if it is obvious, that most household income is spent on medical expenses. Some individuals reach out to their pockets or mobilize family, friends, and relatives to cope with the sudden increase in treatment costs. Sustainable Development Goal 1 is to "eradicate all forms of poverty by 2030." This goal cannot be achieved if households spend all their savings and assets on health care. Due to the current non-infectious epidemic and the COVID-19 pandemic, many households using self-paying medicines are economically devastated.

Research done by (Gharibi et al. 2021) on *"the increase of healthcare costs of multiple sclerosis patients in Iran"* show that hospitalized patients have HEC and 44% live in poverty due to OOP. While occupation status, having additional health insurance, and living in Tabriz all had a significant impact on direct health spending (OPS).

This study identifies evidence on factors influencing the increase in out-of-pocket health costs in developing countries from the reviewed literature. As there is little relevant information available, the findings will help identify research gaps and provide the basis for future studies on rising healthcare costs in developing countries.

2 Materials and Methods

The study conducted systematic reviews of the publication of heath care payments (from 2000 to 2022). The 22 years was purposely selected because they involve MDGs and SDGs. In fact, low and middle-income countries were among the few countries that significantly achieved the MGDs (Koomson et al. 2021). From 2016 and onwards, the country embarked on the SDGs implementation. Therefore, the selection of 22 years aims at capturing changes that may have happened to update knowledge to inform actors in the sector of health specifically in the area of out-pocket medical payment in selected countries. Even though in developing countries is half-way implementing

SDGs, pro-active lessons from this study are needed to provide critical advice to remedy shortfalls.

2.1 Inclusion and Exclusion Criteria

The following are criteria basing on which paper to review were selected

i. Scope of the research: papers focusing on out of pocket medical expenditure
ii. The setting: Selected countries in developing countries.
iii. Type of publications: the publication targeted concerned papers produced from the period selected. It involves editorials in the area of out of pocket medical expenditure.
iv. Language: English was language in which publications were considered

2.2 The Process of Searching and Selecting the Studies

The systematic review was done through the University of Gothenburg Library popular databases such as CINAHL, GUPEA, JUNO, Mediearkivet, NE.se, PubMed, and Scopus and PubMed was selected as the one contains the papers under study and Rayyan software was used as tool of selection of appropriate papers. In all search engines, the study used search strings such as "out of pocket" AND "Out of pocket payment" OR medical expenditure* OR healthcare Payment AND in low and middle income countries. Search scopes were peer reviewed journals and other scholarly publications from 2000 to 2022.

2.3 Data Extraction

Data extraction form was created to collect information about the four key attributes.

a) Paper identification: study citations, study locations, participant characteristics
b) Methodology: Research design such as research type, sampling method, sample size.
c) Concept: How to conduct the research concepts of individual with medical expenses and definition of healthcare
d) Challenges in out of pocket medical payments: Retained attributes are attributes with the increased out of pocket health expenditure to investigate.

2.4 Analysis Approach

To identify Factors related to direct payment of health care services, the study examined all variables identified in previous studies. Rayyan software was used to group them according to their similarity rating. To look into groups that cited numerous studies on similar topics such as (Bredenkamp et al. 2011; Brinda et al. 2014; Cylus et al. 2018; Flores et al. 2008; Mahumud et al. 2020; Woldemichael et al. 2016). To decide which paper to retain, we interactively worked in the interface of the Rayyan software guided by attributes such as "included", "decided", and 'undecided".

2.5 Search Strategy

A systematic literature search was used to identify relevant studies on direct payment of health care expenditure papers published between 2000 and 2022 were included. Data was obtained from the University of Gothenburg Library through popular databases such as CINAHL, GUPEA, JUNO, Mediearkivet, NE.se, PubMed, and Scopus. Policy guidelines were searched on various websites, including government and WHO websites, regarding compliance with out-of-pocket healthcare costs in developing countries. We searched for relevant literature using the following keywords: out-of-pocket, and medical expenses. In research, Boolean terms are used to separate keywords (AND; OR). In our search for eligible studies, we used Medical expenditure. We have included studies in English (Fig. 1).

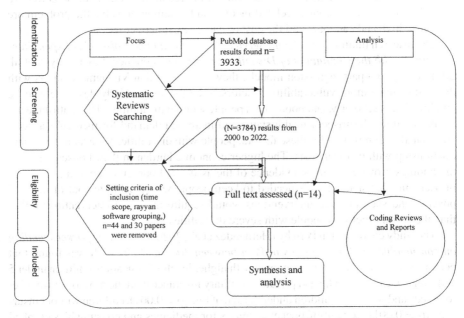

Fig. 1. Diagram of the PRISMA process flow for searching and selecting literature

3 Results

The search yielded 3933 articles from the PubMed database retrieved documents from 2000 to 2022. Only 3784 articles remain based on the criteria for the selected starting year of 2000, and only 14 articles remain based on the inclusion criteria, titles and summaries of all 14 articles to comply with the inclusion criteria. 14 articles were critically analyzed to understand the phenomenon of factors related to direct payment of health expenditure in the said countries. The last 14 scientific journal article publications are still available and all 14 papers were quantitative studies.

Research done by (Schmid 2017) on *"out-of-pocket spending and financial protection in the Chilean healthcare sector: a systematic review"*. Their findings show that household health care costs exceed 30% of households in 2012, Less than 1% of the total population lives in poverty due to lack of money spent on health care services, and 4% of Chilean households are demolished due to medical cost. Their findings highlight the most vulnerable groups and recommend additional research to better understand insurance mechanisms and make reforms for better health services for citizens (Schmid 2017).

(Dhankhar et al. 2021) conducted a study on *"out-of-pocket spending, disaster health spending, and funding for Non-communicable diseases in India: a systematic review"* with meta-analysis. Their results estimated that the proportion of individuals experiencing HEC was 62.7%. The most common methods of funding cancer treatment are borrowing money and selling assets, and one of the recommendations made is to have control policies, cancer control-related income and treatment to solve the problem are obviously high and cannot afford the cost of cancer treatment costs.

The research findings on *"Health Service Utilization and Out-of-Pocket Expenditure Associated with the Continuum of Disability in Vietnam"* conducted by (Nguyen et al. 2021) Using two-part regression models, they discovered that Vietnamese adults with disabilities have more vulnerabilities factors, such as being elderly, less likely to be employed, less scholarly, and poorer than people without disabilities. These attribute are connected through dirt-poor health status and increased demand for medical services, but even after controlling for these factors, people with disabilities have an independent relationship with medical costs. The higher economic burden in their households, and the findings provide empirical evidence of the associated burden of ongoing disability in Vietnam. Decisive action is needed to keep people with disabilities out of medical poverty, and such targeted interventions would require moderate interventions rather than the current focus on people with severe disabilities.

The study conducted in Peru by (Hernández et al., 2020) on *"Out-of-pocket spending on consumables and medications in Peru between 2007 and 2016"*. It was found that persons in coastal areas, women, people with higher levels of education, children under 5 and over 60, and those who had private or military insurance paid the most out-of-pocket for goods and services. Children under the age of five ($p = 0.001$), children without insurance ($p = 0.001$), and out-of-pocket expenses for medicines and consumables climbed significantly among children with insurance between 2007 and 2016, making them one of the quintuplets with the greatest per capita spending. People without chronic diseases ($p = 0.001$), Seguro Integral de Salud ($p = 0.001$), or members of the armed forces ($p = 0.035$), as well as those who live in urban and rural areas (both $p = 0.001$).

The study conducted in Brazil by (Faustino et al. 2020) on *"the income and out-of-pocket health expenses of families with older individuals in Brazil"* demonstrate that In comparison to households without an elderly head of household and even more so in households without elderly members, the proportion of out-of-pocket health expenses by income group per capita is higher for families with elderly members, households with elderly members, or husband and wife as head of household.

The study conducted in Saudi Arabia (Hanawi 2021) on *"Decomposition of inequalities in out-of-pocket health expenditure burden in Saudi Arabia"* their study's goal was to identify and analyze the factors of OOP health cost inequality in Saudi Arabia. Their findings revealed that in Saudi Arabia, relative OOP health care costs are concentrated in the poor, putting a heavy burden on the poor and those aged 60 and over, as well as those with low education, while factors that increase the burden on the wealthy include gender, secondary education, higher education, and health for men under 60. To ensure the overall well-being of the poor, efforts to reduce OOP health care costs must take into account certain factors that burden the poor, such as the elderly and a lack of education.

According to the research conducted by (Herberholz and Phuntsho 2021) on "*Medical, transportation and spiritual out-of-pocket health expenditure on outpatient and inpatient visits in Bhutan*" Demographic, socioeconomic, geographic, and morbidity-related factors were found to influence the likelihood of OOP due to medical, transport, and psychological reasons, as well as hospital costs. Economic burden is most strongly perceived by respondents in rural areas and those with a high need for health services.

Study conducted by (Koomson et al. 2021) on "*Empirical information on the relationship between financial inclusion and out-of-pocket health expenditure*" is scarce, and it was found that the majority of studies conducted in Ghana to date did not use multidimensional measures of financial inclusion. The results show that an increase in the standard deviation of financial inclusion is associated with an increase in the standard deviation. in household out-of-pocket health care costs, ranging from 0.1367 to 1.7608. This conclusion is particularly prominent in homes with female heads and in metropolitan areas.

The findings from (Thanh et al. 2019) on "*Out-of-Pocket Medical Expenses Among Patients in Vietnam with Insurance and Without Insurance*". According to the estimated coefficients of the insurance state variables in the ordinary least squares model, health insurance contributed to the reduction of direct payment of health care services for those with health insurance by 31.1 percent for outpatient care and 31.5% for inpatient care when compared to those without health insurance. And outpatients care with health insurance reduced direct payment by 42.3 percent and 20.2 percent for those enrolled for the community health and health facilities of the districts, respectively. Insurance reduced direct payments expenses for inpatient care by 34.9 percent when compared to those who did not have insurance (Thanh et al. 2019).

The research conducted by (Yusuf and Leeder 2020) on "*estimates of out-of-pocket health care spending in Australia*", the average annual OOP expenditure was A$4290 per household, accounting for 5.8 percent of total spending on goods and services. Despite not being a direct health-care expense, private health insurance (PHI) premiums made up 40.6 percent of all OOP costs. Nearly half of the remaining 59.4% went to physicians and other healthcare providers, and about one-third went to pharmaceuticals. Dental procedures and specialist consultations cost the most in terms of out-of-pocket costs, while visits to general practitioners cost the least. Compared to those without insurance, households with private health insurance spent four times as much on medical expenses.

According to (Roger Muremyi et al. 2020) on "*predicting the out of pocket health expenditure in Rwanda*" they found that the direct payment of health-care services has increased from 24.46 percent in 2000 to 26 percent in 2015 and The treenet model predicted out-of-pocket health expenditure with high accuracy, and the variables total consumption, ages of the head of households, and members of the households all contributed significantly to the increase in out-of-pocket in Rwanda.

According to the research findings of (Barennes et al. 2015) on *"Evidence of High Out of Pocket Spending for HIV Care Leading to Catastrophic Expenditure for Affected Patients in Lao People's Democratic Republic"*. The findings indicated that the main causes of increase of OOPs were transportation, travel time, and the distance to healthcare facilities. However, the households attended the provincial hospital were linked to the use of lower money on healthcare services compared to those attended other health facilities.

According to the research conducted by (Briesacher et al. 2010) on *"Out-of-pocket burden of health care spending and the adequacy of the Medicare*: Part D low-income subsidy" their findings revealed that the proportion of households around 26% 26% had high medical costs.

4 Discussion

This is, as far as we know, the first scoping review that focuses on it direct payment of health services in developing countries. Concerning the factors influencing the rise of direct payments of medical expenses in low and middle-income countries. Several factors have been identified as having an impact on the rise in direct payment of medical costs. However, vulnerable groups who do not understand insurance systems, occupational status, supplemental health insurance, rural residents, people with disabilities, people with chronic illnesses, families with elderly people, and female-headed households are all factors that contributed to the increase of "out of pocket health expenditure" (Dhankhar et al. 2021; Faustino et al. 2020; Gharibi et al. 2021; Hernández et al. 2020; Karim et al. 2021; Koomson et al. 2021; Nguyen et al. 2021; Lee 2021; Yusuf & Leeder 2020). As previously stated in previous studies on out-of-pocket health care costs, a lack of education, transportation, total consumption, age of heads and members, and length of stay all contributed to an increase of out-of-pocket health expenditure(Al-Hanawi 2021; Herberholz & Phuntsho 2021; Roger Muremyi et al. 2020; Shin & Lee 2021; Thanh et al. 2019). A systematic literature search was used to identify relevant studies on "out-of-pocket health expenditure" in order to answer our research question. Data was obtained from the University of Gothenburg Library through popular databases such as CINAHL, GUPEA, JUNO, Mediearkivet, NE.se, PubMed, and Scopus. Relevant literature have been searched by using the following keywords: out-of-pocket and medical expenses. In research, Boolean terms are used to separate keywords (AND; OR). In our search for eligible studies, we used health expenditure. The World Bank index is constantly updated and the country classification is changing. For example, you can see that a country that was classified as a low- and middle-income country in 2000 was classified as a high-income country in 2022. Due to this limitation, it is important to constantly investigate these countries whose categories have changed.

5 Conclusions

Results showed that household sociodemographic characteristics, household location, and health insurance all influenced direct payment of health care costs in low- and middle-income countries. Despite the identification of the aforementioned factors, studies on direct payment of health care costs in low- and middle-income countries are limited. This study highlights the importance of addressing the factors that have led to sustained increases in out-of-pocket medical costs in general.

References

Adam, W., et al.: Progress on catastrophic health spending in 133 countries: a retrospective observational study. Lancet Global Health **6**, 169–179 (2018)

Aimable, T.: Sharing the burden of sickness: Mutual health insurance in Rwanda. World Health Organazation **86**, 823–824 (2008)

Hanawi, M.K.: Decomposition of inequalities in out-of-pocket health expenditure burden in Saudi Arabia. Soc. Sci. Med. (2021). https://doi.org/10.1016/j.socscimed.2021.114322

Barennes, H., Frichittavong, A., Gripenberg, M., Koffi, P.: Expenditure for affected patients in Lao People's Democratic Republic. PLoS ONE (2015). https://doi.org/10.1371/journal.pone.0136664

Bredenkamp, C., Mendola, M., Gragnolati, M.: Catastrophic and impoverishing effects of health expenditure: New evidence from the Western Balkans. Health Policy Plan. **26**(4), 349–356 (2011)

Briesacher, B.A., et al.: Out-of-pocket burden of health care spending and the adequacy of the Medicare Part D low-income subsidy (2010). https://doi.org/10.1097/MLR.0b013e3181dbd8d3

Brinda, E.M., Andrés, R., Enemark, U.: Correlates of the out-of-pocket and catastrophic health expenditure in Tanzania: Results from a national household survey. BMC Int. Health Hum. Rights **14**, 5 (2014)

Cylus, J., Thomson, S., Evetovits, T.: Catastrophic health spending in Europe: Equity and policy implications of different calculation methods. Bull. World Health Organ. **96**, 599–609 (2018)

Dhankhar, A., Kumari, R., Yogesh, B.: Out-of-pocket, catastrophic health expenditure and distress financing on non-communicable diseases in India: a systematic review with meta-analysis. Asian Pacific J Cancer Prevent. APJCP 22(3), 671–680 (2021). https://doi.org/10.31557/APJCP.2021.22.3.671

Faustino, G, Levy, RB, Canella, DS, Oliveira, C, Novaes, HMD: Income and out-of-pocket health expenditure in living arrangements of families with older adults in Brazil. Cad Saude Publica (2020). https://doi.org/10.1590/0102-311X00040619

Flores, G., et al.: Coping with health-care costs: implications for the measurement of catastrophic expenditures and poverty. Health Econ. **17**(12), 1393–1412 (2008)

Garg, C.C., Karan, A.K.: Reducing out-of-pocket expenditures to reduce poverty: a disaggregated analysis at rural-urban and state level in India. Health Policy Plan **24**, 116–128 (2009)

Gharibi, F., Semnan, A., Dalal, K.: The catastrophic out-of-pocket health expenditure of multiple sclerosis patients in Iran. BMC Health Serv. Res. (2021). https://doi.org/10.1186/s12913-021-06251-4

Herberholz, H., Phuntsho, S.: Medical, transportation and spiritual out-of-pocket health expenditure on outpatient and inpatient visits Bhutan. Soc. Sci. Med. (2021). https://doi.org/10.1016/j.socscimed.2021.113780

Hernández, H.-V., Varga-Fernández, R., Magallanes, L., Bendezu, G.: Out-of-pocket expenditure on medicines and supplies in Peru in 2007 and 2016. Medwave (2020). https://doi.org/10.5867/medwave.2020.02.7833

Karim, M.A., Singal, A.G., Ohsfeldt, R.L., Morrisey, M.A., Kum, H.C.: Health services utilization, out-of-pocket expenditure, and underinsurance among insured non-elderly cancer survivors in the United States, 2011–2015. Cancer Medecine (2021). https://doi.org/10.1002/cam4.4103. Epub 2021 Jul 30

Koomson, I., Abdul, A., Abbam, A.: Effect of financial inclusion on out-of-pocket health expenditure: Empirics from Ghana. Eur. J. Health Econ. (2021). https://doi.org/10.1007/s10198-021-01320-1

Liu, K., Subramanian, S.V., Lu, C.: Assessing national and subnational inequalities in medical care utilization and financial risk protection in Rwanda. Int. J. Equity Health (2019). https://doi.org/10.1186/s12939-019-0953-y

Mahumud, R.A., Sarker, A.R., Sultana, M., Islam, Z.: Distribution and determinants of out-of-pocket healthcare expenditures in Bangladesh. Pan Afr. Med. **37**, 55 (2020)

Nguyen, L., Lee, J.T., Hulse, S.G., Hoang, M.V., Le, D.B.: Health service utilization and out-of-pocket expenditure associated with the continuum of disability in Vietnam. International Journal of Environmental Research and Public Health (2021). https://doi.org/10.3390/ijerph18115657

Muremyi, R., Haughton, D., Kabano, I., Niragire, F.: Prediction of out-of-pocket health expenditures in Rwanda using machine learning techniques. Pan Afr. Med. 37, 357 (2020). https://doi.org/10.11604/pamj.2020.37.357.27287

Schmid, A.: Out-of-pocket expenditure and financial protection in the Chilean health care system-a systematic review. Health Policy (2017). https://doi.org/10.1016/j.healthpol.2017.02.013

Shin, S.M., Lee, H.W.: Comparison of out-of-pocket expenditure and catastrophic health expenditure for severe disease by the health security system: Based on end-stage renal disease in South Korea. Int. J. Equity Health (2021). https://doi.org/10.1186/s12939-020-01311-3

Thanh, N.D., Anh, B.T.M., Xiem, C.H., Van Minh, H.: Out-of-pocket health expenditures among insured and uninsured patients in Vietnam. Asia-Pacific J. Public Health (2019). https://doi.org/10.1177/1010539519833549

WHO: Naming the coronavirus disease (COVID-19) and the virus that causes it (2019). www.who.int/emergencies/diseases/novel-coronavirus-2019/technical-guidance/naming-the-coronavirus-disease-(covid-2019)-and-the-virus-that-causes-it

Woldemichael, A., Daniel, Z.G., Shimeles, A.: Community-based health insurance and out-of-pocket healthcare spending in Africa: evidence from Rwanda. African Development Bank Group, 9922 (2016)

Yang, X.: Health expenditure, human capital, and economic growth: an empirical study of developing countries. Int. J. Health Econ. Manag. 163–176 (2020). https://doi.org/10.1007/s10754-019-09275

Yusuf, F., Leeder, S.: Recent estimates of the out-of-pocket expenditure on health care in Australia. Aust. Health Rev. (2020). https://doi.org/10.1071/AH18191

Are They Really that Warm: A Thermal Assessment of Kiosks and Metal Containers in a Tropical Climate?

L. A. Nartey[✉], M. Agbonani, and M. N. Addy

Department of Construction Technology and Management,
Kwame Nkrumah University of Science and Technology, Kumasi, Ghana
nlordaaron@gmail.com

Abstract. Purpose: In Ghana, makeshift structures such as kiosk and metal containers have been increasing in their share of use as dwellings and currently makes up to 20% of dwellings. The growing urbanization in Ghana points to a further increase in these structures. It is generally known that such structures are of poor quality however, little is known about their thermal comfort conditions neither is there any data on them. To this end, this study sought to determine the thermal comfort of residents of makeshift structures.

Design/Methodology/Approach: Field measurement and questionnaire survey were adopted for the study. Field measurements of the study included indoor and outdoor thermal parameters across six selected makeshift structures in Kumasi. A humidity/temperature datalogger was mounted in the various structures to collect data regarding the environmental factors of thermal comfort. A handheld anemometer was also used to record one of the variables. Another set of data was collected using a questionnaire.

Findings: The results from the field measurements show an average temperature of 29.73 °C and an average relative humidity of 66.7% indoors in the selected study structures. Also, survey participants indicated a general dissatisfaction of the conditions and expressed preference for cooler conditions. The high dissatisfaction expressed towards thermal conditions in this study demonstrates that occupying makeshift structures may be thermally unhealthy as temperatures are almost 2 degrees higher than recommended.

Implications/Research Limitations: This data collection period for this study was for a relatively short period. Further data can be collected for longer periods for assessment. Also, subsequent studies can explore thermal comfort using adaptive models and their health implications on the occupants.

Originality/Value: The findings of this study have brought to light statistics surrounding thermal comfort in the makeshift structures. With this information, alternative forms of affordable and more comfortable infrastructure can be explored. This paper provides empirical evidence on thermal conditions in makeshift structures in a tropical warm-humid climate.

Keyword: Climate · Heat assessment · Indoor environmental quality · Makeshift structures thermal comfort

© The Author(s), under exclusive license to Springer Nature Switzerland AG 2023
C. Aigbavboa et al. (Eds.): ARCA 2022, *Sustainable Education and Development – Sustainable Industrialization and Innovation*, pp. 451–463, 2023.
https://doi.org/10.1007/978-3-031-25998-2_33

1 Introduction

One of the principal and primary functions of a building is to provide comfort to occupiers without compromising their health or performance (Thapa and Panda 2015). Indoor thermal comfort has always been very important in building design but recently, it has become an even more critical component of a building's function (Van Hoof 2008; De Dear et al. 2020; Ozarisoy 2022). Trends in global warming and heatwave occurrence warn of an increased discomfort indoors buildings designed for the present (Heracleous and Michael 2018; Chapman et al. 2019: Perkins-Kirkpatrick and Lewis 2020). Unless some drastic measures are taken to curb the anthropogenic effects on the climate, global temperatures will only get warmer (Intergovernmental Panel on Climate Change (IPCC) 2021; IPCC 2022). There's sufficient evidence that indoor thermal environment of a building is greatly influenced by outdoor environmental conditions (Kownacki et al. 2019). An implication of a warmer climate will be the need for further cooling to make indoor environment comfortable in the warmer regions of the world (Ahmed et al. 2021). What this means is that, buildings will be required to be more efficient in filtering the harsh outdoor conditions, hence the requirement of climate responsive building designs (Cui and Overend 2019; Osman and Sevinc 2019). Many forms of low-income houses however may fall short in this regard. According to Akçabozan and Demir (2015), low-income houses have some factor of incompleteness. They are produced with cheap and readily available materials as opposed to the require materials (Godswill et al. 2016). In African urban slums, a very popular form of low-income houses are makeshift structures (Danso-Wiredu and Midheme 2017).

Makeshift as a term is synonymous to "temporary" or "substandard". Makeshift structures such as kiosks and metal containers are shabby in nature (Godswill et al. 2016; Mbiggo and Ssemwogerere 2018). Poor design as well the materials used for constructing them render them inefficient in serving as shields against the climate. However, makeshift structures as forms of informal housing serve as places of abode and trade for a lot of people, mostly the poor in slum settlements (Ochieng 2011; Danso-Wiredu and Midheme 2017). In Africa, the rate of urbanization is higher than the global average (World Bank 2020). The consequence of lack of adequate infrastructure to match urbanization is the rise of slums in the urban centers (Güneralp et al. 2017), slums predominantly constituted of makeshift structures. In Ghana's instance, makeshift structures constitute a fifth of dwellings and that number is still rising (Ghana Statistical Service (GSS), 2021). The occupants of these structures are often people earning low incomes (UN-HABITAT 2007; Mberu et al. 2017). Their economic disposition makes it difficult to afford the appropriate building services such as air conditioning to make the dwellings comfortable (Singh 2016).

In the middle of the current climate crisis with warmer thermal conditions, there's the insinuation that these dwellings will become uncomfortably warmer and might even pose considerable challenges to human health (Vardoulakis et al. 2015). However, there is a lack of documentation on the actual comfort conditions of occupiers of makeshift structures in literature as majority of research regarding thermal comfort in sub-Saharan Africa focuses on formal forms of housing. This research therefore is aimed at uncovering the reality of thermal comfort conditions of people living in makeshift structures in a warm climate.

2 Literature Review

Thermal comfort is a broad satisfaction among the occupants of a particular space with the thermal environment (CIBSE 2006). The approaches to thermal comfort research are usually based on either of two bases; heat balance models such as Fanger's PMV using laboratory studies and adaptive models using field studies (Yao et al. 2009; Muhammad et al. 2022). Heat balance models are useful for assessing air-conditioned spaces while considering outdoor conditions indirectly (Fanger 1970). The adaptive models are suitable for naturally conditioned spaces (Muhammad et al. 2022). In the heat balance models, the subjects usually have no control over their immediate indoor thermal environment while the adaptive models factor behavioural and psychological adaptations and the interaction between outdoor and indoor environments (Shrestha et al. 2021). In conducting thermal comfort surveys, researchers collect data about the thermal environment (physical measurements) and the simultaneous thermal response of subjects going about their everyday lives (comfort vote survey) (Nicol and Humphreys 2002; Wu et al. 2019). This allows comparisons to be made between the thermal environment and how the occupants of that environment actually feel.

As thermal comfort is increasingly becoming critical, several thermal comfort studies have been carried out in the tropical regions of the world over the past decade. Appah-Dankyi and Koranteng (2012) for instance made an assessment of the thermal comfort condition of students and teachers in a school building. The study primarily investigated participants' perception of comfort as well as the thermal conditions of the rooms using subjective assessment questionnaires and physical measurements. One of the conclusions of the study was that, occupants were satisfied with their thermal environment in spite of high temperatures recorded. This according to the study demonstrated a higher tolerance of inhabitants of tropic regions. In more recent times, Koranteng et at. (2019), Guevara et al. (2021), Rodríguez et al. (2021), Adekunle (2021) and Gangrade and Sharma (2022) have all conducted thermal comfort studies on various forms of educational buildings including libraries, halls of residence and classrooms. Simons et al. (2012) also conducted a study in high-rise office buildings to assess comfort in glass buildings. The study confirmed high energy consumption in glass box buildings in attempt to make indoor conditions comfortable. The study also found that during instances of power outages, the offices became uncomfortably warm due to lack of natural ventilation and suggested that glass boxes were not suitable in warm humid climates. Similarly, Rahman et al. (2022) explored comfort conditions in naturally ventilated public hospitals in Malaysia using onsite measurements, field survey and simulation analysis. At the conclusion of the study, it was established that the hospital wards in the region were uncomfortably warm as study participants indicated warm sensations and dissatisfaction during the survey.

However, all these studies on thermal comfort in the region are based on formal forms of housing; offices, educational facilities, hospitals, etc. Appah-Dankyi and Koranteng (2012) for instance stated clearly that the study unit was chosen because of its sustainable features with proper ventilation and orientation. There is a void in literature concerning comfort conditions in makeshift structures specifically in sub-Saharan Africa. The explorations of this study will therefore address that gap.

From existing literature, most indices used to assess thermal comfort comprise two main classes of factors; environmental conditions such as air temperature and humidity and personal factors such as activity levels.

3 Methodology

Six different makeshift structures were considered for this study. This number was informed by the number of dataloggers available for the study. The general approach to data collection was quantitative. Data collection was in two main parts; physical measurement of environmental conditions and a comfort survey of occupants.

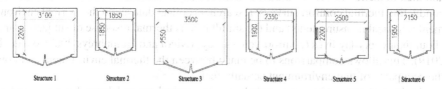

Fig. 1. Plans of selected structures

3.1 Selection of Study Structures

The criteria used for selecting the structures was by the different purposes they served. The purposes these structures served presented some variation in the heat gain and loss patterns in the spaces being studied. The structures were selected within the city of Kumasi. Details of the selected structures are illustrated in the table and figure below (Table 1 and Fig. 1).

Table 1. Characteristics of selected structures

	Occupancy type	Structure Type	No. of Occupants	Occupancy Hours
Structure 1	Trading Point (Grocery Shop)	Metal Container	1	6am–7pm
Structure 2	Trading Point (Cobbler's workshop)	Metal Container	2	6am–7pm
Structure 3	Trading Point (Hairdressing salon)	Wooden Kiosk	3	6am–7pm
Structure 4	Trading Point (Mobile money shop)	Metal Container	2	6am–7pm

(*continued*)

Table 1. (*continued*)

	Occupancy type	Structure Type	No. of Occupants	Occupancy Hours
Structure 5	Squatting Residence	Metal Container	1	6pm–7am
Structure 6	Trading Point (ECG Prepaid Vending)	Wooden Kiosk	3	6am–7pm
Total			12	

3.2 Physical Measurements

The physical measurement entailed mounting dataloggers in the study structures and recording indoor environmental factors i.e., air temperature and air humidity for a period of two weeks. These were done using temperature and relative humidity dataloggers (RHT20 Temperature/Humidity Datalogger). The dataloggers were programmed to take readings every 2 min. This configuration was to guard against exceeding the maximum data storage capacity of the devices (16,000) and ensure uninterrupted recording. They dataloggers were mounted close to the centre of each of the structures to collect uninterrupted data for the entire 14-day period. At the end of this period, each device had logged 10,080 for each parameter. Outdoor air temperature and relative humidity were also recorded in the process. Windows interface software for the various devices were used to retrieve the recorded data which was then exported to Microsoft Excel for analysis.

3.3 Comfort Survey

A point-in-time (PIT) comfort survey was conducted alongside field measurements. Conducting this survey is important as the thermal sensations of individuals within the same environment can vary (Djongyang et al. 2010). The survey was conducted using close-ended questionnaires informed by the samples in ASHRAE Standard 55 - 2017. This survey was aimed at collecting data regarding the following;

i) Occupants' state of health at the time of data collection
ii) Occupants' thermal sensation
iii) Occupants' thermal acceptance
iv) Occupants' thermal preference
v) Thermal environment control systems used in the structures

4 Results and Discussion

4.1 Environmental Factors of Thermal Comfort

Table 2 below shows a summary of readings taken in each structure. During computation and analysis, data logged from 7pm to 6am for Structures 1, 2, 3, 4 and 6 were excluded

Table 2. Readings for indoor environmental conditions

	Air temperature (°C)			Relative humidity (%)		
	Min	Max	Avg	Min	Max	Avg
Structure 1	26.2	32.4	**29.01**	40.8	71.3	**65.05**
Structure 2	26.3	33.2	**29.61**	43.1	70.2	**62.5**
Structure 3	26.1	30.5	**28.20**	40.9	74.5	**68.8**
Structure 4	26.5	32.7	**29.40**	41.2	71.7	**63.3**
Structure 5	26.1	30.1	**28.30**	50.9	82.2	**63.7**
Structure 6	26.4	31.8	**28.96**	41.0	71.6	**64.25**

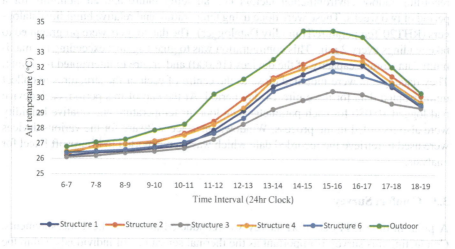

Fig. 2. Average air temperature readings for structures occupied in the day

Table 3. Readings for outdoor environmental conditions

	Air temperature (°C)			Relative humidity (%)		
	Min	Max	Avg	Min	Max	Avg
Outdoor readings	21.3	34.5	**30.55**	52.04	85.55	**68.80**

since the structure are unoccupied during this period. Structure 5 serving a residential purpose unoccupied most of day until 6pm through to 7am the next morning.

According to ASHRAE standard 55, the acceptable comfort range for summer is 25 °C to 28 °C. Comparing this standard to the values recorded and presented in Tables 2 and 3 would suggest that the structures should be uncomfortable for humidity levels from 30% to 70%. From Fig. 2, it can be observed that indoor air temperatures remain within acceptable limits till noon when it gets warmer and temperatures cross the threshold.

Table 4. Participants' State of health during PIT survey

	Frequency	Percent
Well	12	100.0
Unwell	0	0
Total	12	100.0

Temperature readings illustrated in Fig. 3, indicate relatively cooler conditions for most of the night in structure 5.

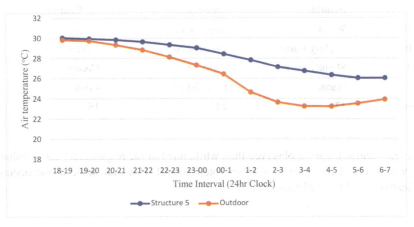

Fig. 3. Average air temperature readings for structures occupied at night

4.2 Participants' State of Health During PIT Survey

A two-point scale was used to record occupants'/respondents' state of health as well or unwell at the moment they were being interviewed of their thermal sensations. It is a well-documented fact the human brain shifts its "internal thermostat" to fight sickness (Székely and Garai 2018; Opp, 2019). This could well interfere with the findings of this study hence the need to document this data and exclude responses by unwell respondents. From Table 4 however, 12 out 12 respondents were in good health and had no fevers or condition that could pose some irregularity to their body temperatures when data was being collected regarding their thermal sensation.

4.3 Comfort Perception of Occupants

4.3.1 Thermal Sensation

Occupants ranked their point-in-time thermal sensation on 7-point scale ranging from −3 to +3 with −3 being cold and +3 being hot. The votes of the occupants are represented

Table 5. Summary of Occupants' Thermal Comfort Perception

Occupant	Thermal sensation	Thermal satisfaction	Thermal preference
001	Slightly warm	Dissatisfied	Cooler
002	Warm	Dissatisfied	Cooler
003	Warm	Dissatisfied	Cooler
004	Slightly warm	Dissatisfied	Cooler
005	Neutral	Satisfied	No Change
006	Slightly warm	Dissatisfied	Cooler
007	Neutral	Satisfied	No Change
008	Neutral	Satisfied	No Change
009	Warm	Dissatisfied	Cooler
010	Slightly warm	Dissatisfied	Cooler
011	Slightly warm	Satisfied	Cooler
012	Warm	Dissatisfied	Cooler
Sum	**13**	**20**	**15**

on Fig. 4 below. It can be observed that, while most of the respondents chose values on the warm side of the scale, 3 of them voted neutral. However, none of the occupants indicated a cool thermal sensation.

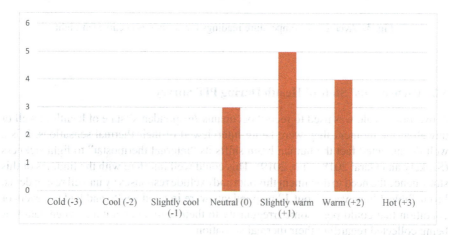

Fig. 4. Statistics for thermal sensation votes of occupants

Thermal Satisfaction

Occupants expressed whether or not the thermal environmental conditions are acceptable or not. All 12 of the study of the participants responded to this survey therefore meeting the 80% requirement of ASHRAE standard 55- 2017. The findings from the survey represented in Fig. 5 below indicates that majority of occupants voted the thermal conditions as unacceptable while only 33.3% were satisfied of the thermal environment.

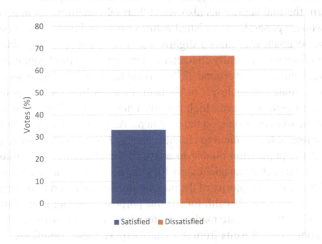

Fig. 5. Thermal Acceptance votes of participants

4.3.2 Thermal Preference

This was measured using a simple 3-point scale. Occupants indicated which condition they prefer; cooler conditions, warmer conditions or no change in thermal condition. The results of this survey are illustrated in Fig. 6. It is clear that majority of occupants prefer cooler conditions while 3 of them are comfortable with the indoor thermal conditions of their structure. No participant prefers conditions warmer than they already are.

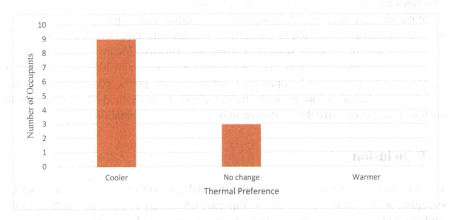

Fig. 6. Thermal Preference Distribution of Occupants

Most participants of this survey clearly are dissatisfied with the thermal conditions within the structures. Their dissatisfaction is understandable as the temperature and humidity values recorded do not fall within the prescribed temperature ranges for comfort. From Table 5 above, it can be observed that most occupants who voted warm or slightly warm thermal sensations also voted thermal conditions as unacceptable. Also, occupants who are generally dissatisfied with thermal conditions prefer cooler conditions including an occupant who finds a slightly warm environment acceptable.

Ghana's climate is generally characterized by hot and humid conditions. The annual average temperature rests at around 28 °C (World Bank 2020). These hot conditions will naturally necessitate higher levels of element filtration from buildings being put up in our part of the world. However, high poverty rates and appalling economic conditions characterize the lives of people on the continent. As reported by Hope Sr (2009), Africa is the most vulnerable continent to climate change. This is mainly due to poverty which renders most the population unable to adapt to the changing conditions. This means proper infrastructure that can adequately filter the harsh outdoor conditions is not within the affordable means of majority of the population. People resort to putting up structures which may not foster good health for various purposes ranging from shops to places of residence. These structures are made of materials that cannot properly shield occupants from harsh conditions. Results from this study expose these conditions. The average air temperature within enclosures is too high for comfort. A similar argument was made in Zheng et al. (2021) which reviewed thermal comfort in temporary structures. The study found out that most users of temporary buildings often face thermal comfort problems and that this situation is not localized to just a single climatic zone but is common in different climatic zones. These structures tend to be too cold when outdoor temperatures are cold and too hot when outdoor temperatures are high.

Another important observation from this study was that air circulation within these spaces is also very poor. This further establishes alarming indoor humidity. This creates ideal conditions for mold growth, asthma and increased vulnerability to infections. Introduction of more openings may help to this effect. However, increasing theft cases may hinder this development since the reason for reducing the number openings on these structures mostly is property protection from theft.

With all these findings, there's a clear indication that some alternative forms of low-cost infrastructure should be considered. Suggestions have been made by previous studies on low-cost and more sustainable building forms with locally available materials and easy construction methods, see for example Rincón et al. (2019) and Zhao et al. (2015). These options can be optimized further and the public educated on adopting such building forms. Adopting such strategy could improve the quality of life for the occupants and even drive the built environment towards sustainability.

5 Conclusion

The aim of this study was to assess the thermal comfort of occupants of makeshift structures within a tropical climate, in this case Ghana. By relating the findings of this study to ASHRAE standards, the study discovered that levels of indoor air temperature and humidity do not meet the comfort requirements for human comfort. The survey conducted for the study also revealed that people inhabiting these makeshift structures are generally dissatisfied with indoor conditions and prefer cooler conditions. With this information, alternative forms of affordable infrastructure should be explored. Also, measures should be taken against excessive heat gain in these structures. The envelope of these structures had no mechanism of reducing thermal radiations into indoor spaces.

One major limitation of the study is the relatively short period that was used in conducting the thermal assessment. This limitation does not invalidate the results but rather provides greater insight to be gleaned from the study. It is expected that further studies will be conducted over a longer period of time. This paper forms part of a larger study aimed at assessing the effect of indoor building environment of makeshift structures on the health of the occupants.

References

Adekunle, T.O.: Thermal performance and apparent temperature in school buildings: a case of cross-laminated timber (CLT) school development. J. Build. Eng. **33**, 101731 (2021). https://doi.org/10.1016/j.jobe.2020.101731

Ahmed, T., Kumar, P., Mottet, L.: Natural ventilation in warm climates: the challenges of thermal comfort, heatwave resilience and indoor air quality. Renew. Sustain. Energy Rev. **138**, 110669 (2021). https://doi.org/10.1016/j.rser.2020.110669

Akcabozan, A., Demir, Y.: A comparative parametric evaluation of informal and formal housing: Maltepe/Istanbul case study. Archit. Urban Plan. **10**(1), 21–26 (2015). https://doi.org/10.1515/aup-2015-0003

Chapman, S.C., Watkins, N.W., Stainforth, D.A.: Warming trends in summer heatwaves. Geophys. Res. Lett. **46**(3), 1634–1640 (2019). https://doi.org/10.1029/2018GL081004

Chartered Institution of Building Services Engineers (CIBSE). KS6: Comfort. London: CIBSE (2006). https://www.cibse.org/knowledge-research/knowledge-portal/ks06-comfort-ks6

Cheung, T., Schiavon, S., Parkinson, T., Li, P., Brager, G.: Analysis of the accuracy on PMV–PPD model using the ASHRAE global thermal comfort database II. Build. Environ. **153**, 205–217 (2019). https://doi.org/10.1016/j.buildenv.2019.01.055

Cui, H., Overend, M.: A review of heat transfer characteristics of switchable insulation technologies for thermally adaptive building envelopes. Energy Build. **199**, 427–444 (2019). https://doi.org/10.1016/j.enbuild.2019.07.004

Danso-Wiredu, E.Y., Midheme, E.: Slum upgrading in developing countries: lessons from Ghana and Kenya. Ghana J. Geography **9**(1), 88–108 (2017). https://www.ajol.info/index.php/gjg/article/view/154657

De Dear, R., Xiong, J., Kim, J., Cao, B.: A review of adaptive thermal comfort research since 1998. Energy Build. **214**, 109893 (2020). https://doi.org/10.1016/j.enbuild.2020.109893

Fanger, P.O.: Thermal comfort. Analysis and applications in environmental engineering. Thermal comfort. Analysis and applications in environmental engineering (1970)

Gangrade, S., Sharma, A.: Study of thermal comfort in naturally ventilated educational buildings of hot and dry climate-a case study of Vadodara, Gujarat, India. Int. J. Sustain. Build. Technol. Urban Dev. **13**(1), 122–146 (2022). https://doi.org/10.22712/susb.20220010

Godswill, O.C., Ugonma, O.V., Ijeoma, E.E.: The determinants of squatter development in Southern Aba Region of Nigeria. Afr. J. Environ. Sci. Technol. **10**(11), 439–450 (2016). https://www.ajol.info/index.php/gjg/article/view/154657

Guevara, G., Soriano, G., Mino-Rodriguez, I.: Thermal comfort in university classrooms: an experimental study in the tropics. Build. Environ. **187**, 107430 (2021). https://doi.org/10.1016/j.buildenv.2020.107430

Güneralp, B., Lwasa, S., Masundire, H., Parnell, S., Seto, K.C.: Urbanization in Africa: challenges and opportunities for conservation. Environ. Res. Lett. **13**(1), 015002 (2017). https://doi.org/10.1088/1748-9326/aa94fe

Heracleous, C., Michael, A.: Assessment of overheating risk and the impact of natural ventilation in educational buildings of Southern Europe under current and future climatic conditions. Energy **165**, 1228–1239 (2018). https://doi.org/10.1016/j.energy.2018.10.051

Hope, K.R., Sr.: Climate change and poverty in Africa. Int J Sust Dev World **16**(6), 451–461 (2009). https://doi.org/10.1080/13504500903354424

Intergovernmental Panel on Climate Change, IPCC: Climate Change 2021: The Physical Science Basis, IPCC Sixth Assessment Report; Summary for Policymakers, p. 14 (2021). https://www.ipcc.ch/report/ar6/wg1/

Intergovernmental Panel on Climate Change, IPCC: Climate Change 2022; Impacts, Adaptation and Vulnerability. IPCC Sixth Assessment Report; Summary for Policymakers, pp. 22–29 (2022). https://www.ipcc.ch/report/ar6/wg2/

Koranteng, C., Simons, B., Essel, C.: Climate responsive buildings: a comfort assessment of buildings on KNUST campus, Kumasi. J. Eng. Des. Technol. **17**(5), 862–877 (2019). https://doi.org/10.1108/JEDT-03-2019-0054

Kownacki, K.L., Gao, C., Kuklane, K., Wierzbicka, A.: Heat stress in indoor environments of scandinavian urban areas: A literature review. Int. J. Environ. Res. Public Health **16**(4), 560 (2019). https://doi.org/10.3390/ijerph16040560

Mberu, B., Béguy, D., Ezeh, A.C.: Internal migration, urbanization and slums in Sub-Saharan Africa. In: Groth, H., May, J.F. (eds.) Africa's Population: In Search of a Demographic Dividend, pp. 315–332. Springer, Cham (2017). https://doi.org/10.1007/978-3-319-46889-1_20

Mbiggo, I., Ssemwogerere, K.: An investigation into fire safety measures in Kampala slums. A case of katanga-wandegeya. Civil Environ. Res. **10**, 30–34 (2018). 20.500.12281/7336

Ochieng, E.: Factors influencing the implementation of Kenya slums upgrading Programme: a case of Kibera Slums in Nairobi County. PhD diss., University of Nairobi, Kenya (2011). http://erepository.uonbi.ac.ke/handle/11295/4166

Opp, M.R.: Fever, body temperature, and levels of arousal. In: Handbook of behavioral state control, pp. 623–640. CRC Press (2019). https://www.taylorfrancis.com/chapters/edit/10.1201/9780429114373-39/fever-body-temperature-levels-arousal-mark-opp.

Osman, M.M., Sevinc, H.: Adaptation of climate-responsive building design strategies and resilience to climate change in the hot/arid region of Khartoum Sudan. Sustain. Cities Soc. **47**, 101429 (2019). https://doi.org/10.1016/j.scs.2019.101429

Ozarisoy, B.: Energy effectiveness of passive cooling design strategies to reduce the impact of long-term heatwaves on occupants' thermal comfort in Europe: climate change and mitigation. J. Clean. Prod. **330**, 129675 (2022). https://doi.org/10.1016/j.jclepro.2021.129675

Perkins-Kirkpatrick, S.E., Lewis, S.C.: Increasing trends in regional heatwaves. Nat. Commun. **11**(1), 1–8 (2020). 10.1038%2Fs41467-020-16970-7

Rincón, L., Carrobé, A., Martorell, I., Medrano, M.: Improving thermal comfort of earthen dwellings in sub-Saharan Africa with passive design. J. Build. Eng. **24**, 100732 (2019). https://doi.org/10.1016/j.jobe.2019.100732

Rodríguez, C.M., Coronado, M.C., Medina, J.M.: Thermal comfort in educational buildings: the classroom-comfort-data method applied to schools in Bogotá Colombia. Build. Environ. **194**, 107682 (2021). https://doi.org/10.1016/j.buildenv.2021.107682

Székely, M., Garai, J.: Thermoregulation and age. Handb. Clin. Neurol. **156**, 377–395 (2018). https://doi.org/10.1016/B978-0-444-63912-7.00023-0

Thapa, S., Panda, G.K.: Energy conservation in buildings–a review. Int. J. **5**(4), 95–112 (2015). https://doi.org/10.5963/IJEE0504001

UN-HABITAT: Sustainable Urbanization: local action for urban poverty reduction, emphasis on finance and planning. Twenty First Session of the Governing Council, GC21, 16–20 April, Nairobi, Kenya (2007). https://www.preventionweb.net/files/1700_462551419GC202120What20are20slums.pdf [accessed on June 24, 2022]

Van Hoof, J.: Forty years of Fanger's model of thermal comfort: comfort for all? Indoor Air **18**(3), 182–201 (2008). https://doi.org/10.1111/j.1600-0668.2007.00516.x

Vardoulakis, S., Dimitroulopoulou, C., Thornes, J., et al.: Impact of climate change on the domestic indoor environment and associated health risks in the UK. Environ. Int. **85**, 299–313 (2015). https://doi.org/10.1016/j.envint.2015.09.010

World Bank: World Bank staff estimates based on the United Nations Population Division's World Urbanization Prospects: 2018 Revision. Data: Urban Population Growth (2020). https://data.worldbank.org/indicator/SP.URB.GROW. Accessed 24 June 2022

Zhang, Y., Zhou, X., Zheng, Z., Oladokun, M.O., Fang, Z.: Experimental investigation into the effects of different metabolic rates of body movement on thermal comfort. Build. Environ. **168**, 106489 (2020). https://doi.org/10.1016/j.buildenv.2019.106489

Zhao, Z., Lu, Q., Jiang, X.: an energy efficient building system using natural resources-superadobe system research. Procedia Eng. **121**, 1179–1185 (2015). https://doi.org/10.1016/j.proeng.2015.09.133

Zheng, P., Wu, H., Liu, Y., Ding, Y., Yang, L.: Thermal comfort in temporary buildings: a review. Build. Environ. **109262** (2022). https://doi.org/10.1016/j.buildenv.2022.109262

Fashion Transformational Synthesis Model for Beauty Pageants in Ghana

S. W. Azuah[1,2]([✉]), K. S. Abekah[1,2], and B. Atampugre[1,2]

[1] Department of Fashion Design and Technology, Faculty of Applied Art and Technology, Takoradi Technical University, Takoradi, Ghana
schazuah@yahoo.com

[2] Department of Marketing and Strategy, Faculty of Business Studies, Takoradi Technical University, Takoradi, Ghana

Abstract. Purpose: The purpose of this study was to develop A fashion concept transformational syntheses model for local pageants like "Ghana's Most Beautiful" Pageant for the designing of activities. To achieve the aim of the study, the objectives sought to determine the standard of beauty exhibited in the pageant, assess viewers' perceptions of the pageant, and further determine the fashion concept and cultural relevance of the pageant.

Design/Methodology/Approach: The study employed the exploratory sequential type of mixed-method design. The process employed enabled the researcher to explore pageants to ascertain in-depth pageant practices and participant views on the cultural relevance. The targeted population consisted TV3 staff, Producers and organizers, Judges, Fashion designers, Council for National Art and Culture officials, Traditional rulers, and viewers of the pageant in the three sectors of Ghana. A proportionate stratification was calculated for each group of participants who formed a Sample size of 1051. Simple random was used in selecting the sample size. Close-ended questionnaires developed from observation, interviews and content analyses were administered to participants. SPSS tool was used in data analyses where inferential statistics, mean, standard deviation and ANOVA were engaged. The data obtained were presented in tables and diagrams and findings employ in developing a model.

Findings: Findings revealed that beauty standards of Ghana's Most Beautiful Pageant were both of Ghanaian and foreign relevance. The fashion concept was also a mixture of both cultures. There was a significant difference between the fashion concept and cultural relevance of the pageant suggesting the adoption of foreign cultures into the pageant. it also revealed main objectives of the pageant was not the only source of inspiration for designing pageant activities which result in occasional deviation.

Research Limitation/ Implications: the study was based on a local pageant in Ghana hence its concentration mainly on Africa/Ghanaian pageantry.

Practical Implication: The knowledge advanced in this study informs organizers of pageants on the need to consider the main objective of pageants, community, cultural relevance and viewers as the source of inspiration for designing pageants

© The Author(s), under exclusive license to Springer Nature Switzerland AG 2023
C. Aigbavboa et al. (Eds.): ARCA 2022, *Sustainable Education and Development – Sustainable Industrialization and Innovation*, pp. 464–473, 2023.
https://doi.org/10.1007/978-3-031-25998-2_34

activities. The employment of the fashion concept would guide against deviation and the introduction of foreign practices.

Social Implication: The knowledge advanced by this study will help policy-makers in the beauty industry to review existing concepts and geared towards helping local pageants in achieving positive results in the communities.

Originality/ Value: The novelty of this study lies in the model created for beauty pageants, it educates on the inclusion of the main focus, communities and viewers of the pageant in the designing of activities for better results especially within the beauty pageant industry.

Keyword: Beauty pageants · Cultural relevance · Fashion concept · Ghana · Transformational model

1 Introduction

Throughout modern society, the tendency towards continuous change can be found through beauty pageants. Beauty pageants are events organisied for females with desirable qualities of a society based on their cultural values and norms. They are celebrated with the purpose of bring out beauty of a community, nation or continent through women. Beauty pageants come in a form of performance art. In such events women are normally selected to represent their collective identity to a wider audience as a symbolic representation.

Research indicates that beauty is always a craft with its products and backgrounds being very local. Thus, society's shared tastes and lifestyles collectively shape and reflect its people's tastes and lifestyles.

Beauty pageants, a vibrant place for creating and questioning cultural definitions uses fashion as a tool to achieve its aims due to fashion's potentials of causing change. As defined by Karunaratne (2016), Fashion is a culturally endorsed form of expression in a specific material or non-material phenomenon that can be discerned. Beauty pageants whether locally or internationally exhibits culture and fashion of its society and portrays a form of competition determined by societal values and norms within its location.

According to Venkatasamy (2015), the individual appearance in society communicates non-verbally such as ones' cultural values, and lifestyle. The employment of fashion for beauty pageants is very much appropriate especially when it has to do with transforming a nation through redefining beauty. Similarly, in beauty pageants, fashion is portrayed to advocate for desired cultural practices such as the form of dressing, talking, dancing, singing, and behaviours. A model that is structured based on the fashion concept for a pageant would therefore facilitate the adoption of its values. Every society tries to preserve its culture by portraying its beauty which is occasionally affected by international styles of fashion. Crawford et al. (2008) stated that nationalism, morality, modernization and globalization is not only reflected through beauty pageants, but also represent the social constructions of gender. In Africa, there has been a long struggle, particularly in urban environments, over the ability to maintain cultural identity and tradition among ethnic groups. In the sense of the ethnic makeup, the pressures from the western world most often have a major effect on efforts to preserve some elements of

African culture (Oster-Beal 2013). There have been many attempts to promote culture through beauty pageants by practitioners over the world, however, research has not been conducted in Ghana to this effect.

The perceptions of the public on the fashion concept and cultural relevance of activities of the pageant have not been analyzed by researchers. For example, every year GMB pageant crowns its queens amid several complaints from the public. According to Balogun (2012), Pageantry in Africa is a borrowed phenomenon, which sometimes, faces criticism due to its concepts not being geared towards promoting African cultural values. Most academic research work on pageantry focused primarily on pageants within the restrictive Western feminist lens. Research conducted by Hansen (2010) revealed that, though beauty pageants have a positive impact on society, sometimes there is the potential of it contributing negatively (Bell 2002). The purpose of the current study was to develop the fashion concept transformational syntheses model for local pageants based on findings of Ghana's Most Beautiful Pageant for designing activities of local beauty pageants that consider local content. In achieving this aim, the study's objectives are to 1) determine the standard of beauty exhibited in Ghana's Most Beautiful pageant 2) assess viewers' perceptions of the pageant, and 3) determine the fashion concept and cultural relevance of the pageant in Ghana.

Theories Underpinning the Study

Research revealed that most beauty pageants both local and international have their central focus; however, the designing of activities to fulfill these main objectives comes with the introduction of foreign materials disrupting the main idea. The beauty ideals were framed in diverse ways yet resulted to that of international beauty ideals (Balogun 2012). This was as a result of the absence of a model directing the designing of standards required for a local beauty pageant that reflects the community.

Through the adapting of Karunaratne, (2016) "Generalized Fashion Model" (GFM), and integrating it with Cholachatpinyo et al. (2002) "Fashion Transformational Process Model" (FTPM) facilitated the designing of a culturally friendly model. This results for achievement of desired purpose in the organization of local pageants. Predictions emerging from these studies formed a concept for the designing of fashionable themes and ideas for the organization of beauty pageants.

1.1 The Fashion Concept Theory

In establishing a beauty pageant, its values and purpose necessitating its establishment must be reflected clearly in its theme or activity. The fashion concept ensures the pageant achieve the desired goals through glamorous presentations. This was accomplished through the combination and adoption of fashion principles to style of the pageantry purpose. Fashion art is the combination of material, concept, and culture with the human figure into becoming one. Thus designing based on the fashion concept model included the process of analyzing pageant design brief, studying pageantry trends and interpreting ideas in line with culture as this forms the main structure of a society. The concept particularly in all activities must have a number of unique characteristics similar to the following:

1. Activities of a pageant are non-permanent; subject to change, and can eventually be replaced by "newer" ideas within main scope. This means undesirable cultural practices is modified to suit demands of current tastes. Designing of activities especially contestants' costumes from various regions must be subjected to change while still upholding pageant values since culture is dynamic.
2. A pageant must contain various characteristics of functionality taking into consideration the general objectives. As illustrated by Chatterjee (2017), a pageant's elements must be purposeful, aesthetically appealing, and pleasurable to viewers. A pageant for example should reflect these in its drama, music, storytelling, costumes, styling, and other aesthetics qualities for social acceptability.
3. Similarly, the practice of a pageant to some extent should satisfy major categories of people in community, institution or nation; even though, research indicates that every fashion product at any given period is either accepted or rejected (King-O Riain 2008).
4. Additionally, the pageant must be possessed by complete originality and novelty when compared to acceptable ones. Therefore pageants should be dynamic and in agreement with current practices of the society in other to command societal adaption and acceptance.
5. Pageant must directly have personal and social physical characteristics, such as societal identification, prominence, status and high recognition. The critical motivations for the acceptance a pageant must be found in its social characteristics. The presented outfits/activity must offer for attraction, education, imitation, expression or instrumental to meet the differentiating and socializing forces within a society.

Karunaratne's view on fashion apparel fit indicated that costumes must be found fitting to main purpose of the pageant. A pageant which seeks to redefine beauty for national development must have all its activities gearing toward achieving the main purpose (Karunaratne, 2016). It is on this basis that a fashion object becomes worthy of acceptance in a community like Ghana. The pageant model is based on the fashion generalized concept due to its numerous attributes to society.

Cholachatpinyo et al. (2002) added that the fashion objects and processes should be considered in the design of fashionable activities to facilitate their acceptance. The designing of a pageant based on the fashion concept requires the consideration of all these factors. The differentiation and socializing effect of individuals must also be considered as the pageant deals with people, regions, communities, tribes, religious groups, the political class and all clusters of lifestyles. An individual contestant representing her region strives to be unique but also social in all her presentations to be relevant.

In studying the fashion object and process, Karunaratne (2016) stated that a shift in the desire of an existing object to a newly emerging one ultimately results from its manifestation. The alternative object exposed to potential adopters cause dimensions of change. Individual adaptors also go through some changes in the cause of constant exposure of the fashion object and process. Change is implicit and critical to fashion objects and must be considered when designing its products. Fashion has the transformational power to change a society into a desired or undesired lifestyle. It is an indispensable tool in achieving societal related visions due to its attractive nature. "With regard to beauty, fashion gives pleasure in a feature that happened to be adaptive and survived in the minds of people" (Chatterjee, 2017). Every pageant has a purpose to fulfil and the development of a model based on fashion is the key as fashion produces pleasure that drives humans to act in desired direction.

Clothing fashions represent the development of time and reflection of cultural drivers to societal changes (Inglessis 2008). A beauty pageant such as Ghana's Most Beautiful has its main purpose of promoting and enhancing national unity. It therefore needs to engage fashion for its transformational agenda since fashion could be a benchmark. Another reason for adopting fashion is its ability to satisfy the need for differentiation and individuality thus, fulfilling the social, psychological, and cultural needs. All these attributes are needed in a beauty pageant for a contestant to express self and her community. A pageant is therefore incomplete without communicating fashion.

The fashion concept facilitates achievement of pageants' objectives and has the power to transform any cultural practices into a contemporary ideas for possible subsequent acceptance and adoption by the society. The present of a model in planning pageant activities facilitates in-depth understanding of pageant practices and possible adoption of its values within the society.

The combination of past and present, popular culture, social issues and fashion trends in the designing of pageant activities and costumes give more meaning to pageantry. Organizers must therefore ensure relevant points are central in organization from the advertising stage through to the crowning and subsequent activities. Again, continues sustenance and modification of culture through beauty pageants should be advocated to promote local culture and its globalization by organizers.

2 Methodology

The study employed the exploratory sequential type of mixed method design. This approach often used in investigating and surveying on research cases that requires research instruments such as observation, content analyses and questionnaires (Creswell 2016; Terrell 2015). The process employed, enabled the authors explore the pageant to ascertain an in-depth pageant practices and traditional rulers' views on the cultural relevance of the pageant. Target population consisted TV3 staffs, Producers, Judges, Council for National Art and Culture officials, Traditional rulers, and viewers of the GMB Pageant in the three sectors of Ghana. A proportionate stratification was calculated for each group of participants who formed the Sample size of 1051. Close ended questionnaires were administered to participants after items were developed from observation, interviews and content analyses. Data analyses was done through SPSS using descriptive and inferential statistics with mean, standard deviation and ANOVA.

3 Findings and Discussion

Activities underlined in the GMB pageant are full of excitement, education, tradition, culture and beauty in all totality. However, a lot needs to be done to relate its concept of beauty more to the Ghanaian cultural values. At face value, most of the activities are cultural-based but some foreign ideas occasionally find their way into the pageant

Table 1. Standard of beauty pageant

Item	M	SD
1. Ghanaian society has a standard of beauty	3.93	1.063
2. During the beauty pageant type of clothing worn is accepted by all stakeholders	4.33	0.913
3. The culture of society was reflected in the types of clothing worn by contestants within the pageant	4.22	1.020
4. The African beauty of a women was exihbitted in costumes worn	3.54	1.242
5. Imported standards were engaged in choosing pageant contestants	**2.70**	1.261
6. Contestants wore make ups appearing more like foreign ladies	**3.53**	1.168
7. International Style of Grooming contestants was applied for GMB contestants	**3.07**	1.223
8. Both GMB and foreign pageants put on similar styles of clothing	2.78	1.391
9. Global application of make ups	2.23	1.167
10. In all activities of GMB pageant, clothing of contestants were culturally accepted	3.89	1.250

Source: (Field work, 2021)

practices. An example is the crowning of an African queen in foreign clothing instead of a traditional costume.

Regarding standard of beauty exhibited at GMB pageant, the survey revealed that though the African beauty of a women was exihbitted in costumes worn (M-3.54, SD-1.242) foreign standards were engaged in choosing pageant contestants (M-2.70, SD-1.261). It was recorded that Both GMB and foreign pageants put on similar elegances regarding their appearance. Viewers again observed that both GMB and international pageants were same in beauty ideals. Intercontinental style of Grooming contestants was applied for GMB contestants as it recorded a high figure of (M-3.07, SD-1.223). What then makes a pageant locally based? A contradictory finding was when respondents indicated that in all activities of GMB pageant, clothing of contestants were culturally accepted (M- 3.89 SD-1.250). This means if clothing were similar to both local and international, and the same time accepted in Ghana then there still remains a gap. A revelation which suggest that practices of the pageant as against its cultural transformational agenda in the country still requires attention. The pageant's objectives of redefining beauty a strict policy governing the direction of all activities: hence the development of a model for the designing of local pageants activities.

Regarding perceptions on the GMB pageant, it was indicated that Ghanaian style of clothing was portrayed at GMB pageant with a score of (M-4.93, SD-0.913). It was also revealed that the culture of each region was reflected in their clothing choices (M-4.22 SD- 4.020). The response of the two variable indicated a higher response to both statements suggesting that participants believed even though style of clothing were to some extent culturally friendly they ocassionally fall short regarding local standard as stated in the earlier discussion on Table 1. The result is consistent with the study of

Balogun (2012) that body features, styles of clothing and make-ups in African beauty pageants most times are not inline with expected local standards. Organizers and Fashion designers must strive to define beauty standard that are acceptable in their communities. Designer should ensure high quality local designs attractive enough to penetrate international markets and possibly create trends are produced. This would go a long way to ensure the economic development of local markets while projecting local cultures.

Table 2. Significant difference between fashion concept and cultural relevance

ANOVA		Sum of Squares	Df	Mean Square	F	Sig
Fashion Concept	Between Groups	71.969	2	35.985	0.680	0.507
	Within Groups	55426.116	1047	52.938		
	Total	55498.085	1049			
Cultural Relevance	Between Groups	1407.845	2	703.922	5.885	0.003
	Within Groups	124387.275	1040	119.603		
	Total	125795.120	1042			

Source: (Field work 2021)

There was significant score on participants' views of GMB pageant regarding the fashion concept (F = 0.680, p = 0.507) and cultural relevance (F = 5.885, p = 0.003) with the cultural relevance rated higher than the fashion concept as indicated in Table 2. An indication that even though GMB pageant is culturally sensitive and related its ability to motivate the general public to the adoption of its attributes and practices, it is however low in comparison to its relevance. There is significant difference on perceptions about the fashion concept and cultural relevance of the GMB pageant (Table 3).

Table 3. Perception rating on fashion concept against cultural relevance test

Items	M	SD	SIG
Fashion Concept and Cultural Relevance	−25.6721	9.09084	0.000

Source: (Field work, 2021)

Comparing the rating of participants' views on the fashion concept and its cultural relevance in the Ghanaian context, it was observed that there was significant difference between the two variables with a record value of p = 0.000 and SD = 9.09084. The results further indicated viewers' perception regarding fashion concept and culture relevance mean rating suggested cultural relevance (−25.11818) is higher as compared to that of the fashion concept (−26.22608) with significance score of (p = 0.005) for cultural relevance and (p-.003) for the fashion concept. There was significance difference on viewers rating regarding the fashion concept and cultural relevance on the GMB pageant. The finding is consistent with that of Sproles and burns fashion theory (1996), which averred that a fashion object must have social acceptability, self-fulfillment, status symbolism and other psycho-social qualities leading to its adoption and acceptability; a finding supporting the work of Karunaratne (2016).

Below is a designed model outlining guidelines for the designing of pageant activities to achieve maximum results, based on relevant literature reviewed and the study's findings.

3.1 A Transformational Fashion Concept Synthesis Model for Beauty Pageants in Ghana

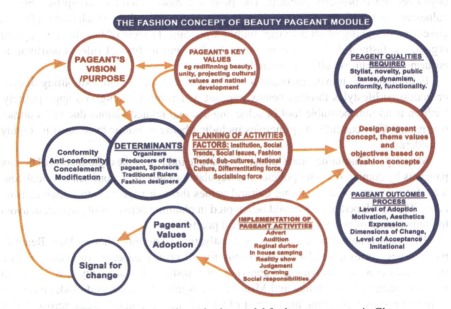

Fig. 1. Wompakeah-Azuah synthesize model for beauty pageants in Ghana

Figure 1 depicts a cultural customized model developed to ensure the achievement of pageant objectives and maximization of ultimate results. The model is drawn with central focus on a pageant objective as shown in upper left circle of the model. When organizing a pageant, the key consideration such as the community (target participants), pageant's key elements (its focus, contestants) and activities are determined which finally become the main source of inspiration for designing of pageants activities. During the designing of pageant's activities, the fashion concept becomes a tool introduced to ensure dynamism, novelty, colourful event, functionality and inoviation of all its activities. Organizers must always consider the expected outcomes of a pageant to avoid activities that result in unwanted results. However when unwanted outcomes results the model provides the possibility of revisiting previous processes of events and possible corrections of subsequent ones for ultimate results. However there is always the possibility of some members of the public conforming, anti- conforming, concealing, modifying and adopting values of the event based on their assessments as observed by Cholachatpinyo, et al (2002).

4 Conclusion

Some pageants attempt to homogenize the concepts of beauty that previously were diverse and culturally based, a practice that compels women to meet the standards set by other beauty pageants. The exhibition of desired behaviours in a pageant brings about general acceptance of the queen and adoption of pageant's practices in the community. A beauty pageant becomes a tool in the hands of the organizers in achieving desired objectives when properly planned. The proposed model therefore comprised beauty values and its key elements inspired by traditional cultural values which formed factors governing designing of all activities within a pageant. In view of this, beauty pageant organizers, fashion designers, community leaders and traditional rulers contribute to ensuring a pageant is culturally friendly.

Fashion has the transformational power to change a society into exihbitting desired or undesired lifestyles through beauty pageant which must be engaged appropraitely. Fashion is an indispensable tool in achieving societal related visions due to its attractive nature. Consequently its principles including being of public taste must be highly considered.

Ministry of Chieftaincy and Culture should take interest in the organization of beauty pageants by coming out with policies that regulate organizers to conform to 'local content' for purposes of maintaining cultural values through pageant rather than causing harm. The developed model should be adopted in planning activities of pageant to avoid the importation of non ideal foreign cultural practices.

The study is not without a limitation. Firstly, the use of only Ghana's Most Beautify pageant begs from generalizing the results for all beauty pageants in Ghana. Future studies may consider studying and applying the transformational beauty pageant model to others such as Miss Malika and Miss Ghana. Again, the developed model based on a studied results was within the context of Ghana; a cross country study across Africa can give a better appreciation of the applicability of the model.

References

Balogun, O.M.: Cultural and Cosmopolitan: Idealized femininity and embodied nationalism in Nigerian beauty pageants. Gend. Soc. **26**(3), 357–381 (2012)

Barnard, M.: Fashion as Communication. Routledge, New York (1996)

Bell, R.H.: Understanding African Philosophy: A Cross-Cultural Approach to Classical and Contemporary Issue. Routledge, London (2004)

Burke, S.: Fashion Designer. From Concept to Collection. Burke

Chatterjee, A., Vartanian, O.: Neuroscience of aesthetics. Ann. N. Y. Acad. Sci. **1369**(1), 172–194 (2016)

Cholachatpinyo, A., Padgett, I., Crocker, M., Fletcher, B.: A conceptual model of the fashion process–part 2: An empirical investigation of the micro-subjective level. J. Fash. Mark. Manage. Int. J. **6**(1), 24–34 (2002)

Crawford, M., Kerwin, G., Gurung, A., Khati, D., Jha, P., Regmi, A.C.: Globalizing beauty: Attitudes toward beauty pageants among Nepali women. Fem. Psychol. **18**(1), 61–86 (2008)

Creswell, J.W.: Research Design: Qualitative, Quantitative, Mixed Methods Approaches. Sage publications (2016)

Hansen, P.H.: 4 Cobranding Product and Nation. Trademarks, Brands, and Competitiveness, p. 77 (2010)

Karunaratne, P.V.M.: Meanings of fashion: context dependence. Int. J. Multidisciplinary Stud. **3**(2) (2016)

Inglessis, M.G.: Communicating through clothing: The meaning of Clothing among Hispanic Women of different levels of Acculturation. The Florida State University (2008)

KingO'Riain, R.C.: Making the perfect queen: the cultural production of identities in beauty pageants. Sociol. Compass **2**(1), 74–83 (2008)

Oster-Beal, M.: Preserving Tradition: Analyzing the Commoditization of Cultural Identity through Beauty Pageants Among Ethnic Minority Groups in Kathmandu (2013)

Sproles, G.B., Burns, L.D.: Changing appearances: Understanding dress in contemporary society. Fairchild publications (1996)

Terrell, S.R.: Writing A Proposal for Your Dissertation: Guidelines and examples. Guilford Publications (2015)

Venkatasamy, N.:. Fashion trends and their impact on the society. In: Conference: International conference on textiles, Apparels and Fashion

Design Creativity and Clothing Selection: The Central Focus in Clothing Construction

S. W. Azuah[✉], A. S. Deikumah, and J. Tetteth

Department of Fashion Design and Technology, School of Applied Art and Technology,
Takoradi Technical University, Takoradi, Ghana
schazuah@yahoo.com

Abstract. Purpose: It looked into how the application of design elements and principles influence final appearance of clothing and their possible selection by the customer. The objective was to study design and construction variables that must be prioritized in garment assembly. The aim was to study outfits designs by the Ghanaian dressmaker and examine proportion, emphasis, rhythm, balance and colour coordination in designs constructed.

Design/Methodology/Approach: The qualitative method was used in studying designs selected from three regions of Ghana namely Accra, Takoradi and Kumasi. This was so because most clothing activities were concentrated mainly in these three cities of Ghana. Design and style produced were studied to discover the assembled design features in finished apparels. Instruments such as observation and content analysis were engaged for data collection. Designed garments selected from the manufacturing houses were examined against design principles, nine outfits were sampled for the study from each city. The data obtained from the clothing constructed were analysed and presented for discussion.

Findings: Findings revealed that most garments had colour combination inappropriately applied and motifs incorrectly placed. Common faults were on design creation and construction processes which affects customer satisfaction and clothing selection.

Research Limitation/Implications: The study focused on the design construction and outer appearance of garments in selected fashion shops of the three cities in Ghana that affect clothing selection. Elements and principles of design, construction and selection of clothing were most times not designed to meet customer selection.

Practical Implication: The knowledge acquired from this study would help in the training of designers in Ghana especially in the informal sector. This would also prompt designers to pay more attention to the construction of clothing for customer satisfaction. Key points realized would significantly influence the mode of training in the clothing industry.

Social Implication: The knowledge advanced by this study would help in clothing designing/ selection, improve productivity in the clothing construction industry to enhance the Ghanaian economy.

Originality/Value: The novelty of this study was the blending of knowledge in fashion design, construction and customer satisfaction in the Ghanaian fashion industry.

© The Author(s), under exclusive license to Springer Nature Switzerland AG 2023
C. Aigbavboa et al. (Eds.): ARCA 2022, *Sustainable Education and Development – Sustainable Industrialization and Innovation*, pp. 474–482, 2023.
https://doi.org/10.1007/978-3-031-25998-2_35

Design Creativity and Clothing Selection 475

Keyword: Clothing selection · Creativity · Design · Garment construction · Ghana

1 Introduction

It is critical to accomplish a goal of dressing distinction during clothing construction at all times. Clothing has become the ultimate advocator in achieving the best results. Clothing can help the body perform better in a variety of situations. Although all clothing is beneficial, functional clothing design focuses on garments that have specific functions and must be designed as such. In the context of this study, the distinction means guaranteeing quality at every stage of garment assembly to wind customer satisfaction. Dressing in a way that flatters a figure and expresses a wearer's personality is very significant and require attention. A designer who knows how to make the most of a person's beauty gets to know the person and studies their physical and psychological characteristics to create clothing that flatters them.

According to studies, Most people want to dress nicely, but they are unsure of what will look good on their body type and measurements. The body's capacity to function under varied circumstances can be enhanced by clothing. Despite the fact that all clothing has a purpose, functional clothing design focuses on clothing with particular purposes. Technical know-how on design creativity, which ensures that apparel has a unique look and better fit, appears to be a challenge in its application. (Omoavowere and Lily 2011; Foster and Ampong 2012).

Clients are increasingly becoming dissatisfied with the fit and style of their clothing and increasingly demanding better products (Dove 2016). Complaints like "the colour combination are inappropriate", and "the motif in the design is incorrectly placed," are common criticisms of the wearer in Ghana. Complaints such as these prove the design and production of the garments were not done properly, affecting clothes fitting and selection. According to Fischer (2009), clothing construction provides a clear introduction to the core skills and knowledge required to create a successful outfit that satisfies the wearer and the designer has a responsibility to create the appropriate outfit.

The ideal silhouette is a one-of-a-kind system for examining figure proportionality and determining appealing, pleasing and fulfilling styles. Blunders can make a client appearance uncertain and disturbed. The first step in dressing for success is to make sure your clothes fit properly. The size, shape, style, and colour of one's attire should all be appropriate and free of construction flaws. Knowing one body shape and deciding the physical attributes to stress or ignore are important considerations.

Clothing has the capacity to conceal, conceal, generate, and harm one's image. (Corley 2007). The first impression a person has of another is based on their clothing, whether it is appropriate or not. The remainder, such as intellectual admiration, comes in second. According to Azuah (2014), by creating fashion illusions that allow the eye to see beautiful appearances while suppressing physical faults, designers can create an ideal appearance of regular proportion. While Ghana has worked to improve the quality of clothing design for local consumption and worldwide export, there are still design concerns in local clothing manufacture that need to be solved. While the local market complains about flaws with apparel design, the international market through AGOA initiative continue to hunt for quality in clothing construction.

Design creativity and assembly principles help improve productivity and relieve bottle-neck situations in the clothing manufacturing industries. However, few fashion designers works address the manufacturing/design from a systematically holistic perspective, implying that literature has yet to explore in detail on the interaction between product design and the satisfaction of the wearer from integral perspective. Engaging the right elements and principles in designing also fosters satisfaction to customers, which are a significant driver and contribution to the build-up of demand for ones products.

2 Theories Underpinning the Study

The study adopted Kano model which portrays attributes that are present in a product leading to customer satisfaction or rejection. The model has five characteristics of a new product that affect customer satisfaction and clothing selection.

Adopted Source: Szymczak and Kowal cited Gurbuz (2018)

The Kano model attributes are categorized into five different groups: the first point "Must-be" explains that in a design there is always a type of feature that satisfies the fundamental requirements of an attire that, when satisfied, are not noticed by customers but which, if unmet, have a significant impact on the design selection. In garment construction some vital processes are carried out which may not be noticed by customers but tend to affect shape when not properly done.

The model also indicates that there is a 'one-dimentional' characteristic which asserts that customers will feel satisfied with a product when some qualities of a design are present but on the other hand would be dissatisfied if their expectations are not fulfilled. Garment designers must always ensure the right features are present in a designs.

The "attractive" characteristics of the model are referred to as "bonuses" which when present in a product, boost customer satisfaction. Nevertheless, their absence has no appreciable impact e.g. some decorative techniques on garments though without any specific major function would enhance the beauty of a dress and consequentely affect the desire for its selection.

There also some attributes that have no effect on customer satisfaction and will have no impact in costumer's purchase decision when included or not.

There are also some "reverse" attributes of the model which affect customer satisfaction negatively and need to be avoided by fashion designers, e.g. fabric defects and seam puckering.

Every designer must always examine the importance of customer needs against their designs. It is important to know how a new product will compare to competing designs that are already on the market. Colour ordinations and style features must be in line with created designs.

Finally the "quality function deployment" concept seeks to incorporate client needs into the design and specs of new products, making them more appealing to consumers.

All textiles and clothing products, including fabrics for interior spaces, are developed and created using design concepts as a foundation. Artisans and other designers have used design components in a number of ways to achieve a specific look throughout history. Every designer uses the elements of design; line, space, form, texture, and colour, regardless of discipline (Anyakoha, 2015).

Clothing design is without a doubt the most apparent manifestation of one's appearance in society, and therefore must be taken seriously. Designers utilize colour, balance, line, shape, proportion, and harmony among other elements to communicate unity, harmony, and a sense of wholeness in their work.

2.1 Colour Influence in Clothing

According to Bell & Ternus, colour is likely the most essential design element, and 2014 is also a personal experience. Our physiology and the viewing environment both influence what we see at any given time and in any particular place. As a result, colour selection is mostly based on personal taste reason why the need for monochromatic colour schemes to be employed to create collections of shades for the consumer.

2.2 Colour Combination in Clothing

To make clothes selection easier, the designer uses colour to create eye-catching combinations. Gupta and Sharma (2021) indicated that colour harmony is a mixture of colours that appear attractive collectively. The choice of fasteners, trims, notions, and fabric colours must all work together to achieve the best results. Three colours are used in a split complementary scheme, with one core colour and two different ones. Double complementary scheme, on the other hand, consists of four colours: two primary colours plus their complementary colours. They must therefore be given adequate attention during clothing construction.

Colours may have different implications in the students' world, which could impact their wardrobe choices. Colour employed to portray emotions and the wearer's identity can be one of the most significant and intriguing aspects of ones design choice. Depending on the garment chosen, colour can make a variety of effects on the wearer. For example, black in Ghana is a typical Ashanti colour for festive occasions which could be used for a variety of purposes in the educational setting. Adolescents may also associate various colours with different meanings. These factors must, of course, be considered by the society, trend of the international world or local fashion designers.

Colours, according to Bell and Ternus (2014) may develop their own names through time. This is usually accomplished by associating with something universally recognized by persons of the same race speaking the same language. Colour is a significant visual element that may set a tone, emphasize characteristics, and draw attention to products on sale. According to studies, colour boosts the power of cloth to attract attention.

2.3 Proportion as an Influential Factor

The proportion in fashion design could be refer to an oversized sleeve worn with a regular-sized shirt or the contrast (severe difference in scale) between large and little pieces that are prominence on an outfit. According to Bell and Ternus (2014), proportion plays a role in fashion coordination as well. A slim lady who wants to display her top, on the other hand, is likely to request such a dress, even if the style is unusual. It is up to the designer to make sure the design is proportioned to a wearer's size.

2.4 Texture as an Influential Factor

It's crucial to remember that texture may both be seen and felt. Fashion designers can influence the appearance of clothing by using textural contrasts in dress design. Remember that the more exposure customers have to certain things, the more likely they are to be persuaded. The selection of texture must therefore be a focus in clothing design (Bell and Ternus 2014).

3 Methodology

The descriptive research method was used for studying designs produced. The population was dressmakers with their produced garments were selected from Accra, Takoradi and Kumasi in Ghana. This was so because most clothing activities were concentrated mainly in these three cities of Ghana. Five clothing items were picked from three garment manufacturing shops in each of the three cities of Ghana. The total clothing items examined for the study were 27, with each of the three shops submitting 3 outfits from each city. Purposive and Simple random sampling was used in selecting fashion houses and designed garments respectively. Designed garments selected from the manufacturing houses were examined against the principles of design and analysed based on content analyses. Principles of the design were used as a guide in determining appropriately designed products and the possibility of their selection. The researcher and assistants visited clothing designing houses and observed designs against the demands of clothing construction and principles/ elements of design.

4 Findings and Discussion

Clothing designs were examined to ascertain the design lapses found in constructed clothing of the fashion houses understudied. Findings were related to colour combinations, stitching, application of elements and principles of designs etc. other key factors in garment construction were also discussed (Fig. 1).

Fabrics and clothing presented below were examined in relation to the demands of each design.

Fig. 1. Colour combination in clothing Source: Field work 2022

Colour proportion application by dressmakers during study was to some extend unsuitably. While the textiles design create a design with a specific statement in mind it is the duty of the fashion designer to bring out clearly his message through the arrangement of design without interruption. As much as the extroverted person prefers warm colours, the restrained person prefers the polar opposite in order to conceal 'him or herself. According to Gbadamosi (2012), several factors such as social media, parental factors, and so on have an impact on clothing colour selection. Furthermore, through daily interactions, about attractions, and clothing designers turn to choice colour based on personal tastes (Johnson and Kim 2013). A practice that does not fit a client's choice.

4.1 Balance Used in Clothing Definition

The term "balance" is used to describe the method of determining the type of clothing (symmetric or assymetric). The phrase composition refers to balancing numerous aspects and elements of a garment in an artistic format in the art world, particularly in fashion design. Stylists use what they know about art to create a garment, but balance must be the guiding principle throughout the process.

As a result, balance is employed as a merchandising tactic. Dressmakers must use notions such appropriate zip fasteners in a skirt, buttons and buttonholes in an outerwear piece to achieve a harmonic effect.

4.2 Rhythm in Clothing Design

The combination of these elements may elicit a desire for such apparel of which designer must beware. Stitches are also used by designers to create a sensation of movement by following the patterns established by a design's composition (Fig. 2).

Fig. 2. Rhythm Source: Takoradi Technical University

Fig. 3. Style emphasizing African print in men's outfits

4.3 Emphasis in Clothing Design

4.4 Emphasis as a Feature in Clothing Selection

In this research work emphasis is placed last for a reason. It plays a very important role in the arrangement of a design of which its absence can the product selection in fashion designing. Bell and Ternus (2014) argued that any of the aesthetic concepts and aspects covered thus far could be used to specific regions. The collection in Fig. 3 is claim to emphasize the usage of African prints in designing men's wear. However, not all outfit were properly designed to portray designers aim. Second outfit from left is an example. When the accent principle is used, a specific component of the clothing is highlighted in a pleasing way. The message the clothing is intended to send when worn determines what the designer emphasized. If a new fashion colour is important, the focus will be on that colour. However too many emphasis disorganizes a design.

4.5 Key Factor in Clothing Selection

Design of clothing is one of the main factors in the selection of clothing. In order to create a magnificent shape that fits the body, fashion designers must grasp and understand the procedures involved in making a three-dimensional garment from a two-dimensional piece. The construction of lines, pockets, collars, finish edges, and how to produce volume and structure to give the wearer a unique look and feel in garment design is the ultimate choice of a designer. The designer determines the most important process of garment construction and provides a foundation on which to build it. It covers pattern cutting and mannequin draping, as well as many methods for turning a flat design drawing into a three-dimensional apparel. Basic sewing skills are discussed, as well as how to use darts, sleeves, collars, pockets, and fabric cuts to change up your designs.

4.6 Cutting/Sewing Details and Clothing Selections

One of the most significant and crucial components for a textile to achieve the established standards in terms of performance and comfort is a garment that corresponds to the human figure and the specific features of the particular garment (Dabolina et al. 2018).

Designers must understand how a garment transforms from a two-dimensional design to a three-dimensional form as soon as possible. Before cutting out and sewing a garment, a pattern is a flat paper or card template designed to specify the features of the garment. It's critical to have a firm grasp on body shape and how body measurements relate to pattern components. To guarantee that the fabric pieces fit together properly and exactly once produced, the pattern cutter must be precise. Sewing specific seams can be utilized to provide decorative detailing, according to Vasbinder (2014) in the book Complete Sewing. The unusual double stitching on the run and fell seam adds flair and should be noted. Top stitching gives definition and sharpness to garments, especially coats which could be a source of clothing selection.

4.7 Defects

The look of a garment determines its beauty. The buyer is more likely to approve of an outfit that is devoid of flaws. Cooklin et al. (2006) reinforced the need to minimize clearly visible cloth faults during spreading in order to avoid rejections due to defective parts. Pattern coordination once again necessitates design alignment during manufacturing; this should be achieved with the least amount of machine manipulation. A distorted outfit along with incorrect shading would almost probably result from poorly positioned components.

Mixed-up colours in clothing might cause key challenges in its marketing. When purchasing garments, someone who values accuracy considers all of these factors. Cutting accuracy is crucial since it has an influence on the dynamics of style especially the silhouete. Cooklin et al. (2006) said again that some finished items could be rejected due to flaws if quality is assured in the manufacturing room.

4.8 Assembling

A badly designed clothing has a limited lifespan. A seam that joins two or more parts of fabrics using series of stitches should really be appropriate for the garment's intended use. The kind of seams is determined by the ultimate use of product. It must meet the appropriate aesthetic and performance standards.

5 Conclusion

Constructional factors are very key to the selection of clothing and appearance. As clothing is chosen because of its beauty, so should every stage in its production be adequately planned to achieve a favourable response in the marketing space. A poorly considered stage in the designing process could cause a company some financial losses and possible closure. The fashion designer needs to apply all relevant creativity skills to produce an excellent outfit that meets individual tastes in garment construction. The clothing manufacturing process is a very vital tool in the selection and satisfaction of an outfit. It is recommended that fashion designers pay more attention to the designing and production process.

References

Adebisi, T., Abdulsalam, A.: Factors influencing colours in clothing selection. J. Asian Reg. Assoc. Home Econ. 24(2), 56–63 (2017)
Bell, J., Ternus, K.: Strategies in Retail Promotion (2014)
Corley, C.F.: Dress for Success, Career Services. Sacramento City College, New Zealand (2007)
Dabolina, I., Silina, L., Apse-Apsitis, P.: Evaluation of clothing fit. In: IOP Conference Series: Materials Science and Engineering, vol. 459, No. 1, p. 012077. IOP Publishing, December 2018
Gurbuz, E.: Theory of New Product Development and its Applications. In (Ed.), Marketing. IntechOpen(2018). https://doi.org/10.5772/intechopen.74527
Dove, T.: Stretch to fit–made to fit. Int. J. Fash. Des. Technol. Educ. 9(2), 115–129 (2016)
Tribalbyn.com/products/egochi-african-print (2021)
Foster, P., Ampong, I.: Pattern cutting skills in small apparel industries and teacher education universities in Ghana. Int. J. Vocat. Tech. Educ. 4(2), 14–24 (2012)
Fisher, C.: Black, Hip and Primed (to shop). American Demographics Inc, New York (1996)
Gbadamosi, A.: Acculturation: an exploratory study of clothing consumption among Black African women in London (UK). J. Fashion Mark. Manage. Int. J. (2012)
Gupta, M.K., Sharma, S.: Significance of basic design elements in spatial and cultural environment of built forms. Significance 5(4) (2021)
Johnson, K., Kim, J.: Clothing functions and use of clothing to alter mood. Int. J. Fashion Design; Technol. Educ. 6(1), 43–52 (2013)
Mahmood, S.: On Demand Product Development Customized For Production (2012)
Omoavowere, E.O., Lily, G.: Analysis of adult female clothing made with adapted patterns and free hand cutting: constraints and prospects. J. Educ. Soc. Res. 1(2), 135 (2011)
Vanderhoff, M.: Clothes, Clues and Career. Ginn and Company, New York-USA (1884)
Vasbinder, N.: Super Stitches Sewing: A Complete Guide to Machine-sewing and Hand-stitching Techniques. Clarkson Potter (2014)
Vikalp, K.: How to apply monochromatic color scheme in design? Many shades of one colour (2021). https://uxplanet.org/how-to-apply-monochromatic-color-scheme-in-designLalji

Assessment of Climate Change Mitigation Strategies in Building Project Delivery Process

A. Opawole[✉] and K. Kajimo-Shakantu

Department of Quantity Surveying and Construction Management, Faculty of Natural & Agricultural Science, University of the Free State, P.O. Box 339, Bloemfontein 9301, South Africa
tayoappmail@gmail.com, kajimoshakantuk@ufs.ac.za

Abstract. Purpose: The purpose of this study is to assess the climate change mitigating strategies adopted at the stages of the building project supply chain towards enhancing the sustainable construction process.

Design/Methodology/Approach: Primary data used for the study were collected through questionnaires administered on 102 construction professionals comprising; architects, quantity surveyors, engineers, and builders in Lagos State, Nigeria. These are major construction professionals that represent various organizations in the supply chain of the building delivery process. Data retrieved were analyzed using both descriptive and inferential statistics.

Findings: Respondents indicated their opinion that the top four stages of the building project supply chain that contribute most to climate change ranked in the order of; the processing of raw materials with means score (MS = 4.14), extraction process (MS = 4.12), and transportation processes (MS = 3.83). The assembling process (MS = 3.40), maintenance operations (MS = 3.32), and other operations such as heating, lighting, etc. (MS = 3.26) were also significant on the scale of assessment used in the questionnaire but they stayed in the category of lesser ranked variables. Findings also indicated that mitigating strategies including waste reduction, waste re-use, and waste recycling (MS = 3.72), more efficient construction processes (MS = 3.62), use of sustainable building materials (MS = 3.59), increased use of local materials (MS = 3.59), were rated more important than those of retrofitting of existing buildings (MS = 3.16), carbon sequestration (MS = 3.03), and carbon mitigation offsets, emissions trading, and carbon tax (MS = 3.01).

Research Limitations: A larger sample size and inclusion of case study projects could bring the findings to a broader perspective.

Practical Implications: More awareness about climate change and its impact should be created by government and professional bodies through the regular advertisement, workshops and seminars to update the knowledge of clients and construction professionals.

Social Implications: Government should enact laws that will regulate indiscriminate extraction of raw materials to minimize its contribution to climate change.

© The Author(s), under exclusive license to Springer Nature Switzerland AG 2023
C. Aigbavboa et al. (Eds.): ARCA 2022, *Sustainable Education and Development – Sustainable Industrialization and Innovation*, pp. 483–492, 2023.
https://doi.org/10.1007/978-3-031-25998-2_36

Originality/Value: The study provided implications for climate change mitigating measures and sustainability in building project delivery.

Keyword: Building process · Climate change · Project delivery · Sustainable construction

1 Introduction

The discussion on climate change has been multidisciplinary. Popular definition of the concept had been a long term and permanent alteration in the global weather resulting from greenhouse gases caused by human activities and natural systems (Ayoade 2003; IPCC 2013; Fawzy et al. 2020). Building supply chain processes contribute at different rate to climate change. These include processes not limited to; extraction of raw materials, manufacturing processes, fabrications, transportation and assemblage of construction materials and components (Ametepey and Ansah 2015). Building processes require energy at varying degree for heating, cooling, drilling, fabrication, welding, drilling, among others, and consequently result in carbon dioxide (CO2) emissions. Building processes put together were identified to responsible for about 40% of CO2 emissions (Lasin et al. 2016).

Review of the processes of building has become necessary to mitigating climate change contribution from the various stages. Recent studies in building delivery processes have emphasized sustainability through green processes and use of alternative construction materials and components (Ngxito et al. 2019). For example, Klufallah et al. (2014) have reported that green buildings could achieve up to 35% reduction in carbon emissions. A number of climate change mitigation initiatives had focused on sustainable construction and these include; renewable energy generation, energy efficient technologies, green buildings, sustainable transportation, and rational use of energy. These initiatives have been well implemented in some developed countries, though the extents of implementation of the strategies have varied across countries and remain unknown in some contexts including the Nigerian construction industry.

One major impediment to sustainable construction is that the delivery process is very complex (Olanipekun 2015) and the favorableness of key project objectives (mostly the cost implication) have not been justified empirically. Kang et al. (2013) had indicated that the use of new technologies and sustainable materials could be very expensive, and significantly push project budget higher (Masia et al. 2020). Whereas studies on climate change mitigating measures have increased significantly in recent times (e.g. Fawzy et al. 2020; Tunji-Olayeni et al. 2021, Olapade 2022, etc.), most of the studies in Nigeria have rather focused on climate change mitigation in other sectors than construction.

Tunji-Olayeni et al. (2021) which had focused on climate change mitigation strategies in the built environment had best centered on manufacturing companies neglecting sources of emissions from other aspects of building process. In another study, Tunji-Olayeni et al. (2019) investigated climate change mitigation and adaptation strategies for construction activities within planetary boundaries. However, the study was limited to review of existing literature on climate change mitigation strategies without conducting an empirical study. The purpose of this study is therefore to assess the climate change mitigating strategies adopted for the stages of building project supply chain towards enhancing sustainable construction process. The study will raise the awareness level of construction professionals on the contribution of the building project supply chain to climate change and its detrimental effects. Consequently, findings will suggest climate change mitigating strategies that are amenable to the Nigerian construction industry. Ultimately, the study will strengthen the body of knowledge on the climate change mitigating strategies in construction project delivery and sustainable building.

2 Literature Review

Climate change is mostly defined as the shift in climate patterns mainly caused by greenhouse gas emissions. United Nations Framework Convention on Climate Change (UN 1992) defined climate change as a change of climate, attributed directly to human activity. According to Yue and Gao (2018), natural systems which had resulted in climate change were identified to include; earthquakes, oceans, permafrost and wetlands, etc.. On the other hand, activities of energy generation, industrial activities and construction are major human activities that have contributed to climate change (Fawzy et al. 2020).

Various production chains of building contribute to climate change at different magnitude. These include; extraction of raw materials, manufacturing and fabrication of materials and components, transportation, on-site assemblage/ construction activities, post-construction operation, and maintenance of buildings. All these processes involve extensive use of energy which results in emission of carbon dioxide into the environment. Climate change mitigation strategy thus refers to the action taken to reduce emissions of CO2 emissions via application of alternative carbon practices and processes. A number of studies have indicated various approaches through which climate change had been mitigated.

Some of these researches have emphasized advancement in green building; adoption of sustainable building and construction; and adoption of alternative materials and components (Ngxito et al. 2019; Oke et al. 2019; Olapade 2012). Other mitigation measures such as; use of sustainable building materials, sustainable design, creation of policy and regulations by government, retrofitting of existing buildings, waste recycling, use of local materials, carbon mitigation offsets, emissions trading, and carbon tax, use of prefabricated elements/off-site manufacturing have also been identified. (Cardez and Czerny 2016; Masia et al. 2020; Fawzy et al. 2020; Tunji-Olayeni et al. 2021). Although these earlier studies indicated robust findings, they had either focused on different contexts of climate change or study areas. Therefore, the aim of this study is to assess the climate change mitigation process in building project delivery with a view to ensuring sustainable construction process. The specific objectives of the study are to; determine the contribution of the various stages of building project supply chain to climate change; and evaluate the climate change mitigating strategies adopted for the stages of building project supply chain.

3 Research Methods

This study adopted a quantitative approach. The study area is Lagos state, Southwestern Nigeria. Respondents were 80 construction professions which were randomly selected from the study area. Lagos is a Southwestern State in Nigeria that hosts the highest number of construction companies operating in Nigeria. Primary data were collected through structured questionnaires administered to construction professionals who were randomly selected from consulting and contracting organizations in the study area. A total number of 138 questionnaires were administered out which 80 copies were found to be properly filled and returned for the analysis. This is a response rate of 58.0%. The valid copies were completed by 16 architects, 42 quantity surveyors, 25 engineers, and 19 builders. The questionnaire was designed in other to capture the research objectives in which the respondents ranked their perception based on the questions asked. The first section of the questionnaire focused on the profile of the respondents which include- type of project undertaken, type of organization, year of experience, academic and professional qualifications. The second section focused on questions relating to the objectives of the study. A 5-point likert scale with intervals between1–5 was be used in obtaining response from the respondents, with 5 representing highest score, and 1 the lowest score. The scales of the questionnaire was considered reliable and internally consistent at Cronbach's alpha value of $0.864 \leq \alpha < 087$. Data retrieved were analyzed using both descriptive and inferential statistics.

4 Results and Discussion

4.1 Profile of the Respondents

This section presents the background information of respondents which includes highest academic qualifications of the respondents, official designations, and respondents' years of experience. The result of the analysis of academic qualifications indicated that more than half of the respondents, 68(65.4%) were First Degree holders, about 17(16.3%) were Master's Degree holders. The least proportion of the respondents, 4(3.8%) were PhD holders. On the professional affiliations of the respondents, majority of the respondents, 42(40.4%) were quantity surveyors. Followed were engineers, 25(24.0%), architects 16(15.4%), and 19(18.3%). The average year of experience of the respondents was 14.5 years. The profile of the respondents revealed that they possess adequate work experience and academic qualifications to provide reliable data for this study.

4.1.1 Contribution of the Stages of Building Project Supply Chain to Climate Change

This section determines the contribution of the various stages of building project supply chain to climate change. A comprehensive literature review was conducted to identify the stages of building project supply chain that contribute to carbon emission. Eight major stages were identified and assessed by respondents based on their level of significance. The data retrieved were analyzed using mean and Kruskal-Wallis test. The result is presented in Table 1. The perceptions of the respondents on the stages of building project supply chain indicated the top four as; processing of raw materials with mean item score of (MS = 4.14), extraction of raw materials (MS = 4.12), transportation of raw materials to factories (MS = 3.83), and transportation of building materials to sites (MS = 3.73).

The stages of building project supply chain with lesser contribution to climate change were assemblage of building components (MS = 3.40), maintenance of the building (MS = 3.32), daily operation of the building (MS = 3.26). Processing of raw materials contributes significantly to climate change. For example, cement is a major construction material with extensive contribution of CO2 in all processes associated with its extraction, processing and final incorporation into building. Similarly, production of steel consumes and releases much gas into the atmosphere in its smelting process. In the extraction of raw materials, large amount of energy is consumed which in turn releases CO2 into the atmosphere. The use of heavy machines that run on fossil energy for mining and extraction of raw materials releases significant amount of CO2 which contributes to CO2. Most of vehicles used in transporting raw materials to factories as well as finished materials to site consume fossil fuels.

Table 1. Stages of building project supply chain that contribute to climate change

Stages	Overall M	SD	R	Architect M	SD	R	Quantity Surveyor M	SD	R	Builder M	SD	R	Engineer M	SD	R	Kruskal Wallis H Test
Processing of raw materials	4.14	0.75	1	4.00	0.56	1	4.36	0.66	1	4.05	0.85	2	3.92	0.86	4	0.114
Extraction of raw materials	4.12	0.65	2	3.88	0.62	2	4.14	0.57	2	4.26	0.65	1	4.12	0.78	1	0.353
Transportation of raw materials	3.83	0.90	3	3.13	1.41	7	3.93	0.75	3	4.05	0.71	2	3.96	0.68	3	0.127
Transportation of materials and components to site	3.73	0.90	4	3.38	1.15	6	3.69	0.95	4	3.74	0.87	5	4.00	0.58	2	0.254
Decommissioning of building	3.54	0.78	5	3.63	0.72	3	3.57	0.67	5	3.63	0.90	6	3.36	0.91	5	0.771
On-site construction/ Assemblage of building components	3.40	0.76	6	3.63	0.50	3	3.29	0.81	6	3.79	0.63	4	3.16	0.80	7	0.030
Maintenance of the building	3.32	0.92	7	3.50	1.03	5	3.26	0.91	8	3.32	0.89	8	3.32	0.95	6	0.767
Daily operation of the building	3.26	0.81	8	3.13	0.96	7	3.29	0.81	6	3.58	0.69	7	3.08	0.76	8	0.102

M = Mean, R = rank, SD = Standard Deviation.

Thus, significant amount of CO2 is emitted. As indicated by Sattary and Thorpe (2012), all processes related to the production of buildings consume significant amount of energy, as such strategy such as bioclimatic principles may be employed in material selection and at construction stages as steps towards reducing embodied energy in the processes.

Kruskal-Wallis test was employed to test the agreement in the opinions regarding the ranking of the stages among the group of respondents. The result showed that assemblage of building components (p-value = 0.030) had a significance value less than 0.05. This implies that there is statistically significant difference in the perception of architects, quantity surveyors, builders, and engineers on this stage. Consequently, there is no significant difference in the perception of the respondents on the remaining seven stages (p-value > 0.05).

4.1.2 Climate Change Mitigating Strategies Adopted for the Stages of Building Supply Chain

The second objective of this study identified the climate change mitigating strategies adopted for the stages of building project supply chain. A review of literature was conducted to identify climate change mitigation strategies and they were assessed by the respondents based on their level of adoption and usage as well as suitability for the processes. The result presented in Table 2 showed that the top five mitigating strategies were waste reduction, waste reuse, and waste recycling (MS = 3.72), more efficient construction processes or techniques (MS = 3.62), use of sustainable building materials (MS = 3.59), increased use of local materials (MS = 3.59), and people-driven change driven by strong demand (MS = 3.58). However, the least five climate change mitigating strategies were retrofitting of existing buildings (MS = 3.16), creation of policy and regulations by government (MS = 3.14), demolition and rebuild (MS = 3.09), carbon sequestration (MS = 3.03), and carbon mitigation offsets, emissions trading, and carbon tax (MS = 3.01).

Waste reduction, waste reuse, and waste recycling is rated high as it would reduce unregulated disposal and burning and thus become an effective mitigating strategy with respective to materials supply. The finding conforms with Tunji-Olayeni et al. (2021) with respect to reuse, and waste recycling as also commonly used climate change mitigation strategy adopted by manufacturing companies. Moreover, Intini and Kuehtz (2011) considered the use of recycled materials (e.g. plastic bottles) as alternative for better thermal insulation which can reduce environmental impact as much as 46%. The use of alternative materials would also suggest reduced processing and transportation energy (Chou and Yeh 2015), as they are mostly embraced where they are locally available. This will consequently result in significant lesser emission of CO_2. Pomponi and Moncaster (2016) had indicated that indicated savings of 9.6 MtCO2e/annum by using wood as an alternative to concrete- and steel-based building systems in the United State building sector. Use of new tools, methods and methodologies in achieving processes like raw materials extraction, manufacturing, fabrication and assemblage would also achieve low-carbon built environment as indicated by (Pomponi and Moncaster 2016).

The result of Kruskal-Wallis test showed that there is a statistically significant difference (sig at $p < 0.05$) in the perception of the respondents on the level of adoption of two of the mitigating strategies; extending the building's life (p-value = 0.005) and demolition and rebuild (p-value = 0.026).

Table 2. Climate change mitigating strategies adopted for stages of building project supply chain

Strategies	Overall M	SD	R	Architect M	SD	R	Quantity Surveyor M	SD	R	Builder M	SD	R	Engineer M	SD	R	Kruskal Wallis H Test
Waste reduction, waste reuse, and waste recycling	3.72	0.89	1	4.00	0.89	1	3.50	0.86	3	3.84	1.07	1	3.80	0.76	1	0.202
More efficient construction processes or techniques	3.62	0.70	2	3.88	0.62	2	3.48	0.74	4	3.63	0.76	3	3.68	0.63	4	0.206
Use of sustainable building materials	3.59	0.69	3	3.50	0.73	9	3.48	0.74	4	3.68	0.48	2	3.76	0.72	2	0.196
Increased use of local materials	3.59	0.80	3	3.75	0.86	5	3.74	0.77	1	3.32	0.67	10	3.44	0.87	8	0.216
People-driven change driven by strong demand from (key role of all BE stakeholders in the built environment)	3.58	0.83	5	3.88	0.62	2	3.60	0.80	2	3.47	0.84	6	3.44	0.96	8	0.539
Sustainable design	3.47	0.93	6	3.25	0.86	13	3.36	0.96	8	3.63	1.01	3	3.68	0.85	4	0.289
Use of innovative tools, methods, and methodologies	3.45	0.83	7	3.63	0.72	7	3.38	0.83	7	3.42	0.90	8	3.48	0.87	7	0.837
Extending the building's life	3.45	0.91	7	3.88	1.09	2	3.10	0.66	12	3.47	0.84	6	3.76	1.01	2	0.005
Increased use of prefabricated elements/off-site manufacturing	3.40	0.81	9	3.50	1.03	9	3.48	0.71	4	3.21	0.79	13	3.36	0.86	10	0.588
Reduction, re-use and recovery of embodied energy and carbon intensive construction materials	3.36	0.85	10	3.75	1.00	5	3.33	0.87	9	3.37	0.83	9	3.16	0.69	15	0.211

(*continued*)

Table 2. (continued)

Strategies	Overall M	Overall SD	Overall R	Architect M	Architect SD	Architect R	Quantity Surveyor M	Quantity Surveyor SD	Quantity Surveyor R	Builder M	Builder SD	Builder R	Engineer M	Engineer SD	Engineer R	Kruskal Wallis H Test
Use of renewable energy	3.34	0.93	11	3.25	0.68	13	3.12	0.89	11	3.63	1.01	3	3.56	1.00	6	0.158
Policy and regulations by the construction sector	3.17	0.88	12	3.13	0.81	16	3.17	0.91	10	3.05	0.71	17	3.28	1.02	11	0.866
Retrofitting of existing buildings	3.16	0.82	13	3.38	0.89	11	2.95	0.76	13	3.26	0.93	12	3.28	0.74	11	0.284
Creation of policy and regulations by government	3.14	0.99	14	3.25	1.13	13	2.93	1.05	14	3.32	1.00	10	3.28	0.74	11	0.467
Demolition and rebuild	3.09	1.02	15	3.63	1.26	7	2.76	0.93	17	3.21	0.92	13	3.20	0.91	14	0.026
Carbon sequestration	3.03	0.90	16	3.37	0.72	12	2.93	0.92	14	3.21	0.79	13	2.84	0.99	17	0.102
Carbon mitigation offsets, emissions trading, and carbon tax	3.01	1.00	17	3.13	0.81	16	2.90	1.08	16	3.11	0.99	16	3.04	1.02	16	0.878

5 Conclusion

The various stages of building project supply chain contributing to climate change have been identified from the literature review and they were assessed by respondents as regards the significance of their contribution to climate change. The study revealed that the mean score ranking of the eight identified stages of building project supply chain contributing to climate change. Five of the identified stages had mean score greater than 3.50 on a scale of 5.00. This implies that these stages of building project supply chain contribute significantly to climate change viz processing of raw materials, extraction of raw materials, transportation of raw materials to factories, transportation of building materials to site, and decommissioning of buildings. Furthermore, the study identified climate change mitigating strategies adopted for the various stages of building project supply chain and assessed the level of their adoption. The mean score ranking of the top five strategies ranged from 3.58 to 3.72. The implication is that these five strategies have been adjudged by the respondents to be often used in the stages of building project supply chain. Based on the findings of this study, the following recommendations are proposed. In the first place, government should enact laws that will regulate indiscriminate extraction of raw materials to minimize its contribution to climate change. Manufacturing companies should embrace technologies that ensure minimal release of

CO2 in the processing of raw materials. More awareness about climate change and its impact should be created by government and professional bodies through the regular advertisement, workshops and seminars to update the knowledge of clients and construction professionals. Courses on climate change and sustainable construction should be integrated into the curriculum of built environment studies to increase the knowledge level of built environment students on the subject. Provision of adequate funding and support is required from the government to subsize and make available sustainable construction materials. The study provided implications for climate change mitigating measures and sustainability in building project delivery.

References

Ametepey, S.O., Ansah, S.K.: Impacts of construction activities on the environment: the case of Ghana. J. Const. Project Manage. Innov. **4**(1), 934–948 (2014)

Ayoade, J.O.: Climate Change: A Synopsis of its nature, causes, effects and management. Vantage Publishers, Ibadan (2003)

Cadez, S., Czerny, A., Letmathe, P.: Stakeholder Pressures and Corporate Climate Change Mitigation Strategies. Bus. Strateg. Environ. **28**, 1–14 (2019)

Chou, J.S., Yeh, K.C.: Life cycle carbon dioxide emissions simulation and environmental cost analysis for building construction. J. Clean. Prod. **101**, 137–147 (2015)

Fawzy, S., Osman, A.I., Doran, J., Rooney, D.W.: Strategies for mitigation of climate change: a review. Environ. Chem. Lett. **18**(6), 2069–2094 (2020). https://doi.org/10.1007/s10311-020-01059-w

Intini, F., Kuehtz, S.: Recycling in buildings: an LCA case study of a thermal insulation panel made of polyester fiber, recycled from post-consumer PET bottles. Int. J. Life Cycle Assess. **16**, 306–315 (2011)

IPCC. IPCC factsheet: what is the IPCC? (2013). https://www.ipcc.ch/site/assets/uploads/2018/02/FS_what_ipcc.pdf

Kang, Y., Kim, C., Son, H., Lee, S., Limsawasd, C.: Comparison of preproject planning for green and conventional buildings. J. Const. Eng. Manage. **139**(2013), 1–10 (2013)

Klufallah, M., Nuruddin, M., Khamidi, M., Jamaludin, N.: Assessment of carbon emission reduction for buildings projects in Malaysia-a comparative analysis. In: E3S Web of Conference 3. p. 01016 (2014). https://doi.org/10.1051/e3sconf/20140301016

Masia, T., KajimoShakantu, K., Opawole, A.: A case study on the implementation of green building construction in Gauteng Province SouthAfrica. Manage. Environ. Qual. **31**(3), 602–623 (2020)

Ngxito, B., Kajimo-Shakantu, K., Opawole, A.: Assessment of alternative building technologies (ABT) for pre-tertiary school infrastructure delivery in the Eastern Cape Province, South Africa. Manage. Environ. Qual. **30**(5), 1152–1170 (2019)

Oke, P.R., Roughan, M., Cetina-Heredia, P., Pilo, G.S.: Revisiting the circulation of the East Australian Current: Its path, separation, and eddy field (2019). https://doi.org/10.1016/j.pocean.2019.102139

Olanipekun, A.: Successful delivery of green building projects: A review and future directions. J. Const. **8**(1), 31–40 (2015). https://www.asocsa.org/documents/JoC-March2015-vol8-no1.pdf

Olapade, O.J.: Application of alternative construction materials and components in building projects in Lagos State, Unpublished B.Sc, dissertation, Department of Quantity Surveying, Obafemi Awolowo University, Ile-Ife, Nigeria (2021)

Pomponi, F., Moncaster, A.: Embodied carbon mitigation and reduction in the built environment – What does the evidence say? (2016). http://dx.doi.org/10.1016/j.jenvman.2016.08.036

Tunji-Olayeni, P.F., Omuh, I.O., Afolabi, A.O., Ojelabi, R.A., Eshofonie, E.E.: Effects of construction activities on the planetary boundaries. J. Phys. **1299**, 012005 (2019). 1–7

Tunji-Olayeni, P., Osabuohien, E., Oluwatobi, S., Babajide, A., Adeleye, N., Agboola, M.: Climate change mitigation strategies: a case of manufacturing companies in Ota, Nigeria, IOP Conf. Ser. Earth Environ. Sci. **655** (2021). https://doi.org/10.1088/1755-1315/655/1/012068

Sattary, A., Thorpe, D.: Optimizing embodied energy of building construction through bioclimatic principles. In: Smith, S.D. (ed.) Proceedings of 28th Annual ARCOM Conference, Edinburgh, UK, Association of Researchers in Construction Management, pp. 1401–1411 (2012)

UN: United Nations Framework Convention on Climate Change (1992). https://unfccc.int/files/essential_background/backgroundpublications_html.pdf/

Yue, X.L., Gao, Q.X.: Contributions of natural systems and human activity to greenhouse gas emissions. Adv. Clim. Change Res. **9**, 243–252 (2018)

Prior Knowledge of Sustainability Among Freshmen Students of University of Professional Studies

N. A. A. Doamekpor[✉] and E. M. Abraham

Department of Business Administration, University of Professional Studies, Accra, Ghana
naaadjeleyashiboe@yahoo.co.uk

Abstract. Purpose: This study investigates UPSA students' prior knowledge of sustainable development (SD) as well as its association with students' gender.

Design/Methodology/Approach: The study is descriptive with data obtained using a cross-sectional survey of freshmen. Data were collected using convenient sampling during orientation sessions for the students. A total of 235 responses were received. The collected data were analysed using relative index to rank items related to students' knowledge of various aspects of SD. Mann-Whitney U test was used to test if there was a significant difference between males and females with regard to the top three items with the highest level of knowledge.

Findings: The findings showed low levels of knowledge on environmental legislation, policy and standard and no significant difference in knowledge of SD between males and females.

Implications/Research Limitations: The results highlight the need for including and emphasizing content on SD aside environmental issues. Particularly legislation, policy and standards should be discussed. Although the participants of the study were first year students from one public university it provides empirical evidence for SD knowledge amongst business and management students. Qualitative approaches are required to further explore the nature and depth of the knowledge students have.

Practical Implications: Understanding student knowledge will inform what is lacking and hence be the focus of subsequent education for sustainability so as to provide valuable grounding in the concepts and principles of sustainability.

Originality/Value: Past studies have investigated sustainability education around the world and across different disciples and levels but have not adequately studied prior knowledge. The findings of this study provide evidence of the knowledge students have of sustainability prior to the start of higher education which is relevant to facilitating further learning on the subject and curriculum development.

Keyword: Education · Ghana · Prior knowledge · Sustainable development · Tertiary

1 Introduction

Education is a critical element for promoting and ensuring a drive toward a more sustainable world. For individuals and organisations to make appropriate choices and decisions towards sustainability, they must learn how to consider the long-term future implications on ecology, economy and society in decision-making. Education for sustainability, therefore, seeks to incorporate knowledge of sustainable development into all courses, at all levels of education (Buckler and Creech, 2014; Cebrián et al., 2019) so as to improve the capability of individuals and organisations to tackle environmental and developmental concerns (UNESCO, 1992).

Education for Sustainable Development (ESD) is "an approach to teaching and learning based on the ideals and principles that underlie sustainability" (Kopnina and Meijers 2014, p. 189). The aim of ESD is to advance the students' knowledge and skills essential for involvement in development issues (Kopnina, 2012). It recognizes that education improves and sustains the aptitude for evaluating and making preferences for Sustainable Development of individuals, associations, institutions and countries and is, therefore, a precondition for SD (UNECE, 2003).

Several studies have investigated sustainability education around the world and across different disciples and levels. In Ghana, however, sustainability education-related research is limited. Research at basic, secondary and tertiary levels has been conducted (Abaidoo, 2016) but has focused on narrow issues of sustainability (Debrah et al. 2021). Prior studies have suggested investigations on the status of EE and ESD in higher education levels since students at this level normally represent the age groups that are likely to be affected by environmental problems of current human activities. In Ghana, such studies at the tertiary level, emphasise curriculum analyses of sustainability content (Etse and Ingley, 2015; Witoszek, 2020). These studies have raised the need for intentional and increased efforts to integrate sustainability in our curriculum.

Many higher education institutions are therefore including sustainability-related input. The University of Professional Studies, Accra (UPSA) is one of the Universities that has included sustainability-related issues in its programmes and activities. UPSA requires all its freshmen to undertake a mandatory course in Environmental management and sustainability. The success or otherwise of such courses to develop and encourage sustainability practices, values and principles in its students is predicated on various precedent factors. One such factor is the prior knowledge of students.

Prior knowledge is the knowledge available in a person's long-term memory at the outset of learning (Simonsmeier et al., 2018). Inadequate or fragmented prior knowledge may hamper learning from the start of studies and may account for between 30 and 60% of the variance in study results (Dochy, 1996). Several studies (Azapagic et al., 2005; Summers et al., 2004; Emmanuel and Adams, 2011; Tuncer, 2008) have investigated tertiary students' knowledge of sustainable development – the aspects of knowledge (social, environmental and/or economic), environmental standards, legislations and policies. Some such studies investigated gender differences in knowledge with mixed results (Chawla, 1988; Loughland et al., 2003; Kagawa, 2007). It is however unclear whether the knowledge investigated is prior knowledge or not.

Prior knowledge should be an important focus in Education for Sustainability since it has an effect on successful learning of sustainability. The effect of prior knowledge has several determinants such as whether prior knowledge is activated (i.e. retrieved from memory); relevant for the learning task at hand and is congruent or incongruent with the to-be-learned content (Brod, 2021). When prior knowledge is appropriately ascertained, teaching can then be done accordingly. Consequently, further education on sustainability must be guided by investigations of prior knowledge. Such investigations remain limited particularly at the tertiary level, amongst business and management students and in the African context.

This study, therefore, investigated the sustainability knowledge UPSA level 100 students have prior to starting their higher education. Understanding student knowledge will inform what is lacking and hence be the focus of subsequent education for sustainability. Students can then develop a broad understanding of sustainability grounded in the concepts and principles of sustainability and to apply these principles in their professional fields (AASHE 2014).

2 Method

The study population was first year UPSA undergraduate students of the 2019/2020 academic year. The total number of students admitted was 7652. Data were collected using convenient sampling during orientation sessions for the students. A total of 235 responses were received. The questionnaire used for the study was adapted from Al-Naqbi and Alshannag (2018). It consisted of a 45-item scale used to measure student knowledge of environmental issues; environment legislation, policy and standards; environmental tools, technologies and approaches and sustainable development. Students were also asked to rate the level of importance attached to sustainable development.

3 Results

3.1 Respondent Characteristics

The sample was made up of students from four different programmes with students of Business Administration being in the majority (Table 1). The females (51%) were slightly more than males (49%) with the average age of the students being 19 years.

3.2 Student Prior Knowledge

Amongst the four categories of environmental and sustainability-related items, students' knowledge of environmental issues were highest whilst their knowledge of Environmental legislation, policy and standards were lowest. Of the top ten items with the highest knowledge, 8 were environmentally related items (Table 3). Table 2 presents the Average Index for SD categories. The results show that both Environmental issues and Sustainable Development had high average relative indices (0.692 and 0.583, respectively).

Table 3 and Table 4 present student knowledge of items ordered according to the top ten and the lowest ten respectively.

Mann–Whitney U (Table 5) was used to examine if there was a significant difference between males and females with regard to the top three items with the highest level of knowledge. The males generally have higher ranks suggesting that they have higher values of knowledge. Given that the significance values are greater than 0.05, males and females do not have significant difference in their perceptions.

Table 1. Demographic Characteristics

PROG

	Frequency	%	Valid %	Cumulative %
Business Administration	150	63.8	63.8	63.8
Business Economics	25	10.6	10.6	74.5
Public Relations	44	18.7	18.7	93.2
Real Estate Management and Finance	16	6.8	6.8	100.0
Total	235	100.0	100.0	

SEX

Female	120	51.1	51.1	51.1
Male	115	48.9	48.9	100.0
Total	235	100.0	100.0	

Table 2. Average Relative Index for SD Categories

Ref	SD Category	Average RI	NO	YES
EN	Environmental issues	0.691	31%	69%
SD	Sustainable Development	0.583	52%	48%
ETTA	Environmental Tools, Technologies and Approaches	0.422	80%	20%
ELPS	Environment Legislation, Policy and Standards	0.285	97%	3%
	TOTAL	0.528	60%	40%

Table 3. Student Knowledge: Top Ten Items

Item	RI	Rating	Ref	NO	YES	N
Deforestation	0.839	1	EN	4%	96%	235
Water pollution	0.834	2	EN	7%	93%	233
Air pollution	0.827	3	EN	7%	93%	235
Ecosystem	0.791	4	EN	10%	90%	233
Desertification	0.788	5	EN	11%	89%	234
Depletion of natural resources	0.787	6	EN	14%	86%	235
Climate change	0.751	7	EN	16%	84%	233
Global warming	0.748	8	EN	19%	81%	233

(*continued*)

Table 3. (*continued*)

Item	RI	Rating	Ref	NO	YES	N
Population growth	0.725	9	SD	25%	75%	228
Sustainable development – definition and the concept	0.717	10	SD	23%	77%	224

Table 4. Student Knowledge: Lowest Ten Items

Item	RI	Ref	NO	YES	N
Kyoto Protocol	0.273	ELPS	99%	1%	233
ISO 14001	0.274	ELPS	98%	2%	232
Montreal Protocol on CFCs	0.276	ELPS	98%	2%	232
The Florence Convention	0.283	ELPS	98%	2%	230
Rio Declaration	0.285	ELPS	98%	2%	222
EU EMAS	0.299	ELPS	96%	4%	231
Intergovernmental Panel on Climate Change (IPCC)	0.308	ELPS	94%	6%	232
Eco-labelling	0.353	ETTA	90%	10%	229
Photochemical smog	0.366	EN	88%	12%	231
Tradable permits	0.372	ETTA	89%	11%	228

Table 5. Gender differences in Knowledge

	SEX	N	Mean	SD	RI	Asymp. Sig (2-tailed
Deforestation	Male	115	2.40	0.660	0.850	0.115
	Female	120	2.32	0.550	0.829	
	Total	235			0.839	
Water pollution	Male	115	2.35	0.750	0.837	0.521
	Female	118	2.32	0.678	0.831	
	Total	233			0.834	
Air pollution	Male	115	2.34	0.736	0.835	0.173
	Female	120	2.28	0.608	0.819	
	Total	235			0.827	

4 Discussion

The results indicated that UPSA freshmen showed overall high levels of understanding of SD. The results are comparable to Al-Naqbi and Alshannag 2018. However, this knowledge was mostly higher for environmental issues but very low for issues related to environmental legislation, policy and standards. Their levels of knowledge in environmentally related issues such as deforestation were high, which is consistent with Nicolaou and Conlon (2012) and Doamekpor and Duah (2022), who found that knowledge of sustainability was focused more on environmental rather than economic and social dimensions. These results may be due to a history of Sustainability that has emphasized environmental issues Edum-Fotwe and Price (2009). Strengthening students' knowledge in sustainability education prior to entering the University is important and should emphasise other dimensions as well as tools and approaches to SD.

Unlike findings from Al-Naqbi and Alshannag (2018) and Levinine and Strube (2012) there was no difference in knowledge of SD by males compared to females. Whereas Al-Naqbi and Alshannag (2018) found females to be more knowledgeable, Levinine and Strube (2012) reported that females showed less knowledge compared to males.

5 Conclusion and Recommendations

The study investigated the prior knowledge of UPSA freshmen towards sustainable development. The association of gender with knowledge was also considered. The findings showed low levels of knowledge on environmental legislation, policy and standard emphasizing what knowledge may be lacking (gaps) in sustainability education. The results also highlight the need for including and emphasizing content on SD aside environmental issues. Particularly legislation, policy and standards should be discussed. This directs the focus of subsequent education for sustainability on these gaps so that students can develop a broad understanding of sustainability issues. Students are therefore better prepared to apply sustainability in their professions which further helps promote sustainability practices in society. Although the participants of the study were first year students from one public university it provides empirical evidence for SD knowledge amongst business and management students. The study highlights the need for further investigations into student knowledge of sustainability. Qualitative approaches are required to investigate the nature of the knowledge students have and the depth.

References

Abaidoo, C.: Assessing students' understanding of, and responses to, climate change in Ghana: A study at the University of Cape Coast (Doctoral dissertation, University of Cape Coast) (2016)

AlNaqbi, A.K., Alshannag, Q.: The status of education for sustainable development and sustainability knowledge, attitudes, and behaviors of UAE University students. Int. J. Sustain. Higher Educ. **19**(3), 566–588 (2018)

Association for the Advancement of Sustainability in Higher Education (AASHE) (2014). "Stars technical manual: version 2. www.aashe.org/files/documents/STARS/2.0/stars_2.0_technical_manual_-_administrative_update_two.pdf. Accessed 4 May 2015

Azapagic, A., Perdan, S., Shallcross, D.: How much do engineering students know about sustainable development? The findings of an international survey and possible implications for the engineering curriculum. Eur. J. Eng. Educ. **30**(4), 349–361 (2005)

Brod, G.: Toward an understanding of when prior knowledge helps or hinders learning. Sci. Learn. **6**(24), 1–3 (2021)

Buckler, C., Creech, H.: Shaping the future we want: UN Decade of Education for Sustainable Development; final report. UNESCO (2014)

Cebrián, G., Segalàs, J., Hernández, À.: Assessment of sustainability competencies: a literature review and pathways for future research and practice. Central Eur. Rev. Econ. Manage. 3(3), 19–44 (2019)

Chawla, L.: Children's concern for the natural environment. Children's Environ. Quart. 5(3), 13–20 (1988)

Debrah, J.K., Vidal, D.G., Dinis, M.A.P.: Raising awareness on solid waste management through formal education for sustainability: a developing countries evidence review. Recycling 6(1), 6 (2021)

Dochy, F.J.: Assessment of domain-specific and domain-transcending prior knowledge: Entry assessment and the use of profile analysis. In: Alternatives in Assessment of Achievements, Learning Processes and Prior Knowledge, pp. 227–264. Springer, Dordrecht (1996). https://doi.org/10.1007/978-94-011-0657-3_9

Doamekpor, N.A.A., Duah, D.: Conceptions of sustainability amongst post graduate (MSC) construction management students. In: Laryea, S., Essah, E. (eds.) Proceedings of 10th West Africa Built Environment Research (WABER) Conference, pp. 305–318. Accra, Ghana (2022)

Edum-Fotwe, F.T., Price, A.D.: A social ontology for appraising sustainability of construction projects and developments. Int. J. Project Manage. 27(4), 313–322 (2009)

Emanuel, R., Adams, J.N.: College students' perceptions of campus sustainability. Int. J. Sustain. Higher Educ. 12(1), 79–92 (2011)

Etse, D., Ingley, C.: Higher education curriculum for sustainability: course contents analyses of purchasing and supply management programme of polytechnics in Ghana. Int. J. Sustain. Higher Educ. (2016)

Kagawa, F.: Dissonance in students' perceptions of sustainable development and sustainability: implications for curriculum change. Int. J. Sustain. Higher Educ. 8(3), 317–338 (2007)

Kopnina, H.: Education for sustainable development (ESD): the turn away from 'environment' in environmental education? Environ. Educ. Res. 18(5), 699–717 (2012)

Kopnina, H., Meijers, F.: Education for sustainable development (ESD): exploring theoretical and practical challenges. Int. J. Sustain. Higher Educ. 15(2), 188–207 (2014)

Levine, D.S., Strube, M.: Environmental attitudes, knowledge, intentions and behaviors among college students. J. Soc. Psychol. 152(3), 308–326 (2012)

Loughland, T., Reid, A., Walker, K., Petocz, P.: Factors influencing young people's conceptions of environment. Environ. Educ. Res. 9(1), 3–20 (2003)

Nicolaou, I., Conlon, E.: What do final year engineering students know about sustainable development? Eur. J. Eng. Educ. 37(3), 267–277 (2012)

Perrault, E.K., Clark, S.K.: Sustainability in the university student's mind: are university endorsements, financial support, and programs making a difference? J. Geosci. Educ. 65(2), 194–202 (2017)

Simonsmeier, B.A., Flaig, M., Deiglmayr, A., Schalk, L., Schneider, M.: Domain-specific prior knowledge and learning: a meta-analysis. Research Synthesis 2018a, Trier, Germany (2018a)

Simonsmeier, B.A., Flaig, M., Deiglmayr, A., Schalk, L., Schneider, M.: Domain-specific prior knowledge and learning: a meta-analysis. Research Synthesis 2018b, Trier, Germany (2018b)

Summers, M., Corney, G., Childs, A.: Student teachers' conceptions of sustainable development: the starting-points of geographers and scientists. Educ. Res. 46(2), 163–182 (2004)

Tuncer, G.: University students' perception on sustainable development: a case study from Turkey. Int. Res. Geographical Environ. Educ. 17(3), 212–226 (2008)

Weehen, H.V.: Towards a vision of sustainable University- a case of University of Amsterdam. IJSHE 1(1), 20 (2000)

Wiernik, B.M., Ones, D.S., Dilchert, S.: Age and environmental sustainability: a meta-analysis. J. Manage. Psychol. 28(7–8), 826–856 (2013)

Witoszek, N.: Teaching sustainability in Norway, China and Ghana: challenges to the UN programme. Environ. Educ. Res. **24**(6), 831–844 (2018)

UNECE: Draft UNECE strategy for education for sustainable development, CEP/AC.13/2004/3 (2003)

UNESCO: UN Conference on Environment and Development: Agenda 21, UNESCO, Switzerland (1992)

Conceptual Framework on Proactive Conflict Management in Smart Education

P. Y. O. Amoako(✉)

School of Computing, College of Science, Engineering and Technology,
University of South Africa, P.O.Box 392, Pretoria UNISA 0003, South Africa
`57640777@mylife.unisa.ac.za`

Abstract. Purpose: Smart education has been the new paradigm in open distance electronic learning or virtual learning. Through the use of intelligent technologies, smart education, an emerging educational paradigm, tends to promote and develop students into future world leaders. The challenge is that many conflicting issues affect the implementation of smart education. This research aims at providing a proactive method of resolving conflicts in such a promising educational system.

Design/methodology/approach: It presents a comprehensive empirical survey on smart education models and reconstructs them to consider a new approach to managing conflict in such smart educational models. The proposed framework relies on the classified critical conflict factors in the smart education environment and the possible resolution strategies.

Findings: The findings can be employed as a reference guide to understanding smart education models and the effectiveness of ensuring a conflict-free smart education environment.

Research Limitation: The study constitute educational models but limited to those that are concerned with open distance electronic learning. Many educational models are in literature, however, the focus of this paper is on only those that provide smart education.

Practical implication: The study presents the following recommendations; a need for establishing standards for constructing educational institution's systems with conflict management, build the capacity of quality service in open distance electronic learning institutions.

Social implication: The survey adds literature to the ongoing research in smart education. In addition, it unfolds new knowledge to the policy makers to amend policies to provide standardized scheme for smart education. This will ensure quality of service in open distance electronic learning environment.

Originality/Value – This research is relevant to conceptualizing smart education implementation with a focus on conflict management and developing a novel proactive automatic conflict management in smart education as a reference guide to understanding conflict management.

Keyword: Conflict classification · Education · Management · Paradigm · Smart

1 Introduction

Typically, Smart education aids teachers in providing learners with effective and efficient learning experiences by integrating innovative learning analytics technology and fine-grained domain knowledge that moves current learning models to universal and contextualized education. Smart education platforms apply technology to guarantee non-venue-based education and incorporate models to improve the "smartness" content and pedagogy by addressing matters such as the scope of customization, the measure of ubiquity, and the level of self-regulation or co-regulation.

Information technology, which has become an indispensable component of human life and society, has caused the evolution of new network generation people whose lives are worthless without emerging technologies. Another notable factor causing every person to rely on technology, specifically virtual activity-supported technologies, is the pandemic that has caused uncertainties in face-to-face operation. Despite this, modern education systems retard sufficient support in advancing the human capital in favorable conditions in virtual environments.

There is a clear trend in the contemporary educational environment on the movement towards smart education, which provides the opportunity to build up smart personnel for the current smart society. It is evident by the emerging new smart universities employing smart technologies in their educational process and management (Semenova et al. 2017).

Conflict is currently a factor in academic life, and learning institutions frequently experience tension in some cases. There exist varying forms of conflict due to its source and has revealed different thoughts in its definition. Thomas defines conflict as "the process that commences when one agent perceives that the other has frustrated, or is about to frustrate some of his interest" (Thomas 1976). Conflict is inevitable and inherently good because it raises and addresses challenges, energizes operations on the most relevant issues, makes agents "be real," and motivates participation. This aids agents learn to identify the differences and how to make them beneficial.

According to (Lyapina et al. 2019), the operating system's compatibility to various smart programs of the educational process of varying teaching directions ensures the establishment of new fundamental science based on mobility and speed of flow of information. Hence, this demand creating new innovational and technological dimensions of research to support diversified teaching and learning agent to prevent their conflicting interest in smart education.

A systematic review on the research landscape of smart education based on a bibliometric analysis revealed the complete silence of researchers on the concept of conflict issues hindering the advancement of smart education (Li and Wong 2021). It was noted that the majority of the study on smart technologies for teaching and learning has to do with hot topics like the Internet of Things, big data, flipped learning, and gamification.

This paper presents a proactive conflict management model for smart education to consider the possible sources of conflict and appropriate resolution techniques in smart education.

2 Method

A systematic literature review aims to identify, evaluate and interpret all relevant available publications suitable under specified criteria to respond to the research question. The work of Kitchenham and Charters presents a detailed plan to specify the quality measurement for literature selection to ensure the review quality (Kitchenham et al. 2010). This research employs such a plan to ensure the review quality. The likelihood of bias is reduced significantly in the review process to advance research efforts with a balanced analysis and interpretation of findings.

This research follows a three-step strategy; the first step is to identify the review requirements, establish the research questions, deduce, and evaluate a review procedure. The second step is to conduct the search policy, choose the criteria and articles' quality, and perform extraction processes to obtain the relevant models. Finally, the relevant models are discussed and reconstructed to answer the research questions.

2.1 Research Questions

The review aims to elicit the state-of-the-art smart education models, the conflicts, and the resolutions required for quality of service in open distance electronic learning. For this purpose, extraction, and analysis are based on the following questions:

RQ1: What are the publication trends and contexts of smart education models?
RQ2: What are the conflicts in smart education of open distance electronic learning?
RQ3: What is the common resolution approach to eliminate conflict in smart education?

2.2 Databases and Search String

The study concentrated on the journals and conference proceedings from January 2012 to June 2022. The selected academic databases as primary sources were Scopus, Science Direct, ACM Digital Library, IEEE Xplore, and SpringerLink. These relevant academic databases were selected based on the advancement in smart education concepts and open distance electronic learning.

The chosen search string for limiting the number of articles obtained and ensuring their relevance by applying keywords of Boolean operators like OR and AND representing synonyms and associative words respectively. Keywords related to the research topic, synonyms, and alternative words were identified manually. The keyword criteria employed to search in the database indexes include; "smart education", "smart model", "smart learning", and "e-learning conflict", "virtual conflict". Diverse combinations of these keywords applying "AND" and "OR" commands are used as the search string.

2.3 Inclusion and Exclusion Criteria

This stage was carried out to determine which article is relevant for further review. We defined the inclusion and exclusion criteria in Table 1.

Table 1. Inclusion and exclusion criteria

Inclusion criteria	Exclusion criteria
Published between 2012 and 2022	book chapters, abstract publications, technical reports, and dissertations
Written in English	Do not mention keywords in the research area
Discussed smart education or e-learning models, e-learning conflict, or virtual conflict	Discussed general education and conflict issues
Included in scientific conferences and journals	do not include smart education models, and strategies for conflict management

2.4 Search and Selection

A quality assessment filter was used to ascertain the articles that passed the inclusion and exclusion criteria for smart education models, conflict identification, and resolution. Table 3 shows the outcome of the search and selection process.

Selection process	Number of articles
Identification with the search string	754
Exclusion & inclusion criteria	62
Remove duplicated studies	58
Screened titles, abstracts, and keywords	27
Eligibility assessment	7

3 Conflict Management in Smart Education

Conflict management is a critical issue in every establishment and smart universities providing smart education need to pay much attention. This section discusses smart education and the concept of conflict and its management.

3.1 Smart Education

By analyzing smart education projects conducted by some developed countries, (Zhu et al. 2016) established some general concepts on smart education. They indicate that smart education aims to foster the workforce to master 21st-century knowledge and skill for societal needs and challenges. Again, (Zhu et al. 2016) define 'smart' regarding the educational environment as engaging, intelligent, and scalable. In order to support customised and individualized learning, the environment must include features like context awareness, content adaptability, collaborative and interactive tools, quick evaluation, real-time feedback, and eventually handling every dispute. Soaring to engage the learner in meaningful learning with effectiveness and efficiency. It must be possible to integrate cutting-edge interfaces, smart devices, and a variety of learning data in a conflict-free operating learning environment, which is one of the prerequisites for open system design.

3.2 Conflict in Smart Education

In the perspective of (Ozturk and Simsek 2012), conflict is the perceived overt or hidden events generated from a different group of agents, triggered by dynamics of the conflict having a role in the smart education process. Systems must be able to resolve a variety of conflict types on behalf of agents, or else they will have to accept their limitations and tell agents that the conflict has been addressed directly, which is not a proactive solution. Resolving a conflict between multiple agents is more difficult than one agent. As a result, the number of agents depends on how challenging context inference and conflict resolution are.

Several conflicts can happen in different dimensions depending on the source, intervenient, detection time, and the ability to solve them. Concerning the source, according to (Carreira et al. 2014), conflicts can occur when multiple agents use the same resource, when multiple agents use the same application, when multiple agents use the same set of policies, when multiple agents use the same set of policies in different contexts, or when multiple agents use the same set of agent profiles and preferences. They once more proposed, based on intervenient, that conflict situations can either be single-agent—where one agent's intentions are at odds with one another—or agent vs. agent (multi-agent), where more than one agent contends for control of a given resource, application, or environment state. Agent vs. bandwidth—where an agent's actions are at odds with any established bandwidth allocation policies—is regarded as a resource-related conflict.

A dispute identified as priori, projected to happen, in their submission on the moment of detection as another means of classifying conflicts. It can also be a possible potential conflict if the likelihood of occurrence is lower than that of the specific probable conflict, or it can also be a definite potential conflict, meaning a conflict that will happen if the agent is in the right situation. Additionally, if a conflict arises, it can be identified (also known as an actual conflict), either by using the context-awareness capabilities of the structure or by obtaining agent feedback. The last situation is when a conflict is detected only after a sufficient amount of time has passed to resolve it, maybe as a result of limitations on context awareness or sensing delays (Lee 2010).

They contend that, in the best-case scenario, conflicts that are settled before they are even noticed as occurring are thought to be avoided. If the problem is not discovered before it occurs, the resolution approach is used. In most cases, a conflict is detected when it really occurs, at which point the system either tries to resolve it or recognizes that it is unable to do so. There is also a chance that the system will miss a conflict and have to deal with it later, possibly due to a sensor's response time being delayed. Figure 1 illustrates the conflict classification in relation to the aforementioned..

3.3 Conflict Management Model in Smart Education

Meng et al. proposed a smart pedagogy framework based on the SMART key elements model (situated learning (S), mastery learning (M), adaptive learning (A), reflective learning (R) and thinking tools (T)), curriculum design method and detailed teaching strategy (Meng et al. 2020). A significant result revealed that applying smart pedagogy promotes learning outcomes failed to address the conflict between teaching and learning agents and their environment. To respond to the technological commitment required

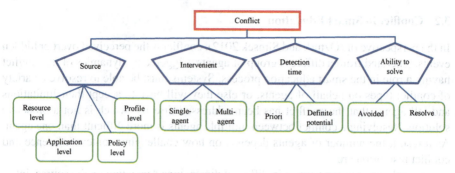

Fig. 1. Conflict Classification in smart education

by Education 4.0, (Hartono et al. 2018) proposed a smart hybrid learning technique and architectural framework based on a three-layer architecture approach. Although the framework points out solutions to essential issues in smart learning, explicitly providing an all-inclusive learning platform, but does not consider the inevitable management of conflict in such an all-inclusive platform. The model is, therefore, reconstructed with the inclusion of conflict management presented in Fig. 2.

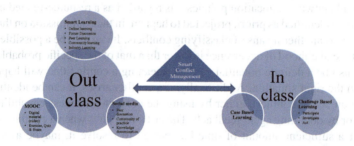

Fig. 2. Enhanced smart hybrid learning method model

Apart from improving the smart hybrid learning method to resolve conflicting situations, the proposed conceptual framework of the smart hybrid learning method is enhanced, as shown in Fig. 3.

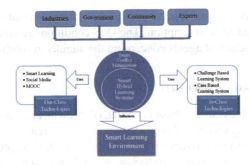

Fig. 3. Conceptual Framework of Smart Hybrid Learning Method

4 Resolving Conflict Proactively in Smart Education

Many researchers have said that conflict is not destructive in itself. Because more options are generated and analyzed before making a decision, conflict is a crucial emergent state that enables agents to make better decisions (Jehn and Mannix 2001). Depending on the nature of the conflict that may surface in smart education require different resolution approaches.

4.1 Goal-Based Conflict Resolution

Each agent in smart education aims to achieve a goal that varies across. These different goals bring conflict and require resolution at their goal or interest level.

The approach by (Tuttlies et al. 2007), primarily on conflicts resulting from disparate user interests, focuses on a priori conflict detection and resolution through avoidance. The method was based on the pervasive computing component model, which uses contracts that include the specifications and functions of components to choose services and devices and assess how well they meet the demands at hand. Applications must disclose how they affect other applications and agents in order to avoid conflicts and promote a resolution, possibly by adopting the incompatible applications. It has a conflict manager module that's connected to a database that holds a context model describing the state of the environment, and another database that keeps track of the situations that are deemed to be conflicting. This allows the conflict management to recognize conflicting circumstances in real time.

Another approach close to the above is proposed by (Armac et al. 2006), which employed a different infrastructure. They introduced a rule-based conflict detection technique that works under the premise that all resources and performance are measured as weighted automata and combined with a set of rules that keep an eye on the various components to spot conflicts as they arise.

Concerning a goal-based resolution of conflict, (Silva et al. 2010) state that when analyzing collective contexts, a shared application may arrive at an inconsistent state where it is unable to satisfy the goals of several individuals simultaneously. They suggest an approach for conflict identification and resolution that uses a client-server architectural paradigm to pick and set up the most effective conflict resolution algorithm currently available. This choice-making depends on the application's demands for quality of services (QoS) and resource consumption. The QoS conditions realized in their work are the collective satisfaction of agents concerning the attained resolution results.

4.2 Resource-Based Conflict Resolution

In an attempt to resolve resource-based conflict, (Retkowitz and Kulle 2009) proposed a technique that investigates dependencies between services realized as bindings. It highlights resource concurrency issues and introduces a method to prevent them through more dynamic and effective resource sharing. A configuration process prioritizes previously established binding policies and constraints while attempting to provide a service configuration that concurrently fits agent requirements, device environment, and service dependencies. To examine service bindings and handle access control taking priority groups into account, they used service method tagging and communication interception. This method is consequently not totally automatic but does provide agents with a tool to visualize and manually change the system status. Another resource-based conflict resolution was proposed by (Huerta-Canepa 2008) which focused on an ad hoc interaction-based management strategy for smart spaces. They agreed that only two jobs could be conducted at once, and that conflicts could be avoided through area device management and resource allocation based on the priorities of the agents, the duration of the jobs, and other similar pre-defined parameters.

4.3 Policy-Based Conflict Resolution

Considering policy-based conflict resolution, (Syukur et al. 2005) investigated methods for conflict identification and avoidance based on context changes. They developed a policy-based application model where application actions are explicitly governed by policy. As suggested by (Kawsar 2007), This research was able to distinguish clearly between policy-based conflict in terms of their causes, detection, and resolution procedures, and different temporal approaches for resolution. Rules with priority schemes among agents are the most prevalent techniques for resolving conflict. The best functioning strategies, according to their findings, were a hybrid of reactive and proactive-based conflict recognition and proactive instantaneous conflict resolution. They contend that the suitability of each method depends on the specifics of the system.

4.4 Authorization-Based Conflict Resolution

Every agent, even the overall administrator of the smart education system, will require authorization before access can be allowed to use the system. Therefore, this can sometimes pose some conflict; for instance, when the system allows an agent access to a single login account, if another mistakenly uses one's credential and reset it, it will cause the actual agent's inability to access. Hence this is much related to policy-based conflict, as supported by (Masoumzadeh et al. 2007). They claimed their method's strength is the static conflict recognition and conflict resolution limited to run-time. In essence, a practical resolution for this is creating precedence among conflicting policies.

4.5 Automatic Conflict Resolution

The conflict resolution approaches discussed above are significant measures useful in a smart education environment to improve conflict management. However, they are not automatically proactive concerning the meeting and averting possible conflicts. There is, therefore, the need to fashion a model that integrates them into an automatic conflict resolution framework.

According to (Hasan et al. 2006), a system that conducts spontaneous context-aware conflict resolution receives input about the context of the environment, scans it for inconsistencies, and then creates a new context in the environment to satisfy incongruous demands or provides information about the incongruous circumstance. The factors surrounding smart conflict management are modeled in Fig. 4.

Fig. 4. Factors of smart conflict management

5 Model for Automatic Smart Conflict Management in Smart Education

Coping with conflict proactively in Smart Education, automatic conflict resolution principles are employed to propose a model. Since real-time responses serve as management operations requirements for real-time systems, the automatic detection and resolution of conflict require the integration of all the classifications of conflicts with the implementation of their resolution mechanisms.

All the conflict detection and resolution techniques are executed in parallel in real-time to detect and resolve. Therefore, this can be achieved through multithreading in order for them to run concurrently. Figure 5 below depicts the model for proactively managing conflict in smart education.

When the conflict is automatically detected, the following operation is to classify them depending on the dimension: source, intervenient, detection time, and ability to solve. However, if it does not relate to any of these classes, it is considered non-classified. These may be those conflicts that occur due to unspecified activities in the smart education environment.

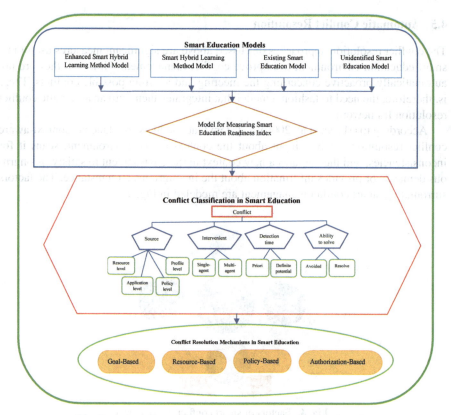

Fig. 5. Automatic conflict resolution model for smart education

The resolution phase is activated with every conflict classified to meet a specific resolution mechanism to salvage the situation. These resolution mechanisms include goal-based, resource-based, policy-based, and authorization-based. It is possible to apply different resolution mechanisms to a single conflict. Hence, other constraints such as capacity and traffic on the system will be enormous due to the implementation of parallel processing and, in effect, cause system inefficiency.

Conflict in smart education is not limited to the conceptual framework for the implementation and at the start but can evolve at any point in the teaching and learning processes. These depend on several factors, as indicated by (Hasan et al. 2006). The framework proposed in this research evaluates the various environmental factors and classifies them as conflict patterns in smart education. Classification as source conflict may be from a resource of smart education system not satisfying all agents. Applications running on the smart education systems may not be the best source of requirements, some smart education policies may hinder the system's smooth running, and some profile specifications or limitations such as the agent's sticking to his preference to win. Intervenient conflict could be single-agent having personal conflicting issues such as possible adverse health effects of long-period technology usage or multi-agent between teacher and student, staff and students, student and learning system, teacher and learning system and other conflict involving two or more agents. Another classification is the conflict detection time prior to its occurrence or at a definite potential period in the teaching and learning process. Conflict can also be classified based on its ability to provide a solution. Avoidance results in not pursuing the conflicting issue, and the second is resolving, which is a conflict that suggests the ability to dig beneath the pressing issue and propose an appropriate solution. The other ways of classification under the ability to solve could be yielding by neglecting or sacrificing some benefits to keep the system moving or compromising by playing safe with identified middle ground amid the conflicting issues.

Identifying and categorizing the various conflicts in smart education provides a springboard spanning into a resolution mechanism. The automatic smart education conflict resolution framework identifies four resolution approaches. Single or multiple approaches may help in resolving the conflict. Section 4 discussed the concepts of these four approaches. Every stage in smart education's teaching and learning process has specific goals; hence, conflicting goals may arise from each stage. A goal-based conflict resolution mechanism is applied to resolve any conflict related to conflicting goals in the smart education environment. Conflicts that occur due to resource limitations, such as bandwidth constraints, are resolved through resource-based conflict resolution mechanisms in smart education. There is a need for an appropriate response to resource distribution to all agents in a smart education environment. Policy issues such as smart course scheduling requirements present some smart education conflict and are resolved through policy-based conflict resolution techniques. Finally, authorization is one of the critical areas of a smart learning environment, especially in smart assessment. Authentication, privacy, and other security-related conflicts are considered under authorization and addressed by authorization-based conflict resolution mechanisms.

6 Conclusion

The advancement in smart education is made paramount by the COVID-19 outbreak, but the major issue has been the various conflict that retards the growth of smart education. Therefore, this requires developing a smart conflict management system to be incorporated into the entire smart campus management system to provide comprehensive support to all stakeholders in the smart education environment.

This paper presents the concept of smart education conflicts and resolution mechanisms. A discussion on smart education points out the basic requirements. The general and specific research perspectives on conflicts in smart education relate to the characteristics of an institution that provides smart education. The conflicts identified are classified based on the source, intervenient, detection time, and the ability to solve. Smart education or smart learning models are enhanced to incorporate conflict management. Finally, a proactive automatic smart education conflict management framework is proposed to provide a comprehensive concept of resolving conflict in smart education.

The future research direction will focus on developing a smart conflict management system based on the various frameworks by employing deep learning techniques.

References

Masoumzadeh, A., Amini, M., Jalili, R.: Conflict detection and resolution in context-aware authorization. In: 21st International Conference on Advanced Information Networking and Applications Workshops, AINAW 2007, pp. 505–511. IEEE (2007)

Carreira, P., Resendes, S., Santos, A.C.: Towards automatic conflict detection in home and building automation systems. In: Pervasive and Mobile Computing (2014). https://doi.org/10.1016/j.pmcj.2013.06.001

Retkowitz, D., Kulle, S.: Dependency management in smart homes. In: Senivongse, T., Oliveira, R. (eds.) DAIS 2009. LNCS, vol. 5523, pp. 143–156. Springer, Heidelberg (2009). https://doi.org/10.1007/978-3-642-02164-0_11

Syukur, E., Loke, S., Stanski, P.: Methods for policy conflict detection and resolution in pervasive computing environments. In: Policy Management for Web Workshop in Conjunction with WWW 2005 Conference, pp. 10–14. ACM (2005)

Kawsar, F., Nakajima, T.: Persona: a portable tool for augmenting proactive applications with multimodal personalization support. In: Proceedings of the 6th International Conference on Mobile and Ubiquitous Multimedia, MUM 2007, pp. 160–168. ACM (2007)

Huerta-Canepa, G., Lee, D.: A multi-user ad-hoc resource manager for smart spaces. In: Proceedings of the International Symposium on a World of Wireless, Mobile and Multimedia Networks, IEEE, pp. 1–6. IEEE (2008)

Hartono, S., et al.: Smart hybrid learning framework based on three layer architecture to bolster up education 4.0. In: International Conference on ICT for Smart Society (ICISS), IEEE [Preprint] (2018)

Armac, I., Kirchhof, M., Manolescu, L.: Modeling and analysis of functionality in ehome systems: dynamic rule-based conflict detection. In: 13th Annual IEEE International Symposium and Workshop on Engineering of Computer Based Systems, ECBS 2006 (2006)

Jehn, K.A., Mannix, E.A.: The dynamic nature of conflict: a longitudinal study of intragroup conflict and group performance. Acad. Manage. J. **44**(2), 238–251 (2001)

Kitchenham, B., et al.: 'Systematic literature reviews in software engineering-A tertiary study. Inf. Software Technol. 792–805 (2010). https://doi.org/10.1016/j.infsof.2010.03.006

Lee, J.: Conflict resolution in multi-agent based Intelligent Environments. Building and Environment [Preprint] (2010). https://doi.org/10.1016/j.buildenv.2009.07.013

Li, K.C., Wong, B.T.M.: 'Research landscape of smart education: a bibliometric analysis. Interactive Technol. Smart Educ. [Preprint] (2021). https://doi.org/10.1108/ITSE-05-2021-0083

Lyapina, I., et al.: Smart technologies: perspectives of usage in higher education. Int. J. Educ. Manage. **33**(3), 454–461 (2019). https://doi.org/10.1108/IJEM-08-2018-0257

Hasan, Md..K., Anh, K., Mehedy, L., Lee, Y.-K., Lee, S.: Conflict resolution and preference learning in ubiquitous environment. In: Huang, D.-S., Li, K., Irwin, G.W. (eds.) ICIC 2006. LNCS (LNAI), vol. 4114, pp. 355–366. Springer, Heidelberg (2006). https://doi.org/10.1007/978-3-540-37275-2_45

Meng, Q., Jia, J., Zhang, Z.: A framework of smart pedagogy based on the facilitating of high order thinking skills'. Interactive Technology and Smart Education **17**(3), 251–266 (2020). https://doi.org/10.1108/ITSE-11-2019-0076

Ozturk, H.T., Simsek, O.: Of Conflict in Virtual Learning Communities in the Context of a Democratic Pedagogy: A paradox or sophism? 1 (2012)

Semenova, N.V., et al.: The realities of smart education in the contemporary Russian universities. In: ACM International Conference Proceeding Series. Association for Computing Machinery, pp. 48–52 (2017). https://doi.org/10.1145/3129757.3129767

Silva, T., Ruiz, L., Antonio, A.L.: How to conciliate conflicting users' interests for different collective, ubiquitous and context-aware applications? In: IEEE 35th Conference on Local Computer Networks, LCN 2010, pp. 288–291 (2010)

Thomas, K.W.: Conflict and Conflict management. In: Dunnette, M.D. (ed.) Handbook of industrial and organizational psychology, pp. 889–935. Rand McNally, Chicago (1976)

Tuttlies, V., Schiele, G., Becker, C.: COMITY - conflict avoidance in pervasive computing environments. In: Meersman, R., Tari, Z., Herrero, P. (eds.) OTM 2007. LNCS, vol. 4806, pp. 763–772. Springer, Heidelberg (2007). https://doi.org/10.1007/978-3-540-76890-6_2

Zhu, Z.-T., Yu, M.-H., Riezebos, P.: A research framework of smart education. Smart Learn. Environ. 3(1), 1–17 (2016). https://doi.org/10.1186/s40561-016-0026-2

A Gated Recurrent Unit (GRU) Model for Predicting the Popularity of Local Musicians

O. O. Ajayi[✉], A. O. Olorunda, O. G. Aju, and A. A. Adegbite

Department of Computer Science, Faculty of Science, Adekunle Ajasin University, Akungba-Akoko, Ondo State, Nigeria
{olusola.ajayi,omojokun.aju,adewuyi.adegbite}@aaua.edu.ng

Abstract. Purpose: Popular musicians are among the most admired people in the world, and music is one of the features of modern culture that is most universally appreciated. Why one musician is well-liked while others are not is frequently a very tough question to answer. Survey shows most local yet talented musicians were not explored but lost due to unwillingness of music promoters to promote them. This study aims at using Gated Recurrent Unit (GRU) model in predicting the rise and popularity of local musicians by considering some established metrics for the evaluation of local musicians, alongside some characteristics/features.

Design/Methodology/Approach: Using the Kaggle dataset of local musicians' performance characteristics and historical data, the study analyzes certain features of local musicians and predicts their possible future rise, using a GRU Deep Learning Model.

Findings: The result shows a high degree (70.1%) of accuracy with a Mean Square Error (MSE) of 0.0069, between the predicted and actual performance. The work shows that the developed model can predict the possible future rise and popularity of local musicians.

Research Limitations: Only a few available original data (relating to local unpopular musicians) were extracted and due to time constraints in implementing the work, the authors could not get down to the community to extract more from local musicians around.

Practical Implication: The model can be used by music promoters to decide on sealing contract agreements with local/upcoming musicians both for their sustainability and promotability.

Originality/Value: While many predictive works were done focusing on music popularity, the song hits rate, and so on, the few that centered their aim on musicians' popularity only analyzed established musicians. This work however ventures into the analysis of the possible elevation of local musicians by considering certain parameters that relate to their songs.

Keyword: Music · Popularity · Prediction · Promotability · Sustainability

© The Author(s), under exclusive license to Springer Nature Switzerland AG 2023
C. Aigbavboa et al. (Eds.): ARCA 2022, *Sustainable Education and Development – Sustainable Industrialization and Innovation*, pp. 514–521, 2023.
https://doi.org/10.1007/978-3-031-25998-2_39

1 Introduction

One may obviously look at publicly available music charts, like the "Billboard Hot 100," which is produced each week by the Billboard Magazine, to evaluate the popularity of local singers for a particular nation or cultural area of the world. However, when attempting to widen the reach to include the entire planet, this simple technique is seldom feasible. However, in certain nations, sales of digital music (through internet shops) are included. These differences in homogeneity between nations, i.e., the inclusion or removal of specific distribution routes, make it difficult to compare such data across various nations in the world. Given that international laws on this subject are so inconsistent, it is also important to take into account any potential significant distortions brought on by (illegal) music sharing methods. In actuality, the vast majority of music distribution today centers their attention and marketing to foreign music and musicians.

However, countries without reliable historical records provide an obvious issue when trying to compile a list of the most well-known musicians ever. After summarizing these difficulties, it is determined that investigating the genres of music and musicians that are well-liked in a certain nation or cultural area requires looking more closely at various data sources and distribution routes (Marcus et al. 2010).

Although the idea of music as a contemplative art gained widespread acceptance in Nigeria through the Church, it took the work of formally trained composers and musicologists to create new idioms and styles in their compositions to create a modern tradition of Nigerian music. The first of these, T.K.E. Phillips, began composing in the 1920s, followed by Fela Sowande in 1940. Given that they form the main structural foundation of this work, concepts of music and musicians are appropriate to discuss in this context.

People frequently only appreciate music for what they perceive it to be (an entertainment medium) and performers for what they perceive themselves to be (entertainers). Music is the science or art of sequencing, combining, and arranging sounds in time to create compositions that are cohesive and continuous. Music is a force, a living force that can be given many different meanings and interpretations. It is more than just an entertainment phenomena (Umezinwa 2009). It is simply not just one thing, but rather a variety of things examined through a prism that unifies all sounds, whether they be mechanical, instrumental, or vocal, that have rhythm, melody, or harmony. As a result of these many definitions and interpretations, a musician is someone who plays an instrument, especially professionally, or who has musical talent (Okafor 2005). The author adds that these musicians "have couched many of our social commentaries, moral rules, and guides in music and songs."

The psychology of music consumption (Berns et al. 2010) and the complex network theory have both been examined (Zanin et al. 2009). In fact, certain businesses, like ReverbNation1 and Next Big Sound2, have begun to study the presence of music artists in social media due to its great marketing potential in order to measure the success of individual artists and local musicians. Popularity indices are frequently used in actual applications to rank artists in terms of their popularity as well as to compare and contrast the temporal dynamics of web-based and service-dependent popularity indices in terms of the rankings that these indices produce.

Several research attempts have been made to address the question of the possibility of predicting if a song/songwriter will be popular. Survey however shows that most of these works only predicted already famous artists. Therefore, this research work explores and established the capability of GRU-Model (Gated Recurrent Unit) in Deep Learning to predict the possible future rise and popularity of local music artists.

2 Related Work

As of the time this study was being conducted, there is very little actual literature on this topic available online.

Estimating Country-Specific Artist Popularity was the focus of Marcus et al. (2010). A Peer-to-Peer network, web search engines, Twitter tweets, meta-data of users' shared folders, and playcount data from last.fm were all employed in the study that sought to compile a list of the most well-known musicians in history. The study found that it is successful when it shows that different, largely homogeneous data sources may be used to estimate artist popularity. The study's hybrid techniques, which take into account the various information outputs from the four heuristics in terms of quantity and quality, are its one flaw.

Barajas (2011) identifying which of these characteristics was most crucial in influencing song popularity was the goal of Predicting Song Popularity. While there are additional external features that we would like to examine, such as radio airtime and the amount of money spent on promotion, they are a setback because they provided numerous unexpected discoveries regarding what makes music popular using Machine Learning (ML) techniques.

Bellogin et al. (2013). Do Web and Social Music Services Agree on the Popularity of an Artist? Web and social music services: Do They Share the Popularity of the Artist? The use of music-related social media platforms (EchoNest3, Last.fm4, and Spotify). They discovered that the service-dependent popularity indices exhibit almost no temporal patterns, most likely because they are not updated on a daily basis. In addition, these indices are similar but not equivalent, neither in terms of ranking nor in the genres or tags associated with their most popular artists. The shortcoming of the web-based popularity index is that it does not take into consideration how an artist is discussed in the context.

Furgiuele (2017). Popularity Prediction, aim to enumerates and analyzes a number of factors assumed to be an indicator of popularity, using a cohost of Evaluation metrics, and state of the art baselines to predict popularity, investigated multiple definitions of popularity, and shown their differences is the achievement obtained. Attention is to improve the accuracy of their proposed method.

Lee and Lee (2018). Metrics, Characteristics, and Audio-based Prediction for Music Popularity. In this study, the Gated Recurrent Unit (GRU) Neural Network is the model/approach utilized to forecast music popularity based on auditory information utilizing a large-scale data set. The success of this effort is measured by the characteristics of the popularity metrics using actual chart data for 16,686 songs that appeared on the Billboard Hot 100 list between 1970 and 2014. Each popularity metric displayed a distinctive distribution from which important findings could be drawn; the accuracy of predictions made, for instance, with deep neural networks.

A study on optical music recognition utilizing deep neural networks was conducted by Jiri (2020). The goal of the project was to investigate an end-to-end method for optically reading handwritten music notation. A Deep Neural Networks model was used to identify handwritten music. By developing the straightforward engraving technology known as the Mashcima engraving system, the study was able to produce cutting-edge results. But not many symbols could yet be engraved by the system (chords, trills, repeats, dynamics, and text).

3 Problem Statement

Finding out why one musician is well-liked when others are not is frequently really challenging. This study analyzes certain features of local musicians and predict their possible future rise, using a GRU Deep Learning Model.

While many predictive works were done focusing on music popularity, songs hits rate, and so on, the few that centered their aim on musicians' popularity only analyzed established musicians. Survey showed most local yet talented musicians were not explored but lost due to unwillingness of music promoters to promote them, and their future rise were adjudged blink. This study aimed at predicting the rise and popularity of local musicians by considering some established metrics, alongside some characteristics/features.

4 Methodology

This study adopts Gated Recurrent Unit (GRU)-Model Recurrent Deep Learning Technique. The following procedure shall be followed in the successful implementation of the work.

Data Collection: Table 1 shows the data collected from kaggle.com platform. Features/Attributes extracted from the data source include artist name, genre, tempo, energy, danceability, liveness, length of songs, acoustic, speech and popularity.

The collected data were information from the Top artists as well as Local unpopular artists from various types of music, such as Gospel, Hip hop, Alive, Juju etc. (Figs. 1 and 2).

Table 1. Source of data collected

S/N	Dataset	Type	Source	Date
1	Musicians' Data	CSV	www.kaggle.com	July 20th, 2021

Data Pre-processing: This stage was executed to ensure the removal of unwanted data (Fig. 3).

```
import pandas as pd
import numpy as np
rawdata=pd.read_csv("C:/users/venerate/desktop/voa/400 level/2nd sem/csc400/project/main/auston/musicians_features.csv")
rawdata.head(50)
```

	S/N	ARTIST NAME	GENRE	TEMPO	ENERGY	DANCEABILITY	LIVENESS	LENGTH	ACOUSTICNESS	SPEECHINESS	POPULARITY
0	1	Frank Edward	Gospel	117	75	65	8	191	10	63	88
1	2	Moses Bliss	Gospel	105	74	70	8	180	12	48	78
2	3	Tim Godfey	Gospel	190	78	55	6	175	15	58	69
3	4	Bola Are	Gospel	93	73	50	10	166	56	80	90
4	5	Tope Alabi	Gospel	150	80	85	12	178	18	68	90
5	6	Nathaniel Bassey	Gospel	88	77	75	14	144	11	87	86
6	7	Steve Crown	Gospel	175	85	70	9	106	4	54	79
7	8	Dunsin Oyekan	Gospel	100	71	45	6	104	34	40	70
8	9	Shola Allyson	Gospel Folk	104	70	75	11	120	14	66	86

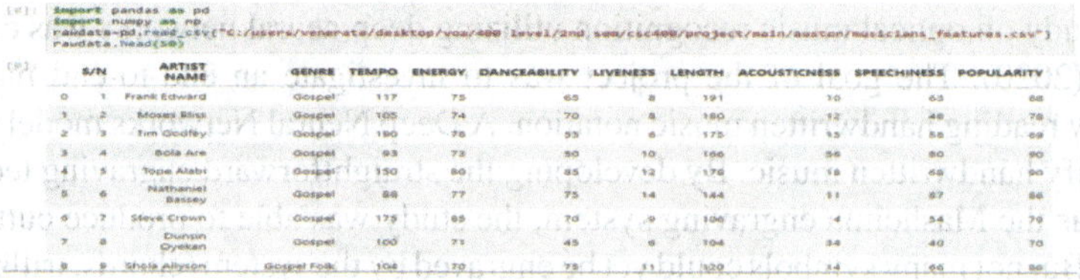

Fig. 1. Popular musicians data

```
import pandas as pd
import numpy as np
rawdata=pd.read_csv("C:/users/venerate/desktop/voa/400 level/2nd sem/csc400/project/main/auston/local_musicians_
rawdata.head(5)
```

	S/N	ARTIST NAME	GENRE	TEMPO	ENERGY	DANCEABILITY	LIVENESS	LENGTH	ACOUSTICNESS	SPEECHINESS	POPULARITY
0	1	Koloflow	Hip Hop	65.0	50.0	65.0	5.0	132.0	34.0	80.0	NaN
1	2	Oladiplenty	Hip Hop Rap	58.0	47.0	60.0	4.0	126.0	30.0	68.0	NaN
2	3	Chrisno	Hip Hop Afrobeats	52.0	45.0	45.0	4.0	134.0	25.0	57.0	NaN
3	4	Roszy	Gospel	88.0	69.0	72.0	6.0	240.0	40.0	64.0	NaN
4	5	Odion	Gospel	80.0	65.0	69.0	5.0	220.0	45.0	60.0	NaN

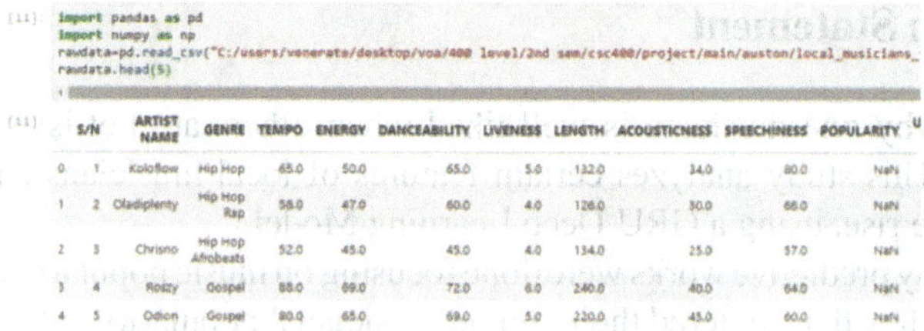

Fig. 2. Local musicians data

```
In [2]: import numpy as np
        import pandas as pd
        rawdata=pd.read_csv("C:/Users/VENERATE/Desktop/VOA/400 level/2nd Sem/CSC400/Project/Main/Auston/Musicians_Features.csv")
        rawdata.head()
Out[2]:
```

	S/N	ARTIST NAME	GENRE	TEMPO	ENERGY	DANCEABILITY	LIVENESS	LENGTH	ACOUSTICNESS	SPEECHINESS	POPULARITY
0	1	Frank Edward	Gospel	117	75	65	8	191	10	63	88
1	2	Moses Bliss	Gospel	105	74	70	8	180	12	48	78
2	3	Tim Godfey	Gospel	190	78	55	6	175	15	58	69
3	4	Bola Are	Gospel	93	73	50	10	166	56	80	90
4	5	Tope Alabi	Gospel	150	80	85	12	178	18	68	90

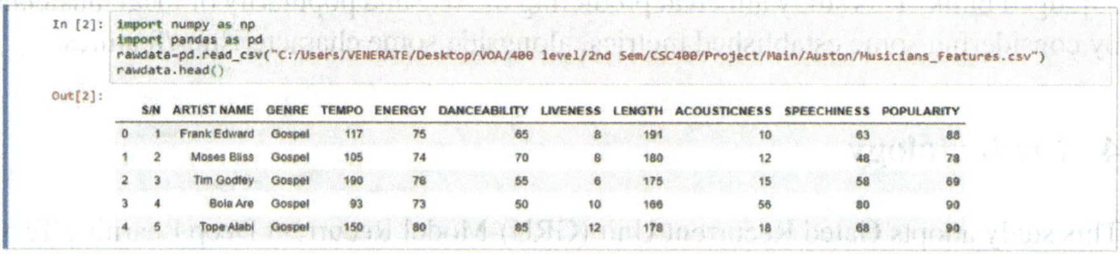

Fig. 3. Data preprocessing using python

Model Building: A gate recurrent unit (GRU) is a type of Recurrent Neural Network (RNN) that has a simpler architecture compared to Long-Short Term Memory (LSTM). For this study, Gated Recurrent Unit (GRU) of Recurrent Neural Network will be used (Fig. 4).

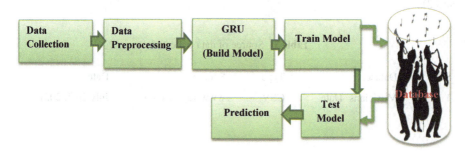

Fig. 4. System Model

Train Model: Training of a model involves applying the algorithm selected on the prepared datasets. The training set consists of 70% of musicians' data that will be used to train the model so it can predict the outcome (Fig. 5).

Fig. 5. Training Data

Test Model: The testing set is a subset of the data set used to test a model. The testing set consists of 30% of musicians' data that will be used to measure the accuracy and efficiency of the algorithm used to train the machine (Fig. 6).

Fig. 6. Testing Data

Database: Database is a collection of organized information in a regular structure. Here, the retrieved data of past and local musicians were stored for the study's analysis purpose.

Prediction: At this stage, the study establishes the capability of the model to predict the future popularity of local musicians.

5 Result And Discussion

Results
The study which concentrated on performance of the local artists to determine their popularity, gave accuracy estimates of 89% (Tempo), 89% (Energy), and 94% (Danceability) (Figs. 7 and 8).

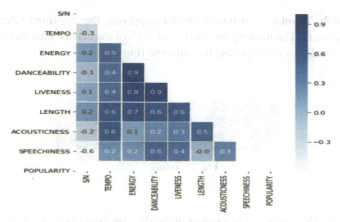

Fig. 7. Musicians features chart

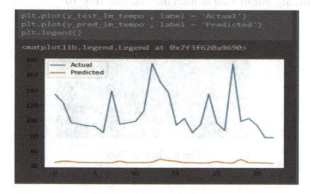

Fig. 8. Actual vs Predicted

6 Conclusion

This study presented a model for predicting the popularity of local musicians (upcoming artists) using some features of popular musicians (genre, tempo, danceability etc.) to determine the popularity/future rise of the particular local musicians using gated recurrent unit (GRU) algorithm. The data used for the analysis were extracted from kaggle.com platform. From the data extracted (100 for popular musicians, and 20 for local musicians; totaling 120 data), 70% were used for the training of the data while 30% were used for testing of the model. The result of the analysis gave an accuracy of 89% (Tempo), 89% (Energy), and 94% (Danceability) proving the metrics to be suitable for measuring the performance and popularity of local musicians.

However, the Actual vs Predicted plot shows a model that is not fit. The reason for this is not far-fetch; Deep Learning Techniques/Models are expected to be used/deployed on big data. For future perusal on this work therefore, it is recommended that more data be used for the analysis to further validate our results. Also, newer metrics could be deployed in the extensive evaluation of the analysis and results. Usage of another approach other than GRU is also welcome.

References

Berns, G.S., Capra, C.M., Moore, S., Noussair, C.: Neural mechanisms of the influence of popularity on adolescent ratings of music. Neuroimage **49**(3), 2687–2696 (2010)

Okafor, R.C.: Music in Nigerian Society. New Generation Books, Enugu (2005)

Marcus, S., Pohle, T., Koenigstein, N., Knees, P.: What's hot estimating country-specific artist popularity. In: ISMIR, pp. 117–122 (2010)

Umezinwa, O.E.: Music and democracy: a tale of nature and contrivance. UJAH: Unizik J. Arts Hum. **11**(1), 122–139 (2009)

Zanin, M., Lacasa, L., Cea, M.: "Dynamics in scheduled networks" chaos: an Interdisciplinary. J. Nonlinear Sci. **19**(2), 023111 (2009)

Bellogin, A., de Vries, A., He, J.: Artist popularity: do web and social music services agree? In Proceedings of the International AAAI Conference on Web and Social Media, vol. 7, no. 1, pp. 673-676 (2013)

Jiri, M.: Optical music recognition using deep neural networks (2020). Retrieved from dspace.cuni.cz, 2020

Junghyuk, L., Lee, J.-S.: Music popularity: Metrics, characteristics, and audio-based prediction. IEEE Trans. Multimedia **20**(11), 3173–3182 (2018)

Furgiuele, A.: Popularity Prediction. Scholarsarchive.library.albany.edu (2017)

Novel Cost-Effective Synthesis of Copper Oxide Nanostructures by The Influence of pH in the Wet Chemical Synthesis

R. B. Asamoah[1(✉)], A. Yaya[2], E. Annan[2], P. Nbelayim[2], F. Y. H. Kutsanedzie[1], P. K. Nyanor[1], and I. Asempah[1]

[1] Faculty of Engineering, Accra Technical University, Accra, Ghana
rbasamoah@atu.edu.gh

[2] Department of Materials Science and Engineering, CBAS, University of Ghana, Legon, Ghana

Abstract. Purpose: This study aims to provide an optimum pH to achieve a highly crystalline CuO nanostructures using a cost-effective facile single-pot wet chemical synthesis approach.

Design/Methodology/Approach: The experiment was undertaken by varying the pH of a single-pot wet chemical precipitation synthesis of copper oxide using NaOH. Different pH of 8,10 and 12 were studied in synthesis B, C, and D respectively where all other conditions remained constant. The resulting synthesized powder samples were analyzed by the TEM and XRD characterization techniques.

Findings: The results revealed that an increase in the pH influenced the morphology and the crystallography of the resulting particles. The formation of nanorod morphology increased respectively in B < C < D (pH 8 < 10 < 12). Consequently, the crystal patterns of CuO were enhanced respectively in B < C < D. Thus, the formation of CuO was found to be pH dependent with pH 12 being optimal for the formation of highly crystalline CuO nanostructures using the single-pot wet chemical precipitation synthesis.

Implications/Research Limitations: The study output highlights the systematic influence of pH in the formation of CuO in the facile single-pot wet chemical synthesis. This synthesis method is the simplest, easy and most cost-effective approach to metal oxide nanoparticle synthesis hence the study is extensible to allied particles such as zinc oxide.

Practical Implications: The identification of the unique pH for highly crystalline CuO nanostructures using the single-pot wet chemical synthesis is significant to eliminating further calcination of nanoparticles which impedes the idea of a facile single-pot synthesis. Hence, producing a cost-effective synthesis route for CuO nanostructures.

Originality/Value: The previous study has not established the influence of different pH on the morphological and crystallographic formation of metal oxide nanoparticles and its associated relative cost efficiency. The findings identified here demonstrate a unique pH for the cost efficient single-pot wet chemical synthesis for obtaining identical morphology and high crystallinity of CuO nanoparticles.

Keyword: Crystallography · Morphology · Nanorods · TEM · XRD

1 Introduction

In recent years, Copper Oxide nanostructures are of significant interest to the industrial sector for production as antibacterial agents. The antibacterial activities of copper oxide are highly recommended for decontaminating infested surfaces (Asamoah et al. 2020a). However, paramount to the production of copper oxide nanostructures is a cost-effective means of synthesis.

Copper Oxide (CuO) nanostructures are monoclinic crystalline particles with narrow bandgap possessing unique properties which are different from their bulk counterparts (Kap et al. 2005). These properties influence their applicability for sensing devices, electronic storage devices, nanodevices for catalysis, anti-bacterial activity among others (Mamani and Gamarra 2014). The properties of nanostructures are determined by their methods of synthesis (Asamoah et al. 2020a). Thermal decomposition, hydrothermal, sol-gel, electrochemical, microwave radiation and wet chemical precipitation synthesis includes the known methods for synthesizing nanostructures (Salavati-niasari et al. 2009). Among these methods of synthesis, wet chemical precipitation synthesis is highly advised because it can be accomplished in a facile manner, simple and easy to undertake, among other factors which makes the method relatively cost effective (Asamoah et al. 2020). The wet chemical precipitation synthesis process for copper oxide typically involves a copper oxide precursor and a precipitating agent for forming copper oxide precipitates in an aqueous media. The precipitating agent reacts with the dissolved copper ions in solution to form precipitates of hydroxides or oxides of copper. The precipitating agent causes precipitation by altering the chemical and physical states of the dissolved copper ions to form solid state precipitates in suspension. Precipitation follows a sequence of steps including; nucleation, growth of nucleus and crystallization. The resulting copper hydroxides and or copper oxides precipitates are insoluble in solution and easily separable from the suspension. Precipitating reagents include; NaOH, CaO, Ca(OH)$_2$, CaCO$_3$, MgO, Mg(OH)$_2$, and NH$_4$OH. The wet chemical precipitation synthesis is highly dependent on the pH of the synthesis solution caused by the precipitating agents. Thus, the formation of copper oxide nanostructures is highly dependent on the concentration of the precipitating agent in the synthesis media. The concentration of precipitating agent is proportional to the pH of the solution (Phiwdang et al. 2013). Hence, the pH is a significant factor in the wet chemical precipitation synthesis. Whiles low pH does not result in the formation of copper oxide, they could however be transformed into the same by thermal dehydration. Thermal dehydration is known for the conversion of copper hydroxide into copper oxide (Cudennec and Lecerf 2003). Thermal dehydration of metastable copper hydroxide into the stable copper oxide is typically accomplished through calcination at high temperatures averagely 400 °C to 600 °C (Wongpisutpaisan et al. 2011). The calcination process for the solid-state transformation into copper oxide is time consuming and demands high energy provided by typically costly sophisticated equipment called furnace. Therefore, the calcination process impedes the facile wet chemical synthesis by increasing the cost and altering its easy and simple single-pot synthesis approach. Moreover, the calcination process may contribute to further particle aggregation thereby altering the morphology (Martis 2011). Contemporarily, the most common synthesis route for copper oxide using the wet chemical precipitation synthesis remains the eventual calcination of the particles to realize crystalline copper oxide. (Phiwdang et al. 2013). Thus, copper hydroxide precipitates are formed during the wet chemical precipitation process. The solid-state precipitates are separated from solution for calcination in a furnace to transform in to copper oxide (Asamoah et al. 2020c). The laborious and multi-staged processes defeat the principle of a facile single-pot wet chemical synthesis.

In order to achieve a facile, simple, easy and cost-effective means of synthesizing metal oxide nanostructures, specifically copper oxide in this case, it remains versatile to eradicate calcination from the series of steps for producing highly crystalline particles by the wet chemical precipitation method. It is therefore significant to identify the pH for realizing highly crystalline copper oxide nanostructures using the wet chemical precipitation synthesis, without further calcination. In this regard, the synthesis of copper oxide nanostructures is investigated by the varying of pH from 8 to 12 in a wet chemical precipitation synthesis. The investigation is determined to establish a suitable pH for the complete formation of high chemically pure and crystalline nanostructured copper oxide without calcination. Sodium hydroxide is used as the precipitating agent to alter the pH from 8 to 12. Hydroxide ions causes the oxidation of copper ions (Cu^{2+}) capped by polyvinylpyrrolidone. The oxidation of copper ions produces copper oxide nanostructures (CuO). The precipitated copper oxide nanostructures are washed severally with ethanol, dried and collected for characterization. The blueprint of this study would be versatile for the influence of pH in the wet chemical precipitation synthesis of highly crystalline copper oxide nanostructures applicable for a range of nanomaterial uses.

2 Experimental

2.1 Materials

The materials used for the synthesis of copper oxide nanoparticles included copper nitrate, sodium hydroxide and polyvinylpyrrolidone. These chemicals were each purchased from Sigma Aldrich. All chemicals were of analytical grade hence required no further purification. De-ionized water used were provided by Milli-Q.

2.1.1 Synthesis

Generally, copper oxide nanoparticles were synthesized in a one pot aqueous wet chemical precipitation synthesis. The nanoparticles were prepared by dissolving 2.9g of copper nitrate and 1.2g of polyvinylpyrrolidone (PVP) in 100ml of distilled water as indicated in the table below. The solution was heated to 60° under magnetic stirring at 800rpm. 2M NaOH aqueous solution was added dropwise at the stated temperature. The mixture continued to stir for a further 1h at 60°. The pH was varied to confirm suitability. In synthesis B, the pH was made 8 whiles in C and D they were 10 and 12 respectively. The synthesis of copper oxide nanoparticles can be summarized according to Eqs. 1 and

2 below (Table 1):

$$CuNO3 + NaOH \rightarrow Cu(OH)2 + NaNO3 \quad (1)$$

$$Cu(OH)2 \rightarrow CuO + H2O \quad (2)$$

Table 1. Quantity of reactants

	CuNO3	PVP	Volume(ml)
Quant.(g)	2.9	1.2	100

2.2 Characterization of the Synthesised Nanoparticles

The influence of synthesis parameters including pH have phenomenal effect on particle morphology (Asamoah et al. 2020b). The phenomenal effects of the synthesis parameter are studied with diverse characterization tools. In this study, the influence of different pH on the synthesis of CuO were analysed by transmission electron microscope (TEM) and X-ray diffractometer. These characterization techniques revealed the morphology and crystallinity of CuO nanoparticles synthesised by different range of pH.

2.2.1 Transmission Electron Microscope (TEM) Analysis of Synthesised Nanoparticles

TEM (JEOL 2010, JEOL Ltd, USA), was used for the verification of the morphologies of the synthesised nanoparticles in this study. The TEM micrographs were obtained by pipetting a minute quantity of the respective nanoparticles onto a carbon grid. The samples were allowed to dry. They were observed by the high resolution TEM at 200kV. The particle size including particle length and width was analysed with an Image J software.

2.2.2 X-Ray Diffractometer (XRD) Analysis of Synthesised Nanoparticles

The crystallinity of the synthesised nanoparticles was verified with an XRD (Empyrean, Malvern Pan-Analytical, UK) to emphasize on the purity of a synthesised sample. X-ray diffraction was obtained of the dry powdered synthesised samples at a wavelength of 1.5418Å at a 2θ range from 0^0 to 110^0. The crystal patterns obtained from the powdered synthesised nanoparticles were analysed relative to the reference XRD crystal pattern of CuO.

3 Results and Discussion

The TEM micrographs of the synthesised nanoparticles by varying pH in the methods B, C and D are shown below in Table 2. The micrographs depicted different kinds of

Table 2. TEM images of synthesis B, C and D.

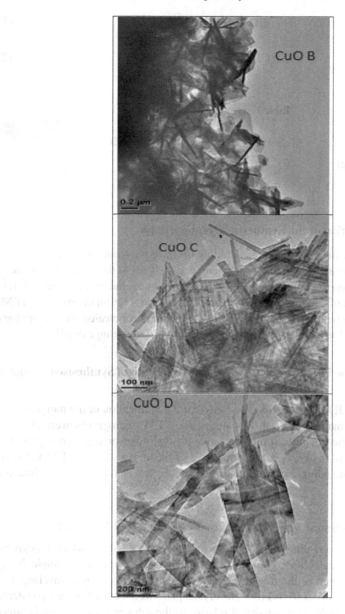

nanoparticle morphologies respective to the pH of the synthesis media. In synthesis approach B which was at a pH of 8, nanorods interspersed among irregular shaped nanostructures are observed. Using Image J, the rather fewer nanorods recorded in the synthesised powered sample of synthesis B were measured to have an average width of approximately 25nm and a length of 350nm. Synthesis C recorded relatively more nanorod nanostructures and a handful irregular shaped nanostructures as compared to those in method B. Synthesis C has nanorods of a width of about 10nm and a length of 340nm. The nanorods in C are partly separated. In synthesis D, nanorods are observed to be core clustered or aggregated. Synthesis D has a length of about 10nm and a length of 397nm. The nanorods in synthesis B, C and D had identical sizes but varying quantity. Thus, the pH of the synthesis is proportional to the nanorod formation. Synthesis D which occurred at a pH of 12 recorded no presence of irregular shaped nanostructures. Thus, all the nanoparticles had a uniform nanorod morphology of one-dimensional nanostructure. In all the synthesis of varying pH, the nanostructures proved to possess some level of the presence of nanorod kind of structures. Wet chemical synthesis of copper oxide are mentioned by previous authors to possess a nanorod morphology (Pandey et al. 2014). The irregular shaped nanostructures such as majorly observed in synthesis B and decreased in C with no occurrence in D are the presence of reactants which mediate the conversion into copper oxide in the transition process (Cudennec and Lecerf 2003). These irregular shaped structures decreased with increasing pH. Thus, the conversion into copper oxide is directly proportional to the pH of the media for the synthesis. The presence of nanorod in the TEM micrograph depict copper oxide in the synthesized sample (Pandey et al. 2014). Hence the formation of copper oxide was pH dependent, in the order B < C < D which had a pH of 8 < 10 < 12 respectively.

The XRD crystal patterns of the synthesised powder samples of synthesis method B, C and D at pH 8, 10 and 12 respectively are illustrated in Fig. 1. The crystal patterns of the powdered samples of synthesis B has low conformation to copper hydroxide and copper oxide. This indicates that at a pH of 8, the wet chemical process synthesis of copper oxide results in the formation of particles with a crystallography that deviates from copper hydroxide and copper oxide. The pH 8 therefore does not lead to the significant formation of copper oxide and copper hydroxide as was also reported by the TEM micrograph. Synthesis C was carried out at a pH of 10. The crystal patterns of the powered samples of synthesis C closely agrees with the crystal patterns of copper hydroxide at a considerable purity. A possible subsequent calcination of copper hydroxide leads to the formation of copper oxide (Cudennec and Lecerf 2003). However, calcination process was not investigated in this study. Synthesis D performed at a pH of 12 recorded a highly crystalline copper oxide particle. The crystal patterns of the synthesised powered samples conformed with coper oxide which is an indication of the sample purity. Therefore pH 12 is optimal for a single-pot synthesis of copper oxide in the wet chemical process. The XRD results confirms the TEM micrographs which recorded significant changes in morphologies as the pH was increased from 8 to 12 indicating a possible different type of elemental formation in the different pH synthesis.

The pH of the synthesis media is a function of its hydroxide ions to influence the precipitation of copper ions in the solution for copper oxide formation. Increase in pH is invariably an increase in the sodium hydroxide concentration in the synthesis media. The hydroxide is directly proportional to the pH. The hydroxides react with the copper ions in solution (Cu^{2+}) according to Eq. 1 above. Lower concentration of hydroxide leaves excess or unreacted copper ions impeding the formation of copper hydroxide which mediate the formation of copper oxide in the wet chemical reaction process. An example of this phenomenon was witnessed in synthesis B which did not result in a significant formation of both copper hydroxide and copper oxide as depicted by both the TEM micrograph and XRD analysis. A considerable equal amounts of copper ions and hydroxide ions in the synthesis media results in the formation of stable copper hydroxide. Synthesis C, at a pH of 8 has a comparably equal amounts of these cation and anion. The XRD analysis revealed an acceptable presence of copper hydroxide evident to this effect. Equation 1 above emphasises this state of the wet chemical reaction process. At high hydroxide concentration resulting in high pH hence excess hydroxides, the excess hydroxides react with hydrides in Cu(OH)$_2$ to form water leading to the complete formation of copper oxide in wet chemical process as observed in Eq. 2 above. Synthesis D, recorded high pH possessing excess hydroxides leading to the formation of copper oxide which were shown by the XRD analysis.

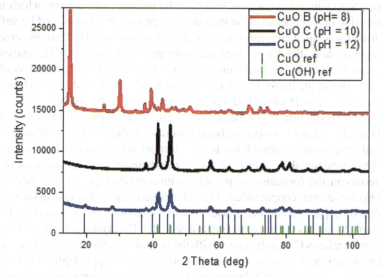

Fig. 1. XRD of synthesized silver nanoparticles.

4 Conclusion

Conclusively, a pH was optimised for the preparation of copper oxide nanostructures using the cost-effective single-pot wet chemical aqueous precipitation synthesis method. Copper oxide nanostructures were found to be synthesized at a high pH of 12. It is expected that the excess hydroxides at high concentration generated at high pH contributes to the formation of crystalline copper oxide. At lower pH as in B and C, there exist no excess hydroxides to influence the formation of copper oxide in the chemical reaction process. Hence, the pH influences the concentration of hydroxide ions for the formation of copper oxide nanostructures in the single-pot wet chemical synthesis. The research output eradicates calcination from the synthesis process hence its associated cost. The resulting copper oxide nanostructures are highly pure, cost efficient and easy to produce with limited skills and resources. The particles generated are applicable for various uses especially as antibacterial agents on contaminated surfaces. The method realized supports sustainable industrial innovation to reduce cost of production of copper oxide nanostructures.

References

Asamoah, R.B., Annan, E., et al.: A Comparative Study of Antibacterial Activity of CuO/Ag and ZnO/Ag Nanocomposites. Advances in Materials Science and Engineering (2020a). https://doi.org/10.1155/2020/7814324

Asamoah, R.B., Mensah, B., et al.: A Comparative Study of Antibacterial Activity of CuO/Ag and ZnO Ag Nanocomposites (2020b)

Asamoah, R.B., Yaya, A., et al.: Synthesis and characterization of zinc and copper oxide nanoparticles and their antibacteria activity. Results Mat. **7**, 100099 (2020c). https://doi.org/10.1016/j.rinma.2020.100099

Asamoah, R.B., Yaya, A. Anan, E.: Applications of nanostructural materials for water remediation (no date)

Cudennec, Y., Lecerf, A.: The transformation of Cu(OH)2 into CuO, revisited. Solid State Sci. **5**(11–12), 1471–1474 (2003). https://doi.org/10.1016/j.solidstatesciences.2003.09.009

Kap, Y., et al.: Size-controlled synthesis of alumina nanoparticles from aluminum alkoxides **40**, 1506–1512 (2005). https://doi.org/10.1016/j.materresbull.2005.04.031

Mamani, J.B., Gamarra, L.F.: Synthesis and characterization of Fe 3 O 4 nanoparticles with perspectives in biomedical applications **17**(3), 542–549 (2014)

Martis, M.: In situ and ex situ characterization studies of transition metal containing nanoporous catalysts, pp. 1–216 (2011). http://discovery.ucl.ac.uk/1334587/

Pandey, P., et al.: Antimicrobial properties of CuO nanorods and multi-armed nanoparticles against B. Anthracis vegetative cells and endospores 789–800 (2014). https://doi.org/10.3762/bjnano.5.91

Phiwdang, K., et al.: Synthesis of CuO nanoparticles by precipitation method using different precursors. Energy Procedia **34**, 740–745 (2013). https://doi.org/10.1016/j.egypro.2013.06.808

Salavatiniasari, M., Fereshteh, Z., Davar, F.: Synthesis of cobalt nanoparticles from [bis (2-hydroxyacetophenato) cobalt (II)] by thermal decomposition Thermal decomposition. Polyhedron **28**(6), 1065–1068 (2009). https://doi.org/10.1016/j.poly.2009.01.012

Topnani, N., Kushwaha, S., Athar, T.: Wet synthesis of copper oxide nanopowder. Int. J. Green Nanotechnol. Mat. Sci. Eng. 1(2) (2010). https://doi.org/10.1080/19430840903430220

Wongpisutpaisan, N., et al.: Sonochemical synthesis and characterization of copper oxide nanoparticles. Energy Procedia **9**, 404–409 (2011). https://doi.org/10.1016/j.egypro.2011.09.044

The Synergestic Effect of Multiple Reducing Agents on the Synthesis of Industrially Viable Mono-dispersed Silver Nanoparticles

R. B. Asamoah[1]([✉]), A. Yaya[2], E. Annan[2], P. Nbelayim[2], F. Y. H. Kutsanedzie[1], P. K. Nyanor[1], and I. Asempah[1]

[1] Faculty of Engineering, Accra Technical University, Accra, Ghana
rbasamoah@atu.edu.gh

[2] Department of Materials Science and Engineering, CBAS, University of Ghana, Legon, Ghana

Abstract. Purpose: The synthesis of industrially viable monodispersed silver nanoparticles was investigated by the combined effect of two reducing agents and variations in the synthesis reaction conditions including temperature.

Design/Methodology/Approach: The microstructural analysis using transmission electron microscope (TEM) of the as-synthesized nanoparticles showed the nanoparticle's morphology. The particle size distribution was evaluated with Image J.

Findings: The synergy of a combined reducing agent produces nanoparticles with; enhanced total surface area, narrow-size distribution, spherical morphology and highly mono-dispersed as compared to a single reducing agent.

Research Limitations: The results obtained reveals a paradigm shift in the synthesis of monodispersed silver nanoparticles. Narrow-size particle distribution with enhanced total surface area and its associated morphology are industrially viable. The study is extensible to feasible applications requiring size-controlled nanoparticles such as antibacterial gent, drug delivery, among others.

Practical Implications: The monodispersed nanoparticles have identical particle sizes significant for size-dependent homogeneous physico-chemical properties. Monodispersity have enhanced total surface area with efficiency hence they are relatively economically viable.

Originality/Value: This study has emphasised on the effect of combined reducing agents on the synthesis of monodispersed silver nanoparticles under different synthesis parameters. The study output demonstrates a novel influence of the synergy of multiple reducing agents on the synthesis of industrially viable silver nanoparticles.

Keywords: Nanoparticles · Reducing agents · TEM · Mono-dispersion

1 Introduction

Industrially viable nanoparticles are expected to be monodispersed. The functional properties are size and shape dependent (R.B. Asamoah et al., 2020). Most frequently, identical sizes exhibit indistinguishable features which are useful for conforming applications.

Preparation of nanoparticles with narrow (uniform) size distribution is therefore important for a variety of scientific and industrial applications (R.B. Asamoah et al., 2020). Controlling the size of nanoparticles (nps) is therefore critical to obtaining a narrow size distribution with identical features appropriate for a specific application (Odularu, 2018). For instance, the anti-bacterial application of silver nanoparticles (nps) increases with decreasing nanoparticle's size (Anderson et al., 2017). The prominent anti-bacterial agent which is the release mechanism of silver ions against bacteria, increases with decreasing size of silver nps (Shah et al., 2019).

The chemical synthesis of size-controlled silver nps is influenced by a number of factors that affect the thermodynamic and kinetic behavior of reactions during nucleation (Asamoah, Yaya and Anan, 2020). These factors include the type of reducing and capping agents (Mott et al., 2007), reaction temperature (Nyathi et al., 2019), pH (Hussain et al., 2011), concentration of reactants. Reducing agents such as sodium borohydride ($NaBH_4$) and.

hydrazine (N_2H_2) have strong reduction capabilities. They exhibit a fast nucleation times resulting in small particle sizes normally less than 20 nm. On the contrary, weak reducing agents including hydroxides and organics such as trisodium citrate (TSC) exhibit rather slow reaction to produce nanoparticles of approximately 50 nm (Hashemipour, Zadeh and Pourakbari, 2011). The effects of reducing agents on nanoparticle size is also affected by the concentration of the reducing agents (Roberson et al., 2014). To produce finer particles, research has demonstrated that it is ideal for the concentration of reducing agent to be in excess of the precursor compound. Furthermore, to complement the actions of reducing agents in nanoparticle synthesis, surfactants are used to encapsulate the nanoparticles. The amphiphilic molecules caps around mostly single individual nanoparticles to prevent their agglomeration (Garni et al., 2017). This effect causes the individually covered particles to disperse within a suspension. The molar ratio between surfactant and precursor is vital to ensuring an efficient surfactant action by the separation of nanoparticle surfaces. Many researchers have resulted to a ratio of 4:1 in favor of surfactant for the micellar concentration. A marginal ratio limits the ability of surfactants to form micelles uniformly protecting nanoparticles (Dang et al., 2011). However, the quantity of aqueous media is also of a prominent consideration. Whiles no set limits have been unanimously set, it is practically essential that in wet chemical synthesis, the aqueous content remains high with respect to all reactants. In the divergent approach, a near equal or greater surfactant(s) content to aqueous media creates a reverse micelle. Reverse micelles defined in microemulsion synthesis is better achieved when the aqueous with relatively smaller mass ratio to surfactant forms an immiscible solution with a non-polar solvent (oil). The reverse micelles form a reactor with an aqueous nano-core surrounding the precursor. In such a case, amidst other factors, surfactant to water ratio is a major contributor to controlling particle size hence narrow size distribution (Davarpanah et al., 2015).

In this study, we highlight the mono-dispersion of nanoparticles specifically metal nanoparticles that is mostly studied using a single strong or weak reducing agent, a surfactant and metal precursor. To enlighten our understanding of the scope, a new paradigm shift which is rarely looked at is explored. The size of silver nanoparticles is controlled for narrow size distribution by the synergy of two strong reducing agents of $NaBH_4$ and N_2H_2; the weak reducing agent tri-sodium citrate (TSC), is used as co-reducing agent and also as the surfactant of the reaction. The results from this synergy are compared to that resulting from a single strong reducing agent of sodium borohydride with TSC.

2 Experimental

2.1 Materials

The synthesis of silver nanoparticles was carried out using a range of materials which was purchased from Sigma Aldrich, UK. Silver nitrate was the precursor used; Sodium borohydride and hydrazine were the reducing agents; Tri-sodium citrate was simultaneously used as a co- reducing agent and a capping agent for the nanoparticle's synthesis. Sodium hydroxide was acquired from Kimix. All chemicals used in the synthesis were of pure analytical grade hence did not require further purification before use. All aqueous solutions in the synthesis were prepared with deionised water from Milli-Q.

2.2 Synthesis

The synthesis of size-controlled silver nanoparticles was carried out in three different procedures as diagrammatically shown in Fig. 1. In procedure 1, the nanoparticles were prepared by dissolving sodium borohydride and trisodium citrate in an aqueous solvent. The solution was kept under stirring at 1400 rpm at a temperature of 60 °C for 30 min. Silver nitrate was then added to the solution under stirring whiles maintaining the temperature constant. Upon the dropwise addition of silver nitrate, there was a change in color. 1M sodium hydroxide was added to adjust the pH to 10.5. The synthesis continued for a further 30 min under stirring at the stated temperature after which it was cooled slowly. Maintaining all other conditions, in procedure 2, hydrazine was also used as a reducing agent simultaneously with sodium borohydride as used in procedure 1. In procedure 3, nanoparticles were prepared as in procedure 1 except that synthesis was performed at a temperature of 25 °C throughout the procedure. The concentration and quantity of chemicals used are listed in Table 1.

x: total volume of $AgNO_3$ which is added dropwise to y: total volume of $NaBH_4$ and TSC or $NaBH_4$, N_2H_4 and TSC.

2.3 Results and Discussion

The synthesis of monodispersed silver nanoparticles was carried out by varying molar concentration of reacting silver precursor and those of reducing agents as well as surfactant. The pH was constantly maintained at 10.5 by the addition of 1M NaOH. The pH was monitored using a pH meter. The constant pH for all synthesis ensured uniformity as pH is observed to also influence the particle size (Mascolo, Pei and Ring, 2013). The concentrations of reactants were varied for each synthesis procedure. The experiments (1a, 1b and 1c) in procedure 1 used a

The Synergestic Effect of Multiple Reducing Agents on the Synthesis   533

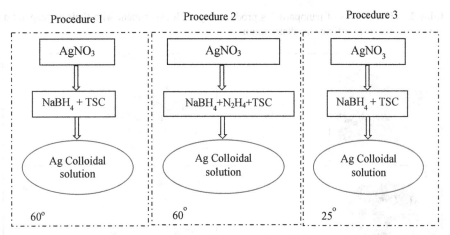

Fig. 1. Procedures for synthesizing the silver nanoparticles at different temperatures and varying reducing agents.

Table 1. Different reaction conditions for the synthesis of monodispersed silver nanoparticles.

Procedure	Exp	AgNO$_3$ mol dm^{-3}	NaBH$_4$ mol dm^{-3}	N$_2$H$_4$ mol dm^{-3}	TSC mol dm^{-3}	Volume of reactants (ml)	pH	Temp °C
1	1a	0.001	0.002		0.004	x5y45	10.5	60
	1b	0.002	0.002		0.004	x5y45	10.5	60
	1c	0.003	0.002		0.004	x5y45	10.5	60
2	2a	0.001	0.004	0.001	0.01	x5y45	10.5	60
	2b	0.002	0.004	0.001	0.01	x5y45	10.5	60
	2c	0.004	0.004	0.001	0.01	x5y45	10.5	60
3	3a	0.001	0.002		0.004	x5y45	10.5	25
	3b	0.002	0.002		0.004	x5y45	10.5	25
	3c	0.003	0.002		0.004	x5y45	10.5	25
	3d	0.002	0.004		0.003	x5y45	10.5	25
	3e	0.002	0.006		0.002	x5y45	10.5	25

single strong reducing agent of sodium borohydride. These experiments produced particles of wider size ranges. These particles produced had sizes with a wide deviation from the average mean particle size which are shown in Table 2 below. The simultaneous use of hydrazine as a co-strong reducing agent in combination with sodium borohydride had an impact on the particle sizes. The deviations of particle sizes (see Table 2) from the average mean were much narrower which is shown in Table 3. This means, the particles produced in this case were much more

534 R. B. Asamoah et al.

Table 2. TEM images of nanoparticles produced by each experiment with their corresponding histogram showing particle size distribution.

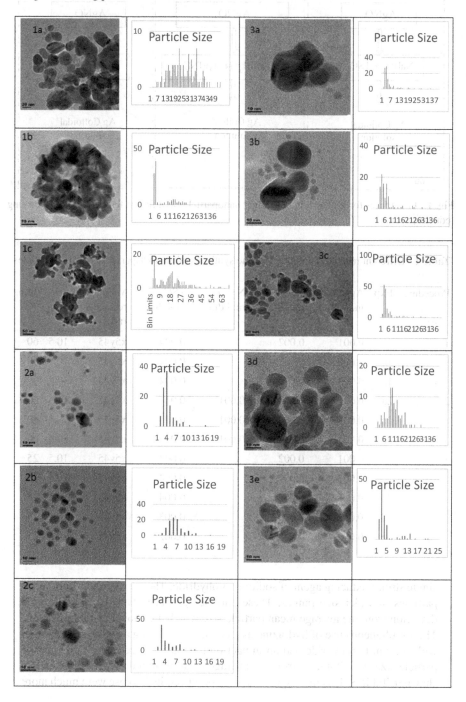

identical in size (see Table 2 and 3). The two strong reducing agents used in this procedure have a rather much faster nucleation rates on the particle formation. They reduce the silver ions (Ag$^+$) at a much faster rate resulting into much smaller particle sizes which are more uniformly distributed. With respect to Table 3, the experiments in procedure 1 have larger standard deviation. This means the particles are rather widely distributed with respect to the average mean particle size. The same trend is observed in the experiments in procedure 3. It is observed that the deviation (see Table 2) is again wide with respect to the average mean particle size which can be verified from Table 3. In these cases, particles were synthesized, ranging from a mixture of smaller and larger particles. In relation to procedures 1 and 3, the particles in procedure 2 have comparably smallest deviation from the mean particle size which is attributed to the synergy of two combined reducing agents as shown in Tables 2 and 3. The experiments in this synthesis procedure produced particles of a uniform sizes (narrowly distributed). The narrow distribution can be attributed to the synergy of the two combined reducing agents. The combined effect of the two strong reducing agents increases the concentration of reducing agents in the wet chemical synthesis media for reducing the silver ions. The synergy also increases the reaction rate to cause the growth of particles that are smaller and more similar in sizes. The narrow distribution among particles increases their characteristic similarities to exhibit identical properties which is suitable for specificity in application such as water decontamination among others. The difference in concentration in precursor (AgNO$_3$) among the different experiments (2a, 2b, 2c) in procedure 2, has different size distributions indicated by their standard deviations (Table 2). Whiles 2a has the most narrow in size distribution, the size distribution increase iwith increase in precursor to reducing agent ratio as in Tables 2 and 3. This effect was due to the increase in the concentration of AgNO$_3$ with respect to other reactants (Table 1). Increasing the concentration of precursor (AgNO$_3$) over other reactants (reducing and capping agents) limits the effectiveness of the reducing and capping agents. Capping the particles protects them from agglomeration. Agglomeration promotes rather larger and uneven size distribution. The effect of increasing precursor concentration over the concentration of reducing agents increases the amount of precursor (silver) nuclei generated. Excess nuclei cause agglomeration during particle growth. The capping agent acts as barrier against particle aggregation during nanoparticle formation. The particles produced in such a case increases in size. The effect of increased concentration in precursor over reducing and capping agents was therefore observed in 2b and 2c as the standard deviation of particles increased leading to the resultant production of wider particle size distribution. It is also observed that in procedure 2, the particles are of more spherical nature as compared to particles in other procedures (1&3). This effect is also attributed to the synergy of the two strong reducing agents used in the nanoparticle synthesis. This also confirms what is in literature that, whiles strong reducing agents such as sodium borohydride produces spherical nanoparticles, weak reducing agents such as trisodium citrate may results in nanoparticles that are less spherical or irregular shapes. In procedures 1 and 3, nanoparticles produced using sodium borohydride alone as a single reducing agent appear to be less spherical relative to those in Procedure 2 which has a synergy of two strong reducing agents, see Fig. 3. Again, the less spherical nature of these nanoparticles is due to the action of trisodium citrate. Therefore, certain nanoparticles within these experiments appear to have a less spherical shape. The combined effects of two strong reducing agents in procedure 2 prevent irregular shape of nanoparticles by trisodium citrate. The nanoparticles in experiments of Procedure 2 shown in Table 1 are therefore more spherical in shape. In procedure 2 which has a synergy of two fast reducing agents the reaction mechanism is much faster and generates smaller particle sizes at a faster rate. The synergy of two strong reducing agents would remarkably increase the fast reaction rates to grow and form particles at a faster rate which prevents the weak reducing agent to have any significant impact on particle formation. The trisodium citrate being a weak reducing agent therefore only acts as a surfactant in such a case to protect the surface of formed particles from agglomeration. However, in the case of using single fast reducing agents as in procedures 1 and 3, the reaction mechanism may not be fast enough. However, in such a case, trisodium citrate which has a higher concentration in the aqueous solution may also contribute to the reduction of silver ions hence affecting the shape of the formed silver nanoparticles. The synergy of the two strong reducing agents in procedure 2 therefore produces faster and comparatively smaller particles. Synthesis Procedure 2 also results in more highly mono-dispersed silver nanoparticles. Procedure 1 and 3 were synthesized at temperatures of 60 °C and 25 °C respectively. The room temperature synthesis of silver nanoparticles demonstrated in Procedure 3 has improved mono-dispersity relative to Procedure 1 though both synthesis employed single strong reducing agent. Comparably, in synthesis Procedure 1,2 and 3, the synergy of two strong reducing seen in Procedure 2 produces much highly mono-dispersed nanoparticles.

Table 3. Nanoparticle mean size for the various experimental synthesis with their corresponding standard deviation.

Procedure	Experiment	Mean Particle Size (nm)	Standard Deviation (nm)
1	1a	24.47	9.92
	1b	6.99	5.77

(*continued*)

Table 3. (*continued*)

Procedure	Experiment	Mean Particle Size (nm)	Standard Deviation (nm)
	1c	20.12	14.54
2	2a	3.9	2.02
	2b	6.05	2.29
	2c	4.39	2.48
3	3a	5.76	5.76
	3b	6.04	4.75
	3c	5.75	5.96
	3d	10.81	4.9
	3e	4.39	3.96

3 Conclusion

The synthesis of industrially viable monodispersed silver nanoparticles was achieved by the combined effect of two reducing agents. The synthesis procedure generates nanoparticles of identical spherical shape with a controlled narrow-size distribution. The identical morphology characterizes a homogeneous physico-chemical property of nanoparticles. Nanoparticles with identical properties can be quantified for their effectiveness in a specific application. Thus, enhancing their efficiency, hence the economic viability of monodispersed nanoparticles. This study contributes to the sustainable development goal of industrial innovation for the production of viable silver nanoparticles.

Acknowledgement. The authors acknowledge financial support from the Africa Regional International Student Exchange (ARISE) and the research support of the Department of Chemical Engineering, University of Cape Town, under the supervision of Professor Patricia Kooyman.

References

Anderson, D.E., et al.: Investigating the influence of temperature on the kaolinite-base synthesis of Zeolite and Urease immobilization for the potential fabrication of electrochemical urea biosensors. Sensors (Switzerland), 17(8) (2017). https://doi.org/10.3390/s17081831

Asamoah, R.B., et al.: A Comparative Study of Antibacterial Activity of CuO/Ag and ZnO/Ag Nanocomposites. Adv. Mater. Sci. Eng., 2020 (2020). https://doi.org/10.1155/2020/7814324

Asamoah, R.B., et al.: Results in Materials Synthesis and characterization of zinc and copper oxide nanoparticles and their antibacteria activity. Results Mater., 7(March), 100099 (2020). https://doi.org/10.1016/j.rinma.2020.100099

Asamoah, R.B., Yaya, A., Anan, E.: Applications of nanostructural materials for water remediation (2020)

Dang, T.M.D., et al.: Synthesis and optical properties of copper nanoparticles prepared by a chemical reduction method. Adv. Nat. Sci.: Nanosci. Nanotechnol. 2(1) (2011). https://doi.org/10.1088/2043-6262/2/1/015009

Davarpanah, S.J., et al.: Synthesis of Copper (II) Oxide (CuO) Nanoparticles and its application as gas sensor. J. Appl. Biotechnol. Rep. 2(4), 329–332 (2015)

Garni, M., et al.: Biochimica et Biophysica Acta Biopores/membrane proteins in synthetic polymer membranes ☆. BBA - Biomembranes 1859(4), 619–638 (2017). https://doi.org/10.1016/j.bbamem.2016.10.015

Hashemipour, H., Zadeh, M.E., Pourakbari, R.: Investigation on synthesis and size control of copper nanoparticle via electrochemical and chemical reduction method. Int. J. Phys. Sci. 6(18), 4331–4336 (2011). https://doi.org/10.5897/IJPS10.204

Hussain, J.I., et al.: Silver nanoparticles: preparation, characterization, and kinetics. 2(3), pp. 188–194 (2011). https://doi.org/10.5185/amlett.2011.1206

Mascolo, M.C., Pei, Y., Ring, T.A.: Nanoparticles in a Large pH window with different bases, pp. 5549–5567 (2013). https://doi.org/10.3390/ma6125549

Mott, D., et al.: Synthesis of size-controlled and shaped copper nanoparticles. Langmuir 23(10), 5740–5745 (2007). https://doi.org/10.1021/la0635092

Nyathi, T.M., et al.: Impact of nanoparticle–support interactions in Co 3 O 4 /Al 2 O 3 catalysts for the preferential oxidation of carbon monoxide. ACS Catal. 9(8), 7166–7178 (2019). https://doi.org/10.1021/acscatal.9b00685

Odularu, A.T.: Metal Nanoparticles: thermal decomposition, biomedicinal applications to cancer treatment, and future perspectives. Bioinorganic Chemistry and Applications 2018 (2018). https://doi.org/10.1155/2018/9354708

Roberson, M., et al.: Synthesis and characterization Silver Zinc Oxide and Hybrid Silver/Zinc oxide nanoparticles for antimicrobial applications. Nano LIFE 04(01), 1440003 (2014). https://doi.org/10.1142/s1793984414400030

Shah, V., et al.: Molecular insights into sodium dodecyl sulphate mediated control of size for silver nanoparticles. J. Molecular Liquids 273, 222–230 (2019). https://doi.org/10.1016/j.molliq.2018.10.042

Perormance and Nutrient Values of *Clarias Gariepinus* Fed with Powdered Mushroom (*Ganoderma Lucidum*) and Tetracycline as Additives

A. M. Adewole(✉)

Department of Animal and Environmental Biology, Adekunle Ajasin University, Akungba Akoko, P.M.B. 001, Ondo, Nigeria
adeyemo.adewole@aaua.edu.ng

Abstract. Purpose: The comparative effects of *Ganoderma lucidum* and Tetracycline as dietary additives on aquaculture indices and nutritive values of African catfish were evaluated.

Design/Methodology/Approach: Juveniles (n = 15, mean weight: 9.00 ± 0.06g) were given diets of 40% crude protein containing:(0.5% TET1- 1% TET2) and 0.75% (GLM1), 1.5% (GLM 2),3% (GLM 3), 6% (GLM 4), 9% (GLM 5) and Control (0.0%), twice a day (morning and evening), at five percent body weight for 84days using a Completely Randomized Experimental Design. The Average Weight Gain (AWG); Specific Growth Rate (SGR), proximate and mineral composition of fish carcasses were evaluated. Data were subjected to statistical analyses using mean, percentages and analysis of variance (ANOVA) at ($\alpha_{0.05}$).

Findings: The maximum AWG (53.99 ± 0.32g) was obtained from fish given GLM2 diet while the minimum(30.52 ± 0.02g) was recorded from fish given the control diet. The SGR (0.96 ± 0.02%/day) was significantly different (P < 0.05) in fish given GLM2 diet in comparison to others. The maximum protein (62.38 ± 0.03%) content was obtained in the carcass of fish given GLM3 diet and the minimum (60.41 ± 0.05%) was recorded in the fish carcass given the control diet. The maximum Na^+, Mg^{2+} ions content which varied significantly (P < 0.05) among the groups were found in the fish flesh given GLM5 diet and minimum content was jointly from the fish flesh given TET2 and GLM4 diets respectively.

Implications/Research Limitation: Inclusion of 3% *G. lucidum* improved the performance and carcass qualities of *C. gariepinus*. Furthermore, research on the organoleptic properties of the fish flesh given these additives are desirable.

Practical Implication: The use *G. lucidum* as a substitute for orthodox drug like tetracycline in the culture of *C.gariepinus* by farmers is recommended.

Originality/Value: Mushrooms are useful for growth and health promotion of animals and humans. Furthermore, antibiotics have been extensively utilised in fish farming systems as growth promoters. These antibiotics are costly and promote multi drug resistance (MDR) by the microbe and consequently have environmental effects. Therefore, this study evaluated the impact of these additives on the growth parameters of African catfish *Clarias gariepinus* fingerlings and it's impact on the nutritive values of the flesh of *C. gariepinus* given the herbal based diets in comparison with the flesh of the fish given the synthetic antibiotic based diets respectively. This is evident based on the fact, that consumers are now increasingly aware and expressing concerns about the quality of farmed produced, especially products from fish, poultry and other livestock given or raised with medicated drugs/chemical substances such as antibiotics, hormones, enzymes and so on, due to their hazardous impact on consumer's health, as well as the farmed animals' health and the environment respectively.

Keyword: Aquaculture · Fish nutrition · Growth performance · Mineral contents proximate composition

1 Introduction

With the increasing world population from the 7.8 billion in 2020 to around 10.9 billion by 2100. The United Nation predicted, that close to 70% of the projected increase in the world population from 2020 to 2050 will occur in Africa (United Nation World Population Prospects, 2017). This translates to more mouths to be given foods, both globally and regionally (this is due to the fact that one out of ten of the world populace are living in abject poverty and struggling to meet daily essential needs such as; education, health, access to water and good sanitation and more importantly food security) (World Bank, 2020).

This development calls for serious attention and effective implementation of the Sustainable Development Goals (SDGs), particularly (SDG) No.1 (End poverty everywhere) and SDG No.2 (Zero/No hunger), which means adequately healthy food must provided for everyone in order to have access to improved nutrition, adequate food security and promoting sustainable food production (UNDP, 2020). If sustainable agriculture is one of the key targets and sustainable aquaculture/ sustainable fish farming are the indicators to the 'means of achieving' SDGs 1 and 2 respectively. Therefore, there is the needs for mobilization of resources to end poverty and hunger respectively as soon as possible.

The aim of sustainable agriculture is to make provision for the societal basic needs such as food and textiles, today and without jeopardizing the ability of future generations to meet their own needs UC Sustainable Agriculture Research and Education Program (2021). Sustainable fish farming is the production of different fish species for commercial purposes by means that have a friendly, if not kind and gentle, but with overall effect on the ecosystem, contributing to livelihood and developing the local area, at the same time, promoting economic benefits (Burger, 2022). Furthermore, sustainable fish farming has proven evidence that capture fisheries are overexploited and increasing quantities of fish species are being threatened with extinction (Burger, 2022). Furthermore, overfishing which is the unsustainable exploitation of the fisheries resources, has really becomes a big problem, that have affected the globe and its inhabitants. To solve this, there is a need for an alternative sustainable practices, which is the production of fish through sustainable aquaculture. However, a reasonable proportion of aquaculturists raised fish in an unsustainable method, that are not environmentally friendly and having deleterious effect on the consumers (Eco Caters, 2022).

Nutritional studies entail manipulation and oral administration of substances hitherto alien to the animals and far from their conventional diets (Azeez, Adah, Adenkola, and Ameen, 2016). Alteration in physiological process of animal was reported by Satchithanandam *et.al.* (1990) based on feeds and feed intake. However, the use of functional food as dietary supplement in nutritional studies can make positive alterations, as it has benefit (Ajibola, 2015). Additive such as antibiotics are used for prophylactic, curative and growth promotional purposes in the course of animal productions such as poultry (Al-Bahry, Al-Mashani, Elshafie, Pathare, and Al-Harthy, 2006), fish (Adewole, 2016).

Despite the positive effects of antibiotics, Ajani (2019) observed that antibiotic produces visible negative impacts such as destruction of the microbial population in the aquatic ecosystem, development of antibiotics – resistance strains of bacteria, suppression of animal immune system and residue in animals' products and its effects on human health are of major concern. Furthermore, some fish farms use additives that were unhealthy such as hormones, aquatic biocides and antibiotics that have potentially hazardous effects on the health status of the people that eat the fish as well as the environment (Eco Caters, 2022).

Evidently, deficiencies from the utilization of synthetic drugs and the aquaculture industry turning into the use of alternative sustainable practices, such as sustainable aquaculture cum fish farming, thus, prompting the removal of in-feed antibiotics among the feed ingredients in some regions of the world (Ratcliff, 2000). The adoption of antibiotic-free production system where natural alternatives replace the use of in-feed antibiotics is the current research focus for animal nutritionists (Jimoh Ayuba, Ibitoye, Raji and Dabai, 2017; Adewole, 2021).

Varieties of mushrooms offered certain advantages such as protection from pathogens, resistance to diseases, improved nutrient utilization, leading to fast growth. Proximate composition and phytochemical contents of mushrooms have revealed the presence of biochemical substances that aid in their functioning as feed substitutes/alternatives and as well as medicines to cure different parasitic diseases, while, improving wound healing and other medical values (Adejumo and Awosanya, 2005).

Ganoderma lucidum is a mushroom well known, as useful source of feed supplement and medicine, thus helping in the inhibition of cancerous cells growth (Borchers, Keen and Gershwin, 2004). *G. lucidum* contain triterpenes group named ganoderic acids, that have similar chemical configuration as steroid hormones.Furthermore, it produces other secondary metebolites that are present in fungi, such as alkaloids, coumarin, mannitol and polysaccharides (Paterson, 2006). These triterpenes found in *G. lucidum* have the potential capacity to supress LPS/D-Galactosamin e-induced liver damage by the reduction of tumor necrosis factor-alpha (TNF-α) and interleukine-6(IL-6) expression respectively (Hu, Du, Xiu, Bian, Ma, Sato, Hattori, Zhang, Liang, Yu and Wang,2020).

Recently, Giannenas Pappas, Mavridis, Kontopidis, Skoufos and Kyrizazkia, (2010) reported the use of both or the combination of Chinese herbs and mushroom extracts as a substitutes for synthetic drug such as antibiotic in the promotion of growth in animals. Furthermore, the sustainable search for alternative to protein sources and the current trends of replacement of fishmeal and soybean meals with other plant sourced meals justified the quest or thirst for the adoption of new and non-conventional feed resources in fish nutrition. Fungal based additives are suitable sources of ingredients based on their ability to convert woody substrates into promising protein resources with little dependence on other factors: water, land and climate change (Lapena,Olsen, Arntzen, Kosa, Passoth,Eijsink,and Horn 2020a;Lapena, Kosa, Hansen, Mydland, Passoth, Horn and Vincent 2020b). However, the sustainable development of aquacultural industry is essential in order to sustain the projected increase in demand for animal source of proteinous foods, due to rapid increase in human population size (Agboola, Øverland, Skrede, and Hansen, 2021).

Furthermore, available researches suggested that varieties of yeast can serve as appropriate alternatives or main proteinious ingredients in animal feeds. Subsequently, most appropriate and most optimum doses or dosages of these yeast/fungal probiotics have not being subjected to biological/medical evaluations (Agboola, Øverland, Skrede and Hansen, 2021). Thus, more researches are necessitated in order to determine appropriate dosage of yeast/fungal probiotics for fish species of aquacultural importance. There is little or no information on the biological evaluation of dried powdered *G. lucidum* meal diets in fish. Hence, the objectives of this study were to determine the effect of different dosages of *G. lucidum* as alternative to tetracycline as functional additives on performance and nutritive values of African catfish *Clarias gariepinus*.

Various studies have shown the fact that different species of animal react at varying degrees to different types of medicinal plants as buttressed by the quotation of Watson, (2001) that "Discriminate use of many herbal products are safe, while indiscriminate use of herbs are unsafe and may be toxic". Furthermore, just as we each have preferences for different foods, we each react best to particular herbs Hawkey and Hayfield (1999). Also Abatan, (2012) reported that different species of animals are poisoned with varying degree of severity and by different types of poisonous plants and Pamplona-Roger (2001) reported the doses must be well calculated in order to remain within narrow therapeutic range of these plants, because their toxic dose is only slightly higher than their therapeutic dose. Therefore, this study was aimed at supplying the missing information about the reactions of African walking catfish *Clarias gariepinus* to medicinal mushroom *(Ganoderma lucidu*m) and synthetic antibiotics Teteracycline based meal diets respectively; as it affects the growth performance and nutritive values of the flesh. Furthermore, the success in catfish farming in Nigeria, allowed the fish feed industry to expand and grown tremendously in both investments and innovations.Thus necessitate the use of alternative antibiotics/therapeutics for possible inclusion in the diet of farmed fish or as a total replacement for commonly used ones where possible.

In Nigeria, unlike most developed countries of the world, there are no records of any programmes for the supervision of drug residues either in animal products or in abattoirs (Dipeolu and Alonge, 2001). While (Dina and Arowolo, 1999; Kanu, 2008) reported gross misuse of animal based drugs in the country, due to uncontrolled and lack of regulation of drug usage, along the importation and sharp practices in the fish feed industry. Coffman (2000) reported that, parts of the conclusions and recommendations of the N.R.C., (1999) was that further research into alternatives to drugs were also recommended as part of the solution to drug resistance. Okeke and Sosa (2003), observed that novel drugs are obtainable from natural products, an ancient source of antimicrobial, while, Doyle (2001) submitted that the efficiency of these additives such as phytogenics, probiotics in animal feed must be able to give equivalent benefits as the synthetic drugs used in animal production.

Natural products are beneficial, less expensive, and environmentally safe for both man and his animals. Plants or medicinal plants (herbs) have been reported to have been used as remedy for human and animal diseases for many ages. All civilizations, took advantages of herbs in order to ease suffering and many times in order to cure and prevent diseases as well. May be the Creator gave human this vegetal world, with all its curative power and potential to make our life easier (Schneider, 2001).

Furthermore, natural products can be obtained from food components and medicinal herbs with antioxidant activities. An antioxidant is a biochemical compound that reduces the rate of oxidation of other molecules. Antioxidants do stop the series of reactions by eliminating free radical groups, and inhibitions of other oxidative reactions. However, while, doing this, antioxidants get oxidized themselves, so antioxidants are often regarded as reducing agents such as polyphenols, ascorbic acid or thiols (Bjelakovic, Nikolova, Gluud, Simonetti and Gluud, 2007).

These reactive oxygen species do play a major role in the occurrence of many diseases: hypertension, atherosclerosis, diabetes melliitus, cancer, neurodegenerative disorders (Parkinson's disease, Alzhemeir), dementia. These also include protein oxidation and lipid peroxidation, DNA damage and reperfusion injury of the brain and liver respectively (Valko,Leibfritz,Moncol,Cronium,Mazu and Telser, 2007; Finkel, 2005).

Also, the idea of producing crops for the health benefit rather than for food or fiber has gradually changed the trends in plant biotechnology and medicine (Abatan, 2012). However, plant biotechnology and medicine have been known to produce new therapuetics such as plant-based pharmaceuticals, functional foods, dietary supplements, multicomponents botanical drugs and plant-produced recombinant protein respectively (Abatan, 2012; Soetan and Abatan, 2008).

The use of'Functional foods' is based on the concept that they supply more than nutrients to the body, they provide extra metabolic advantages to the consumers. Medicinal plants that are richly blessed with abundant levels of vitamins and nutrients, can be a good source antioxidant, which can be effectively utilized as functional foods (Demming-Adams and Adams, 2002; Block, Koch, Mead, Tothy, Newman and Gyllenhaal, 2008).

The idea of knowing the fish nutrient contents is very important in the quest for the proper uitlization of fish as food item. It can also be a reflector of the proximate compositions of the fish products (Moghaddam, Mesgaran, Najafabadi and Najafabadi, 2007; Fagbenro and Akinbulumo, 2005). Furthermore, the proximate analyses show the average percentage nutritive values such as carbohydrate, lipids, protein, water and mineral within the fish tissues. The nutritive values varied extensively among fishes, as well as within the same species or indivudualized member of the same species (Reinitz, 1983; Hoffman, Casey and Prinsloo, 1992). Furthermore, the variation in the body composition of fish has been known to be dependent on various parameters: environment, age, size, feeding habits and species (Reinitz, 1983; Degani, 1988 and Degani 1989). However, most of the past researches on African freshwater food fish species, dealt with mainly on the study of their biology and ecology, thus limited attention has been given to the nutritive values and chemical composition (Fagbenro and Akinbulumo,(2005).

There is dearth of information, especially from the biological evaluations of different feed ingredients and additives like functional foods and growth promotant (Adewole, 2021).

Since mushrooms have been generally regarded as valuable source of feed supplement and good sources of antioxidants that promote growth and prevent diseases in farm animals. Furthermore, the body composition of fish is greatly influenced by nutritional history. Therefore, the use of *G. lucidum* is based on its rich medicinal values as compared to the orthrodox drugs like the tetracycline. This a right approach to exploring novel medicinal plants that are abundantly around us, thus promoting fish medicine and health.

2 Materials and Methods

Experimental designs and procedures: 360 *C. gariepinus* fingerlings with a mean Initial total weight (ITW) of 135.09 ± 0.50g and Initial total length (ITL) of 10.98 ± 0.08cm were obtained from the Hatchery and research unit of the Department of Animal and Environmental Biology, Adekunle Ajasin University, Akungba–Akoko, Nigeria. The fingerlings were stocked in triplicates, after being acclimatized for 14 days in 24 rectangular plastic tanks (49 × 33.5 × 33.5 cm^3) using a Completely Randomized Experimental Design. The fingerlings were given five percentage of the inital mass (g) twice daily (8.00h and 18.00h) for a period of 84 days between October 13, 2020 and January 5, 2021.The routine cleaning and maintenance of the rectangular plastic culture tanks were done as recommended by Adewole, (2014). The different metric measurements were taken after every two weeks and the rations given were adjusted according to the new fish weight attained. At the completion of the experimental periods, the biometric measurements of the fingerlings were taken according to Adewole, (2014).

Experimental site: The research was conducted in the Fish hatchery and research unit of the Department of Animal and Environmental Biology, Adekunle Ajasin University, Akungba–Akoko, Nigeria.

Mushroom collection, identification, diets formulation, processing and preparation: The mushrooms (*G.lucidum*) was gathered in the raining period of September,2020 in the wild forest of Aklungba -Akoko. The mushroom samples were recognized and approved in the Department of Plant Sciences and Biotechnology, Adekunle Ajasin University, Akungba-Akoko, Nigeria. The samples were dried in the sun for 14 days and grounded into powder using hammer mill. The powdered meal was kept in a polythene bag at ambient temperature according to Adewole, (2014).

The various ingredients were obtained from a feedmill store in Ikare -Akoko, Ondo State, Nigeria. Eight isonitrogenous (40% Crude Protein) diets were formulated as shown in Table1. The diets were 0% (Negative Control), 0.5% Tetracycline (TET1), 1%Tetracycline (TET2) as positive controls, and 0.75%, 1.5%, 3%, 6%, 9% dried powdered *G. lucidum* meals (coded as GLM1 - GLM5) respectively. The processing, preparation and storage were done using the methods of Adewole, (2014).

Data collection: The data obtained from the biometric measurements were used to calculate the different aquacultural parameters: Average Weight Gain (AWG), Specific growth rate (SGR) and Relative growth rate (RGR), Condition factors (K) and Survival rate (SR) and Nutrient utilization parameters:Feed conversion ratio (FCR), Protein efficiency ratio (PER), Mean feed intake (MFI), Mean protein intake (MPI) according to Adewole, (2014).

The proximate and mineral elements analyses were determined in accordance to the protocol of the Association of official analytical chemists (AOAC), (2009). All determinations were done in triplicates according to Adewole, (2017).

Statistical analysis: The data obtained from aquacultural and nutrient evaluation indices, proximate, along with the mineral composition were statistically analysed with one-way Analysis of variance (ANOVA) at ($\alpha_{0.05}$), while the test for means at different levels of significance was done using Duncan's Multiple Range Test using Statistical Analysis System (SAS, 2008) model.

Table 1. Percentage of ingredients (g/100g) in Ganoderma lucidum and Tetracycline meal diets

Ingredients	Control 0%	TET1 0.5%	TET 2 1%	GLM1 0.75%	GLM2 1.5%	GLM3 3.0%	GLM4 6.0%	GLM5 9.0%
Fish meal	20.72	20.60	20.49	20.55	20.38	20.04	19.36	18.68
Groundnut cake meal	20.72	20.60	20.49	20.55	20.38	20.04	19.36	18.68
Soya bean meal	20.72	20.60	20.49	20.55	20.38	20.04	19.36	18.68
Yellow maize	15.42	15.35	15.27	15.30	15.18	14.94	14.46	13.98
Rice bran	15.42	15.35	15.26	15.30	15.18	14.94	14.46	13.98
Vitamin and Mineral premix	2.00	2.00	2.00	2.00	2.00	2.00	2.00	2.00
Carboxymethyl cellulose	2.00	2.00	2.00	2.00	2.00	2.00	2.00	2.00
Starch	1.00	1.00	1.00	1.00	1.00	1.00	1.00	1.00
Salt	0.50	0.50	0.50	0.50	0.50	0.50	0.50	0.50
Bone meal	1.00	1.00	1.00	1.00	1.00	1.00	1.00	1.00
Tetracycline	-	0.50	1.00	-	-	-	-	-
Mushroom	-	-	-	0.75	1.50	3.00	6.00	9.00
Total	**100**	**100**	**100**	**100**	**100**	**100**	**100**	**100**

3 Results

Total final weight (TFW) ranged from 601.75 ± 29.84g -861.08 ± 26.72g. The maximum TFW value was from the fish given GLM2 diet and the minimum value was obtained in the fish given GLM5 diet. The TFW was significantly different (P > 0.05) in the fish given GLM2 diet compared to the fish given the various treatments. (Table 2). The maximum values for AWG, RGR and SGR (53.99 ± 0.32g/fish;536.47 ± 19.93%; 0.96 ± 0.02%/day) were recorded from fish given GLM2 diet and the minimum values of

Table 2. Growth performance and nutrient utilization of *Clarias gariepinus* given *Ganoderma lucidum* meal diets in plastic tanks for 84 days

Parameters	Control	TET1	TET2	GLM1	GLM2	GLM3	GLM4	GLM5
Total Initial weight (g)	135.17 ± 0.20a	135.06 ± 0.13a	135.38 ± 0.10a	135.27 ± 0.11a	135.29 ± 0.10a	135.30 ± 0.08a	135.11 ± 0.16a	135.39 ± 0.13a
Mean Initial weight(g)	9.01 ± 0.01[a]	9.00 ± 0.01[a]	9.03 ± 0.01[a]	9.02 ± 0.01[a]	9.02 ± 0.01[a]	9.02 ± 0.01[a]	9.01 ± 0.01[a]	9.03 ± 0.10[a]
Total Final weight(g)	649.09 ± 24.68b	673.08 ± 16.57b	668.19 ± 32.28b	656.9 ± 71.53b	861.08 ± 26.72a	715.02 ± 18.70b	710.52 ± 15.01b	601.75 ± 29.84b
Mean Final weight(g)	39.53 ± 0.42[b]	50.49 ± 6.10[b]	53.76 ± 2.30[b]	51.89 ± 1.70[b]	63.01 ± 6.20[a]	53.63 ± 1.80[b]	54.65 ± 1.52[b]	42.99 ± 2.85[b]
Body weight gain	513.89 ± 24.84[b]	538.02 ± 16.64[b]	532.81 ± 32.37[b]	521.65 ± 71.43[b]	725.79 ± 26.76[a]	579.73 ± 18.7[b]	575.41 ± 15.09[b]	466.36 ± 29.82[b]
Average Weight Gain (g/fish)	30.52 ± 0.02[c]	41.49 ± 0.03[b]	44.73 ± 0.40[b]	42.87 ± 0.41[b]	53.99 ± 0.32[a]	44.61 ± 0.09[b]	45.64 ± 0.65[b]	33.96 ± 0.12[c]
Relative growth rate (%)	380.22 ± 18.85b	398.38 ± 12.52b	393.97 ± 24.20b	385.57 ± 52.57b	536.47 ± 19.93a	428.50 ± 14.02b	425.91 ± 11.43b	344.45 ± 21.95b
Specific growth rate (%/day)	0.81 ± 0.02b	0.83 ± 0.01b	0.82 ± 0.02b	0.81 ± 0.06b	0.96 ± 0.02a	0.86 ± 0.01b	0.86 ± 0.01b	0.77 ± 0.03b
Survival rate (%)	86.67 ± 6.67[a]	88.89 ± 5.88[a]	82.88 ± 4.44[a]	84.44 ± 4.44[a]	91.11 ± 2.22[a]	88.89 ± 2.22[a]	86.67 ± 6.67[a]	93.33 ± 3.85[a]
K_1	0.78 ± 0.77[ab]	0.85 ± 0.27[b]	0.72 ± 0.39[ab]	0.75 ± 1.32[ab]	0.83 ± 0.14[a]	0.74 ± 0.59[ab]	0.77 ± 0.58[ab]	0.73 ± 0.12[ab]
K_2	0.86 ± 0.88[ab]	0.89 ± 0.97[ab]	0.97 ± 0.95[ab]	0.90 ± 0.33[b]	1.00 ± 0.34[ab]	0.81 ± 0.70[ab]	0.90 ± 0.86[ab]	0.88 ± 1.00[a]
Total Feed Intake	1335.69 ± 100[ab]	1553.35 ± 64.78[a]	1256.45 ± 51.08[b]	1344.75 ± 123.49[ab]	1514.61 ± 19.53[b]	1330.56 ± 26.86[ab]	1211.56 ± 30.29[b]	1104.51 ± 101.60[b]
Mean Feed Intake (g)	100.20 ± 0.02[ab]	116.53 ± 0.04[a]	101.08 ± 0.01[ab]	106.14 ± 0.06[a]	110.80 ± 0.05[a]	99.82 ± 0.06[b]	93.20 ± 0.03[b]	78.89 ± 0.01[c]
Food Conversion Ratio	2.60 ± 0.16[a]	2.89 ± 0.10[a]	2.38 ± 0.22[bc]	2.62 ± 0.14[ab]	2.09 ± 0.10[c]	2.30 ± 0.12[bc]	2.11 ± 0.05[c]	2.36 ± 0.14[bc]
Protein Intake	565.24 ± 42.43[bcd]	692.22 ± 28.87[a]	510.67 ± 20.77[cd]	571.39 ± 52.91[bcd]	634.27 ± 8.19[ab]	578.66 ± 11.67[bc]	525.21 ± 13.10[cd]	467.40 ± 43.06[d]
Mean Protein Intake	44.41 ± 0.02[bc]	51.93 ± 0.01[a]	44.12 ± 0.02[bc]	46.15 ± 0.03[b]	48.62 ± 0.02[ab]	43.41 ± 0.03[bc]	40.40 ± 0.02[c]	34.31 ± 0.03[d]
Protein Efficiency Ratio	0.92 ± 0.06[bc]	0.78 ± 0.03[c]	1.05 ± 0.11[ab]	0.90 ± 0.05[bc]	1.15 ± 0.05[a]	1.00 ± 0.05[a]	1.10 ± 0.02[ab]	1.01 ± 0.06[ab]

Table 3. Proximate and mineral compositions of *Clarias gariepinus* given *Ganoderma lucidum* meal diets in plastic tanks for 84 days

Parameters	Control	TET1	TET2	GLM1	GLM2	GLM3	GLM4	GLM5
Crude Protein (%)	60.41 ± 0.05^e	60.56 ± 0.02^d	61.58 ± 0.01^{bc}	61.69 ± 0.02^b	61.47 ± 0.01^c	62.38 ± 0.03^a	61.65 ± 0.09^b	62.37 ± 0.03^a
Crude Lipid (%)	9.50 ± 0.00^d	9.64 ± 0.01^c	7.85 ± 0.01^f	8.05 ± 0.03^e	7.69 ± 0.05^g	10.40 ± 0.01^a	7.78 ± 0.01^f	10.19 ± 0.01^b
Crude Ash (%)	2.49 ± 0.01^a	2.35 ± 0.01^b	2.29 ± 0.01^c	2.26 ± 0.00^{de}	2.25 ± 0.01^{de}	2.25 ± 0.01^e	2.24 ± 0.01^e	2.31 ± 0.01^c
Crude Fibre (%)	0.85 ± 0.01^{ab}	0.83 ± 0.01^{bc}	0.79 ± 0.01^d	0.79 ± 0.02^d	0.83 ± 0.01^{bc}	0.79 ± 0.01^d	0.85 ± 0.00^b	0.87 ± 0.01^a
N F E (%)	25.02 ± 0.33^{ab}	24.45 ± 0.01^{bc}	25.25 ± 0.03^{ab}	25.05 ± 0.07^{ab}	25.48 ± 0.03^{ab}	21.87 ± 0.02^d	26.80 ± 1.23^a	22.95 ± 1.25^{dc}
Dry Matter (%)	97.77 ± 0.01^b	97.81 ± 0.01^a	97.76 ± 0.00^b	97.73 ± 0.01^c	97.83 ± 0.01^a	97.69 ± 0.01^d	97.60 ± 0.01^e	97.49 ± 0.01^f
Na (mg/L)	1.06 ± 0.01^a	0.66 ± 0.08^c	0.92 ± 0.01^b	0.65 ± 0.00^c	0.66 ± 0.00^c	0.85 ± 0.00^b	1.08 ± 0.01^a	1.11 ± 0.00^a
K (mg/L)	4.32 ± 0.00^a	4.26 ± 0.05^a	4.26 ± 0.03^a	4.00 ± 0.02^c	4.10 ± 0.00^b	3.96 ± 0.01^c	4.12 ± 0.01^b	4.31 ± 0.03^a
Mg (mg/L)	0.74 ± 0.00^d	0.77 ± 0.00^c	0.72 ± 0.00^f	0.77 ± 0.00^c	0.74 ± 0.00^e	0.78 ± 0.00^b	0.72 ± 0.00^f	0.79 ± 0.00^a
Ca (mg/L)	3.63 ± 0.00^f	5.24 ± 0.03^c	4.34 ± 0.02^e	5.21 ± 0.01^c	5.10 ± 0.04^d	5.33 ± 0.02^b	5.27 ± 0.00^{bc}	5.68 ± 0.03^a

MGW (30.52 ± 0.02g) was from the fish given the control diet; while the minimum values of RGR and SGR (344.45 ± 21.95%; 0.77 ± 0.03%/day) were from the fish given GLM5 diet. The fingerlings given GLM2 diet was significantly different (P < 0.05) in all growth performance indices when compared with other tested diets and the control diet respectively.

The K2 and SR ranged (0.81–1.00 ± 0.34;82.88% -93.33%) respectively. The maximum K2 was from the fingerlings given GLM2 diet and the minimum was from the fingerlings given control diet, however, best SR value obtained in the fingerlings given GLM5 diet and the minimum from the fingerlings given TET2 diet respectively (Table 2).

The best FCR and maximum PER values (2.09 ± 0.10;1.15 ± 0.05) were obtained from the fingerlings given GLM2 diet, while, poorest cum minimum values (2.89 ± 0.10 and 0.78 ± 0.03) were from the fingerlings given TET1 diet. The FCR and PER were significantly different (P > 0.05) within the treatments (Table3). Furthermore, MFI and MPI varied significantly (P > 0.05) within the fingerlings given the experimental diets. The maximum values for MFI and MPI (116.53 ± 0.04g; 51.93 ± 0.01g) were obtained from the fingerlings given TET1 diet and the minimum values (78.89 ± 0.01g; 34.31 ± 0.03g) were from the fingerlings given GLM5 diet respectively (Table 3).

All the proximate parameters varied significantly (P < 0.05) within the fingerlings given the experimental diets (Table 3). However, percentage crude protein (CP%) contents obtained from the carcasses of *C.gariepinus* varied from 62.38 ± 0.03 - 60.41 ± 60.41%, while the maximum value was obtained from the flesh of fingerlings given GLM3 diet and the minimum value was from the flesh of fingerlings given the control diet. Also, the percentage crude lipid (CL%) contents ranged from 7.69 ± 0.01% to 10.40 ± 0.01%, the maximum value was obtained from the flesh of fingerlings given GLM3 diet and minimum value was obtained from the flesh of fingerlings given GLM2 diet.

Percentage crude ash (CA%) contents varied from 2.24 ± 0.01- 2.49 ± 0.01%.The maximum value was obtained in the carcasses of fish given the control diet and minimum value was obtained in the carcasses of fish given GLM4 diet (Table 3). Percentage crude fibre (CF%) contents of the fish carcasses had the maximum value of 0.87 ± 0.01% from the fish carcass given GLM5 diet, while the minimum values 0.79 ± 0.01/0.79 ± 0.02 were co-jointly or among the carcasses of fish given TET2; GLM1 and GLM3 diets respectively (Table 3).

The mineral compositions of the fleshes of *C.gariepinus* reflected that significant variation (P < 0.05) existed among the groups. Ca contents ranged from 5.68 ± 0.03 to3.63 ± 0.00 mg/L, with the maximum value obtained in the flesh of the fingerlings given GLM5 diet and the minimum value was obtained in flesh of fingerlings given the control diet (Table 3). K contents varied from 3.96 ± 0.00 to 4.32 ± 0.00mg/L and maximum value was obtained in the flesh of the fingerlings given the control diet, and the minimum value was obtained in the flesh of the fingerlings given GLM3 diet (Table 3) respectively. Mg concentrations ranged from 0.72 ± 0.00 to 0.79 ± 0.00mg/L, with the maximum value obtained in the flesh of the fingerlings given GLM5 diet and the minimum value was obtained in the flesh of the fingerlings given TET2 and GLM4 diets respectively (Table 3). Na contents ranged between 0.65 ± 0.00 to 1.11 ± 0.00mg/L. The maximum value was obtained in the flesh of the fingerlings given GLM5 diet and the minimum value was obtained in the flesh of the fingerlings given GLM1 diet (Table 3) respectively.

4 Discussion

Weight gain is an important index in getting the response(s) of fish subjected to biological evaluations of different novel diets and also a valuable parameter for growth measurements (Balogun, Abdullahi, Auta and Ogunlade, 2004; Adewole, 2021). The inclusion of *Ganoderma lucidum and* Tetracycline meal diets in *C. gariepinus* showed improvement from the initial weights to the final weights in all the fish exposed to the experimental diets. However, a better result in fish given GLM2 diet than fish given other diets was observed. This result obtained was similar to Muyideen, Lawal and Aarode, (2013), which given *Sclerotium ptuberregium* to *Orechomis niloticus* fingerlings. They reported that the replacement of up to 12.5% soybean meal (SBM) by *Sclerotium* gave better AWG and SGR as against control diet. Furthermore, result from this study was corroborated with that of Oliva-Tales and Goncalves (2001) who reported that the best growth performances and nutrient evaluation index in Sea bass given diets containing 30% dietary protein from the supplementation of brewer's yeast.

Furthermore, the results obtained from this study are similar to the following studies: Essa, Mabrouk, Mohamed and Michael, (2011) reported that the same fish species used in this study, given with 0–2% *S. cerevisiae* had improved performance and profitability after 186 days of biological evaluations. Also, Fronte, Abramo, Brambilla, De Zoysa and Miragliotta, (2019) observed that Gilthhead sea bream (*Sparus aurata*) given *S. cerevisiae* at 20% replacement of fish meal (which contain 4.6% protein contents) offered for 92 days, could possibly have no hazardous impact on growth partern and gut histological architecture. Jin, Xiong, Zhou, Yuan, Wang and Sun, (2018) given diets containing 1% yeast hydrolysate or yeast biomass as substitute to Pacific white shrimp (*Litopeneaus vannamei*) for 56 days and reported that the 1% addition of yeast hydrolysate or yeast biomass improved aquacultural index such as growth, enhance immunity and promote resistance to ammonia nitrogen stress. However, South African dusky kob (*Argyrosomus japonicus*) given diets that had the addition inactivated *S. cerevisiae* ranging from 0–30% for 42 days by Madibana and Mlambo, (2019) had result that showed that 5% inclusion of *S. cerevisiae* did no compromised aquacultural index such as growth and state of health of dusky kob. Subsequently, the supplementations greater than 5%, there were observed depressions in growth. Furthermore, Abass, Obirikorang, Campion, Edziyie and Skov (2018) reported that supplementation of *S. cerevisiae* as feed ingredient for Nile Tilapia (*Oreochromis niloticus*) between 0–5% for 84 days, discovered that *S. cerevisiae* improved the ability of the fish to tolerate sudden changes in the heat condition and low oxygen potential. The study revealed the additive *S. cerevisiae* enhanced the aquacultural index: growth, reduced stress and increases resistance to disease Tilapia. Similarly, the evaluation of the addition of yeast Mannn oligo saccharide (MOS) as dietary additive on the increment in size and rate of survival in *Amphiprion ocellaris* fingerlings was evaluated for 90 days by Ramanadevi Kaliyan, Thangappan and Muthusamy (2013). The results indicated that the final aquacultural indices evaluated like changes in weight and length, along the specific growth rate in all the groups indicated no statistical significance ($P > 0.05$).

The mortality rate was lowest in fish given GLM5 diet; this reflected the fact that the increasing concentrations of mushroom in the diets were well tolerated by the fish. The increased survival rate of fish given GLM diets were probably caused by increase in innate immunity as reported by Daulloul, Lillehoj, Lee and Chung, (2006), that the secondary metabolites in mushroom such as lectin increase inborn or natural immunity in broilers challenged with *Eimeria acervulina*. The high survival rates experienced during the experimental period, was attributed to stable water qualities and good husbandry practices. Furthermore, this result is similar to Ramanadevi, Kaliyan, Thangappan and Muthusamy,, (2013) that deduced that Mann oligo saccharide (MOS) did not have any impact on the aquacultural index such as changes in the size of the biomass (i.e. length and weight) but it reduces the percentage mortality of *Amphiprion ocellaris* fingerlings.

A major factor limiting fish cultivation is the problem of balancing the effect of rapid increase in the rate of fish growth and effective utilization of the feed supplied (Gockcek, Mazlum and Akyurt, 2003).This work showed different levels of optimum utilization of supplied diets by *C.gariepinus*. The results of total feed intake (TFI) revealed that the maximum TFI was obtained by the fish given TET1 diet and was not significantly different from the fish given GLM2 diet, while the minimum TFI was from the fish given GLM5 diet respectively. However, lower TFI by the fish given GLM5 diet could be attributed to the reduction in the taste of the supplied diet, thus, resulting in appetite reduction, and this may be attributed to the bitter taste of *G. lucidum*. The bitter taste of the mushroom was reported by Smart *et al.*,(2001). However, Jackson, Capper and Matty, (1982) reported a decrease in TFI as increment in the quantity of mango peel diet given to *O. niloticus* was observed.

The minimum value for FCR was obtained in the fish given GLM2 diet, while maximum value was obtained in fish given TET1 diet respectively. This observation could be related to the presence of phytonutrients and metabolites like saponins, tannis, oxalates in *Ganoderma* (Soetan and Oyewole, 2009) that may provoke some physiological responses in animals. The values for FCR in the study were lower than 4.80 to 5.03 reported by Analyn *et al.*, (2008) for *O. niloticus* given *G.lucidum* and higher than Ogunji, Kloas, Wirth, Neumann, and Pietsch, (2008), that reported FCR ranging from 1.2 to 1.5 for Nile Tilapia (*O. niloticus*) given maggot meal.The differences in the results may be attributed to nutritional values of *G. lucidum* which contain more quanties of protein, carbohydrates and the lower quantities of fibre, fat and ash that gave a better FCR and also the differences in the species of fish used.

Several studies have been reported on the neutriceutical values of yeast cell wall and its derivatives in fish (Meena, Das, Kumar, Mandal, Prusty, Singh, Akthar, Behera, Kumar, Pal and Mukherjee, 2013; Torrellas, Montero, and Izquierdo, 2014). However, diverse factors such as post-fermentation processing, yeast strain and fermentation media, differences among the fish varieties or species and diet preparation, processing and formulation were deduced to be attributable to the reduction in the biomass (length and weight) and nutritional values as increment in the quantity of *S.cerevisiae* in a few species of fish evaluated (Øverland and Skrede, 2017).

PER is an index measuring the efficiency of the protein sources within the diet, and possibly make provision of essential amino acid needed by the animal and may also favor fat deposition (Singh, Maqsood, Samoon, Phulia, Mohd, and Chalal, 2011).

The PER in this study ranged from 0.78 to 1.15, which is similar to Singh, Maqsood, Samoon, Phulia, Mohd, and Chalal, (2011). These authors observed the possibility of fat deposition in common carp given of papain as growth promotant. Excess protein beyond the required level needed by the fish will be excreted as ammonia by the gills as a form of nitrogenous by products (Lovell, 1981). The fish given GLM2 diet has the minimum FCR and maximum PER values. This result is in agreement with Olukunle, (2011) who given *C. gariepinus* with different dietary oil source and Siddhuraju and Becker (2000) given *Cyprinus carpio* with *Mucuna pruriens*. These authors reported that a diet with minimum FCR indicates more efficient utilization of diets by fingerlings. The better result obtained in treatments with GLM diets could be as a result of increased protein retention which made it more available for the fish growth.

The inclusion of powdered *G. lucidum* diets in *C. gariepinus* at varying levels had significant effects on the body chemical contents. The analyses indicated better improvement in the tissue chemical compositions of fish given the diets with either oxyteteracycline or the mushroom when compared with fish given the control diet. However, for body fat composition, there was a slightly increase in the flesh of *C. gariepinus* given GLM5 and GLM3 diets, while in the fish gven GLM2 diet, the diet had no pronounced impart on the lipid composition of the fish flesh. For all the fish fed with tetracycline based diets, there was significant increment in the tissue fat deposition. The fat as a nutrient serve as a good source of energy and essential fatty acids, which are useful for fish wellbeing.

The Na and Ca contents were maximum in flesh of fish the fish given GLM5 diet, while K content followed an opposite trend of Na concentration. The Mg^{2+} varied significantly among the groups, with the maximum value from flesh of the fish given GLM5 diet and the minimum was jointly from the carcasses of the fish given TET2 and GLM4 diets respectively. This showed that there are large quantities of Mg, Na and Ca in *G.lucidum*, however, lower values of Ca (0.40%) and P (0.30%) in wild *G. lucidum* were reported by Ogbe *et al.*, (2012). These minerals elements have played significant physiological roles and helped in several biochemical activities such as cellular enzymatic reactions. They are useful for the development of skeletal and musclar activities, along with the provision of normal growth, particularly calcium Ogbe and John (2012). Furthermore, sodium and potassium are essentially useful for the maintenance of osmotic and ionic balance, while potassium and calcium are necessary for the stimulation of action potential across the synapses, along with the strengthening of the heart muscle contractile rate (Muhammad, Dangoggo, Tsafe, Itodo and Atiku, 2011). Therefore, the presence of these minerals within the carcasses of the fingerlings given the experimental meal diets, particularly the mushroom based diet supported the reported nutritive value of *G. lucidum* to contain reasonable quantity of fibre (35%), protein (13.3%),fat (2.6 ± 0.3), calcium (0.4%) and phosphorus (0.3%) respectively as observed by Ogbe, Ditse, Echeonwu, Ajodoh, Atawodi and Abdu, (2009).Other benefits of yeast culture is that it helps in chelating many essential and useful minerals elements. Peterson, Streeter and Clark, (1987) observed a gradual increment in the retentive capability of K,Cu and Zn in farmed animals given yeast culture based diets.

5 Conclusion

Potentially, the results obtained in this study indicated necessity for fish feed producers and nutritionists to make use of medicinal herbs in the fish industry.

In conclusion, the use of fungal additive especially, medicinal mushrooms like *G.lucidum* in the culture of *C. gariepinus* can provide profitable, safe and sustainable fish production system.

Furthermore, inclusion of mushroom at 3% gave the best performance and thus established the baseline dosage for the production of African catfish *C. gariepinus*, without compromising the growth performance and nutritional values.

The adoption of the use of this medicinal mushroom by the fish farmers should be encouraged. This will translate to the production of healthier fish food by the fish farmers, since the synthetic antibiotics have been negatively implicated to produce fish products with drug residues and other untowards consequences.

Also, there should be awareness campaign by the government agencies and other stakeholders such as the fish nutritionists, fish feed producers and marketers, fish farmers that are involved in the regulation and monitoring of the fish feed production and its quality, and marketing, to stop the use of in-feed antibiotics due to its short comings.

Further research on other Clariid species using *G.lucidum* and other medicinal/edible mushrooms/plants are advocated.

The laboratory culture and production of the medicinal/edible mushrooms should be giving a trial, in order to faciltate all year round availability of the mushroom/plant.

It will also be an avenue for more income generation by the farmers through the adoption of integrated fish farming (i.e.fish farming cum mushroom farming).

Also, the production of the mushroom by the farmers will help in the environmental control and disposal, and conversion of some agrowastes (non-food lignocellulose) products such as maize stovers that have constituted menace in the environment.

Furthermore, this is in furtherance of the position (UN General Assembly,2015) that stated that: "changing our world: the 2030 Target for Sustainable Development"; supporting the policy statements of the following SDGs: Goal No.1:Ending of hunger, to achieve food security and improvement in nutirtion and sustainable farming system; Goal No.2: Ending of poverty in all itsramifications; Goal No.3: Provision of good health and wellbeing and Goal No12: Responsible consumption and production respectively.

The implication of the mushroom *G. lucidum* as functional additive on the health and immunity of the fish should be considered as area of future research.

Also, a look into the overall quality assessment and acceptability, safety of the fish flesh through the organoleptic properties of the fish flesh given these additives are desirable.

References

Abass, D., Obirikorang, K., Campion, B., Edziyie, R., Skov, P.: dietary supplementation of yeast *(Saccharomyces cerevisiae)* improves growth, stress tolerance, and disease resistance in juvenile Nile tilapia *(Oreochromis niloticus)*. J. Eur. Aquaculture Soc. **26**, 843–855 (2018)

Abatan, M.O.: Poisons and poisoning the nightmarish of a society without diagnostic tools. An Inagural Lecture, 2011/2012. University of Ibadan. Ibadan University Press, Publishing House, University of Ibadan, Ibadan, Nigeria (2012)

Adejumo, T.O., Awosanya, O.B.: Proximate and mineral composition of four edible mushroom species from South western Nigeria. Afr. J. Biotech. **4**, 1084–1088 (2005)

Adewole, A.M.: Performance and production economics of African catfish (Clarias gariepinus) given Allium sativum meal diets. Niger. J. Fish. 18 (2), 2364–2680 (2021)

Adewole, A.M.: Production economics and nutritive values of clarias garieprinus given two dietary additives in organic aquaculture. In: Proceeding of the 5th annual conference of school of science (SOS 2017), pp. 85–95 (2017)

Adewole, A.M.: Growth performance, blood profile, organosomatic indices and histopathology of *Clarias gariepinus* given Amoxicillin as dietary. Int. J. Aquac. **6**(15), 1–13 (2016)

Adewole, A.M.: Effects of Roselle as dietary additives on growth performance and production economy of Clarias gariepinus. J. Emerg. Trends Eng. Appl. Sci. (JETEAS) 5 (7):1–8 (2014)

Agboola, J.O., Øverland, M., Skrede, A., Hansen, J.Ø.: Yeast as a major protein rich ingredient in aquafeeds: a review of the implications for aquaculture production. Rev. Aquac. **13**, 949–970 (2021)

Ajani, E.K.: Engaging the fish: elixir for Nigeria's self sufficiency in fish production, p. 130. University of Ibadan, Nigeria, An Inaugural Lecture (2019)

Ajibola, A.: Physico-chemical and physiological values of honey and its importance as a functional food. Int. J. Food Nutr. Sci. **2**(6), 1–9 (2015)

Al-Bahry, S.N., Al-Mashani, B.M., Elshafie, A.E., Pathare, N., Al-Harthy, A.H.: Plasmid profile of antibiotic resistant *E.coli* isolated from chickens intestines. Alabama J. Acad. Sci. **77**, 152–159 (2006)

AOAC: (Association of Official Analytical Chemists), 2009. Official methods of analysis, Washington, DC (2009)

Azeez, O.M., Adah, S.A., Adenkola, A.Y., Ameen, S.A.: Changes in erythrocyte membrane properties following exposure to premium motor spirit (petrol vapour) and modulatory effect of *Moringa oleifera* and vitamin C in wistar rats. J. Afr. Ass. Physiol. Sci. **4**(2), 102–108 (2016)

Balogun, J.K., Abdullahi, A.S., Auta, J., Ogunlade, O.P.: Feed conversion, protein efficiency, digestibility and growth performance of *Oreochomis niloticus* given *Delmixregia* seed meal. In: Proceedings of the National Conference of Fisheries Society of Nigeria (FISON), 838- 842 (2004)

Bjelakovic, G., Nikolova, D., Gluud, L.L., Simonetti, R.G., Gluud: Mortality in randomized trials of antioxcidant supplements for primary and secondary prevention: systematic review and meta-analysis. JAMA **297**(8), 842–857 (2007)

Block, K.I., Koch, A.C., Mead, M.N., Tothy, P.K., Newman, R.A., Gyllenhaal, C.: Impact of antioxidant supplementation on chemotherapeutic toxicity: a review of the evidence from randomized controlled trials. Int. J. Cancer **123**(6), 1227–1239 (2008)

Borchers, A.T., Keen, C.L., Gershwin, M.E.: Mushrooms, tumors and immunity: an up-date. Exp. Biol. Med. **229**, 393–406 (2004)

Burger, A.: What is Sustainable Aquaculture? Last Modified 29 Mar 2022

Coffman, J.: Regulation of antibiotic resistance the US. *AgBio Forum*, volume 3, November 2 and 3, pages 141–147 (2000)

Degani, G.: Body composition of African catfish *(Clarias gariepinus)* at different ages. Israeli J. Aquac.-Bamidgeh **40**, 118–121 (1988)

Degani, G.: The effect of different protein level and temperatures on feed utilization, growth and body composition of *Clarias gariepinus* (Burchell 1822. Aquaculture **76**, 293–301 (1989)

Demming-Adams, B., W.W.: Antioxidants in photosynthesis and human nutrition. Sci. **298**(5601), 2149–2153 (2002)

Daulloul, R.A., Lillehoj, H.S., Lee, H.S., Chung, K.S.: Immunopotentiating effect of a *Fomitella fraxinea*- derived lectin on chicken immunity and resistance to coccidiosis. Poult. Sci. **85**, 446–451 (2006)

Dina, O.A., Arowolo, R.O.A.: Some considerations on veterinary drug use and supply in Nigeria. Revue – d' Elevange et-de-medicine veterinaries –des – pays – Tropicaux **44** (1), 29 – 31 (1991)

Dipeolu, M.A., Alonge, D.O.: Residues of tetracycline antibiotics in cattle meat marketed in Ogun and Lagos states of Nigeria. Asset Series A **1**(2), 31–36 (2001)

Doyle, M.E.: Alternatives to antibiotic use for grwoth promotion in animal husbandary. Food Research Institute Briefings, 1-17. University of Wisconsin-Madison (2001)

Caters, E.: What is sustainable fish farming? Accessed 30 Mar 2022

Essa, M.A., Mabrouk, H.A., Mohamed, R.A., Michael, F.R.: Evaluating different additive levels of yeast, *Saccharomyces cerevisiae*, on the growth and production performances of a hybrid of two populations of Egyptian African catfifish, *Clarias gariepinus*. Aquaculture **320**, 137–141 (2011)

Fagbenro, A.O., Akinbulumo, M.O.: Flesh yield, proximate and mineral composition of four commercial West African food fish. J. Anim. Vet. Adv. **4**(10), 848–851 (2005)

Finkel, T.: Radical medicine: treating ageing to cure disease. Nat. Rev. Mol. Cell Biol. **6**, 971–976 (2005)

Fronte, B., Abramo, F., Brambilla, F., De Zoysa, M., Miragliotta, V.: Effect of hydrolysed fifish protein and autolysed yeast as alternative nitrogen sources on gilthead sea bream *(Sparus aurata)* growth performances and gut morphology. Ital. J. Anim. Sci. **18**, 799–808 (2019)

Giannenas, I.S., Pappas, S., Mavridis, G., Kontopidis, J., Skoufos, S., Kyrizazkia, I.: Performance and antioxidant status of broiler chickens supplemented with dried mushrooms (*Agaricus bisporus*) in their diet. Poult Sci. **89**, 303–311 (2010)

Gockcek, C.K., Mazlum, Y., Akyurt, I.: Effects of feeding frequency on the growth and survival of *Himri barbell* and *Barbus luteus* fry under laboratory conditions. Pak. J. Nutr. **7**(1), 66–69 (2008)

Hu, Z., et al.: Protective effect of triterpenes of *Ganoderma lucidum* on lipopolysaccharide induced inflammatory responses and acute liver injury. Cytokine **127**, 154917 (2020)

Jackson, A.J., Capper, B.S., Matty, A.J.: Evaluation of some plant proteins in complete diets for the the tilapia, *Sarotherodon mossambicus*. Aquaculture **27**, 97–109 (1982)

Jimoh, A.A., Ayuba, U., Ibitoye, E.B., Raji, A.A., Dabai, Y.U.: Gut health maintenance in broilers: comparing potential of honey to antibiotic effects on performance and clostirdial counts. Niger. J. Anim. Prod. **44**, 106–113 (2017)

Jin, M., Xiong, J., Zhou, Q.C., Yuan, Y., Wang, X.X., Sun, P.: Dietary yeast hydrolysate and brewer's yeast supplementation could enhance growth performance, innate immunity capacity and ammonia nitrogen stress resistance ability of Pacifific white shrimp (*Litopenaeus vannamei*). Fish and Shellfifish Immunology **82**, 121–129 (2018)

Kanu, S.N.:. Sanitizing Nigeria's fish feed industry. African Aquaculture and Fisheries Digest. March/April, 2008. Vol. 1(3, 5–8 (2008)

Lapena, D., et al.: Spruce sugars and poultry hydrolysate as growth medium in repeated given-batch fermentation processes for production of yeast biomass. Bioprocess Biosyst. Eng. **43**, 723–736 (2020)

Lapena, D., et al.: Production and characterization of yeasts grown on media composed of spruce-derived sugars and protein hydrolysates from chicken by-products. Microbial Cell Factories **19**, 1–14 (2020a). https://doi.org/10.1186/s12934-12020-11287-12936

Lovell, R.T.: Escalating feed costs require more efficient fish feeding. Aquac. Mag. **7**(15), 38 (1981)

Madibana, M.J., Mlambo, V.: Growth performance and hemobiochemical parameters in South African dusky kob *(Argyrosomus japonicus*, Sciaenidae) offered brewer's yeast *(Saccharomyces cerevisiae)* as a feed additive. J. World Aquac. Soc. **50**, 815–826 (2019)

Meena, D.K., et al.: Beta-glucan: an ideal immunostimulant in aquaculture (a review). Fish Physiol. Biochem. **39**, 431–457 (2013)

Moghaddam, H.N., Mesgaran, M.D., Najafabadi, H.J., Najafabadi, R.J.: Determination of chemical composition, mineral contents and protein quality of Iranian Kilka fish meal. Int. J. Poult. Sci. **6**, 354–361 (2007)

Muhammad, A., Dangoggo, S.M., Tsafe, A.I., Itodo, A.U., Atiku, F.A.: Proximate mineral and anti-nutritional factors of *Gardenia aqualla (Gautandutse)* fruit pulp. Pakistani J. Nutrition. **10**(6), 577–581 (2011)

Muyideen, O., Lawal, A.Z., Aarode, O.: Growth and economic performance of Nile Tilapia, *Oreochromis niloticus* (L.) fingerling given diets containing graded levels of Sclerotium. AACL BIOFLUX Aquaculture, Aquarium, Conservation and Legislation. Int. J. Bioflux Soc. 6 (3), 180–187 (2013)

N.R.C.: National Research Council. The use of drugs in food animals: benefits and risk. National Research Council Washington, D.C, National Academy Press (1999)

Ogbe, A.O., Ditse, U.,Echeonwu, I., Ajodoh, K., Atawodi, S.E., Abdu, P.A.: Potential of a wild medicinal mushroom, *Ganoderma* sp., as feed supplement in chicken diet: effect on performance and health of pullets. Int. J. Poultry Sci. 8 (11): 1052–1057 (2009)

Ogbe, A.O., John, P.A.: Proximate study, mineral and anti-nutrient composition of *Moringa olifeira* leaves harvested from Lafia, Nigeria: potential benefit in poultry nutrition and health. J. Microbiol., Biotechnol. Food Sci. **1**(3), 296–308 (2012)

Ogunji, J. O., Kloas, W., Wirth, M., Neumann, N., Pietsch, C.: Effect of housefly maggot meal (magmeal) diets on the performance, concentration of plasma glucose, cortisol and blood characteristics of *Oreochromis niloticus* fingerlings. J. Animal Physiol. Animal Nutr. 92 (4), 511–518 (2008)

Okeke, I.N., Sosa, A.: Antibotic resistance in Africa – discerning the enemy as plotting a defence. Features, African Health **25**(3), 10–15 (2003)

Oliva-Tales, A., Gonçalves, P.: Partial replacement of fishmeal by brewer'syeast *Saccharomyces cerevisae* in the diets for sea bass *Dicentrachus labrax* juveniles. Aquaculture **202**, 269–278 (2001)

Olukunle, O.: Evaluation of different dietary oil sources on growth performance and nutrient utilization of *Clarias gariepinus* juveniles. Niger. J. Fish. **8**(1), 189 (2011)

Øverland, M., Skrede, A.: Yeast derived from lignocellulosic biomass as a sustainable feed resource for use in aquaculture. J. Sci. Food Agric. **97**, 733–774 (2017)

Pamplona – Roger, G.D.: Education and Health Library. Editorial Safehz, Pradillo, 6 – Poligno Industrial La Mina E- 28770 Colmenar Viego, Madrid Spain. Fourth Print in English, p. 398 (2001)

Paterson, R.R.: *Ganoderma* – a therapeutic fungal biofactory. Phytochemistry 67(18), 1985–2001 (2006)

Peterson, M.K., Streeter, C., Clark, C.K.: Mineral availability with lambs given yeast culture. Nut. Rep. Int. **36**, 521 (1987)

Ramanadevi, V., Kaliyan, M., Thangappan, A., Muthusamy, T.: The efficacy of dietary yeast mannan oligo saccharide on growth and survival rate in *Amphiprion ocellaris* fingerlings. Eur. J. Biotechnol. Biosci. **1**(2), 12–15 (2013)

Ratcliff, J.: Antibiotic bans- a European perspective. In: Proceeding of the 47th Maryland nutrition conference for food manufacturers. March, 22–23, pp. 135–152 (2000)

Reinitz, J.: Variations of body composition and growth among strains of rainbow trout. Trans. Am. Fisheries Soc. **108**, 204–207 (1983)

Hoffman, L.C., Casey, N.H., Prinsloo, J.F.: Fatty acid, amino acid and mineral contents of African catfish *(Clarias gariepinus)* fillets. South African J. Food Sci. **4**, 36–40 (1992)

SAS: SAS Institute Inc. 2008.ASA/STAt user's Guide version 9.2. for windows.Carry. North Carolina, U.S.A (2008)

Satchithanandam, S., Apler, M.V., Calvert, R.J., Leeds, A.R., Cassidy, M.M.: Alteration of gastrointestinal mucin by fibre feeding in rats. J. Nutr. **120**, 1179–1184 (1990)

Siddhuraju, P., Becker, K.: Preliminary nutritional evaluation of Mucuna seed meal *(Mucuna pruriens* var. *utilis)* in common carp *(Cyprinus carpio* L.): an assessment by growth performance and feed utilization. Aquaculture **196**, 105–123 (2000)

Singh, P., Maqsood, S., Samoon, M.H., Phulia, V., Mohd, D., Chalal, R.S.: Exogenous supplementation of papain as growth promoter in diet of fingerlings of *Cyprinus carpio*. J. ssInt. Aquatic Res. **3**, 1–9 (2011)

Soetan, K.O., Oyewole, O.E.: The need for adequate processing to reduce the anti –nutritional factors in animal feed. A Rev. African J. Food Sci. 3 (9),223- 232 (2009)

Torrecillas, S., Montero, D., Izquierdo, M.: Improved health and growth of fish given mannan-oligosaccharides: potential mode of action. Fish Shellfifish Immunol. **36**, 525–544 (2014)

UC Sustainable Agriculture Research and Education Program: What is sustainable agriculture? UC Agriculture and Natural Resources (2021). <https://sarep.ucdavis.edu/sustainable-ag>

UNDP. Goal 1: No poverty. UNDP. Retrieved 30 December 2020. UNDP. Archived from the original on 30 December 2020. Cited in What is Sustainable Aquaculture? Andrew Burger Last Modified on March 29, 2022 (2020)

United Nations. World Population Prospects (2017): The 2017 Revision, key findings and advance tables; United Nations: New York, NY, USA (2017)

Valko, M., Leibfritz, D., Moncol, J., Cronium, M.T., Mazur, M., Telser, J.: Free radical and antioxidants in normal physiological functions and human disease. Int. J. Biochem. Cell Biol. **39**, 44–48 (2007)

World Bank: Decline of global extreme poverty continues but has slowed. World Bank. Cited in what is sustainable aquaculture? Retrieved 26 August 2020. Cited in Andrew Burger and Last Modified on March 29, 2022 (2020)

Impact of Age Distribution and Health Insurance Towards Sustainable Industrialization

H. T. Williams[1(✉)], T. S. Afolabi[1], and J. N. Mojekwu[2]

[1] Department of Finance, Redeemer's University, Ede, Osun, Nigeria
{williamsh,afolabis}@run.edu.ng

[2] Department of Actuarial Science and Insurance, University of Lagos, Lagos, Nigeria
mojekwejn@unilag.edu.ng

Abstract. Purpose: This paper demonstrates quantitative evidence for estimating the optimal contribution of life table and health insurance towards sustainable industrialization and innovation in Nigeria. In achieving the objectives, the study developed a period life table and derives the values for age distribution. The Malthus theory was tested for support or decline on sustainable industrialization and innovation.

Design/Methodology/Approach: A stratified sampling was used. Nigeria population data of 95,000,000, death rate of 18.55% (as at 1990) and life expectancy growth rate of 0.06% were used to compute a period life table. The data were analyse with ARDL and Cobb Douglas Production Function model.

Findings: The results of the study shows that health insurance reverse the inverse relationship that exist between age distribution and sustainable industrialization and innovation.

Research Limitation/Implications: The age's distribution values are derived from a periodic life table based on assumptions.

Practical Implication: The quantitative analysis in this study would avail information to professionals in health insurance that persons of a specified age are potent variable for attainment of sustainable industrialization and innovation.

Social Implication: The ability of the researchers to integrates, life table, health insurance and sustainable industrialization and innovation demonstrates knowledge policy makers to review health insurance policy, give attention to an applied life table as a medium to developed sustainable industrialization and innovation.

Originality/Value: The uniqueness of study lies in the mathematical models and variables used. Life table, health insurance and the Cobb Douglas was first integrated by the researchers to examine sustainable industrialization and innovation.

Keyword: Age distribution · Health insurance · Sustainable industrialization · Cobb-Douglas model · Life table

1 Introduction

Driving sustainable development, industrialization and innovation with population statistics and healthcare have been one of the core agenda of the BRICS (Brazil, Russia, India,

China and South Africa) (Zachy 2022). India and China have leverage on population to drive and sustain economic growth and sustainable industrialization and innovation goals. Before the unveiling of BRICS, population statistics have been essential variables of both conventional and modern demographic (Williams & Mojekwe, 2021). Within the year 2010 and 2021, Canada have used age distribution statistics in her migration policy to drive and sustain economic growth and sustainable industrialization and innovation (Zachy 2022). Sustainable development has a driving tool to meet the needs of man and also to sustain natural system, it is not only depended on policies and strategies but also on some demographic factors (Mensah, 2019). Crenshaw (1997) stated that there is a fundamental linkage between demographic variables and economic growth. Therefore, the fundamental factors of demographic variables such as age distribution of a population is essential for both economic growth and sustainable industrialization and innovation.

Driving sustainable development, industrialization and innovation with population statistics and healthcare have been one of the core agenda of the BRICS (Brazil, Russia, India, China and South Africa) (Zachy 2022). India and China have leverage on population to drive and sustain economic growth and sustainable industrialization and innovation goals. Before the unveiling of BRICS, population statistics have been essential variables of both conventional and modern demographic (Williams & Mojekwe, 2021). Within the year 2010 and 2021, Canada have used age distribution statistics in her migration policy to drive and sustain economic growth and sustainable industrialization and innovation (Zachy 2022). Sustainable development has a driving tool to meet the needs of man and also to sustain natural system, it is not only depended on policies and strategies but also on some demographic factors (Mensah, 2019). Crenshaw (1997) stated that there is a fundamental linkage between demographic variables and economic growth. Therefore, the fundamental factors of demographic variables such as age distribution of a population is essential for both economic growth and sustainable industrialization and innovation.

Tjarve and Zemīte (2016), Mensah and Enu-Kwesi (2018) stated that the concept of sustainable industrialization and innovation may connote the act of improving and sustaining a healthy economic, ecological and social system for human development. Mensah (2019) stated that sustainable industrialization and innovation revolves round three essential elements environment, economy, and society. For the sustainability of these three elements, the need for a growing and stable population and industrialization and innovation. This is because population drive sustainable industrialization and innovation and sustain the environment. Crenshaw (1997) underpinned a linear relationship between the continuous existence of the environment and the natural resources with a potential population. The International Institute for Sustainable Development (IISD) (2020) 'Sustainable industrialization and innovation meets the needs of the present without compromising the ability of future generations to meet their own needs'. The United Nations (UN) pinpoint many variables that made up its Sustainable Development Goals (SDGs) in which decent work and economic growth are stated as the 8th Development Goals. Williams and Mojekwe (2021) stated that China and India uses population to drive sustainable industrialization and innovation. The issue of sustainable industrialization and innovation is a complex concept in contemporary development discourse.

In development literature, Basiago (1999) stated that sustainability is the act of continuous maintainers of an entity, an outcome or a standard process over a specific time. Maintaining sustainable industrialization and innovation in Nigeria posed a difficulty challenges admitting the impact of the Coronavirus Disease (COVID-19), the insurgencies and economic uncertainty facing the country. The impact of the COVID-19, the lack of standard live table and a structured health system undermine sustainable industrialization and innovation in Nigeria. Therefore, this study focus on how age distribution and health insurance drive sustainable industrialization and innovation.

2 Theories Underpinning the Study

This study underpinned the Malthus theory of population. The theory stated that the population grows more rapidly than the basic needs of man such as food, and diseases may decline the population. We tested the Malthus theory on COVID-19 which decline population through dealth and proposed that the Malthus theory is sufficient in this study. However, modern literature (Williams & Mojekwe, 2021) postulated that good healthcare through health insurance can increase and stabilize population growth. Bedford (2020) states that Age distribution statistics shows that older adults were affected more by COVID-19 than younger adults hence driving sustainable industrialization and innovation is slow. On the premise of this, demographic factors such as age becomes an essential parameter to forecast economic progress and sustain industrialization and innovation. Diep and Hoai (2015) stated that economic growth both in the period of economic stability and period of uncertainty affect sustainable industrialization and innovation. The DESA-UN (2018) stated that to differential between older adults and younger adult, age is the parameter hence age distribution is a fundamental variable for policy decision.

Lobo, Pietriga, and Appert (2015) stated economic sustainability, sustainable industrialization and innovation entails creating a production system that aimed at satisfying a nation current consumption levels without any effects of compromising the nation future needs while Du and Kang (2016) stated that most economists assuming that most natural resources are limited for the total consumption of a specified population and hence call for a structured model to be used by a standard market to effectively and efficiently allocate resources to persons with the market. Du and Kang (2016) further stated that an increasing scale of an economic system has overstretched the natural resource base and thus required a healthy population to rethink of the traditional economic postulations. Cao, (2017) concluded that three activities such as production, distribution and consumptions are the main three activities used in accounting framework to guide evaluate the economy. The implication of this is that these activities obviously alters values and this does not augur well for society and the environment. Most times, persons with disability, age distribution that are dependent creates economic burden for the population. Campagnolo, Carraro, Eboli, Farnia, Parrado and Pierfederici (2018) postulates that in achieving the sustainable development goals, the need for the vital systems to work together. These system include, the population and the population characteristics, the economic and the environment.

Health insurance and age distribution are essential variables for sustaining economic growth, the need to harness these variables for proper output. Age distribution and healthcare are critical variables to be considered in the period of pandemic. Investigating age distribution and health in Nigeria has been intriguing to researchers over the years. Most of the investigation on healthcare and demographics were done mainly with an objective or for the motive of assessing the effect of income and age on economic growth. No focus study had looked into age distribution and health insurance on economic sustainability in West Africa. Bloom and Williamson (1998) demonstrated a substantial progress in economic growth and development in Asia in the period from 1965 to 1990 using healthcare variables and age distribution. In as much as age distribution and healthcare play a significant role in forecasting population and sustaining economic growth, their impacts has not been seen or measured on economic indicators like human capital development. It is on the basis of this premise that the hypotheses for this study is being carried out on age distribution, health insurance and economic sustainability in Nigeria.

Population growth and age structure on a nation's growth and development from the Malthus's theory shows that demographic factors are essential for sustainable industrialization and innovation while the endogenous growth theory states that investment in knowledge is a significant contributor to growth and development. The Malthus's theory and the endogenous growth theory are the theories in which the researchers anchor the idea of this study. However there are some literatures that provide supportive evidence for pessimistic view. Pintu and Predip (2020) states that demographic factor and health care are major public health problem, they further stress that adequate utilization of health care facility in terms of maternal health care services could be an effective means for tumbling maternal mortality when demographic factors are considered. Tosun and Leininger (2017) explained the linkages between sustainable development and governing policies with focus on population statistics and economic variables. International Institute for Population Sciences (IIPS) (2018) states the importance of demographic distribution in population planning in Indian and its corresponding effect on economic activities. Coale and Hoover (1958) formulated a hypothesis based on an Indian database and found out that there is negative connection between demographic factor such as age in young dependence and economic savings in a short run which lower the standard of living. Williams, Abiola and Ojikutu (2021) stated that the act of minimizing cost for health care for staff in tertiary institution is an act of sustaining the University system and the healthcare of staff has a linear relationship with research outputs.

Bloom (2003) stated that a stable and rapid growth of population requires tested and efficient technology in health care and until the year 1800, the gap of age distribution in population between the countries with high technology and low-technology countries was very small. Bloom and Williamson (1998) stated that age distribution was essential as information for implied policy in East Asia as it boost economic growth from 1965 to 1990. Azomahou and Mishra (2008) used age distribution of 0–14, 15–64, 65 and above from OECD 110 countries and concluded that young working age group are fundamental to the building of sustainable industrialization and innovation.

3 Methodology

This study adopts both statistical and econometric method of analysis. Data were collected from existing documents and records and was further transformed by assumptions to fit in to the appropriate quantitative models adopted in the study. The first step was to test for the stationarity of the data using unit root and the Augmented Dickey-Fuller test statistic. We applied a research design that examine age and health on sustainable industrialization and innovation. We identify the Cobb-Douglas production model as the optimal model. The study used the Cointegration and the Auto Regressive Distributive Lag (ARDL) techniques before adopting the Cobb-Douglas model.

A standard Cobb–Douglas production is:

$$Q = A^{K^\alpha L^\beta}$$

where:

Q = total production (the real value of output), L = labor input, K = capital input. A = measure efficiency of the production, α and β are the output elasticities or measure the factor intensity.

In line with the Cobb–Douglas model stated above, the following variables were used as proxy:

Q = Sustainable industrialization and innovation (SII), L = ADV = Age Distribution Value, K = HIP = Health Insurance Premium, $A = R^2$ = coefficient of determination for Human Capital Development. α and β are proxied with standard errors.

The Cobb-Douglas model for the study is hereby specified as follows:

$$Q = AK^\alpha L^\beta \qquad (1)$$

We substitute economic sustainability and other variables used in the study in the Cobb–Douglas model:

$$HCD = R^2 ADV^\alpha HIP^\beta \qquad (2)$$

To obtain $A = R^2$ of the ARDL result

$$LOGHCD = \beta_0 + \beta_1 LOGADV + \beta_2 LOGHIP \qquad (3)$$

4 Findings

We used the information below to build a period life table with a bound of year 0–100.

Population Statistics (Nigeria- 1990) = 95,000,000 (year 0), Death rate Statistics (Nigeria-1990) = 18.55%., Life Expectancy Constant Growth Rate of 0.060% as at the year 1994.

From this table it can deduce that at 5% level of significance that there are statistical evidence to prove the acceptance of the alternative hypotheses that individual series are stationery at their first I(1) and levels I(0) of differences.

Table 1. Period life table data presentation

Age interval	d_x	f_x	q_x	p_x	L_x	T_x	e_x
1990 (0)	17622500	95000000	0.1855	0.8145	133688750	800776069.1	8.42922178
1	14353526.3	77377500	0.1855	0.8145	108889486.9	667087319.1	8.621205378
2	11690947.1	63023973.75	0.1855	0.8145	88690487.06	558197832.2	8.85691268
3	7982285.64	51333026.62	0.1555	0.8445	73008397.11	403015618.9	7.851000524
4(1994)	8015552.01	43350740.98	0.1849	0.8151	61018335.47	330007221.8	7.61249322
5(1995)	6533476.44	35335188.97	0.1849	0.8151	49736045.24	268988886.3	7.612493215
100	0.02399823	0.129790309	0.1849	0.8151	0.129790309	0.129790036	0.999997894
						734284342.8	

Source: computed with excel using

Table 2. Extracted data from the period life table data and the life insurance premium

Age distribution	T_x (ADV)	HIP
15–19	120566255.9	3,680,090.19
20–24	43379067.57	6,941,383.41
25–29	15607545.22	9,147,417.17
30–34	5615505.476	10,157,021.18
35–39	2020424.984	10,660,071.84
40–44	726934.9585	14,649,276.46
45–49	261544.4198	15,751,837.52
50–54	94099.4847	17,161,904.95
55–59	33853.72556	18,678,544.87
60–64	12177.63549	21,083,121.06
65–69	4378.698478	21,982,812.23

Source: computed

Table 3. Summary of unit root test and the augmented dickey-fuller test statistic

Variables	ADF	Probability	Critical values	Order of integration
LOGHCD	-7.11246	0.0295	-2.90084**	I(1)
LOGADV	-3.00423	0.0051	-2.90084**	I(0)
LOGHIP	-4.10049	0.0026	-2.90084**	I(0)

*** significant at 1%, ** significant at 5% and * significant at 10%.

The (**) indicates the rejection of the hypothesis at 5% significance level. Since the Likelihood Ratio (LR) is greater than the Critical Value (CV) at 5% level of significant are co-integrated.

Table 4. Summary of the Johansen Co-integration test

Included Observations: 55 (Age 15–69)
Series: LogHCD, LogADV, LogHIP

Eigen value	Likelihood Ratio	5% Critical Value	Hypothesized No
0.755	113.25	97.93	None**

Source: computed

Table 5. ARDL summary of result

Dependent Variable: HCD
Method: ARDL

Variable	Coefficient	Std. Error	Prob.
C	0.77110	0.45003	0.0050
ADV	1.00569	0.12563	0.0100
HIP	2.11375	0.00012	0.0030
R-squared	0.740022	2.01190	
Adjusted R-squared	0.730950	0.08117	

Source: Computed using Eview9, 2022

Table 6. Extracted from Table 4 showing the p-value of health insurance premium.

Variable	Coefficient	Std. Error	Prob
HIP	2.11375	0.00012	0.0030

Source: Extracted from ARDL result in Table 4

Table 4 shows the positive relationship age distribution, health insurance premium and sustainable industrialization and innovation. The coefficient of determination shows that about 76 % of age distribution value and the health insurance premium are captured by sustainable industrialization and innovation. To test the hypotheses which states that Health insurance practice does not affect sustainable industrialization and innovation, the loglinear regression was carried out and the results shown in Table 4.

The P-value was used to test hypothesis on Health insurance Premium (HIP) against sustainable industrialization and innovation (SII). The P-value calculated score was compared with the 0.05 standard value. The HIP which is a proxy for health insurance premium is 0.0030 which is the P-value, and is therefore compared with 0.05 i.e. 0.0030 < 0.05 we conclude that health insurance premium affect sustainable industrialization and innovation in Nigeria.

In an attempts to test the hypothesis of age distribution and sustainable industrialization and innovation, we proxy the total person-years of life contributed after attaining age x (T_x) derived from the life table as age distribution value (ADV). The results was shown in Table 4.

Table 7. Extracted from Table 4 showing the p-value of Age distribution.

Variable	Coefficient	Std. Error	Prob
ADV	1.00569	0.12563	0.0100

Source: Extracted from ARDL regression result

Table 8. Cobb-Douglas Production Function Output (Q) with Eq. 7

Age Group	(Tx)ADV	HIP	Q (SII)
15–19	120566255.9	3,680,090.19	4506672739
20–24	43379067.57	6,941,383.41	4243270217
25–29	15607545.22	9,147,417.17	3187368007
30–34	5615505.476	10,157,021.18	2149321498
35–39	2020424.984	10,660,071.84	1398787375
40–44	726934.9585	14,649,276.46	1078824454
45–49	261544.4198	15,751,837.52	712898530
50–54	94099.4847	17,161,904.95	475011488
55–59	33853.72556	18,678,544.87	316287930
60–64	12177.63549	21,083,121.06	215474518
65–69	4378.698478	21,982,812.23	139614752

Source: computed with excel

The p-value for ADV is 0.0100 and is therefore compared with 0.05 i.e. 0.0100 < 0.05 it is concluded that the null hypothesis should be rejected hence age distribution values affects sustainable industrialization and innovation.

The Cobb Douglas Production Function.

The Cobb-Douglas model for the study is hereby specified as follows:

$$Q = AK^\alpha L^\beta \quad (4)$$

The study Model is thereby written in Cobb–Douglas form:

$$HCD = R^2 ADV^\alpha HIP^\beta \quad (5)$$

where $A = R^2 = 82$ $\alpha = 0.45$ $\beta = 0.63$

$$Q = 76 ADV^\alpha HIP^\beta \quad (6)$$

$$Q = 76 ADV^{0.45} HIP^{0.63} \quad (7)$$

5 Discussions

The Period Life Table was the focal point in which the data of the study was used as proxies for age distribution and health insurance after adjustment. The Table 7 shows that as the age distribution increase the output declines, indicating an inverse relationship between age distribution and sustainable industrialization and innovation. However, with the presence of health insurance in the model, the rate in which the output decline due to age distribution increase was cushioned with health insurance. The result of this study also corroborates with the result that of Basiago (1999), Campagnolo et al. (20181999) and Mensah (2019) that populations statistics are potent variables to sustainable industrialization and innovation. Also, Cobb-Douglas production function and the econometrics analysis used in the study shows that that there a long run relationship exist among the variables proxy for age distribution, health insurance and sustainable industrialization and innovation. This result corroborates with the works of Azomahou and Mishra (2008) and Zachy (2022) that nations uses population statistics such as age distribution variables to drive sustainable development goals. The study test of hypothesis using the p-value confirmed that age distribution and health insurance affect sustainable industrialization and innovation while the Cobb-Douglas production model outputs display an increasing return to scale, this result tailored towards the results of Cao (2017) and that as the age of a person increases the economic production declines. However, the presence of Health insurance in the Cobb-Douglas model indicates a supporting variable that aid sustainable industrialization and innovation. The coefficient of determination R^2 of 76% from the ARDL result shows that substantial proportion of health insurance and population statistics are explained on the growth of sustainable industrialization and innovation.

6 Conclusion

This study ascertains empirically the effect of age distribution and health insurance premium on sustainable industrialization and innovation by applying Period Life table, error correction, cointegration, ARDL and the Cobb Douglas production model. The study found out that there is an inverse relationship between age distribution and sustainable industrialization and innovation. To reverse this relationship, the need for health insurance. Also, the combination of a particular age distribution and a standard health insurance scheme would boost both sustainable industrialization and innovation in Nigeria. The study shows that age is one of the sole aims of health insurance premium consideration and most economic activities are dependent on the age of the individual hence the study recommend that persons within the age bracket above 69 can still contribute to sustainable industrialization and innovation if considered for special health insurance scheme. The Implication of this study is that quantitative analysis used in this study would informed health insurance professionals and persons of a specified age that age are significant elements for sustainable industrialization and innovation and the knowledge advance and demonstrated in this study would assist the Nigerian policy makers to review health insurance policy, give attention to an applied life table that is geared towards policy that would achieve sustainable development goals with emphasis on industrialization and innovation. The uniqueness of this study lies in the study use of period life and the Cobb Douglas function model.

Data link for Nigeria population for the year 1990: https://data.worldbank.org/indicator/SP.POP.TOTL?locations=NG.

Data link for Nigeria Death Rate for the year 1990: https://www.macrotrends.net/countries/NGA/nigeria/death-rate.

Data link for Nigeria life expectancy growth rate.
https://www.macrotrends.net/countries/NGA/nigeria/life-expectancy.

References

Basiago, A.D.: Economic, social, and environmental sustainability in development theory and urban planning practice: The environmentalist. Boston: Kluwer Academic Publishers

Bedford, S.: The End of World Population Growth in the 21st Century: New Challenges for Human Capital Formation and Sustainable Development. Routledge (2020)

Bloom, D.E., Williamson, J.G.: Demographic transitions and economic miracles in emerging Asia. The World Bank Economic Review, 12(3), 419–455 (1998)

Coale, A.J., Hoover, E.M.: Population growth and economic development in low income countries: a case study of Indias prospects (1958)

Campagnolo, L., Carraro, C., Eboli, F., Farnia, L., Parrado, R., Pierfederici, R.: The Ex-Ante Evaluation of Achieving Sustainable Development Goals. Soc. Indic. Res. 136(1), 73–116 (2018). https://doi.org/10.1007/s11205-017-1572-x

Cao, J.G.: Trading contract and its regulation. Journal of Chongqing University (Social Science Edition) 23, 84–90 (2017)

Collste, D., Pedercini, M., Cornell, S.E.: Policy coherence to achieve the SDGs: using integrated simulation models to assess effective policies. Sustain. Sci. 12(6), 921–931 (2017). https://doi.org/10.1007/s11625-017-0457-x

Crenshaw, S.: Population dynamics and economic development: Age-specific population growth rates and economic growth in developing countries, 1965 to 1990. Am. Sociol. Rev. 974–984 (1997)

David, A.: Demographic structure and economic growth: Evidence from China. J. Compar. Econ. 38(4), 472–491 (2018)

De la Croix, D.: Demographic change and economic growth in Sweden: 1750–2050. J. Macroecon. 31(1), 132–148 (2009)

DESA-UN. (2018, April 4). The Sustainable Development Goals Report 2017. https://undesa.maps.arcgis.com/apps/MapSeries/index.html

Diep, M.A., Hoai, N.T.: Demographic Factors and Economic Growth: The Bi-Directional Causality In South East Asia. Paper submitted to the ninth Vietnam Economists Annual Meeting. Da Nang City, 11 – 12th August (2015)

Du, Q., Kang, J.T.: Tentative ideas on the reform of exercising state ownership of natural resources: Preliminary thoughts on establishing a state-owned natural resources supervision and administration commission. Jiangxi Soc. Sci. 6, 160 (2016)

Lobo, M.-J., Pietriga, E., Appert, C.: An evaluation of interactive map comparison techniques. In Proceedings of the 33rd Annual ACM Conference on Human Factors in Computing Systems - CHI '15 (pp.3573–3582). New York, USA: ACM Press (2015). https://doi.org/10.1145/2702123.2702130

Mensah, J., Enu-Kwesi, F.: Implication of environmental sanitation management in the catchment area of Benya Lagoon, Ghana. J. Integr. Environ. Sci. (2018). https://doi.org/10.1080/1943815x.2018.1554591

Mensah, J.: Sustainable development: meaning, history, principles, pillars, and implications for human action: Literature review. J. Cogent Soc. Sci. 5(1), 1–21 (2019)

Pintu, P., Pradip, C.: Socio-demographic factors influencing utilization of maternal health care services in India. Clinical Epidemiology and Global Health. (8), 666–670 (2020)

Tjarve, B., Zemīte, I.: The Role of Cultural Activities in Community Development Acta Universitatis Agriculturae et Silviculturae Mendelianae Brunensis, **64**(6), 2151–2160 (2016). https://doi.org/10.11118/actaun201664062151

Tosun, J., Leininger, J.: Governing the interlinkages between the sustainable development goals: Approaches to attain policy integration. Global Challenges (2017). https://doi.org/10.1002/gch2.201700036

Williams, T.H, Abiola, B., Ojikutu R.K.: Minimizing Healthcare Cost in Selected Tertiary Institutions in Nigeria. J. Innov. **64**(1), 544–554 (2021)

Williams, T.H., Mojekwe, J.N.: Examine the Effect of Age Distribution and Health Insurance Business on Nigerian Economic Growth. Unilag J. Math. Appl. **1**(2), 154–171 (2021)

Zachy, J.C.: Achieving sustainable development: the Centrality and multiple facets of integrated decision making. Indiana Journal of Global Legal Studies **10**, 247–285 (2022). https://doi.org/10.2979/gls.2003.10.1.247

Determination of Overall Coefficient of Heat Transfer of Building Wall Envelopes

E. Baffour-Awuah[1,2], N. Y. S. Sarpong[1,3](✉), I. N. Amanor[1], and E. Bentum[4]

[1] Mechanical Engineering Department, Cape Coast Technical University, Cape Coast, Ghana
{serwaah.sarpong,ishmael.amanor}@cctu.edu.gh
[2] Agricultural Engineering Department, University of Cape Coast, Cape Coast, Ghana
[3] Department of Agricultural and Biosystems Engineering, Kwame Nkrumah University of Science and Technology, Kumasi, Ghana
[4] Director of ICT Service, University of Cape Coast, Cape Coast, Ghana

Abstract. Purpose: The purpose of the study was to determine the rate at which heat is transferred through the walls of building envelopes with particular reference to building structures in Ghana.

Design/Methodology/Approach: The accidental sampling technique was used to select uncompleted building structures, supported with physical measurements; and interviews were used to gather data in the Kumasi Metropolis. A computer language, C + + for Visual Studios, was employed to design a source code to compute the overall heat transfer coefficient of the walls. Both primary and secondary data were employed in the study. The data obtained using the designed program was analyzed using descriptive analysis in SPSS Statistics 20.

Findings: Results indicated a total of sixteen thousand nine hundred and forty (16,940) OCHT values of wall envelopes. Values ranged between 0.09295 W/(m^2.K) and 3.4411 W/(m^2.K). The walls were categorized and coded concerning the various homogeneous layers making the building envelope. Twelve wall categories in terms of OCHT were obtained. A total number of 4840 concrete walls, 7260 brick walls and 4840 sandcrete walls were identified. Thus, thermal resistances and OCHT values were computed for a total of 16940 building wall envelopes.

Research Limitation: The study assumed that contact resistances between wall interfaces were insignificant. Other assumptions include the fact that individual layers were homogeneous; there was equilibrium or steady-state heat transfer; effects of heat storage were disregarded. Secondary data such as thermal resistance, thermal conductance, thermal transmittance and thermal conductivity were largely relied upon even though such data could be fraught with errors which could have a negative rippling effect on the final results.

Practical Implication: The values of OCHT computed could be used in the design of air conditioning rooms, cold rooms and warm rooms in temperate regions. The source code could be easily modified to compute OCHT values for more than five layered wall envelopes. Though the source code was specifically designed for walls, it can as well be used to compute OCHT values of any material of multiple layers provided the required secondary data are available.

Social Implication: The application of cold rooms in the preservation of fresh meat, cooked food and plant-based crops could reduce food spoilage thus indirectly enhancing the incomes of individuals and households.

Originality/Value: The uniqueness of this study lies in the fact that OCHT data on individual walls in Ghana were not available in literature or elsewhere for quick and easy reference. The study has introduced a coding and classification system of wall envelopes in Ghana based on OCHT.

Keyword: Brick · Concrete · Sandcrete · Temperature · Transmittance

1 Introduction

The factors of human comfort depend largely on dry-bulb temperature, and as such, dry-bulb temperature is the single most important factor of human comfort. However, the rate at which heat is transmitted through the walls of habitable rooms, to a large extent, contributes to the dry-bulb temperature of the room. The overall heat transfer coefficient, U-value, which is also defined as the rate at which heat is transmitted through the wall per unit area per unit temperature difference depends on the dry-bulb temperature-difference between the outside air layer and the internal air layer of the wall (ASHRAE, 2013; Boregowda et al., 2012; Jones, 2007).

In Ghana, which lies in the tropics, with average diurnal dry-bulb temperature of 27 °C, heat gain from sunlight, electric lighting and appliances create unpleasant conditions in building envelopes, unless windows are opened. Even with moderate speeds of wind, open windows cause excessive droughts, which become worse on the upper floors of buildings which are tall. Furthermore the presence of noise and dirt make the situation worse. The effect of objectionable noise is much felt on the lower floors of buildings, especially in urban and industrial areas when windows are opened. Unfortunately the reduction of anxiety and distress given by airflow through natural means by opening windows only produces the intended results for a depth of about six meters inward from the wall. It is obvious that the real advantage derived from opening of windows is not in the least beneficial to the inner portions of deep buildings. Ironically, such portions require high intensity electric lighting continuously. Hence lack of ventilation will give rise to a great deal of discomfort to the people occupying the building. It is also clear that mechanical ventilation alone cannot solve the problem of human comfort in habitable rooms. However, closing windows will also demand full air-conditioning (Arens et al., 2012; Boregowda et al., 2012; Jones, 2007; Krajcik et al., 2012).

Full air-conditioning implies the automatic control of an atmospheric environment for the comfort of human beings, animals or for the proper performance of some industrial and scientific process such as food and crop storage in terms of air movement, temperature, purity, pressure and relative humidity. Within the limits of design specifications, air-conditioning is usually associated with refrigeration. Refrigeration accounts for the high cost of air conditioning. It accounts for four-fifth of the cost of only heating the room. Air-conditioning therefore seeks to overcome heat gains arising from both the outside sources and those originating from the conditioned space. The heat gains are of two categories – sensible and latent heat (ASHRAE, 2013; Jones, 2007).

Overcoming sensible and latent heat gains through air conditioning must be done efficiently and effectively in order to minimize the cost involved in refrigerating the room. Achieving this all-important goal may be possible if the heat gain can be determined as efficiently and effectively as possible. The heat gain through the walls of buildings is the sum of the relative steady-state heat flow that takes place as a result of the temperature difference between the external and internal air of the wall and the unsteady-state heat gain as a result of the variation of solar radiation intensity on the outside surface of the wall (Arens et al. 2012; Boregowda et al., 2012; Jones, 2007).

In calculating the heat gain of the wall it is important to know the rate at which heat is transmitted through the wall per unit area per unit temperature change across the air films on both surfaces of the wall. It has been given various nomenclatures such U-value, U-factor, thermal transmittance, or the overall coefficient of heat transfer. In the computation of heat gain or loss by building wall structures, whether one is dealing with steady-state or unsteady-state heat flows, overall coefficient of heat transfer values are significant. Hence it is necessary that they are established and documented for various wall envelopes.

Building wall materials used in Ghana are manifold. The most common materials used are concrete, sandcrete, brick, wood, stones, terrazzo, shingles and mortar. Composite wall structures in Ghana are generally made of concrete, sandcrete, shingles, tiles, bricks, mortar and wood. The major materials in terms of volume in these walls are layers of concrete, sandcrete and brick, each layer being homogeneous. In spite of this, data on individual walls are not available in literature or elsewhere for quick and easy reference. It is in this light that this study sought to determine heat transfer characteristics of concrete, brick and sandcrete wall envelopes such as overall coefficient of heat transfer, specifically, in Ghana. The main aim of the study was, thus, to determine the overall coefficient of heat transfer of various concrete, bricks and sandcrete wall envelopes in Ghana. The objectives, however, were to introduce a coding and a classification system of wall envelopes in Ghana based on OCHT; develop a computer program for the evaluation of the overall coefficient of heat transfer for concrete, brick and sandcrete wall envelopes available in Ghana; and document the OCHT values of these envelopes.

2 Method and Materials

2.1 Theoretical Considerations

Generally, in most steady-state heat transfer problems, a combinations of heat transfer modes may be involved, namely, conduction, convection and radiation. The wall of a building (which may consist of inside plaster, a row of blocks, and perhaps cement rendering on the outside surface) is constructed in layers to form a composite wall. In such situations the various heat transfer coefficients may be combined into an overall heat transfer coefficient so that the total heat transfer due to conduction, convection and radiation can be calculated from the terminal temperatures, though the effect of radiation is quite insignificant. The analysis to the problem is much simpler if the concept of thermal circuit is employed.

According to Fourier's law the rate of flow of heat through a single homogenous solid is directly proportional to the area A of the section at right angles to the direction of heat flow and to the change of temperature with respect to the length of the path of the heat flow, dt/dx. This law is illustrated in Fig. 1 in which a thin slab of material of thickness dx and surface area A has one face at a temperature t and the other at a lower temperature (t - dt). Heat flows from the high-temperature face to the low-temperature face and the temperature change in the direction of the heat flow is –dt. Applying Fourier's law, the rate of heat flow;

$$Q \alpha A \frac{dt}{Dx} \qquad (1)$$

$$=> Q = -kA \frac{dt}{Dx} \qquad (2)$$

where k is the thermal conductivity of the material (refer Fig. 1).

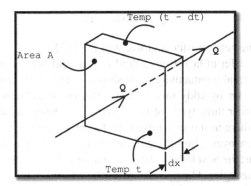

Fig. 1. Heat transfer through a solid slab

Considering the transfer of heat through the slab shown in Fig. 1(b) at section x – x,

$$Q = \frac{-kAdt}{dx} \qquad (3)$$

$$=> Qdx = -kAdt$$

Upon integration.

$$D \int_0^x Qdx = -\int_{t_1}^{t_2} kAdt$$

$$=> Qx = -\int_{t_1}^{t_2} kAdt$$

Assuming steady head flow, Q.

For most solids such as concrete, brick and sandcrete, the thermal conductivity is approximately constant over a wide range of temperatures. Hence k can be taken as constant,

$$\text{Thus} \quad Qx = -Ak \int_{t_1}^{t_2} dt$$

$$\text{Hence} \quad Q = \frac{-kA}{x}(t_2 - t_1) \qquad (4)$$

$$= \frac{kA}{x}(t_1 - t_2)$$

The area A in the direction at right angles to the heat flow Q remains constant through the slab. Another law which needs to be considered when dealing with heat transfer through solids is the Newton's law of cooling which states that the heat transfer from a solid surface of area A, at a temperature t_w, to a fluid of temperature t, is given by;

$$Q = hA(t_w - t); \qquad (5)$$

Where h is called the heat transfer coefficient of the fluid.

Consider the transfer of heat from a fluid A to a fluid B through a dividing wall of thickness x and thermal conductivity k, as shown in Fig. 1. Thermal convection may involve energy transfer by eddy mixing and diffusion in addition to conduction. For large Reynolds number three types of flow may exist. Closely adjacent to the wall is a laminar sub layer where heat transfer exists by thermal conduction; outside the laminar layer is the transition layer, a buffer region, where both eddy and conduction effects are significantly present; the next layer outside the buffer zone is a total eddy mixing which is predominant in the region. Outside and inside the wall the main flow is turbulent while the laminar layer exists at the solid wall only: thus the main body of heat transfer around the wall is the turbulent region where conduction occurs only at the surface of the wall. Hence the thermal resistance of the film is given by:

$$R = \frac{1}{hA} \qquad (6)$$

Assuming a steady state heat transfer, the heat flowing from fluid A to the wall is equal to the heat flowing through the wall, which is also equal to the heat flowing from the wall to fluid B, the temperatures t_A, t_1, t_2 and t_B remaining constant with time. Hence from Eqs. (1) and (2).

$$\begin{aligned} Q &= hA(t_A - t_1) \\ &= k/x(t_1 - t_2) \\ &= hA(t_2 - t_B) \end{aligned} \qquad (7)$$

dA and dB are small thicknesses close to the solid wall on either sides where there is heat conduction in those thin film of layers. Equation 7 can therefore be rewritten as,

$$t_A - t_1 = \frac{q}{ht}; \quad (t_1 - t_2) = \frac{qx}{k}; \quad (t_2 - t_B) = \frac{q}{hB} \qquad (8)$$

Adding the corresponding sides of the three equations and for a unit sectional area,

$$(t_A - t_1) + (t_1 - t_2) + (t_2 - t_B) = q(\frac{q}{ht} + \frac{qx}{k} + \frac{q}{hB})$$

$$\Rightarrow (t_A + t_B) = q(\frac{1}{hA} + \frac{x}{k} + \frac{1}{hB}).$$

$$\Rightarrow q = \frac{tA - tB}{\frac{1}{h} + \frac{x}{k} + \frac{1}{hB}}$$

or $q = U(t_A - t_B)$ for a unit area.

For a given area,

$$Q = UA(t_A - t_B); \tag{9}$$

where A is the cross-sectional area through which heat is passing, and

$$\frac{1}{u} = (\frac{1}{hA} + \frac{x}{k} + \frac{1}{hB}). \tag{10}$$

Wall structures in Ghana are constructed using different materials in layers to form composite walls. For example, a composite wall structure could be made of outside terrazzo layer, a plaster of mortar, a concrete layer and an inside plaster of mortar. Consider the general case of a composite wall shown in Fig. 2. There are n layers of material of thickness x_1, x_2, x_3 etc. and of thermal conductivity k_1, k_2, k_3 etc.

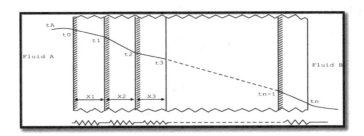

Fig. 2. Heat transfer through composite walls

On one side of the composite wall is a fluid of air at temperature t_A, and the heat transfer coefficient from fluid to wall is h_A; on the other side of the composite wall there is fluid of air B and the heat transfer coefficient from wall to fluid is h_B. Let the temperature of the wall in contact with fluid A be t_o and the interface temperatures are then t_1, t_2, t_3 etc. as shown in Fig. 2. The heat flow is caused by a temperature difference with a corresponding thermal resistance. The flow of heat can therefore be considered to be analogous to the electric current. In this case the temperature difference is analogous to potential difference while the thermal resistance is analogous to electrical resistance. Ohm's Law states that for an electrical circuit,

$$V = IR; \tag{11}$$

where, V = potential difference,
I = Current and
R = Electrical resistance.
Comparing Eq. (5) with Eq. (1),

$$Q = \frac{kA}{x}(t_1 - t_2). \qquad (12)$$

Hence the thermal resistance.

$$R = \frac{x}{kA}; \qquad (13)$$

where
Q is analogous to I and $(t_1 - t_2)$ is analogous to V.

The composite wall is analogous to a series of resistances. Resistances in series can be added to give the total resistance to form the 'thermal circuit'. Comparing Eq. (11) with Ohm's laws, the thermal resistance of a fluid film is given by;

$$R = \frac{1}{hA},$$

where

Q is analogous to I and
$(t_w - t)$ is analogous to V.

Referring to Fig. 2 for the thermal circuit we have.

$$R_A = \frac{1}{h1A},$$
$$R_1 = \frac{x}{k1A},$$
$$R_2 = \frac{1}{k2A} \quad \text{etc.}$$
$$R_n = \frac{xn}{kn-A} \quad \text{and}$$
$$R_B = \frac{1}{hB}$$

The total resistance to heat flow is therefore given by.

$$R_T = R_A + R_1 + R_2 + \cdots + R_n + R_B$$
$$= \frac{1}{hAA} + \frac{x_1}{k1A} + \cdots + \frac{xn}{knA} + \frac{1}{hBA}.$$

For any number of layers,

$$R_T = \frac{1}{hAA} + \sum_1^n \frac{x}{kA} + \frac{1}{hBA}, \qquad (14)$$

Hence $U = \dfrac{1}{\frac{1}{h_A A} + \sum_1^n \frac{x}{kA} + \frac{1}{h_B A}}$. (15)

For a unit surface area,

$$U = \dfrac{1}{\frac{1}{h_A} + \sum_1^n \frac{x}{k} + \frac{1}{h_B}}.$$ (16)

U is called the overall heat transfer coefficient and the unit is Watt/per meter square per Kelvin (W/m$^2 \cdot$K). It is the time rate of heat flow per unit area under steady state conditions from the fluid on the warmer side of the barrier to the fluid on the cold side, per unit temperature difference between the two fluids (Fig. 3).

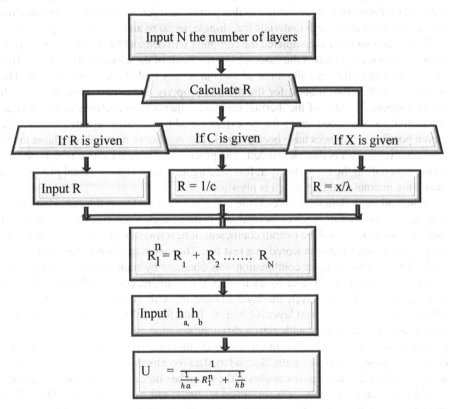

Fig. 3. Flow chart of source code

The code takes input from the user and gives an output. Input data required include the number of materials in each layer, the value of the thermal property of the material, as well as the thickness of the layer as the demand may be. Material, here, refers to the value of the different thermal properties that are available for each layer (Table 1, Table 2, Table 3 and Table 5). A given layer in a wall structure may have different possible materials. For example, there are four possible different materials that may be used as external cladding for a given wall. These materials are stone, shingle, terrazzo and clay tiles. Each has a different thermal property such as thermal resistance, thermal conductance or thermal conductivity.

For instance, for the five-layered concrete wall, the computer will demand the number of materials in the first layer which is the external cladding. Upon imputing the number of possible materials for the first layer the program will then demand which of the thermal property of the first material is available. There are options-the value of the thermal resistance, thermal conductance or thermal conductivity. The first option is the thermal resistance. If it is available an input of the value is made. If it is not available then the thermal conductance is input as the second option. If the thermal conductance is also not available then the thermal conductivity is input as the last option. If the last option is made the program will further demand for the thickness of the layer since both are required for computing the resistance of that particular layer. The process is repeated for the second, third and fourth materials for, shingle, terrazzo and clay tiles respectively.

When the first layer is completed the computer will demand the number of materials in the second layer which is the external plaster. Each of the eleven different (In terms of resistance since they are all mortar plaster) materials could be used for the wall. The program will therefore request for the thermal property of the first material of which the user inputs the value of the thermal resistance, thermal conductance or the thermal conductivity as described above for the external cladding. This is repeated until all the eleven possible resistances have been entered. The number of materials available to the next layer (concrete) is then demanded. The procedure described above is repeated for all the materials in the concrete layer, followed by the fourth layer (internal plaster) and finally the internal cladding which is plywood.

When all input data were entered the computer then performs the computations to generate the number of layer combinations; code of each wall; thermal resistance of each composite wall; and the overall coefficient of heat transfer values of the composite walls. The results were then stored in a text file. They were used for the analysis in the study. The number of layer combination was obtained by multiplying the number of possible materials of each layer by each other. For example, consider the three-layered concrete walls. The first layer, the layer outside the wall is mortar plaster. It has 11 different resistances. The next layer is concrete. It has four different resistances. The last layer is also mortar plaster with eleven different possible resistances. Hence the layer combinations is (11x4x11) 484. In this category, therefore, there are four hundred and eighty-four different types of walls. Secondary data were used for the computation in this regard. Tables 1, 2, 3 and 4 are examples of the secondary data employed. These examples were for five-layered walls each for concrete, brick and sandcrete walls. Other data employed were for four- and three-layered walls, also for concrete, brick and sandcrete.

2.2 Data Analysis

The data obtained using the designed program was analyzed using descriptive analysis in SPSS Statistics 20. Tables were entirely used to display the summarized output in the form of minimum and minimum values of overall coefficient of heat transfer of the various building wall envelopes in Ghana.

Table 1. Resistance of concrete walls with external cladding, internal cladding, external plaster and internal plaster

Resistances of external claddings, $(m^2.K)/W$	Resistances of external plaster, $(m^2.K)/W$	Resistances of concrete layer, $(m^2.K)/W$	Resistances of internal plaster, $(m^2.K)/W$	Resistance of internal cladding, $(m^2.K)/W$
8.801×10^{-3}	0.0176	0.125	0.0176	0.164
8.800×10^{-3}	0.0194	0.195	0.0194	
1.408×10^{-2}	0.0211	0.266	0.0211	
0.165	0.0229	0.352	0.0229	
	0.0246		0.0246	
	0.0264		0.0264	
	0.0282		0.0282	
	0.0299		0.0299	
	0.0317		0.0317	
	0.0335		0.0335	
	0.0352		0.0352	

Source: ASHRAE Handbook Fundamentals, 2013 (Converted from imperial to SI system)

Table 2. Resistance of brick walls with external cladding, internal cladding, external plaster and internal plaster

Resistances of external claddings, $(m^2.K)/W$	Resistances of external plaster, $(m^2.K)/W$	Resistances of brick layer, $(m^2.K)/W$	Resistances of internal plaster, $(m^2.K)/W$	Resistance of internal cladding, $(m^2.K)/W$
8.801×10^{-3}	0.0176	0.261	0.0176	0.164
8.800×10^{-3}	0.0194	0.22	0.0194	
1.408×10^{-2}	0.0211	0.185	0.0211	
0.165	0.0229	0.155	0.0229	
	0.0246	0.13	0.0246	
	0.0264	0.1	0.0264	
	0.0282		0.0282	
	0.0299		0.0299	
	0.0317		0.0317	
	0.0335		0.0335	
	0.0352		0.0352	

Source: ASHRAE Handbook Fundamentals, 2013 (Converted from imperial to SI system)

Table 3. Resistance of layers of sandcrete walls with external cladding, internal cladding, external plaster and internal plaster

Resistances of external cladding, (m².K)/W	Resistances of external plaster, (m².K)/W	Resistances of sandcrete layer, (m².K)/W	Resistances of internal plaster, (m².K)/W	Resistance of internal cladding, (m².K)/W
8.801 x 10⁻³	0.0176	0.0908	0.0176	0.164
8.800 x 10⁻³	0.0194	0.1136	0.0194	
1.408 x 10⁻²	0.0211	0.1362	0.0211	
0.165	0.0229	0.182	0.0229	
	0.0246		0.0246	
	0.0264		0.0264	
	0.0282		0.0282	
	0.0299		0.0299	
	0.0317		0.0317	
	0.0335		0.0335	
	0.0352		0.0352	

Source: ASHRAE Handbook Fundamentals, 2013 (Converted from imperial to SI system)

Table 4. Resistances (R) and heat transfer coefficients (h) of air

Position of surface	Direction of heat flow	h of Surface (non-reflective) (W/m².K)	Resistance (m².K/W)
Still air (Vertical)	Horizontal	hi = 8.293	R = 0.12
Moving air 12.075km/h wind for summer	(Any position)	ho = 22.72	R = 0.044

Source: ASHRAE Handbook Fundamentals, 2013 (Converted from imperial to SI system)

3 Results and Discussion

3.1 Introduction

The results obtained from the evaluation of the overall coefficients of heat transfer of concrete, brick and sandcrete walls in Ghana have been summarized in this chapter. For convenience the results of the study have been categorized into twelve cases. The first four cases, cases I, II, III and IV are the results obtained for concrete walls. Cases V, VI VII and VIII are the corresponding results obtained for brick walls while cases IX, X, XI and XII are those obtained for sandcrete walls. Table 5 shows the category of walls that fall under each case. The computer program was used to generate the results. The data used were obtained from AHRAH Handbook Fundamentals (ASHRAE Handbook Fundamentals, 2013). The heat transfer coefficient of inside and outside wall shown in Table 5 were also obtained from the same source. The results are discussed based on the cases stated above.

Table 5. Categories of building wall structures in Ghana

Case I	Concrete walls with external cladding, internal cladding, external plaster and internal plaster
Case II	Concrete walls with external cladding, external plaster and internal plaster
Case III	Concrete walls with internal cladding, external plaster and internal plaster
Case IV	Concrete walls with internal and external plaster
Case V	Brick walls with external cladding, internal cladding, external plaster and internal plaster
Case VI	Brick walls with external cladding, external plaster and internal plaster
Case VII	Brick walls with internal cladding, external plaster and internal plaster
Case VIII	Brick walls with internal and external plaster
Case IX	Sandcrete walls with external cladding, internal cladding, external plaster and internal plaster
Case X	Sandcrete walls with external cladding, external plaster and internal plaster
Case XI	Sandcrete walls with internal cladding, external plaster and internal plaster
Case XII	Sandcrete walls with internal and external plaster

Source: Study data, 2020

3.2 Concrete Walls with External Cladding, Internal Cladding, External Plaster and Internal Plaster

Walls under this category are composite concrete wall structures made of five composite layers. The outer layer may be of stone, shingle, tile or terrazzo. The next layer, plaster, may vary in resistance from $0.10(m^2.K)/W$ to $0.20(m^2.K)/W$, in intervals of $0.01(m^2.K)/W$. The thickness of the mortar is taken to be constant at 2.54 cm since the variation in thickness was found to have no significant effect on the overall heat transfer coefficient. The mortar thickness ranged between 2.04 cm and 3.04 cm. The difference in the resistances results from the different densities used in the Ghanaian environment. The next layer inwards the building is the main building wall layer, concrete. Four types of concrete layers are used in Ghana. The concrete may vary in terms of weight and thickness. There are two types of weight. These are light weight concrete and heavy weight concrete. In terms of thickness, there are either a 10.16 cm or 20.32 cm of concrete. The next layer adjacent to the concrete layer is the internal plaster layer of eleven different types, varying in terms of density from 1.682 g/cm3 to 2.162 g/cm3, is similar to the external mortar plaster described above. The next adjacent layer is the internal cladding of resistance $0.164(m^2.K)/W$. Under this category one thousand, nine hundred and thirty six (1936) walls are identified. The maximum overall coefficient of heat transfer value obtained was 2.0096554 $W/(m^2.K)$ while the minimum value was 0.929543 $W/(m^2.K)$. The range was calculated to be 1.0801124 $W/(m^2.K)$ for the walls in this category.

3.3 Concrete Walls with External Cladding, External Plaster and Internal Plaster

Wall structures under this category are also concrete walls. All the layers are also similar to those in case I except that the walls in this category have no internal cladding. The maximum U-value is 2.99762 W/(m^2.K) while the minimum value is 1.2161 W/(m^2.K). The absence of the internal cladding, plywood, is seen to have a significant effect on the overall coefficient of heat transfer values obtained. It increased the maximum and minimum overall coefficient of heat transfer values of the corresponding walls in case I by 32.96% and 23.6% respectively. These four-layered walls have an overall coefficient of heat transfer value range of 1.78152 W/(m^2.K). One thousand nine hundred and thirty-six (1936) different walls were also identified in terms of their thermal properties.

3.4 Concrete Walls with, Internal Cladding, External Plaster and Internal Plaster

These are walls that fall under case III. They are four-layered walls with the main layer also being concrete. The outside layer may be mortar plaster with eleven different densities. The resistances vary from 0.10(m^2.K)/W to 0.20(m^2.K)/W also in the interval of 0.1(m^2.K)/W. The next layer from outside to the room is the concrete layer. There are four different types of concrete in terms of their thermal properties, similar to the two cases above. The thermal resistances of the concrete layers are 0.125(m^2.K)/W, 0.195(m^2.K)/W, 0.266(m^2.K)/W and 0.352(m^2.K)/W according to their weight and thickness. This is also similar to the two cases I and II described above. The third layer is mortar plaster made of 11 thermal resistances, similar to the outside layer described above. And finally the inside layer is plywood of thermal resistance 0164(m^2.K)/W. The overall coefficient of heat transfer values obtained from these layer-combinations vary from a maximum of 2.04584 W/(m^2.K) to a minimum of 1.09794 W/(m^2.K), the range being 0.9479 W/(m^2.K). Compared to cases I and II, the walls in case III have no external cladding. As many as four hundred and eighty-four (484) different wall structures were identified under this category.

3.5 Concrete Walls with External Plaster and Internal Plaster

These walls are the simplest in construction as far as the concrete walls are concerned. This is because they are made of only three layers. These layers are the outside layer which is made of mortar plaster layer, the main layer which is the concrete and the inside mortar plaster. Practically these are the commonest concrete construction walls that are built in Ghana. Walls under this group have neither internal cladding nor external claddings.

In addition to the other concrete walls described above, they are used to build private storey buildings and public storey structures. Scarcely in Ghana will one find non-storey building wall made of concrete. Research has found that though they are stronger, firm and durable, the capital cost could be relatively beyond the reach of the average Ghanaian. Few upper and middle class people who engage in building storey structures use concrete.

Four hundred and eighty four (484) building wall structures are also identified. The maximum overall coefficient of heat transfer value is 3.07884 W/(m^2.K) and the minimum 1.33905 W/(m^2.K), giving a range of 1.73979 W/(m^2.K). Comparing these groups of walls with those described above it can be seen that heat can pass through them more readily than the others. For instance, it will be more difficult for heat to pass through the wall made up of shingle, mortar, concrete, mortar and plywood (the min overall coefficient of heat transfer value for case I) by 30.58% than it will do in a plaster-concrete-plaster wall (the min overall coefficient of heat transfer value) for case IV. This is due to the absence of external and internal claddings in the three-layered wall structures under this case IV. Below is also a summary of the results in cases I, II, III and IV for concrete walls.

Table 6. Maximum and Minimum U-values of concrete walls in Ghana

CATEGORY	Max. U-value W/(m^2.K)	Min. U-value W/(m^2.K)
CASE I	2.00965554	0.929543
CASE II	2.99762	1.2161
CASE III	2.04584	1.09794
CASE IV	3.33905	1.33905

3.6 Brick Walls with External Cladding, Internal Cladding, External Plaster and Internal Plaster

Two thousand nine hundred and four (2904) building wall structures are found. They are brick walls made of five layers. The outside layers are made of external claddings composing of stone, clay tile, terrazzo or shingle.

Stone and clay tile are those with the least thermal resistances, 8.801x10^{-3}(m^2.K)/W, while shingle has the largest thermal resistance of 0.165(m^2.K)/W respectively. The next layer is made of mortar plaster with similar thermal properties as those described above. The thermal resistances vary from 0.1(m^2.K)/W to 0.20(m^2.K)/W in the interval of 0.01(m^2.K)/W. The third layer of the wall is brick. Though the brick thickness used in Ghana is 10.16 cm, the density may vary. Six types of brick materials of different density ranges with six corresponding different range of thermal conductivities were used. However, this study used the average conductivity values of these ranges to calculate the resistances of the brick. The fourth layer into the room is also made of mortar plaster similar to those described above. The last and innermost layer of this group of walls is plywood of thermal resistance of 0.164(m^2.K)/W. The maximum overall coefficient of heat transfer value obtained was 2.11596 W/(m^2.K) while the lowest value obtained is 1.0872 W/(m^2.K) with a range of 1.02876 W/(m^2.K). It must be noted that the wall structures under case I (concrete walls) are comparable to those in case V in that they are also made up of five layers. However their overall coefficient of heat transfer values were quite different. For instance, the maximum values are 2.0065554 W/(m^2.K) and

2.11596 W/(m^2.K) respectively. It must be noted that these two walls have everything in common except that the main component is concrete for case 1 and brick for case V. The difference in values is therefore attributed to the difference in the thermal properties of concrete and brick.

Generally concrete is a better conductor of heat due to the ionic bonding in stones and sand which constitute the main materials in concrete as a result of the presence of granite in stones and silica in sand in their material make-up. However density also plays a very important part in heat transfer properties in solids. This is reflected in the result obtained by showing that some of the brick walls can conduct heat better than some of the concrete walls. About 23% of the brick walls in case V can conduct heat faster than case I. In other words, about 67% of the brick walls in case V can resist heat transfer better than some of the concrete walls, both groups being of five layers.

3.7 Brick Walls with External Cladding, External Plaster and Internal Plaster

The main component of the layers in this group of walls is brick, like case V. They are four-layered walls. They have no internal claddings but inside mortar plaster and outside mortar plaster. The external cladding may be of stone, clay tile, terrazzo or shingle.

The resistances are 8.801x10^{-3} (m^2.K)/W for stone and tile, 1.408x10^{-3} (m^2.K)/W for terrazzo and 0.165(m^2.K)/W for shingle. The other layers, mortar plaster and brick, have the same thermal properties as those in case V above. Two thousand nine hundred and four (1904) wall structures were computed with the maximum value being 3.24046 W/(m^2.K) and the minimum value 1.32311 W/(m^2.K). The range of overall coefficient of heat transfer values is 1.91735 W/(m^2.K). 29% of walls in this group can transmit heat faster than some of their corresponding concrete walls in case II. This is as a result of the "density effect" explained under case V.

3.8 Brick Walls with Internal Cladding, External Plaster and Internal Plaster

Seven hundred and twenty six (726) four-layered walls were also identified under this category. The main wall layer is brick. They have no external claddings but plywood as internal claddings. The exterior layer is mortar plaster, followed by the brick with another mortar plaster layer next to the brick. The internal cladding is the innermost layer of the wall. The mortar plaster has thermal properties similar to those in case V. Overall coefficient of heat transfer values ranged from a minimum of 2.15611 W/(m^2.K) to a maximum of 1.32486 W/(m^2.K), the range being 0.83125 W/(m^2.K). About 8.9% of these walls can transmit heat better than some of their corresponding four-layered concrete walls in case III.

3.9 Brick Walls with External Plaster and Internal Plaster

The building wall structures in Ghana that fall under this category are seven hundred and twenty six (726). They are three-layered walls with the main component being brick. The other two layers are mortar plaster, each on either side of the brick. These structures are the simplest and the commonest type of the walls mainly composed of brick. Among the brick walls they can transmit heat most readily. This is indicated in the available results upon comparison. For example, the maximum overall coefficient of heat transfer value is 3.33558 W/(m^2.K) compared to 2.15611 W/(m^2.K) for case VII, 3.24046 W/(m^2.K) for case VI and 2.11596 for case V. Comparing the minimum overall coefficient of heat transfer values, it has 1.83218 W/(m^2.K) in relation to 1.32486 W/(m^2.K), 1.32311 W/(m^2.K) and 1.0872 W/(m^2.K) for cases VII, VI and V respectively. This can be explained from the fact they have neither internal nor external claddings which can contribute to additional thermal resistance to the walls.

Table 7. Maximum and minimum U-values of brick walls in Ghana

CATEGORY	Max. U-value W/(m^2.K)	Min. U-value W/(m^2.K)
CASE V	2.11596	1.08720
CASE VI	3.24046	1.32311
CASE VII	2.15611	1.32486
CASE VIII	3.33558	1.83218

3.10 Sandcrete Walls with External Cladding, Internal Cladding, External Plaster and Internal Plaster

The building wall structures in this category are five-layered. The main layer is sandcrete. The other layers, from the outside of the wall into the room are external cladding made of either stone, clay tile, terrazzo and shingles, outside mortar plaster, inside mortar plaster and internal cladding made of plywood. The sandcrete is sandwiched by both the outside and inside mortar plaster. One thousand nine hundred and thirty six walls were identified. The overall coefficient of heat transfer values ranged from a minimum of 1.21301 W/(m^2.K) to a maximum of 2.15797 W/(m^2.K), the range being 0.94496 W/(m^2.K). Referring to the five-layered walls of concrete and brick it can be observed that though the maximum value for sandcrete is higher than the corresponding value of brick walls, it is lower than concrete. The presence of stones in concrete should have given an opposite result but this is not the case. This can be explained from the fact that the density of sandcrete can be varied to a very large degree that some of them have thermal conductivities that are higher than the thermal conductivities of some concrete. The thermal conductivities of sandcrete ranged from 0.72 W/(m.K) to 1.512 W/(m.K) for a corresponding density range of 1.682 g/cm^3 to 2.162 g/cm^3. Sandcrete block thicknesses used in Ghana are 10.16cm, 12.7 cm, 15.24 cm, and 20.32 cm. Average thermal conductivity value of 1.116 W/(m.K) was therefore used in the computations. For this reason the 'density–effect' was also instrumental in the anomaly.

3.11 Sandcrete Walls with External Cladding, External Plaster and Internal Plaster

This group of walls is made of external claddings, outside mortar plaster, sandcrete and inside plaster. The resistances of inside and mortar plaster are similar to those used in all the other cases above They have no internal claddings. Hence they are four-layered walls. The external cladding may be stone, clay tile, terrazzo or shingles with resistances $8.801 \times 10^{-3} (m^2.K)/W$ for stone and tile, $1.408 \times 10^{-3} (m^2.K)/W$ for terrazzo and $0.165 (m^2.K)/W$ for shingle respectively. Similar to case IX, an average thermal conductivity of $1.116 W/(m.K)$ is used in the computation of the thermal resistance of the sandcrete layer. A total number of one thousand nine hundred and thirty six (1936) sandcrete wall structures are therefore identified. The maximum overall coefficient of heat transfer value is $3.34004 W/(m^2.K)$. The minimum overall coefficient of heat transfer value is also computed to be $1.51424 W/(m^2.K)$, giving a range of $1.8258 W/(m^2.K)$.

3.12 Sandcrete Walls with Internal Cladding, External Plaster and Internal Plaster

These building wall structures are four hundred and eighty four (484) in number. They are also four-layered walls with no external claddings but with internal cladding of plywood. The other layers are outside and inside mortar plaster that are sandwiching the sandcrete layer. The thermal properties of this layer are therefore similar to those used for the sandcrete walls already mentioned above in cases IX and X. The maximum U-value was $2.19975 W/(m^2.K)$ while the minimum value obtained was $1.51654 W/(m^2.K)$. The range was therefore $0.68321 W/(m^2.K)$.

3.13 Sandcrete Walls with External Plaster and Internal Plaster

Among the sandcrete walls, this category is the simplest and commonest in Ghana. In fact among all the various categories of walls discussed above they are the commonest due to the simplicity involved in the molding of the sandcrete unit block and the reduction in cost in putting up the building wall. While internal cladding requires a carpenter to construct, external claddings also require additional skills to add to the construction. It is therefore relatively cheaper, quite easier and simple to construct these walls. It is therefore not surprising that most construction building walls in Ghana fall under this category. This study has found that they constitute about 95% of urban and sub-urban building structures in Ghana. Four hundred and eighty–four types were identified in terms of their thermal properties. The overall coefficient of heat transfer values ranged from a minimum of $2.01858 W/(m^2.K)$ to a maximum value of $3.44118 W/(m^2.K)$ with a range of $1.4226 W/(m^2.K)$.

3.14 Assumptions

The following fundamental assumptions were made in the study:

(i) The individual layers are homogeneous

Table 8. Maximum and minimum overall coefficient of heat transfer values of sandcrete wall in Ghana

CATEGORY	Max. U-value W/(m^2.K)	Min. U-value W/(m^2.K)
CASE IX	2.15797	1.21301
CASE X	3.34004	1.51424
CASE XI	2.19975	1.51654
CASE XII	3.44118	2.01858

(ii) There is equilibrium or steady-state heat transfer, disregarding effects of heat storage.
(iii) The contact resistances at the interface between layers were negligible.
(iv) There is perfect contact between the layers and that no cracks exist in the wall layers.

Assumption 3 above requires the following explanation: thermal resistance at an interface between two solid materials may be a function of the surface properties and characteristics of the solids, the contact pressure, and the fluid on the surface, if any. Fang, Gou, Chen et al. (2018); and Wang, Wang, Gu et al. (2015) studied the influence of some solid properties on contact pressure and compared it to data for wet and dry coils. Chumak & Martynyak (2012) also investigated the effects of surface pressure, surface roughness, hardness, void material, and the pressure of gases in solid voids. Chumak (2016) has also worked on the resistance of adhesive bonds while Kaspareck (1964) and Clausing (1964) have provided data on the contact resistance in a vacuum environment. Results from these studies show that resistance of interface may be insignificant. For this reason the thermal resistances and overall coefficients of heat transfer obtained in this study did not take into account the effects of contact between adjacent layers.

The overall thermal performance can also be affected significantly by such factors as improper installation and shrinkage, setting or compression of a material. Loose contact between layers and the presence of cracks can also affect the thermal resistance and the U-values of the wall substantially. The quality of contact will depend on the care taken during construction, the type of material, the temperatures involved and the age of the wall. During the survey it was found that some brick walls in Ghana have had their outside plaster detached completely with time and under the influence of the weather. Studies on plate fin coils have also indicated that substantial losses in performance can occur with fins that have cracked collars, but negligible thermal resistance was found in coils with continuous collars and property expanding tubes (Capozzoli et al., 2015; Taler, 2007; Yastrebov, 2015).

4 Conclusion

The thermal resistances and overall heat transfer coefficients (overall coefficient of heat transfer values) of sixteen thousand nine hundred and forty (16940) building wall structures in Ghana have been determined in terms of their thermal properties using existing data. Among these six thousand seven hundred and seventy-six (6776) five-layered building wall structures were identified with their thermal resistances and overall coefficient of heat transfer values computed and documented. A total number of eight thousand, four hundred and seventy (8470) four-layered building walls were also identified in Ghana. Their thermal resistances and overall coefficient of heat transfer values have also been also documented. The three-layered structure gave a total of one thousand six hundred and ninety-four (1694) building walls. Their thermal resistances and overall coefficient of heat transfer values have been computed and documented as well.

Tables 6, Table 7 and Table 8 show the maximum and minimum overall coefficient of heat transfer values of five-, four- and three-layered walls respectively. The maximum overall coefficient of heat transfer value for the five-layered walls is 2.15796 W/(m^2.K) while the minimum is 0.09295 W/(m^2.K). For the four layered-walls the minimum overall coefficient of heat transfer value 1.097 W/(m^2.K) is also found while their maximum overall coefficient of heat transfer value is 3.34 W/(m^2.K). Finally, the minimum overall coefficient of heat transfer value computed for the three- layered walls is 1.339 W/(m^2.K), with their maximum value being 3.4411 W/(m^2.K).

The results computed have been categorized into "cases" for easy identification, quick referencing and clarity. Cases I, II, III, and IV have been categorized for concrete walls. Case V, VI, VII and VIII are grouped for brick walls while the results for sandcrete walls were finally categorized into cases IX, X, XI and XII. The results obtained for cases I, V and IX are for five-layered wall structures of concrete, brick and sandcrete envelopes respectively. Case II and III; VI and VII; and X and XI are the respective results for four-layered concrete, brick and sandcrete building wall structures. Cases IV, VIII and XII are also the results for three-layered concrete, brick and sandcrete walls respectively.

Four thousand eight hundred and forty (4840) building wall structures have been identified for concrete walls; seven thousand two hundred and sixty (7260) building wall structures have also been identified for brick walls while four thousand eight hundred and forty (4840) building wall structures have been identified for sandcrete.

A total number of 4840 concrete walls, 7260 brick walls and 4840 sandcrete walls have been identified. Hence thermal resistances and overall heat transfer coefficients have been computed for as many as 16940 building wall structures were therefore identified. A Building wall code, based on thermal properties, has been established for concrete, brick and sandcrete wall structures in Ghana. A computer program has also been developed to compute the overall coefficient of heat transfer values of any building wall structure in Ghana, from a single layer to an infinite number of layers. A program was developed to compute the thermal resistances and overall heat transfer coefficients of three- four- and five-layered walls. Tables are given for thermal resistances and overall coefficient of heat transfer values of wall structures in Ghana built with concrete, brick and sandcrete. A set of tables for thermal resistances and overall coefficient of heat transfer values have also been established for three-layered, four-layered and five-layered building wall structures in Ghana. The greater the number of layers of a particular wall (concrete/brick/sand) the higher the resistance, and the lower the overall coefficient of heat transfer, though the weight–to–overall coefficient of heat transfer ratio increases.

Recommendations

Based upon the findings, the study recommendations that a unified building wall-code should be developed for all concrete, brick and sandcrete walls in Ghana. The Ghana institution of engineers should take the initiative in this regard. The study also recommends that further research work by academic and research institutions be organized to obtain the thermal resistance and overall coefficient of heat transfer values of building wall structures other than concrete, brick and sandcrete. Besides, the study recommends that the computer program used in this study should be developed so that it can be used to compute thermal resistance and overall coefficient of heat transfer values of any multi-layered solid body other than three-, four- and five-layered bodies. Finally, the study recommends that research work should be carried out to experimentally determine the thermal resistances as well as the overall coefficient of heat transfer values of the building wall structures considered in this study. A comparative analysis in this regard could be of benefit to both researchers and industry players.

Acknowledgement. I want to appreciate Prof. K. O. Kesse of blessed memory, formerly of the Mechanical Engineering Department, Kwame Nkrumah University of Science of Technology, Kumasi, Ghana, for his onerous and lovely supervision of this study. I also wish to appreciate computer analyst Mr. Ernest Bentum of University of Cape Coast, Cape Coast, Ghana for actively and extensively supporting this work during the design of the source code.

References

Arens, E., Zhang, H., Hoyt, T., Kaam, S., Goins, J., Bauman, F., et. al. (2012). *Thermal and air Quality Acceptability in Buildings that Reduce Energy by Reducing Minimum Airflow from Overhead Diffusers.* Draft Thermal Report, ASHRAE Technical Report Committee 2.1 (Physiology and Human Environment)

ASHRAE: ASHRAE Standard 55–2013, Thermal Environmental Conditions for Human Occupancy. American Society of Heating Refrigeration and Air Conditioning Engineers, Inc., Atlanta (2013)

Boregowda, S.C., Choate, R.E., Handy, R.: Entropy generation analysis of human thermal stress responses. ISRN Thermodynamics. Volume 2012 (2012), ID 830103, 11 (2012)

Capozzoli, A., Fantucci, S., Favoino, F., Perino, M.: Vacuum insulation panels: analysis of the thermal performance of both single panel and multilayer boards. Energies 2015 8, 2528–2547 (2015). https://doi.org/10.3390/en8042528

Capozzoli, A., Lauro, F., Khan, I.: Fault detection analysis using data mining techniques for a cluster of smart office buildings. Expert Syst. Appl. **42**(9), pp. 4324–4338 (2015)

Chumak, K., Martynyak, R.: Thermal rectification between two thermoelatic solids with a periodic array of rough zones at the interface. Int. J. Heat Mass Transf. **55**(21–22), 5603–5608 (2012)

Chumak, K.: Adhesive contact between solids with periodically groved surfaces. Int. J. Solids Struct. **78**, 70–76 (2016)

Fang, W.Z., Gou, J.J., Chen, L., Tao, W.Q.: A multi-block lattice Boltmann method for the thermal contact resistance at the interface of two solids. Appl. Therm. Eng. **138**, 122–132 (2018)

Garazzino, S., Montagnani, C., Donà, D., Meini, A., Felici, E., Vergine, G., Bernardi, S., Giacchero, R., Vecchio, A.L., Marchisio, P., Nicolini, G.: Multicentre Italian study of SARS-CoV-2 infection in children and adolescents, preliminary data as at 10 April 2020. Eurosurveillance **25**(18), p. 2000600 (2020)

Jones, K.S.: Automatic summarising: The state of the art. Inf. Process. Manage. **43**(6), pp. 1449–1481 (2007)

Krajcik, M., Simonea, A., Olesena, B.W.: Air distribution and ventilation effectiveness in an occupied room heated by warm air. Ener. Build. **55**(2012), 94–101 (2012)

Martynyak, R., Chumak, K.: Effect of heat-conductance filler of interface gap on thermoelastic contact of solids. Int. J. Heat Mass Transf. **55**(4), 1170–1178 (2012)

Taler, D.: Effect of thermal contact resistance on the heat transfer in plate finned tube heat exchangers. ECI symposium series, volume 105. In: Proceedings of 7th International Conference on Heat Exchanger Fouling and Cleaning - Challenges and Opportunities, Eds, Müller-Steinhagen, H., Malayeri, M. R., Watkinson, A.P. Engineering Conferences International, Tomar, Portugal, (July 1 - 6, 2007)

Van Eemeren, F.H., Grootendorst, R., Johnson, R.H., Plantin, C., Willard, C.A.: Fundamentals of argumentation theory: A handbook of historical backgrounds and contemporary developments. Routledge (2013)

Wang, T.-J., Wang, L.-Q., Gu, L., Zhao, X.: Numerical analysis of elastic coated solids in line contact. J. Central South Univ. **22**(7), 2470–2481 (2015). https://doi.org/10.1007/s11771-015-2775-4

Yastrebov, V.A., Anciaux, J.F., Molinary, J.F.: From infinitesimal to full contact between rough surfaces: evolution of the contact area. Int. J. Solids Struct. **52**, 83–102 (2015)

Post-harvest Losses of Coconut in Abura/asebu/kwamankese District, Central Region, Ghana

E. Baffour-Awuah[1,3], N. Y. S. Sarpong[2,3(✉)], and I. N. Amanor[3]

[1] Agricultural Engineering Department, University of Cape Coast, Cape Coast, Ghana
[2] Department of Agricultural and Biosystems Engineering, Kwame Nkrumah University of Science and Technology, Kumasi, Ghana
[3] Mechanical Engineering Department, Cape Coast Technical University, Cape Coast, Ghana
{serwaah.sarpong,ishmael.amanor}@cctu.edu.gh

Abstract. Purpose: The paper was purposed to assess the perception of stakeholders in the coconut-cultivation industry involving farmers, drivers and retailers in the Abura/Asebu/Kwamankese district in the Central Region of Ghana in terms of the sources of post-harvest losses.

Design/Methodology/Approach: Thirty farmers each were targeted as the study population in Abura, Asebu, and Kwamankese townships and their surrounding communities after a preliminary study of 5 farmers. The study employed both quantitative and qualitative data using a close- and open-ended questionnaire, interview and observation as research instruments. The purposive and snowball sampling methods were employed. Sixty farmers responded to the questionnaire at a response rate of 67%, while 9 drivers and 3 roadside retailers were interviewed. The IBM SPSS Statistics 20 software program was used for the data analysis employing descriptive analysis.

Findings: The outcome of the study revealed that sources of post-harvest losses include loss arising from fresh fruit consumption; method of de-husking; processing and storage of copra. Others include losses arising from harvesting (early or late harvesting); collection and gathering; transportation drying of copra. The rest were losses that arise from seasoning; and insects, pests and fungi. Factors that contributed most to postharvest losses of coconut within the district were found to include copra drying (19.98%); processing (19.98%); storage of copra (18.6%); as well as insects, pests and fungi destruction (16.4%). The factor that contributed least was de-husking (2.8%). Postharvest losses among the respondents ranged between 5 and 25 % with an average value of 14.01% and a standard deviation of 4.2%.

Research Limitation: The sample size was limited considering the population involved in the coconut industry in the district. The relationship between individual parameters could not be considered and analysed.

Practical Implication: Kiln drying instead of sun-drying when extensively introduced could improve drying time and mold infestation thus reducing postharvest losses from this source. Application of weedicides to clear farms must be encouraged.

© The Author(s), under exclusive license to Springer Nature Switzerland AG 2023
C. Aigbavboa et al. (Eds.): ARCA 2022, *Sustainable Education and Development – Sustainable Industrialization and Innovation*, pp. 589–602, 2023.
https://doi.org/10.1007/978-3-031-25998-2_45

Social Implication: The annual earnings of most of the farmers were below the standard poverty level. Poverty reduction interventions should be extended or intensified to the farmers since most of the contributing factors of post-harvest losses were due to poverty and lack of funds to deal with both basic human and industry needs. Education of farmers with regards to the application of technology such as the application of insecticides, pesticides and fungicides must be introduced.

Originality/Value: The study established 10 sources of post-harvest losses of coconut in the district and as a case study for Ghana. Post-harvest losses of coconut could be reduced with the help of agronomic interventions from agricultural extension officers and institutional goodwill from the Ministry of Food and Agriculture.

Keywords: Coconut · Coconut oil · Copra · Harvesting · Processing · Storage

1 Introduction

In sincerity, utilization and beauty, there is no other tree that can bypass the coconut plant. It is the most widely grown nut in the world; the most economical utility palm. It provides individuals their basic necessities such as food, drink, shelter, fuel, furniture, medicine, decorative materials among others. They are a necessity and a luxury. It is sometimes referred as the tree of the heavens; tree of life; tree of abundance; and the supermarket of nature. In spite of these benefits and accolades, its cultivation is bereft with many challenges, including postharvest losses (Pirmansah, 2014).

According to Jabatan Pertanian Malasia (DOA) (2021) out of the global production about 93 % is cultivated in the Asian and Pacific regions. The mean global annual production of coconut between 2010 and 2020 was estimated to be over 61 million metric tons. Among the global production of coconut, over half is processed into copra. The other half is processed into desiccated coconut, edible kernel products fresh nut consumption (Zakaria et. al., 2022).

Being the major coconut product from coconut, copra, the dry coconut kernel, and its derivatives coconut oil and copra cake, plus desiccated coconut constitute the major source of foreign exchange to many coconut producing countries. For many small countries including Ghana and other Pacific nations, copra could be a major source of foreign exchange.

Generally, it takes 12 months for coconut to mature from pollination to maturity. The color of the husk is the most important sign to indicate the maturity of the nut. To obtain better quality coconut products, it is important that the nuts are harvested within the right maturity period. It is therefore important that nuts that are partially or completely brown are harvested. In order to increase copra and oil quantity and quality, nuts harvested at the tenth month or at the color-break condition, must be stored or seasoned within a certain period of time. Thus in order to obtain maximum copra and oil yield, the coconut fruit should be harvested when it is fully ripen. Between 11 and 12 months after pollination the nut should be matured ripen and therefore suitable to be harvested.

Practically, the harvesting cycle varies between 45 and 60 days or 90-day periods. Nevertheless, it is recommended that harvesting cycle should be every 45 days for practical and economic reasons and as a result of hired labor expenses. During harvesting two or three bunches of the nuts are matured enough to be harvested from each palm tree if this cycle is adhered to. It has been found that this cycle produces a substantial quantity of quality mature nuts with high copra and oil production.

Harvesting is followed by collection, ripening and de-husking of nuts. Harvested nuts are generally collected together on a single layer, usually on the bare ground within the farm. A moist soil will cause the nuts to germinate, thereby creating postharvest losses. For this reason, the freshly harvested nuts should not be put on the moist ground for a long period of time, but rather be settled on dry locations. When the fresh coconut is de-husked, the hard but brittle shell is exposed to the atmosphere. The shell is then split open into two halves with the help of a cutlass. The liquid content of the coconut is then drained off, after which the cups with the meat attached, are dried. During the drying process, the meat is reduced both in size and moisture content making it easy to be removed from the cupped shell. These cups of coconut meat are further dried for further removal of moisture and relative increase in the concentration of oil content in the copra that will be obtained.

Copra is produced after drying the coconut kernel. Copra, coconut oil as well as the cake derived from it are a major source of foreign exchange for many coconut growing countries in Asia, the Pacific, and Africa. The quality and quantity of copra and copra cake that will be obtained are affected by how the coconut kernel is dried. Improperly dried copra gives rise to moldy copra, the most common and harmful mold being yellow green mold referred as *Aspergillus flavus* and other aflatoxin producing molds. Aflatoxin is harmful both for man and animals. Poor drying methods, thus, give rise to post harvest losses (FAOSTAT 2020).

From harvesting, through seasoning, drying to storage, the absence of awareness and specific knowledge and skills relating post-harvest technologies of the coconut fruit have caused major postharvest losses. Though wastage and losses occur at various value chain points, it has been found that the stage at which the copra is dried is the most critical since the efficiency of the drying process at the farm level consequently and subsequently influences increased losses in terms of product price and quality (FAOSTAT 2020).

The fact that coconut is a smallholder crop and therefore millions of rural people depend on it for survival cannot be overemphasized. Thus, the development of coconut, specifically in terms of postharvest operations, can therefore be the foundation for rural development in communities where coconut is cultivated. The Abura/Asebu/Kwamankese district in the Central Region of Ghana is no exception.

Postharvest losses of coconut could have a broad range of unfavorable consequences (Thampan, 1996). These include loss of money and reduction of adequate food consumption. Postharvest losses also act as a barrier to provision of coconut transformation from a subsistence to a cash crop situation, specifically in various districts within the coconut producing communities in the country. Fresh coconuts suffer from dual postharvest loss consequences. First, primary internal deterioration resulting from fundamental physiological degradation; and second, deterioration resulting from microbial degradation.

Besides the conspicuous physical postharvest loss of coconut, postharvest deterioration may result in quality reduction, with the negative effect on marketing, and associated consequential price inefficiencies and ineffectiveness. Postharvest losses may also result in diversion of normal product utility. An instance of such a situation is when copra meant for the production of pure coconut oil is rather used to produce crude coconut oil as a result of the copra growing moldy or contaminated with aflatoxins; thus the harvested product being diverted into lower-value product (Westby, 2002). In spite of the importance of postharvest losses to the economy and food security in the nation it appears data relating various crops are not available. Notable among these crops is the coconut fruit. There is apparent scarcity of accurate data that quantifies the magnitude of these losses. Though some studies have observed certain levels of losses in the coconut value chain, there has been no attempt to organize an assessment study (Vowotor et al., 2012). Additionally, it appears these estimates are just rough data based on anecdotal information in certain documents. According to Permansah (2014) postharvest losses could influence between 5% and 30% individuals in a number of South South Asia (SSA) and Asian countries. Nevertheless, where and how the losses occurred within the value chain were not indicated. In the case of Africa in general and Ghana in particular, the least said about available data, the better.

The Food and Agriculture Organization of the United Nations (FAO, 2011) Attempted a Systematic Analysis of Losses for Various Crops Around the Globe in 2011. However, Only Physical Losses Were Included in the Estimates, Excluding Economic Losses. Moreover, Most of the Losses, According to the Study, Occurred at the Processing Stage and Almost Entirely Influencing, Rather, the Processed Products. Other Domestic Research in Ghana Have also Concentrated on Only Postharvest Loses During Traditional *Gari* Processing (Boahen, 2004) and Agbelima (Dziedzoave Et. Al, 1999), but not the Coconut Fruit. This Study Shall Thus, Contribute to the Establishment of Data on Postharvest Losses of the Coconut Fruit in Ghana Using Abura/Asebu/Kwamankese as a Case Study Within the Coconut Value Chain, with Reference to Harvesting Through the Processing of the Commodity. The Aim of the Study Was Therefore to Assess Postharvest Losses of Coconut Fruit in the Abura/Asebu/Kwamankese District in the Central Region of Ghana. The Objective Was to Ascertain the Perception of Stakeholders Concerning the Source and Quantification of Postharvest Losses of Coconut Fruit in the Same District. Mussi-Dias and Freire (2016) Report of Sources of Losses to Include Include Loss Due to Harvesting; Loss Due to Collection and Gathering; Loss Due to Transportation by Foot/feeder Road; and Loss Arising from Copra Production/drying; the Others Were Loss from Fresh Fruit Consumption; Loss Arising from De-husking; Loss from Domestic (Micro-scale) Processing; as Well as Loss Due to Storage of Copra. The Rest Were Losses Arising from End-Stem Rot and Losses Due to Seasoning of Fresh Fruit.

2 Method and Materials

2.1 Target Population

The population of Abura/Asebu/Kwamankese District is about 117,185. It is one of the twenty-two districts in the Central Region of Ghana with a land mass totalling 324 km^2. The population is made up of 55,275 males and 61,910 females; 39,428 urban and 77,757 rural inhabitants. Fifty percent of the population are engaged in skilled agriculture, forestry and fishery activities. Coconut farming is included in this category of employment as private informal employees or self-employees (GSS, 2014). Thirty farmers each was targeted as the study population in Abura, Asebu, and Kwamankese townships and their surrounding communities. A preliminary study of 5 farmers showed that percentage postharvest losses could vary between 0 and 6 on a 0–10 scale at various levels of the value chain.

2.2 Data Source and Collection

The study required data to be used to answer some research questions. Data was therefore collected as a piece of information which was obtained during the research activity. The study employed quantitative and qualitative data to give real and complete meaning to the behaviour of the respondents towards their actions, inactions and observations. Both quantitative and qualitative methods were thus used in gathering data, using questionnaire as the research instrument. The researcher eventually decided to use an interview schedule due to the low educational background of most of the inhabitants in the district. Though an interview schedule was used to gather the information, in few instances, unstructured interviews were employed as supplement to clarify issues for clearer understanding and explanation of respondents. The observational tool was also employed on few instances, particularly the nature and size of coconut farms that were visited. Open-ended and close-ended questions were used in gathering the data of the study. Open-ended questions were employed for gathering information on the meanings and explanations of respondents while close-ended questions were used to solicit information concerning each objective of study and also for statistical analysis. This procedure reduced researcher's bias as much as possible. It also ensured the comparability of individual responses and reasons to responses. Besides, it expedited the collection, organization and analysis of the data obtained.

2.3 Sampling Design

Kumar (2005) presents the various types of sampling design methods and their applications and advantages. This study employed the purpose and snowball sampling methods. Several reasons were considered before selecting these methods. One, the information was to be obtained from a known population. Two, the specific respondents were unknown. Three, the respondents were willing to share the required information after a pilot study was instituted. Four, very little information about post-harvest losses of coconut fruit in the district was known to the general public and finally, the information was required to be confidential for ethical reasons and therefore demanded a one-on-one approach. A preliminary study involving 5 coconut farmers was first organized after which a questionnaire for the structured interview was designed and administered. An unstructured interview was also organized for the 6 drivers and 3 roadside retailers employing the same approach. A period of 20 working days was utilized to finally collect all data within the months of September and October in 2021.

2.4 Sample Size

The study targeted the Abura/Asebu/Kwamankese district for the necessary data. However, emphasis was laid on the coconut farmers who are located in the three main towns and the surrounding communities in the district: Abura, Asebu and Kwamankese. This was done for proximity reasons, nonetheless, some of the farms were located in the rural communities outside the towns. Though 90 farmers were target, the final sample size was 60; 20 from each community. Some of the challenges include reachability, willingness to participate in the study and complete responses to the questions in the structured interview that was employed. A total of 9 long and short distance drivers (6 short distance haulers (*Aboboyaa*); and 3 long distance drivers (trucks) were also interviewed using a one-on-one unstructured interview. Three fresh fruit retailers were additionally interviewed using a similar instrument. While the drivers provided responses relating transportation losses, the retailers provided some information on the effects of insects, pests and fungi.

2.5 Data Processing and Analysis

The IBM SPSS Statistics 20 software program was used to analyse the data obtained for the study. Demographic data such as age, gender, social status, educational background and other behavioural and social characteristics were analysed. Data relating coconut harvesting; transportation and handling; de-husking, storage and processing were also obtained from the questionnaire and interviews analysed. The objective was to obtain information relating post-harvest losses of harvested coconut fruits in the district. A descriptive statistical tool was employed, dwelling largely on frequencies, percentages and tables.

3 Results and Discussion

The results and discussion of the study has been divided into 8 sections based on the various sections of the questionnaire. These are demographic information, harvesting information as well as handling and transportation information. The others were de-husking information and storage information. The rest were processing information and postharvest losses information. The postharvest losses information was based on how the respondents perceived the degree of losses at every stage of the value chain with reference to their own personal experience, knowledge and skill.

3.1 Demographics of Respondents

The demographic variables of the farmers that were interrogated include gender, age, highest educational attainment, annual coconut income, family size, marital status and number of years in coconut cultivation as shown in Table 1. These variables were relevant because they might influence some of the activities, methods and/or procedures employed by farmers in the value chain of the commodity. Out of the ninety interview schedules that were distributed the response rate was 67 %. Thus 60 responses were returned taking into consideration the completely responded schedules and the willingness of respondents to participate in the study.

Table 1. Demographic variables of the farmers.

Gender	Frequency	Percentage %
Male	39	65
Female	21	35
Total	60	100%
Age	Frequency	Percentage %
20–29 years	3	5
30–39 years	4	6.7
40–49 years	46	76.7
Above 50	7	11.6
Total	60	100%
Marital Status	Frequency	Percentage %
single	11	18.3
Married	29	48.3
Divorced	20	33.3
Total	60	100%
Education	Frequency	Percentage %
Informal	10	16.7
Middle	33	55
Secondary	11	18.3
Tertiary	6	10
Total	60	100%
Work Experience	Frequency	Percentage %
< 5 years	17	28.3
> 5 years	43	71.7
Total	60	100%

3.1.1 Age and Gender

Age was considered since a person's age could influence the level of perception of the individual. In all, 39 male and 21 female farmers were interviewed. This is an indication that coconut farming in the district is predominantly male-dominated. The age of the farmers ranged between 28 and 68. Majority of the farmers (76.7%) were between the ages of 40 and 49 years. The mean age and standard deviation were 45 years and 3 months, and 11 years and 5 months respectively. The median age was 46 years and 6 months; while the mode was 49 years.

3.1.2 Highest Educational Attainment

Table 2 shows that 16.7% of the respondents had informal education; 55% had Middle School Leaving Certificate/Junior Secondary/High School; 18.3% had secondary education and 10% had tertiary education. Tertiary education included teacher training college, polytechnic and university education. It is of interest to note that over 95% of the respondents were of the view that there were post-harvest losses as far as coconut value chain activities were concerned.

3.1.3 Annual Income

Since most of the farms were on subsistence basis and crop is a perennial commodity the income on the cultivation of the crop was sought on annual basis. Acreage of farms ranged between 2.0 acres and 20.5 acres. The average acreage was 4.8 acres; with about 75% of the farms between 2 acres and 4.5 acres. The standard deviation was 8.25 acres; with the 20.5-acre farm being an outlier. The farm sizes reflected on the annual coconut income of farmers. Annual income of the farmers was found to range between GH¢ 1200/ac to GH¢ 24000/ac. Average income of farmers was found to be about GH¢ 5200 per acre of farm. This excludes the income obtained from coconut palm fronds, coconut coir and coconut shell. Annual income was sought since it could be used to quantify post-harvest losses in terms of monetary cost.

3.1.4 Marital Status

The study indicated that 18.3% of the respondents were single (not married before); 48.3% were married while 33.3% were divorced though about 80% of the divorced were in serious relationship. About 72% of the households had size ranging between 3 and 5. While about 15% of the households numbered between 1 and 2; 13.3% had 6 or more occupants. Information on marital status and household size could influence some of the activities in the value chain and subsequently affect post-harvest losses of the commodity. For example, household size which could also depend on marital status may contribute to losses during gathering and collection after the harvesting operation. Larger households may have enough hands to support the gathering and collection of the nuts by helping to locate all the harvested nuts without any getting missing on the farm. Again a larger household is more likely to get hands to weed the farm before harvesting of the crop. In a weedy farm, nuts are more likely to get lost in the weeds on the farm thus contributing to higher post-harvest loss.

3.1.5 Number of Years in Harvesting

Over 28.3% of the respondents had been in crop harvesting for about 5 years while the rest (71.7%) had been in the harvesting business for over 5 years. This creates a reliability benefit on the responses from the respondents since they could be considered to be experienced enough in the farming business and therefore abreast with the issues concerning post-harvest losses of the commodity.

3.2 Harvesting Information

Though both hands and mechanical knives were used in the harvesting process, majority of the farmers (78.3%) employed hand harvesting. About 90% of the farmers employed between 1 and 5 hands during the harvesting process. Only 10% employed more than 6 hands during harvesting. Larger numbers were required for larger farms. Sometimes, when there were not enough time for the farm owner, larger number of hands were sought to speed up the harvesting operation. The issue of post-harvest loss of the nuts was usually not the bone of contention in this regard.

Studies have shown that in order to reduce post-harvest losses of the crop, harvesting should be done when the nuts are fully matured, supposedly between 11 and 12 months (Punchihewa & Arancon, 1999). However, the study found that only 16.7% of farmers would harvest the nuts at this maturity age. Thus 83.3% of the farmers harvested their crop before this maturity age. Indeed, as much as 23.3% would harvest before the nuts were 10 months. These nuts were mainly harvested for the consumption of the fresh meat. In some cases, the shell might contain very little or no meat to the extent that the entire nut is lost to the bin. Added to this injury is that the liquid content of such nuts could be tasteless to be consumed thus making the entire edible part of the nut lost to the bin. Some reasons given were early harvesting of the crop. The major reasons were attributed to lack of funds to maintain the basic needs of the household and the family needs such as education, health care and rent for accommodation. Early harvesting of the coconut fruit could therefore be one of the important cause of postharvest loss of the commodity. Nuts that are allowed to fall naturally (late harvesting) might also be lost through germination.

The study showed that less than 19% of farmers applied pesticides between fertilization and harvesting stage. Meanwhile, it is documented that the application of pesticides go a long way to reduce pest infestation of immature nuts and consequently, the control of post-harvest loss due to pest infections (Punchihewa & Arancon, 1999). It can therefore be surmised that the deficiency of pesticide application in coconut farms could be a contributory factor of post-harvest loss of coconut fruit in the district.

As many as 68.3% of farmers did hire labour to harvest their nuts. Thus only 31.7% employ family hands to support harvesting of the crop. Considering the fact that about 52 % of respondents were either single or divorced it is not too doubtful that as many as 68.3 % employed additional hands to support their harvesting operations. Employing external hands in harvesting and associated collection and gathering operations could contribute to losses since external hands may be less likely to be committed and sincere to these operations as compared to that of family members. In a situation where wages may not be satisfactory, the commitment, sincerity and loyalty levels could be far less than what is expected.

As a result of hired labour cost it has been recommended that coconut harvesting cycle should be 45 days (Punchihewa & Arancon, 1999). This is for practical and economic reasons, though 60 or 90-day cycle may be adopted. This implies that anything less than 45 days could lead to harvesting of immature fruits. The study ironically revealed that 55 % of the respondents do harvest the crop before 45 days. This could lead to losses after harvest. Even for the consumption of fresh coconut meat, some of the nuts may have little or no meat, in addition to distasteful liquid content. Harvesting before 45 days also brings forth nuts with low copra and oil recovery if the purpose is to produce copra and coconut oil. As already indicated, harvesting of immature fruits could be as a result of household need for income in terms of education, food, health and other subsistence demands.

It is important that immature meat is seasoned for about two to four weeks under a shed, preferably with concrete or wooden floor or mats. Improper seasoning could lead to insect and mold attacks resulting in losses. The study showed that over 90 % of the respondents adhere to this practice. Thus the contribution of improper seasoning to postharvest losses of the commodity can be said to be minimal within the district.

Since proper management of farms, including clearing of weed, could contribute to postharvest losses, respondents were asked as to the number of times they clear their farms in-between farming seasons. Ideally, farms are supposed to be cleared twice between two farming seasons. Fifty percent of respondents cleared their farms twice every farming season. The rest either sometimes do not do clearing, once a year or more than once annually. It can therefore be concluded that clearing of farms could contribute to postharvest loss of the commodity. Generally, loss due to harvest, collection and gathering could be as high as over 8 %. This is comparable to 10 % value in Malaysian farms (Punchihewa & Arancon, 1999).

3.3 Transport and Handling Information

After Harvesting the Nuts Would Have to Be Collected and Gathered at a Central Point. This May Be Done by Throwing the Nuts or Carrying Them in Baskets or Pans to a Central Point. It Has Been Found that Throwing of Nuts Could Damage Some of the Nuts. Unfortunately, Majority of Farms Experience Throwing of Nuts by Gatherers to the Central Point (58.3%). Thus Collection and Gathering Could Contribute to Losses in the Farms in the District. Some of the Reasons Given to Throwing is the Ability to Gather Large Quantities of Nuts Within a Specific Period of Time and at Less Expense of Human Effort. After Gathering, About 65 % of Respondents Would Allow the Nuts on Their Farms or Road Side Before Finally Sent to Their Homes or Sold to Middle Men or Retailers. This Practice is Considered Beneficial Since It Allows the Immature Nuts to Get Ripen and Matured Before They Get to the Final Consumer Thereby Reducing Losses After the Nuts Are Harvested.

Majority of the Respondents Gathered Fruits Under Shade, (62.3%). This Practice is also Beneficial Since Nuts Exposed to Sunlight Could Have Their Meat and Liquid Content Negatively Affected. Though Packaging Could Reduce Post-harvest Losses, Majority of the Respondents (65%) Did not Package the Nuts During Transportation. Packaging is Usually Done Using Plastic or Vegetable Fibre Sacks.

3.4 De-husking Information

The study showed that though over 90 % of the respondents used cutlass for de-husking, about 87 % were of the view that less than one percent of nuts got damaged during de-husking. It can therefore be asserted that de-husking has little to contribute as far as postharvest losses of the crop along the value chain is concerned.

3.5 Storage Information

Majority of the respondents stored their harvested fruits in three ways: within the farm floor (shaded by the coconut trees); by the roadside (covered with coconut palm fronds); or by the roadside unsheltered. Shaded fruits are protected from postharvest losses since direct contact with the sun's radioactive rays could damage the internal content of the nuts. Since about 81.6 % remain shaded it can be concluded that the contribution of shading to postharvest loss of the fruit could be minimal. Thus, while 38.3 % of the respondents allowed nuts on the farm shaded; 43.3 % cover them with coconut palm fronds by the roadside. The road side storage facilitates the transportation of the nuts either to the house, market centres or processing destination.

Majority of the respondents (90%) would not fumigate the coconut meat before storage though fumigation goes a long way to reduce attacks from pests, insects and microbes, thereby reducing post-harvest losses. The absence of fumigation could therefore be a major contributing factor to postharvest loss of the crop.

It is important that the meat (copra) is washed with clean water before storage. The issue of clean water needs more than desired since in some instances the water may be contaminated. About 55 % of the respondents agreed seeing beetles, cockroaches, moth or earwigs at one time or the other being associated with stored copra. The presence of these pests could reduce the quality of copra and subsequently destroying the product. In one way or the other, these insects/pests in the absence of pesticides, insecticides, and fumigants, could go a long way to contribute to loss of the product. It should be noted that 51.6 % of the respondents would not apply insecticides on the stored copra.

The study further revealed that sun-drying of copra is the most popular method employed for drying (83.3%) as against direct kiln/firewood drying (16.7%). The disadvantage of solar drying is that it is slow. Copra is exposed to rain and dew (moisture) which could reduce the quality of the product and subsequently contribute to damage and loss. The fact that 68.3 % of the respondents would find their copra mouldy and fungal while 55 % would find insects on the drying copra is an indication that sun-drying may not be the best drying method to prevent postharvest loss. Loss due to copra storage was found to be 18.6 % as compared to 5–10% in Malaysian farms. This could be due to difference in fruit variety as well as the absence of application of fumigants, pesticides, insecticides and fungicides, among others, during drying and before storage in the study area.

3.6 Processing Information

Processing of copra was based on that done by the respondents, who happen to be the farmers themselves. The study showed less than 30 % of coconut was sent home and used for edible oil production. The remainder was consumed in the fresh state including the liquid content (about 10%). The oil may be consumed as food or used for soap processing. Crude oil from copra, which is produced from unwholesome copra, would usually be used for soap production. Production of oil is at the micro scale level depending on family which was found to range between 1 and 5 individuals. The rest are sold to retailers or middle men who came from outside the district to purchase the harvested

Table 2. Contribution of activities to post harvest losses (PHL) of coconut in the Abura/Asebu/Kwamankese district

Activity	Contribution to PHL (% mean)	Minimum (%)	Maximum (%)	Mean (%)	Standard Deviation
G1	3.80	0	1	0.53	0.10
G2	4.30	0	1	0.60	0.15
G3	4.30	0	1	0.60	0.14
G4	19.98	1	5	2.80	0.90
G5	3.80	0	1	0.53	0.14
G6	2.80	0	1	0.40	0.14
G7	19.98	1	4	2.80	0.70
G8	18.60	1	6	2.60	0.70
G9	6.04	0	2	0.85	0.13
G10	16.40	1	3	2.30	0.95
Total	100	5	25	14.01	4.20

nuts. Table 2 shows the responses of 75 respondents on a scale of 0 to 10. The minimum and maximum response were 0 and 6 respectively.

Legend: G1 = Loss due to harvesting; G2 = Loss due to collection and gathering; G3 = Loss due to transportation by foot/feeder road; G4 = Loss arising from copra production/drying; G5 = Loss from fresh fruit consumption; G6 = Loss arising from de-husking; G7 = Loss arising from domestic (micro-scale) processing; G8 = Loss due to storage of copra: G9 = Loss due to storage of fresh fruit (seasoning); G10 Loss arising from pests, insects and fungus (Source: Field survey, 2021).

The results of the study is summarized as follows:

1. Majority of the farmers (78.3%) harvested nuts with their hands through climbing of the tree.
2. Only 16.7 % of farmers harvest their nuts between 11 and 12 months of maturity.
3. Less than 19 % of farmers sprayed their plant before harvesting s\eason.
4. Hired labour for harvesting amounted to 68.3 %.
5. Fifty-five percent of farmers harvested their fruits below the minimum 45-day cycle.
6. Ninety percent of farmers practice two-four week seasoning of fresh fruits.
7. Over half of the farmers practice the minimum two-times clearing of farms between every two farming seasons.
8. Over 58% of the farmers gathered harvested nuts by throwing to the central gathering point.
9. After gathering 65 % of the farmers would allow nuts on their farms or roadside, both shaded, before final conveyance.
10. Sixty-five percent of farmers would not package nuts during final conveyance to marketing centres.

11. Though more than 90 % de-husk using cutlass, 87 % were of the view that less than 1 % of nuts got damaged through de-husking.
12. Majority of farmers would not apply fumigants, pesticides and insecticides before storing copra (90%).
13. Majority of the farmers (55%) would find cockroaches, beetles, moths or/and earwigs on stored products.
14. Over 51 % of farmers did not apply insecticides.
15. Over 83 % of farmers employed sun-drying of coconut meat.
16. More than 68 % of farmers would find stored copra mouldy and fungal infected.
17. Less than 50 % of copra is processed into coconut oil. Oil is either eaten or used to manufacture soap.

4 Conclusion

The aim of the study was to assess postharvest losses of coconut fruit in the Abura/Asebu/Kwamankese district in the Central Region of Ghana. The objective was to ascertain the knowledge level of coconut value chain stakeholders with reference to postharvest losses. Respondents' views were sought based on their subjective perception as to the losses of the product along the value chain. A preliminary study among 5 farmers indicated that losses along the supply chain could vary between zero and 11 %. However, the study found that postharvest losses among the respondents ranged between 5 and 25 % with an average value of 14.01 % and a standard deviation of 4.2 %.

Factors that were likely to contribute to postharvest losses of coconut in the district were found to include harvesting (early or late harvesting); loss due to collection and gathering; loss due to transportation and loss due to drying of copra. The other factors include loss arising from fresh fruit consumption; loss arising from method of de-husking; loss due to processing and loss due to storage of copra. The rest were loss due to seasoning and loss arising from end-stem rot. Factors that contributed most to postharvest losses of coconut within the district was found to include copra drying (19.98%); processing (19.98%); storage of copra (18.6%); as well as insects, pests and fungi damage (16.4%). The factor that contributed least was de-husking (2.8%).

5 Recommendations

The study recommends that education should be given by extension officers in terms of good agronomic practices and the use of modern technology in the industry. For example, immature nuts should not be harvested since this reduces the overall annual income of the farmers. Education may include sanitary practices, improperly dried copra, drying under moist conditions and storage time. Social relief interventions should be extended to the farmers since some of the limitations arose as a result of poverty and lack of funds to deal with basic needs. The supply of free/subsidised chemicals (insecticides, pesticides and fungicides) to deal with insects, pests and fungus by the Ministry of Food and Agriculture shall go a long way to prevent some, if not all, the postharvest losses of the commodity in the district.

References

Pirmansah, A.: Review of the coconut statistically 2014. In: Coconut Statistical Yearbook (2014). http://www.apccec.org

Jabatan Pertanian Malasia (DOA). Booklet statistic tanaman, (2021)

Mussi-Dias, V., Freire, M.G.M.: Evaluation of postharvest stem-end rot on coconut fruits. J. Agric. Environ. Sci. **5**(2), 25–35 (2016). https://doi.org/10.15640/jaes.v5n2a4

Dziedzoave, N.T., Ellis, W.O., Oldham, J.H., Osei-Yaw, A.: Subjective and objective assessment of 'agbelima' (cassava dough) quality. Food Control **10**, 63–67 (1999). https://doi.org/10.1016/S0956-7135(98)00153-4

FAO. Continental programme on postharvest losses (PHL) reduction: Rapid country needs assessment – Working Document – Ghana (2011)

Punchihewa, P.G., Arancon, R.N.: Coconut: post-harvest operations. Asian and Pacific Coconut Community (APCC). Food and Agricultural Organization of the United Nations (FAO) (1999). www.apcc.org.sg

FAOSTAT (2020). http://www.fao.org/faostat

Thampan, P.K.: Coconut for prosperity. Peekay tree crops development foundation. Kerala, India (1996)

Vowotor, K.A., Mensah-Bonsu, A., Mutungi, C., Affognon, H.: Postharvest losses in Africa – Analytical review and synthesis: the case of Ghana. Researchgate (2012)

Wenham, J.E.: Postharvest deterioration of cassava: a biotechnology perspective. Food and Agriculture Organization of the United Nations. FAO Plant Production and Protection, Paper 130, Rome, Italy (1995)

Westby, A.: Cassava utilization, storage and small scale processing. J. Biol. Prod. Utilization. (2002). https://doi.org/10.1079/9780851995243.0281

Zakaria, M.H., Dardak, R.A., Ahmed, M.F.: Business potential of coconut-based products in the the global markets. FFTC Agricultural Policy Platform (FFTC-AP). Food and Fertilizer Technology Center for the Asian and Pacific Region (2022). www.ap.ffttc.org.tw

Hazard Assessment and Resilience for Heavy Metals and Microbial Contaminated Drinking Water in Akungba Metropolis, Nigeria

T. H. T. Ogunribido(✉)

Department of Earth Sciences, Adekunle Ajasin University, Ondo State, Akungba – Akoko, Nigeria
thompson.ogunribido@aaua.edu.ng

Abstract. Purpose: This research was to measure the physical, chemical, and microbial parameters of natural water in the Akungba – Akoko for health hazard assessment and resilience.

Design/method/Approach: Twenty water samples collected in clean two-liter polythene bottles in the study area were analyzed using standard procedures. The pH meter was used to measure pH, TDS, temperature, and electrical conductivity; nitrate by titration, chromium, aluminum, lead, iron, copper, and arsenic by AAS. The microbial population was determined by inoculation of cultured agar.

Findings: Results showed that pH ranged from 6.0–7.4, conductivity from 141μs/cm to 961 μs/cm, TDS from 112 to 961 mg/L, temperature from 25 to 25.6°C, nitrate from 1.0 to 20 mg/L, arsenic from 0.01 to 0.116 mg/L while values of copper, chromium, total iron and lead were 0.017 to 0.213 mg/L, 0.176 to 0.629 mg/L, 0.052 to 0.367 mg/L, 0.003 to 0.129 mg/L respectively. The coliform from 0 to 6cfu and Escherichia coli from 0 to 2cfu. Gibb's plot revealed rock weathering as the major source of heavy metals, heavy metals values in the freshwater are above WHO drinking limits; coliform and Escherichia coli isolated is an indication of fecal contamination, cluster analyses showed that the water was contaminated, therefore water samples are not potable.

Research limitations: Reverse – osmosis could not be used to remove contaminants due to financial constraints.

Practical implication: Water would be harmful to people drinking it due to the toxicity of heavy metals and microbial contamination, therefore water has to be treated before use or discarded.

Originality/Value: This paper is unique and is not a replication of any existing content.

Keywords: Contamination · Natural water · Resilience · Cluster analyses · Potable

1 Introduction

Natural water is noted for public health hazards and its associated health-related problems. A constant supply of good quality water will reduce poverty and also prevents the spread of waterborne diseases or any other hygiene problems (Agbo et al., 2019, Cosgrove and Rysberman, 2000, UNICEF, 2005). Rare metals cause ill- health in humans and sources of these metals are industrial and anthropogenic activities and manufacturing processes. There is a need to ascertain the quality of natural water apart from its adequate supply for safe consumption. It is pertinent that water is chemically and biologically safe for consumption and other uses. Water quality may be affected by human activities leading to pollution or contamination. Anthropogenic activity may cause water quality degradation. Such activities include dumping of refuse into waterways and disposal of chemicals into the water which may cause public health hazards when such water is consumed. Groundwater is a major source of drinking water in a suburb in Nigeria. There is an increasing demand for groundwater supply in Nigeria (Adeyemi et al., 2005), due to the increase in population or production. Rivers and streams are easily degraded more than groundwater and if groundwater is degraded, it is very difficult and expensive to clean up its aquifer.to restore its quality (Purandara and Varadarajan, 2003). Degradation of water not only affects freshwater quality but also human health and socioeconomic development negatively (Milovanovic, 2007). Groundwater quality depends on the inflow, rainfall, rivers, and weathering processes around its aquifer. The physical or chemistry of water depends on the amount of organic or inorganic compounds present in them. Some of the compounds pose a threat to the ecosystem, some serve as substrates to aquatic life and others are responsible for the color or appearance of the water body (Eletta and Adekola, 2005). Generally, because of the good quality of groundwater, it serves as the major source for drinking and other uses (Okoro, 2012). It is very important to monitor and protect groundwater quality against heavy metals and microorganisms that can cause serious illnesses and sometimes even death.

1.1 Materials and Method

Study Site: Is between latitudes $7°\ 26^1$ and $7°\ 29^1$ N of the Equator and Longitudes $5°\ 43^1$ and $5°\ 45^1$ E, (Fig. 1). The study area falls under a subequatorial climatic region characterized by two climatic seasons which are drought and rain. The rainy season is from April to September while the drought is from October to March. The study area is accessible through roads and footpaths. Akungba rocks consist of pre-Cambrian Basement Complexes whose age is Pan-African (Rahaman, 1988, Rahaman and Ocan, 1978). The predominant rocks are granite gneiss, grey gneiss, quartzites, and charnokites (Fig. 2). Granite Gneiss forms the largest rock type in the study area. They are highly foliated. They show alternating bands of light and dark minerals. The light-colored minerals are quartzofeldspathic while the dark-colored mineral is made up of ferromagnesian minerals. The major minerals are quartz, feldspar, and biotite. Grey Gneiss forms the second largest percentage of rock types made up of biotite, quartz, and feldspar. They were formed due to the metamorphism of pre-existing rocks. The Quartzite occurs as intrusions in the parent rock body. It has a thickness of about fifteen centimeters wide and its color range from light to brown. The other rock types include charnockitic rock and minor pegmatite intrusion. The major minerals are plagioclase feldspar and biotite. The charnockitic rock occurs along the Global -com Mast in Akungba which is a flat-lying outcrop.

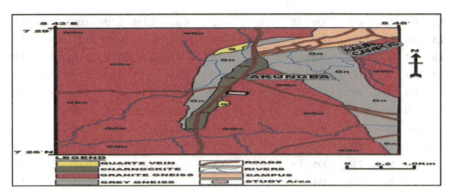

Fig. 1. Geological Map of Akungba-Akoko (Modified After Independent Mapping)

Sample Collection and Laboratory Analyses: Twenty samples were collected in clean 2-L polyethylene bottles. The method of sample collection, preservation, and analyses were done using the standard method (APHA, 2002). Biological and chemical parameters were measured in the laboratory and physical quality was onsite. pH, total dissolved solids, electrical conductivity, and temperature were measured with a pH meter, heavy metals by AAS, and nitrate by titration. Microbes were determined for the total viable or *coliform* count using heterotrophic plate count and MPN method and coliform was isolated from water samples.

1.2 Results and Discussion

The physical, biological, and chemical quality measured are presented in Table 1 below.

Table 1. Results of physical, biological, and chemical analyses in the study area

	pH	Temp °C	EC μs/cm	TDS mg/L	Coli Form cfu	E.-Coli cfu	NO$_3^-$ mg/L	As3+ mg/L	Cu2+ mg/L	Cr3+ mg/L	Fe2+ mg/L	Pb2+ mg/L	Al3+ mg/L
L1	6.2	25	240	252	0	0	1	0.058	0.213	0.368	0.108	0.011	4.62
L2	6.2	25.6	361	280	0	0	1.5	0.024	0.187	0.271	0.081	0.068	1.20
L3	6.4	25	405	168	0	0	0.4	0.007	0.23	0.419	0.063	0.011	1.27

(*continued*)

Table 1. (*continued*)

	pH	Temp °C	EC μs/cm	TDS mg/L	Coli Form cfu	E.-Coli cfu	NO$_3^-$ mg/L	As3+ mg/L	Cu2+ mg/L	Cr3+ mg/L	Fe2+ mg/L	Pb2+ mg/L	Al3+ mg/L
L4	6.0	25	487	336	0	0	1.2	0.01	0.154	0.309	0.052	0.058	1.20
L5	6.0	25	310	217	2	1	4	0.071	0.279	0.514	0.217	0.092	1.40
L6	6.2	25	668	462	0	0	3	0.036	0.212	0.48	0.166	0.044	1.10
L7	6.4	25	112	910	0	0	0.01	0.024	0.152	0.629	0.132	0.076	1.45
L8	6.4	25	227	154	3	0	0.01	0.013	0.17	0.362	0.071	0.014	1.30
L9	6.0	25	141	980	0	0	4	0.096	0.296	0.479	0.367	0.129	1.32
L10	6.2	25	142	980	0	0	1.5	0.027	0.174	0.229	0.262	0.063	1.10
L11	6.2	25	885	616	4	2	2.5	0.116	0.038	0.287	0.123	0.077	1.20
L12	6.4	25	232	336	0	0	1.8	0.048	0.017	0.209	0.103	0.057	1.38
L13	6.2	25	483	161	5	1	0.01	0.028	0.076	0.288	0.078	0.015	4.60
L14	6.0	25	410	154	3	1	3	0.075	0.147	0.322	0.163	0.102	1.41
L15	6.2	25	860	160	1	0	2.2	0.026	0.177	0.183	0.134	0.077	1.35
L16	6.4	25	884	140	0	0	1	0.065	0.077	0.247	0.117	0.096	1.22
L17	6.4	25	450	95	1	0	1	0.055	0.132	0.211	0.144	0.022	1.33
L18	7.4	25.6	961	140	5	1	0.05	0.029	0.092	0.176	0.269	0.031	1.09
L19	6.2	25	345	170	1	0	0.04	0.021	0.186	0.207	0.135	0.047	1.25
L20	6.0	25	453	160	6	2	20	0.045	0.213	0.312	0.086	0.003	1.19

pH: Table 1 shows the various pH range of the water, which ranged from 6.0 to 7.4. The low pH values of water may be due to the application of fertilizers since the inhabitants practice agriculture. pH drinking for fresh water is 6.5 to 8.5 (WHO, 2017). Therefore the pH of the drinking water was slightly acidic except sample in location 18 which was slightly basic.

Total Dissolved Solids: For this study, the value of total dissolved solids ranged from 98 to 980 mg/L. The total dissolved solids in locations 9 and 10 had a value of 980 mg/L which was the highest while location 17 has the lowest value of 95 mg/L. The World Health Organization's permissible limit is 500 mg/L and the maximum contaminant level is 1500 mg/L. Therefore, water samples still fall within the maximum contaminant limits.

Electrical Conductivity: Indicates chemicals in any natural water. Results showed that the electrical conductivity ranged from 112 to 961 micro siemens/cm.

Temperature: This is the measure of the degree of hotness and coldness of the water. Temperature itself does not have a remarkable effect on the water quality but aids some activities in water such as increasing the rate of biological and chemical activities. Low temperatures can reduce the efficiency of treatment processes. The temperature of water samples ranged from 25 to 27°C. WHO, (2011) standard water temperature is 24.4–29.0°C, indicating that the water temperatures of the study area are normal.

Nitrate Concentration: Ranged from 0.01–20mg/L. The lowest value of nitrate was detected in locations 7, 8, and 13 with 0.01 mg/l, location 20 has the highest nitrate concentration of 20 mg/L. Nitrate may come from animal excretion, the atmosphere, and plant. Nitrate in water is derived mainly from agricultural chemicals. Nitrate compounds are highly soluble. However, the nitrate level in water samples from the study area is low when compared with the WHO, (2017) limit of (45 mg/L) for drinking water. Children below six months, that ingest water containing nitrate above the drinking limit of 45 mg/L, could have a shortage of blood (Armenia), known as a blue baby syndrome.'

Aluminum (Al^{3+}): Aluminum concentration ranged from 1.09 to 4.62 mg/L. WHO (2017) drinking water limit is 0.02 mg/L. The value of aluminum was above the drinking water limit of 0.02 mg/L in all the drinking water.

Iron Concentration: Total Iron ranged from 0.052 to 0.269 mg/L. The lowest value occurred at location 4 while the highest was at location 18. Values of total iron were above the World Health Organization's permissible limit of iron in water, which is 0.03 mg/L.

Arsenic Concentration: Arsenic is associated with ores of copper and lead (ATSDR, 2012). Arsenic ranged from 0.007 mg/L to 0.116 mg/L. The lowest occurred in location 3 while the highest in location 11. The maximum contaminant level of Arsenic in freshwater is 0.01 mg/L (WHO, 2011). This showed that arsenic is very high in the water except in sample 3, showing arsenic pollution of water samples in the study area. Long exposure to arsenic may cause nausea, vomiting, abnormal heartbeat, and damage to blood vessels (Manju, 2015, Malik et al., 2014).

Copper Concentration: Ranged from 0.017 mg/L to 0.296 mg/L. These values are higher than the World Health Organization's permissible limit of 0.05 mg/L except in locations 11 and 12 which were within the drinking limit. When copper concentration is higher, this will impair the potability of the water causing a bitter taste (Ogunribido and Kehinde-Philip, 2011). Copper in large concentrations is dangerous to infants and people with metabolic disorders (Salem et al. 2000).

Chromium Concentration: Values ranged from 0.176 mg/L to 0.629 mg/L. The lowest concentration occurred in location 18 while the highest was in location 7. Concentrations of chromium here are all above the World Health Organization's permissible limit of 0.05 mg/L. Chromium can cause cancer, therefore minimum intakes of water with chromium were recommended (WHO, 2011). Drinking water that contains chromium (VI) can cause digestive tract, Kidney, hepatic, or brain problems (Engawa et al., 2018).

Lead: Dissolve easily in water at low pH (Obasi and Akudinobi, 2020). This may be responsible for the high value of lead since the drinking water pH is low. Lead value ranged from 0.003 to 0.129 mg/L. WHO, (1984) proposed a drinking standard of 0.05 mg/L and later changed it to 0.01 mg/L in 1993. This downward review was done due to the toxicity of lead as a result of bioaccumulation in humans. Acute lead poisoning symptoms include headache, abdominal pain, learning difficulties, and central nervous system disorder. (Bala - Chennaiah et al., 2014).

Coliform and Escherichia Coli Bacteria: The presence of these bacteria in the drinking water is an indication of fecal pollution. Drinking water should not contain fecal bacteria. Coliform ranged from 1 to 6 CFU and Escherichia coli ranged from 0 to 2 CFU in the water samples. Escherichia coli is an indicator of enteric viruses and protozoa (WHO, 2017).

Cluster Analyses: Cluster analyses grouped water here into three as stated below and its dendrogram is shown in Fig. 3.

Group 1: This is made up of water in locations 3, 8,14,16,11,12,20,5,6, and 7, this represents 50% of water in the study area, and water in this group was highly polluted. Group 2: consists of 15, 19, 4, 17, and 2, they also represent 25% of the water samples, and they have intermediate pollution. Group 3: while in this group 9, 13, 1, 10, and 18, also represent 25% of water in the study area, they have medium pollution (Fig. 3).

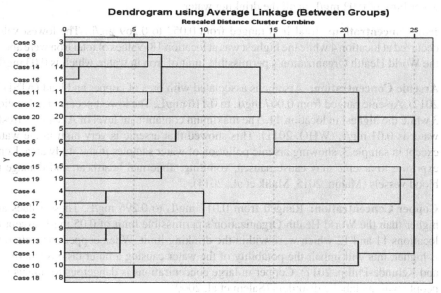

Fig. 3. Dendrogram of cluster analyses of the water samples.

GIBB'S (1970) Diagram: Is used to measure the geochemical evolution of groundwater (Ogunribido, 2018). Gibbs plots in this study indicate rock dominance (Fig. 4). This suggests that the dissolved metals in the freshwater were mainly due to the chemical weathering of the aquifer material.

Resilience Management. Contamination prevention and remediation are usually very expensive, to prevent non–point sources and biological contamination all the wells in the study area must have a platform and sanitary cover. Waste must be disposed of properly, agrochemicals or pesticides should be sparingly used, and good maintenance of the on-site sewage system. For water samples contaminated by the heavy metals in the study area, reverse–osmosis can be used to filter the chemical contaminants. Regular analyses of the water samples should be done to monitor degradation in the water quality.

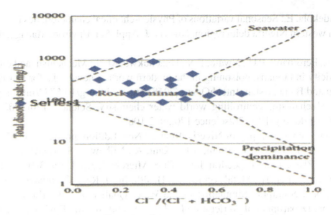

Fig. 4. Gibbs diagram of water samples in the study area

1.3 Conclusion

In conclusion, a comparison of water analytical results with the World Health Organization drinking limits indicated that the drinking water in the study area was contaminated with heavy metals and pathogen microbes. Cluster analyses that grouped the water into three, showed that all the water in each group was contaminated. Gibb's diagram indicated that the main source of dissolved rare metals is the result of weathering of aquifer materials. Continuous drinking of the water may cause learning disabilities in children and kidney and intestinal problems in adults. This may also increase their poverty level because more money would be spent on drugs. For effective management of the water resources in the study area, wells are to be disinfected and be provided with sanitary cover and contaminants are to be removed by Reverse – Osmosis or the drinking water be discarded to avoid public health hazards. This study will provide baseline information for future research studies.

References

Adeyemi, G.O., Ariyo, S., D & Odukoya A. M.: Geochemical characterization of the aquifer to the basement complex sediment transition zone around Ishora southwestern Nigeria. Water Resour. J. **16**, 31–36 (2005)

Agbo, B.E., Ogar, A.V., Akpan1, U.L .,C. I. Mboto1, C.L (2019). Physico-chemical and bacteriological quality of drinking water sources in calabar municipality, Nigeria, J. Adv. Microbiol., **14**(4), 1–22 (2019)

ATSDR: Toxicological profile for Chromium U.S. Department of Health and Human Services, Public Health Service, Division of Toxicology 1600, Agency for Toxic Substances and Disease Registry Atlanta, GA 30333 (2012)

APHA: Standard methods for the examination of water and wastewater, 21st Edition American Publication Health Association, Washington. p. 136 (2002)

Bala Chennaiah, J. Rasheed, M. A, Paul, D.J.: Int. J. Plant, Animal Environ. Sci., **4**(2), 205–214 (2014)

Cosgrove, W.J., Rysberman, F.R.: World water vision: Making water everybody's Business. Earth Scan Publication Ltd; UK. p.108 (2000)

Eletta, M., Adekola, E.: Seasonal variations of Physico-chemical characteristics in water resources quality in western Nigeria delta region, Nigeria. J. Appl. Sci. Environ. Manage. **9**(1), 191–195 (2005)

Engawa, G.A., Ferdinand, P.U., Nwalo, F.N., Unachukwu, M.N.: Mechanism and effects of heavy metal toxicity in humans, poisoning in the modern world—new tricks for an old dog? Ozgur Karcioglu and Banu Arslan, IntechOpen. (2018) https://doi.org/10.5772/intechopen.82511

Gibbs, R.J.: Mechanisms controlling world water chemistry. Science, **170**(3962), 1088–1090 (1970) https://doi.org/10.1126/science.170.3962.1088

Iloeje, N.P.: The new geography of Nigeria. Revised New Edition, pp. 32–45 (1979)

Malik, D., Singh, S., Thakur, J., Singh, R.K., Kaur, A., Nijhawan, S.: Heavy metal pollution of the Yamuna River. an introspection. Int. J. Curr. Microbiol. Appl. Sci. **3**(10), 856–863 (2014)

Manju, M.: Effects of Heavy Metals on Human Health, Int. J. Res.- Granthaalayah, 1–7 (2015)

Milovanovic, N. Geographic Simulation of the Water Quality of Lake Kastoria. Postgraduate Dissertation, Department of Agriculture Ichthyology and Aquatic Environment, University of Thessaly, Greece (2007)

Obasi, P.N., Akudinobi, B.B.: Potential health risks and levels of heavy metals in water resources of lead–zinc mining communities of Abakaliki, southeast Nigeria, Applied Water Science, **10**(184),1–24 (2020)

Ogunribido, T.H.T., Kehinde, P.O.O.: Multivariate Statistical Analysis of the Assessment of Hydrogeochemistry analysis in Agbabu Area, southwestern Nigeria, pp. 425–434. Federal University of Agriculture, Abeokuta, Proceeding of environmental management conferences (2011)

Ogunribido, T.H.T.: Bacteriological and hydrogeochemical investigation of surface water and groundwater in Ikare Akoko, Nigeria, Int. J. Adv. Geosci. **6**(1), 27–33 (2018)

Okoro, R.: The mechanism controlling world river water chemistry. Science **170**, 1088–1090 (2012)

Purandara, P., Varadajan, T.: Evaluation of hydrogeochemistry and water quality inBist-Doab region. Punjab, India, Environ. Earth Sci. **72**(3), 693–706 (2003)

Rahaman, M.A., Ocan, O.O.: On relationships in the Precambrian Migmatitic gneisses of Nigeria, J. Mining Geol. **15**(1), 23–32 (1978)

Rahaman, M.A.: Recent advances in the study of basement complexes in Nigeria.In: Precambrian geology of Nigeria. In: Edited by Oluyide P. O. Mbonu, A.E. Thepotassic granites of Igbetti are further evidence of the polycyclic evolution of the Pan African belt in southwestern Nigeria. Precambrian Res. **22**, 11–92 (1988)

Salem, H.M., Eweida, A.E., Farag, A.: Heavy metals in drinking water and their Environmental impact on human health, ICEHM, 542–556 (2000)

UNICEF: National Rural Water Supply and Sanitation Investment Programme. Final Draft (2005)

WHO: International Standards for drinking water. 3rd guidelines for drinking water quality. Vol. 2: Health criteria and other supporting information. Geneva, World Health Organization, p. 328 (1984)

WHO : WHO's Guidelines for Drinking-water Quality, the International Reference Point for standard setting and drinking-water safety, Geneva, pp. 1–29 (1993)

WHO: Guidelines for Drinking -Water. In: 4th Edition, NLM Classification: WA 657 World Health Organisation, Geneva, Switzerland. p. 433 (2011)

WHO: Guidelines for Drinking -Water, NLM Classification: WA 657, World Health Organisation, Geneva, Switzerland, p. 2624 (2017)